Cellular Drug Delivery

Cellular Drug Delivery

Principles and Practice

Edited by

***D. Robert Lu,* PhD**

*School of Pharmacy, Temple University,
Philadelphia, PA*

and

***Svein Øie,* PhD**

*College of Pharmacy, University of Georgia,
Athens, GA*

HUMANA PRESS TOTOWA, NEW JERSEY

999 Riverview Drive, Suite 208
Totowa, New Jersey 07512

humanapress.com

This publication is printed on acid-free paper. ∞
ANSI Z39.48-1984 (American Standards Institute)
Permanence of Paper for Printed Library Materials.

Production Editor: Robin B. Weisberg.

Cover design: Patricia F. Cleary.

Cover illustration provided by Robert Lu, PhD.

For additional copies, pricing for bulk purchases, and/or information about other Humana titles, contact Humana at the above address or at any of the following numbers: Tel.: 973-256-1699; Fax: 973-256-8341; E-mail: humana@humanapr.com

Printed in the United States of America. 10 9 8 7 6 5 4 3 2 1

1-59259-745-9 (e-book)

Library of Congress Cataloging in Publication Data

Cellular drug delivery : principles and practice / edited by D. Robert Lu and Svein Øie.
p. cm.
Includes bibliographical references and index.
ISBN 1-58829-254-1 (alk. paper)
1. Drug targeting. 2. Cell membranes. I. Lu, D. Robert. II. Øie, Svein.

RM301.63.C45 2004
615'.7--dc22

2003056682

Preface

Cellular drug design and development of pharmaceutical platforms for controlled delivery of bioactive substances to the cellular sites of action have been the focus of considerable research and literature discussions throughout the last decade. To understand the controlling mechanisms of drug delivery (including drug targeting and drug transport) to cells, and to effectively design novel pharmaceutical formulations and drug delivery devices, it is essential to carry out research and development with different cell types and characteristics. Work in this area has been rapidly evolving, creating a critical need to develop a resource devoted to the fundamentals, principles, and practice of cellular drug delivery.

Cellular Drug Delivery: Principles and Practice is intended to serve as an up-to-date reference book covering this broad field. It includes 18 chapters contributed by well-established scientists from US schools of pharmacy and the pharmaceutical industry. The introductory chapter gives an overview of recent progress, as well as the future potential of cellular drug delivery. Part I contains two chapters that describe both the fundamental concepts and the best current thinking in the areas of cellular structure, transport, growth, and defense response, a well as their involvement in cellular drug delivery. Part II addresses the broad spectrum of cellular delivery principles, and ranges from the coverage of cell-mediated immunity, gene delivery, and protein targeting to discussions of cellular drug transport, cellular drug permeability, and a variety of carrier systems related to targeted drug delivery. Part III focuses on technological development in cellular drug delivery. The development of novel formulations for the delivery of DNA and antisense oligonucleotides, as well as drug targeting with immunoglobulin formulations and antibody-mediated approaches are discussed in depth. Finally, research and development on the efficient cellular delivery of bioactive substances through automated processes, which has become a vital part of the modern pharmaceutics, is critically evaluated.

The editors would like to express their sincere appreciation to each of the chapter authors who have contributed their time and expertise to make *Cellular Drug Delivery: Principles and Practice* a comprehensive and valuable resource.

D. Robert Lu
Svein Øie

Contents

Contributors

S. Dean Allison, *P.R. Pharmaceuticals, Ft. Collins, CO*

Mansoor Amiji, *Department of Pharmaceutical Sciences, School of Pharmacy, Northeastern University, Boston, MA*

Thomas J. Anchordoquy, *School of Pharmacy, University of Colorado, Denver, CO*

Taylor K. Armstrong, *School of Pharmacy, University of Colorado, Denver, CO*

Anand Balakrishnan, *School of Pharmacy, University of Maryland, Baltimore, MD*

Antara Banerjee, *OSU Biophysics Program, The Ohio State University, Columbus, OH*

Adrian Bot, *Department of Immunology and Cancer Research, MannKind Corp., Valencia, CA*

Michel Cormier, *Biological Sciences, ALZA Corporation, Mountain View, CA*

Peter E. Daddona, *Macroflux® Technology Development, ALZA Corporation, Mountain View, CA*

Li Di, *Chemical Technologies, Wyeth Research, Princeton, NJ*

Xiang Gao, *Department of Pharmaceutical Sciences, School of Pharmacy, University of Pittsburgh, Pittsburgh, PA*

Daniel J. Hicklin, *ImClone Systems Incorporated, New York, NY*

Rodney J. Y. Ho, *Department of Pharmaceutics, University of Washington, Seattle, WA*

Hans E. J. Hofland, *OPTIME Therapeutics Inc., Petaluma, CA*

Yan Huang, *Department of Pharmaceutical and Biomedical Sciences, University of Georgia, Athens, GA*

Jeffrey Hughes, *Department of Pharmaceutics, University of Florida, Gainesville FL*

William W. Johnson, *OSI Pharmaceuticals Inc., Boulder, CO*

Jeffrey S. Johnston, *Division of Pharmaceutics, College of Pharmacy, The Ohio State University, Columbus, OH*

Edward H. Kerns, *Chemical Technologies, Wyeth Research, Princeton, NJ*

Gary S. Koe, *Valentis Inc., Burlingame, CA*

Sushma Kommareddy, *Department of Pharmaceutical Sciences, School of Pharmacy, Northeastern University, Boston, MA*

Min Jae Lee, *Department of Pharmaceutical Sciences, School of Pharmacy, University of Pittsburgh, Pittsburgh, PA*

Yvonne K. Lentz, *School of Pharmacy, University of Colorado, Denver, CO*

WeiQi Lin, *Biological Sciences, ALZA Corporation, Mountain View, CA*

Dexi Liu, *Department of Pharmaceutical Sciences, School of Pharmacy, University of Pittsburgh, Pittsburgh, PA*

D. Robert Lu, *Department of Pharmaceutical Sciences, School of Pharmacy, Temple University, Philadelphia, PA*

MARION D.C. MOLINA, *School of Pharmacy, University of Colorado, Denver, CO*

DEREK T. O'HAGAN, *Vaccine Research, Chiron Corp., Emeryville, CA*

SVEIN ØIE, *Department of Pharmaceutical and Biomedical Sciences, University of Georgia, Athens, GA*

GUANGLIANG PAN, *Department of Pharmaceutical and Biomedical Sciences, University of Georgia, Athens, GA*

KINAM PARK, *Departments of Pharmaceutics and Biomedical Engineering, Purdue University, West Lafayette, IN*

MAYANK M. PATEL, *School of Pharmacy, University of Colorado, Denver, CO*

JAMES E. POLLI, *School of Pharmacy, University of Maryland, Baltimore, MD*

PAUL R. SEO, *School of Pharmacy, University of Maryland, Baltimore, MD*

MANMOHAN SINGH, *Vaccine Research, Chiron Corp., Emeryville, CA*

DAN SMITH, *Alliance Pharmaceutical Corp., San Diego, CA*

FRANK L. SORGI, *OPTIME Therapeutics Inc., Petaluma, CA*

EDWARD G. SPACK, *PharmaSTART, SRI International, Menlo Park, CA*

PETER W. SWAAN, *Department of Pharmaceutical Sciences, The University of Maryland at Baltimore, Baltimore, MD*

JEFFREY B. ULMER, *Vaccine Research, Chiron Corp., Emeryville, CA*

EMILY WARE, *Department of Pharmaceutical and Biomedical Sciences, University of Georgia, Athens, GA*

ERICA L. WOODAHL, *Department of Pharmaceutics, University of Washington, Seattle, WA*

WU XIAO, *Department of Pharmaceutics, University of Florida, Gainesville, FL*

PREETI YADAVA, *Department of Pharmaceutics, University of Florida, Gainesville, FL*

CHUN ZHANG, *Department of Pharmaceutics, University of Florida, Gainesville, FL*

GUISHENG ZHANG, *Department of Pharmaceutical Sciences, School of Pharmacy, University of Pittsburgh, Pittsburgh, PA*

YE ZHANG, *School of Pharmacy, University of Colorado, Denver, CO*

ZHENPING ZHU, *ImClone Systems Incorporated, New York, NY*

Introduction to Cellular Drug Delivery

Present Realities and Future Prospects

Kinam Park

1. Introduction

Drugs play a central role in modern medicine *(1)*. The total global pharmaceutical market in 2002 was more than $400 billion, and the annual growth rate of the global market has been more than 10%. The large market size allows big pharmaceutical companies to invest heavily in research and development (R&D). In 2002 alone, the combined R&D budget of Pfizer, Pharmacia, Merck, Eli Lilly, and Bristol-Myers Squibb was more than $16 billion. Despite an ever-increasing R&D budget, development of new drugs has been slow, and the number of new drugs introduced has become smaller than the number of drugs that went off patent protection. The real cost of developing a new drug is not well established, but it is generally agreed to be about $800 million. One of the reasons for such a high cost is that many drug candidates are abandoned owing to their unfavorable physicochemical and biochemical properties. It is also becoming more and more difficult to develop new drugs due to the increasing difficulty of reaching their intended targets. One of the ways of making drug development more efficient is to utilize drug-delivery technologies from the early stages onward. The development of new drug-delivery technologies also makes the existing drug more useful.

Drug-delivery technologies contribute significantly in various stages of drug development. For a drug to exert its bioactivity, it has to enter the target cells. This seemingly simple process requires a series of steps that are often difficult to overcome. The steps of drug delivery to the target cells are described in Fig. 1. A drug first has to dissolve in aqueous solution first to have any chance of being bioactive (Step A in Fig. 1). Many newly developed drug candidates are poorly water soluble so a large number of potentially useful drugs are abandoned. Poorly soluble drugs can be dissolved in aqueous solution via polymeric micelles, polymeric nano/micro-particles, or polymeric prodrugs. New protein drugs can increase their bioactivity by protecting themselves in blood using various drug delivery technologies, e.g., PEGylation and microencapsulation. Once a drug is in solution, it has to be delivered to the target cells (Step B in Fig. 1). Commonly, a drug is delivered throughout the body, which causes unwanted side effects. Targeting minimizes such side effects. Targeting may be the only way for some treatments. For example, whether gene therapy will be successful or not depends entirely on whether genes can be delivered to the target site and effectively enter the

From: *Cellular Drug Delivery: Principles and Practice*
Edited by: D. R. Lu and S. Øie © Humana Press Inc., Totowa, NJ

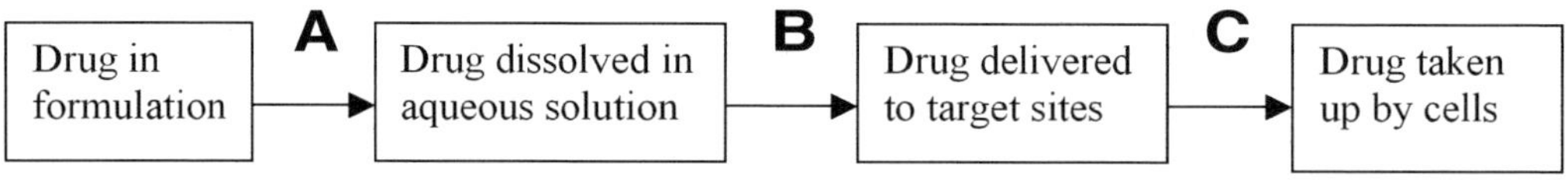

Fig. 1. Drug molecules in different stages.

target cells for expression (Step C in Fig. 1). Entering into the target cell is half of the process. The genes also must escape from endosomes to make themselves available for expression. For most drugs, the ultimate bioactivity is appreciated after they have entered the cells, and thus, cellular uptake can be considered one of the most important steps in drug delivery. In this regard, cellular drug delivery is a very timely topic to consider.

2. Current Technologies in Cellular Drug Delivery

There is a saying in China that you win every battle if you know both your enemy and yourself. Knowing one's enemy is critical in winning the war. In the war against disease, we have to know the enemy, such as viruses and bacteria. At the same time, we have to know ourselves, that is, the cells in our body. This book is about understanding ourselves in preparing for the war against diseases. We have to understand how our body functions and then apply that knowledge in the design of new strategies for efficient drug delivery.

Cellular Drug Delivery: Principles and Practice provides information on both drug uptake into cells and methods of improving cellular uptake. As listed in Table 1, the book discusses the cellular processes, drug delivery to cells, and targeting and formulation. The first three chapters provide background on how cells uptake nutrients and drugs, and other basic information on cell biology. We have a good understanding of transport processes through cell membranes that can be exploited in cellular delivery of various drugs (*see* Chapters 1 and 2). Understanding the biological processes of cell-mediated responses (*see* Chapter 3) should provide strategies for promoting cell-mediated immunity against viral infection, such as influenza virus and, it is hoped, the virus causing severe acute respiratory syndrome (SARS). Two chapters are dedicated to nonviral DNA-delivery systems (*see* Chapters 4 and 5). Although viral vectors are most effective in gene delivery, the inherent danger of using live viruses has been proven to be problematic. There is no doubt that development of nonviral vectors will be essential in routine clinical applications of gene therapy in the future. One of the approaches for efficient cellular delivery may be utilizing naturally occurring peptides that are known to enhance entry into cell membranes (*see* Chapter 6). Transporters not only absorb drugs and nutrients into cells, but also they can efflux absorbed drugs, and thus understanding various transporters in cell membranes is important in optimizing drug therapy (*see* Chapters 7 and 8). Because the majority of drugs are delivered by oral administration, understanding the roles of various transporters in intestinal drug absorption is critical in screening various drugs (*see* Chapter 9).

The two chapters on targeting using colloidal carriers and protein nanocomplexes should be useful for the delivery of various types of drugs (*see* Chapters 10 and 11). The last six chapters discuss formulation-related problems, which often, unfortunately, are neglected, for without proper formulation, no drugs would be clinically useful. Bulk manufacturing of polynucleotides in their stable forms in effective delivery sys-

Table 1
Organization of the Book

Topic	Chapter
Cell biology	1. Transport processes through cell membranes
	2. Cell growth and cell cycle
	3. Cell-mediated immunity
Cellular delivery	4. Nonviral DNA delivery
	5. Cellular delivery of nucleic acid
	6. Membrane permeation promoting peptides
	7. Transporters
	8. MDR modulators (inhibitors)
	9. Intestinal cell permeability
Targeting	10. Colloidal carriers
	11. Protein complexes
Formulation	12. Preservation of DNA integrity
	13. Cationic microparticles
	14. Transdermal delivery
	15. Immunoglobulin formulation
	16. Antibody formulation for cancer treatment
	17. Pharmaceutical profiling

tems is essential in widespread use of nucleotide-based therapeutics (*see* Chapters 12 and 13). One of the first applications in controlled drug-delivery technologies was the development of transdermal products, and one of the first therapies of antisense oligodeoxynucleotides may be by transdermal delivery (*see* Chapter 14). Immunoglobulins are considered to be good vectors to deliver T-cell epitopes to antigen-presenting cells (APCs), and targeted delivery of immunoglobulin-engineered microparticle systems have a great potential (*see* Chapter 15). Antibody-mediated targeted delivery of anticancer agents also holds great promise (*see* Chapter 16). Chapter 17 on the automated pharmaceutical profiling assays is one of the most interesting chapters of the book. Rapid screening of physicochemical and biochemical properties of drug candidates that affect cellular drug delivery provides new opportunities in pharmaceutical research beyond in vivo animal experiments.

3. Opportunities in Cellular Delivery

The concept of controlled drug delivery is more than four decades old, and the drug-delivery area is still evolving. Through the years, numerous novel drug-delivery systems have been developed and a large number of formulations are helping patients. Although the advances made to date are impressive, a number of breakthroughs are necessary for solving various problems facing us today. The main progress to be made is in the precision of targeting. Until a few decades ago, destroying a target in the enemy territory during a war required delivering hundreds of thousands of bombs that literally destroyed the whole town. Nowadays, laser-guided smart bombs or bombs with global positioning capability brought a new concept of one bomb per one target. The ultimate success in drug delivery is expected to bring about a concept of one drug molecule per one target. In fact, all that is required for successful gene therapy is to deliver a gene into a cell as long as the delivered gene is properly expressed. This means that the precision targeting has to be accompanied by effective cellular delivery.

For delivery of macromolecular drugs, such as anticancer drug molecules grafted to polymer chains, protein drugs, and genes, absorption into cells by endocytosis may be the most effective way. Endocytosis occurs when macromolecular drugs or drug carriers interact with the receptors on cell membranes, and receptor-mediated cellular uptake has been exploited quite extensively. In addition to receptor-mediated endocytosis, nonspecific electrostatic interaction with cell membranes is also known to cause endocytosis. One of the possible mechanisms of enhanced cellular absorption by cell penetration peptides *(2,3)* and trans-activating transcriptional activator (TAT) (*see* Chapter 6) is nonspecific electrostatic interactions between positive charges in side chains of those peptides with cell membranes. Effective endocytosis based on nonspecific electrostatic interaction is important for designing the cellular delivery systems that also have the additional function of escaping from endosomes. Drug-delivery vehicles can be designed with high flexibility for effective endocytosis with a minimum number of specific ligand molecules that bind to receptors on cell membranes. Enhanced cellular delivery may also exploit the unusual ability of neutral block copolymers, such as poly(ethylene oxide)-poly(propylene oxide)-poly(ethylene oxide) triblock copolymers, to get inside cells. They are known also to decrease significantly the P-glycoprotein efflux activity in P-glycoprotein overexpressing membranes, probably through incorporating themselves into cell membranes, leading to increased fluidization *(4)*. The cumulated data clearly demonstrated that those block copolymers are taken up inside cells to spread throughout the cytoplasm. Recent studies using fluorescent micelles showed that micelles made from polycaprolactone and poly(ethylene oxide) enter into cells by endocytosis to distribute themselves within several cytoplasmic compartments, but not in the nucleus *(5)*. There is a great potential in improving cellular uptake using various polymeric micelles and in order to control the distribution of the absorbed drug in the cytoplasm.

Opportunity also exists in designing the delivery systems that have the ability to escape from endosomes once they enter cells by endocytosis. One of the main disadvantages of nonviral vectors as compared with viral vectors is their inability to deliver drugs intact from endosomes into the cytoplasm. A number of approaches have been attempted for providing endosome-disruptive properties. Endosome-disruptive hemagglutinin peptides from influenza virus and membrane-active amphipathic peptides have been used to release the delivered agents from endosomes to the cytoplasm *(6,7)*. Although these approaches are promising, they suffer from the shortage of such peptides. One recently developed approach is to utilize synthetic polymers that can disrupt the endosomal membranes at pH around 6.0. Polymeric histidine has shown its ability to disrupt endosomes when pH is lowered to deliver drugs into cytosol *(8–10)*. Combining polycationic nature for endocytosis and polyhistidine for endosomal escape may be a practical combination for enhanced cellular delivery of a variety of drugs ranging from low molecular-weight anticancer drugs to proteins and genes.

Understanding cell biology, membrane interfacial phenomena, protein chemistry, and polymer chemistry are all essential in developing effective cellular delivery systems. Being an expert on any of these areas provides a great potential to open new avenues of cellular drug delivery. If the history of the pharmaceutical industry is any guide, one can easily imagine the ever-increasing values of such research areas and the high rewards accompanying them. Opportunities in the future basically are infinite and only our imagination is a limit. *Cellular Drug Delivery: Principles and Practice* will certainly serve as an excellent resource in expanding our imagination in developing novel drug-delivery systems for the most efficient methods of cellular delivery.

References

1. Drews J. In Quest of Tomorrow's Medicines. New York, NY, Springer-Verlag, 1999.
2. Langel Ü. *Cell-Penetrating Peptides. Processes and Applications.* Boca Raton, FL, CRC Press, 2002.
3. Nielsen PE. Peptide Nucleic Acids. Methods and Protocols. Totowa, NJ, Humana Press, 2002.
4. Batrakova EV, Li S, Alakhov VY, et al. Optimal structure requirements for Pluronic block copolymers in modifying P-glycoprotein drug efflux transporter activity in bovine brain microvessel endothelial cells. J Pharmacol Exp Ther 2003;304:845–854.
5. Savic R, Luo L, Eisenberg A, Maysinger D. Micellar nanocontainers distribute to defined cytoplasmic organelles. Science 2003;300:615–618.
6. Wagner E, Curiel D, Cotten M. Delivery of drugs, proteins and genes into cells using transferrin as a ligand for receptor-mediated endocytosis. Adv Drug Del Rev 1994;14:113–135.
7. Plank C, Zauner W, Wagner E. Application of membrane-active peptides for drug and gene delivery across cellular membranes. Adv Drug Del Rev 1998;34:21–35.
8. Midoux P, Monsigny M. Efficent gene transfer by histidylated polylysine/pDNA complexes. Biconjug Chem 1999;10:406–411.
9. Putnam D, Gentry CA, Pack DW, Langer R. Polymer-based gene delivery with low cytotoxicity by a unique balance of side-chain termini. Proc Natl Acad Sci USA 2001;98:1200–1205.
10. Benns JM, Choi JS, Mahato RI, et al. pH-sensitive cationic polymer gene delivery vehicle: N-Ac-poly(L-histidine)-graft-poly(L-lysine) comb shaped polymer. Bioconjug Chem 2000;11:637–645.

I

Fundamentals of Cell Biology and Cellular Drug Delivery

1

Cellular Structure, Function, and Membrane Transport

Emily Ware, D. Robert Lu, and Svein Øie

Numerous scientific disciplines have contributed significantly to the study of cellular drug delivery. From activity and toxicity screening of potential active ingredients to elucidating the biological mechanisms of site-specific interactions, the science of cellular drug delivery is a complex, integral part of the treatment and prevention of life-threatening illnesses. Every year, new information is made available from many different research areas allowing for continuous improvement through technological advancements. In order to continue along this route, a multidisciplinary background is necessary to understand the process of drug delivery at the cellular level. The purpose of this chapter is to present basic principles of cellular biology and their importance in cellular drug delivery. Subheading 1 deals with the fundamentals of cellular structure and intracellular compartments. Membrane structure, transport inside the cell, passive transport through the membranes, and other membrane properties are discussed in Subheading 2, as well as how these membranes serve as a major barrier to the drug-delivery process. Finally, Subheading 3 describes receptor proteins, how they function, where they are located on a cellular level, how they act as transporters to deliver drugs to the cell, and their role in multidrug resistance.

The effectiveness of a drug therapy is dictated by several factors, including the rate and extent of the drug molecule's penetration into, and permeation through, the body tissues and cells to reach a site of action. Many barriers prevent the drug from reaching the site of action. They can be grouped according to size, including the organism itself, on a 1–2 m scale, an organ representing the 1–100 mm scale, tissue in the 1–100 μm size, and the extracellular and intracellular space of only 1–100 nm *(1)*. Understanding the processes governing a single cell and its makeup is the fundamental objective of this chapter.

1. Internal Organization and General Function

Although there are many different types of cells, every cell has some common components necessary to carry out life processes. In general, the chemical make-up of cells and their internal compartments or organelles are similar *(2)*. From a drug-delivery point of view, the cell or a group of cells is in some way the "site of action," whether the therapy is to destroy cancer cells or produce antibodies through vaccination. There are several traits that an ideal drug carrier should possess in order to deliver the drug to this site of action. First, carriers should target the area or group of cells specific to their

From: *Cellular Drug Delivery: Principles and Practice*
Edited by: D. R. Lu and S. Øie © Humana Press Inc., Totowa, NJ

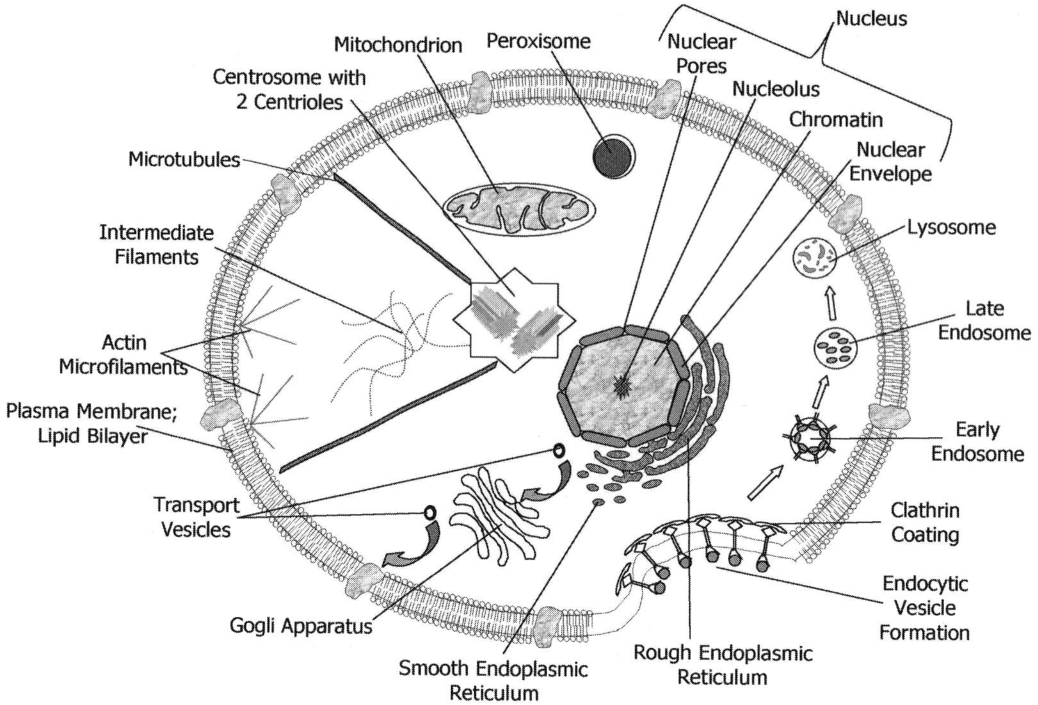

Fig. 1. Schematic of a typical eukaryotic cell, including key internal compartments and endocytic vesicle formation and transformation to lysosomes.

action while avoiding interactions with nontarget cells. They should release therapeutically relevant amounts of bioactive substances at that site for a required duration and frequency. They should also be readily excretable or degradable without causing undesirable biochemical or immune reactions in the body *(3)*. There are a number of cellular compartments common to most eukaryotic cells that a drug carrier may encounter in pursuing its therapeutic goal. A schematic of a typical cell and its compartments is shown in Fig. 1, and Table 1 summarizes how each cellular compartment can act as a drug-delivery barrier.

1.1. The Plasma Membrane

The plasma membrane forms the boundary around the cell and allows it to maintain chemical coordination. The membrane itself is formed from a series of phospholipids that are noncovalently linked together by hydrophobic interactions to form the bilayer around the cell. Imbedded within the membrane are proteins that function to transport molecules from one side to the other. Only specific types of molecules can be recognized, internalized, or transported by these proteins making passage into the cell difficult for noncompatible molecules. There are other means of transporting molecules into the cell, including fluid-phase pinocytosis, phagocytosis, receptor-mediated endocytosis, membrane fusion, passive diffusion, and channel or pore transport, which will be discussed in Subheading 2.1.2.

Table 1
Internal Cellular Organelles Can Act as Barriers to Drug Delivery

Compartment	Drug-delivery barriers
Plasma membrane	Membrane transport protein specificity
	Lipid bilayer penetration
Cytoplasm	Slow macromolecular transport
	Diffusion rate limited
Nucleus	Double lipid bilayer penetration
	Pore complex size and specificity
	Nucleus membrane receptor protein specificity
Endoplasmic reticulum	Metabolic enzymes
Golgi apparatus	Metabolic enzymes
Lysosomes	Presence of hydrolytic enzymes
	Low pH microenvironment
Mitochondria	Double lipid bilayer penetration
Peroxisomes	Presence of catalases
	Presence of membrane-bound oxidative enzymes
Cytoskeleton	Microtubules restrict internal traffic

1.2. The Cytoplasm

After drug molecules cross the plasma membrane, they may encounter the cytoplasm, depending on the mechanism of uptake. The cytoplasm consists of the cytosol and the organelles or compartments of the cytoplasm. These compartments include the nucleus, mitochondria, peroxisomes, endoplasmic reticulum (ER), golgi apparatus, lysosomes, endosomes, and cytoskeletal elements. The cytosol makes up about 54% of the total cell volume and provides the semi-liquid media where proteins are synthesized and degraded and where most intermediate metabolism occurs. The main transport mechanism through the cytosol is diffusion. Macromolecules such as enzymes or high molecular-weight drug molecules diffuse slowly in the cytosol owing to constant interactions with various cellular organelles. On the other hand, smaller molecules diffuse relatively efficiently *(4,5)*.

1.3. The Nucleus

The nucleus is located in the cytoplasm and is bound by the nuclear envelope consisting of a double lipid bilayer membrane. These two bilayers are about 20–40 nm apart and form a significant barrier limiting molecular penetration into the nucleus. The main passage into the nucleus is through pores of about 60–100 nm in diameter formed in the bilayers. This passage is regulated by pore complexes composed of about 50 different proteins called nucleoporins, which are arranged in an octagonal shape. There are approx 3000–4000 pore complexes in a nuclear envelope of the typical mammalian cell transporting many function-essential substances. Within the pore complexes are open aqueous channels that permit small molecules of less than 5000 Daltons to diffuse readily while sufficiently preventing passage of macromolecules of 60,000 Daltons or greater. Macromolecules rely on specific receptor proteins to carry them into or out of the nucleus *(2,5,6)*. If a macromolecule is unrecognized by nuclear receptor proteins, it is prevented from entering the nucleus, making it difficult to use macromolecular targeting carriers. To overcome this impasse, researchers have developed macromolecular carriers, which can be used to target drugs specifically to the nucleus. They are made of *N*-(2-hydroxypropyl)methacrylamide (HPMA) copolymers, containing targeting sites for cell specificity, and degradable drug units. As the carrier enters the cell—probably through endocytosis—and is carried to the lysosomal compartment, the drug unit is cleaved from the polymer backbone by lysosomal enzymes. The free drug molecule then enters the cytoplasm and rapidly diffuses into the nucleus to carry out its effect *(7)*. Other research groups have utilized modular recombinant transporters (MRTs) to target cellular nuclei and deliver photosensitizer agents for anti-cancer therapy. These MRT constructs are made up of several components including α-melanocyte-stimulating hormone, Simian virus (SV40) large T-antigen nuclear localization signal and a translocation domain of the diphtheria toxin, allowing selective drug targeting to melanocyte nuclei *(8)*.

The nucleus is one of the most important components of the cell because it contains the cell's genetic information encoded through deoxyribonucleic acid (DNA). Inside the nucleus, DNA is packaged into chromosomes by associating with proteins called histones to form chromatin. There are a variety of functional enzymes, strands of messenger ribonucleic acid (mRNA) and transfer ribonucleic acid (tRNA), and ribosomal components within the nucleus that are constantly constructed and transported into the cytosol. The ribosomal components (ribosomal RNA or rRNA) are synthesized and assembled in the subunit of the nucleus called the nucleolus, which has a spherical structure composed of 5–10% RNA, protein, and DNA *(5,6)*. Continuous exchange of components across the nuclear membrane is necessary for proper cellular function and can occur through specific transporters in the nuclear pore complex called importins and exportins *(9)*. This nucleocytoplasmic transport pathway could be exploited as a potential mechanism to deliver drugs to the nucleus. Many diseases may be more effectively treated with direct delivery of genetic material into the nucleus to treat or kill target cells. Potential delivery systems, both viral and nonviral vectors, to deliver genes to the nucleus are reviewed in Chapters 5 and 6.

1.4. The Endoplasmic Reticulum

Interspersed throughout the intracellular space and continuous with the outer nuclear membrane is the endoplasmic reticulum (ER). It accounts for more than half of the total membrane systems in the cell. It is a network of flattened sacs with bound ribosomes (the rough ER) and branching tubules (the smooth ER), each with different structures and functions. Inside the ER is the lumen or ER cisternal space, which is separated from the cytosol by the ER membrane. The rough ER is responsible for the synthesis and modification of membrane proteins and phospholipids as well as the synthesis of secretory proteins such as insulin. One of the main functions of the smooth ER is to bud off transport vesicles containing the newly synthesized proteins and lipids. In specialized cells, such as hepatocytes, the smooth ER is the site of drug and lipid metabolism and a site for the storage of calcium ions. The enzymes in the smooth ER are broad acting, such that encountering more than one drug of the same type will cause proliferation of the smooth ER, increasing metabolism and potentially leading to drug tolerance *(2,5)*.

1.5. The Golgi Apparatus

After being released from the smooth ER, many transport vesicles travel to the Golgi apparatus. The vesicle fuses with the Golgi membrane, releasing the contents into the Golgi apparatus. Here the proteins and lipids are modified for specific functions and stored before being repackaged into other vesicles bound for a specific part of the cell or the cell surface. The Golgi apparatus is also the manufacturing site for certain macromolecules, such as glycosaminoglycans and some carbohydrates. It is similar in structure to the rough ER, containing a series of flattened sacs or cisternae stacked upon each other. It is a polar membrane with *cis* and *trans* sides, and thus can receive and export transport vesicles, respectively, at opposite membranes.

1.6. Lysosomes

Lysosomes are membrane-bound organelles whose specific function is digestion or chemical breakdown of various substances. They function to recycle some macromolecules produced in the cell by breaking them down into the constituent monomers. Lysosomes contain about 40 different hydrolytic enzymes and all of them are acid hydrolases. The internal pH of the lysosome is maintained at about 5.0 providing a more effective hydrolyzing environment. The lack of proper lysosomal enzymes can result in cellular damage owing to accumulation of undigested toxic substances. For example, Pompe's disease is a disorder owing to indigestion of glycogen in the liver, and Tay-Sach's disease sufferers are missing a lipid-digesting enzyme in the brain, causing impairment *(2,5)*. Many of these kinds of diseases are treatable with the delivery of specific enzymes into the lysosomal compartment or genetic material into the cell, allowing production of the necessary lysosomal enzymes.

Delivery of drug molecules to cells is often a challenge because most endocytosed drug molecules eventually suffer degradation in the lysosomal compartment. Some researchers have developed a method to circumvent lysosomal degradation using pH-responsive polymers capable of lysing the endosome and releasing the endocytosed drug molecules into the cytoplasm. These polymers are based on α-alkyl acrylic acids such as poly(2-ethylacrylic acid), which, upon exposure to pH 6.3 or below, undergo a conformational change that disrupts the lipid bilayer of the endosome. The main mechanism of action is the polymer permeabilizes phosphatidlycholine membranes, disrupting the lamellar structure of the bilayer and causing lysis *(10)*. This type of technology can be useful in delivery of biomolecules to the cytoplasm; however, little is known of the potential toxic effects of releasing the endosomal components into the cell's cytoplasm.

1.7. The Mitochondria and Peroxisomes

The mitochondria within cells function to extract energy from sugars and fats to synthesize ATP. They are contained by a double lipid bilayer with an internal membrane that has a large surface area owing to complex infoldings called cristae. In the heart of the mitochondria enclosed by the inner membrane is the mitochondrial matrix, where many of the proteins and enzymes reside. The mitochondria contain their own DNA, mitochondrial DNA, which is different from nuclear DNA, and their own ribosomes, which synthesize the proteins and enzymes necessary for specific functions. In a typical mammalian cell, there may be thousands of these 1–10 μm long mitochondria making up about 20–22% of the total cytoplasmic volume *(2,5,6)*.

Peroxisomes are also metabolic centers. They are bound by only a single lipid bilayer and produce hydrogen peroxide as a by-product from reactions of breaking down various substrates. They grow by incorporating cytosol-produced lipids and proteins into their membrane and split into two when they reach a certain size. Contained within the membrane of the peroxisomes are oxidative enzymes that act to break down fatty acids, sugars, and macromolecules for use in other cellular processes. Another important function of the peroxisomes, especially in liver and kidney cells, is the detoxification of substances that enter the blood, such as alcohol.

1.8. The Cytoskeleton

The cytoskeleton provides the structural support necessary for the cell to maintain its shape. It consists of a network of fibers that extend throughout the cytoplasm. There are three main types of fibers of the cytoskeleton: the large microtubules, the intermediate filaments, and the actin filaments or microfilaments. Each serves a slightly different purpose within the cell.

The microtubules are small tube-like structures measuring 25 nm in diameter and 200 nm to 25 μm in length. They are made from the protein tubulin and dictate the positions of various organelles while directing intracellular transport. The intermediate filaments have an intermediate diameter of 8–12 nm. They are constructed of a diverse family of proteins called keratins and may serve as the framework for the entire cytoskeleton. Some intermediate filaments line the inside of the nuclear envelope, providing a strong support and barrier to protect DNA. Microfilaments are not hollow like microtubules. Instead they are solid rods composed of the globular protein actin. Actin filaments are flexible structures of 5–9 nm in diameter and tend to organize themselves in tight linear bundles. Their main purpose is to provide structural support so that the cell holds its shape.

2. Passive Transport Into and Through the Cell

The complexity of cells and their compartments pose tremendous challenges to drug-delivery scientists. Nearly every component in and around the cell can be viewed as a barrier to drug delivery. Targeting specific pathways, receptor proteins, or various cellular transport mechanisms can help scientists overcome those barriers. In this section, the transport mechanisms occurring into and throughout the cell will be elucidated and several examples found in the literature will be given. Mostly passive transport will be talked about here as active transport and the proteins responsible for active transport are reserved for Subheading 3.

2.1. Transport Across the Lipid Bilayer

The lipid bilayer or plasma membrane surrounding cells is composed of amphipathic molecules arranged in a dual-layer configuration that is about 5 nm thick. Fatty lipid molecules make up 50% by weight of the membrane, with the remainder being embedded proteins that serve to actively transport molecules across the membrane, catalyze membrane associated reactions, and provide many other functions. These lipid components act as formidable barriers to aqueous soluble substances. If a drug molecule approaches the cell surface attempting to gain entry, how does it cross the lipid bilayer and obtain access to the internal aqueous cytoplasm? There are several possible mechanisms described below.

2.1.1. Passive Diffusion

One of the main routes for both hydrophilic and hydrophobic drug molecules to enter the cytoplasm is through passive diffusion. The driving force for passive diffusion is a concentration gradient. The molecules move from an area of high concentration to an area of low concentration by random thermal motion. The factors that can affect the diffusion coefficient include solubility and molecular weight. Usually small (<300 Daltons) molecules that are uncharged species and have a moderate oil/water partition coefficient, such that they are somewhat hydrophobic, can diffuse readily. If the partition coefficient is too high, the drug exhibits high lipophilicity and is unable to reach the membrane because it cannot dissolve in the aqueous extracellular compartment. On the other hand, if the drug has a very low partition coefficient and hence is very hydrophilic, it cannot diffuse through the lipophilic membrane to reach the interior of the cell. As molecular weight increases, the rate of diffusion decreases. Larger macromolecules greater than 300 Daltons, such as many oligonucleotides and sucrose, are essentially excluded from the lipid bilayer with a diffusion half-life of 4–10 d *(11)*. Nevertheless, large macromolecules may gain access into the cell by other means.

2.1.2. Endocytic Processes

Many large macromolecules and even small particles can be taken up by the cell through a process called cytosis, or endocytosis. The lipid bilayer forms a cavity or indentation owing to an external stimulus, and through normal fluid motion and rearrangement of the membrane, an envelope, or endocytic vesicle forms engulfing the substance, including any surrounding extracellular fluid, in the cavity. Most cells undergo endocytosis on a regular basis and several different types of endocytosis exist based on the specificity and the size of the material engulfed *(11–13)*. Several types of coating proteins are involved in the process to coat newly formed vesicles. For the internal compartments of the cell, there is a similar mechanism for vesicle formation with a major difference in the protein used to coat the cytosolic side of the membrane. One type of coating protein that has been identified in several types of endocytoses, including pinocytosis, trancytosis, and potocytosis, is caveolae. This protein is unique because it uses part of the lipid bilayer that is very high in concentration of cholesterol to form vesicles. Caveolae have been involved in the internalization of Simian virus (SV40) and cholera toxin, and it is known that caveolae play a role in membrane fusion and intracellular trafficking of transport vesicles. Owing to the high numbers of caveolae in pulmonary capillary epithelial cells, some researchers have investigated macromolecular delivery to the lungs, specifically targeting caveolae membrane domains *(14,15)*.

Pinocytosis is a process in which both small and large molecules dissolved in the extracellular fluid flow into the formed cavity and are pinched off as vesicles within the cell. Because pinocytosis deals with fluids, it is also referred to as fluid-phase pinocytosis, and the vesicle internalized via fluid-phase pinocytosis is called an endosome. Much of the time, the dissolved drug molecule interacts with membrane protein receptors and basically is "caught" in the cavity through specific (receptor mediated pinocytosis) or nonspecific (adsorptive pinocytosis) binding to the receptor when an endosome is formed *(13)*. One fate of the endocytosed material is that its vesicle fuses with an organelle of the cell called an early endosome, which has an acidic environment. Some of the material, especially membrane components and receptor proteins, is then selectively recycled back to the membrane or exocytosed, while the rest becomes a late endosome containing hydrolases and an even more acidic environment with a pH of approx 5.0–6.0. The late endosomes move toward the interior of the cell and become lysosomes containing more digestive enzymes that act to degrade the contents previously endocytosed.

A less destructive endocytic process is transcytosis. Often, the endocytosed material is specifically diverted from the lysosomal route and sent to the plasma membrane, thereby being recycled through exocytosis, or more importantly in drug therapy, transferred to the other side of the cell. Proteins and peptides such as immunoglobulin G (IgG), nerve growth factor (NGF), and epidermal growth factor (EGF) are known to undergo pinocytosis and transcytosis through intestinal mucosa cells to reach the blood vessels on the other side. In some cases, as with ricin, Shiga toxin, pertussis toxin, and cholera toxin, the endocytosed protein toxin is transported to the Golgi apparatus initially, instead of undergoing the normal transitions from endosomes to lysosomes. The toxins then, undergo retrograde transport to the ER *(16)*. Elucidating these types of mechanisms provides insights not only into intracellular transport mechanisms, but also on potential treatments for these toxins.

Another type of endocytosis involves the uptake of large substances such as drug particles, cells, bacteria, or viruses. This is called phagocytosis and occurs in a series of steps similar to pinocytosis. Drug particles or other large substances adhere to the cell membrane. The binding usually occurs at a specific site that activates phagocytic mechanisms within the cell, similar to the process of when an antibody segment bound to a virus particle binds specifically to the Fc receptor on the surface of macrophages initiating uptake *(2)*. The entire particle along with some extracellular fluid is ingested or engulfed by the membrane, forming a detached vesicle or phagosome within the cell. The phagosome then fuses with a lysosome, and the particle and other phagocytosed material is digested while the bilayer materials are recycled. Phagocytosis only occurs in specialized cells called professional phagocytes that exist in specific areas of the body, mainly the reticulo-endothelial system (RES). This includes the liver, bone marrow, and spleen macrophages and circulating blood monocytes. The phagocytes are tremendously important in regulating the body's immune system because they destroy foreign substances and remove particulate antigens. Their importance in drug therapy is limited to those drugs that are delivered in particulate carriers to the cell such as liposomes, erythrocytes, or microspheres. The therapy can be targeted with those drug molecules whose site of action is some region of the RES *(13)*.

Receptor-mediated endocytosis is the most versatile type, involving a high degree of recognition and specificity. Agonists binding to surface receptors activate the endocytic pathway causing invagination of the bilayer and vesicle formation. The receptors themselves are usually recycled back to the cell surface and the agonist either participates in metabolism or is taken back to the cell surface. The size of material that can be endocytosed via receptor-mediated endocytosis ranges from smaller than 100 Daltons to in excess of 1 million Daltons, such as a virus particle. Many drug-delivery efforts focus on targeting various receptors on the cell surface so that particular biochemical pathways can be exploited. Agonists are typically developed to extend the duration of action via altered metabolic pathways in the lysosomal compartment, or to reduce side effects by specific binding to certain receptors on certain cells.

One interesting drug-delivery application developed to exploit receptor-mediated endocytosis is to engineer specific receptors on the cell surface so that targeting can be more specific. This approach, investigated at the University of California, Berkley, involves feeding the cells with chemically modified sugars that can be metabolized and incorporated as the terminal sugar on membrane glycoproteins. The chemical modification adds an electrophilic group that dangles off the cell membrane. When nucleophilic-bound receptor molecules are presented to the cell surface, they bind instantly to the electrophile to provide the cell with newly inserted receptors. Specifically designed drug molecules or virus particles can bind to these receptors to initiate receptor-mediated endocytosis for targeted drug delivery *(17,18)*.

2.1.3. Membrane Fusion

Membrane fusion occurs mostly intracellularly. However, similarities exist between membrane fusion occurring from extracellular space into the plasma membrane and from transport vesicles into the membranes of other internal organelles. Virus particles typically undergo membrane fusion to gain entry into host cells. The mechanism of action is not well-known, but there are several hypotheses. Two of the most commonly accepted ones will be discussed briefly below.

One set of hypotheses are called "proximity models" where a cell and another bilayer surrounded entity come so close together that hydrophilic molecules are displaced and the two lipid bilayers fuse into one membrane spilling the contents of the entity into the cellular cytoplasm *(19)*. There are proteins that apparently catalyze all membrane fusions by anchoring the two membranes together to initiate the process. Inside the cell, these proteins are called soluble *N*-ethylmaleimide-sensitive factor attachment receptor (SNARE) *(20)* proteins and they function to fuse transport vesicles with a target organelle.

A second group of hypotheses are called "fusion pore models," where the space between the two membranes is bridged by a protein or lipid pore complex *(19,21)*. Upon external activation, the pore complex undergoes a conformational change and an aqueous channel is formed linking the internal compartments of the two entities. The lipid layers then migrate inward opening up the channel until fusion is complete. This second hypothesis seems to explain many exocytic pathways that have been experimentally observed, and some authors suggest that this is the most likely theory *(19,21)*.

2.1.4. Pore Transport

The passive transport of solutes, small molecules, or ions across the lipid bilayer can occur through aqueous pores or ion channels. Membrane transport proteins in the lipid bilayer form aqueous pores connecting the extracellular space with the cytosol. Owing to the narrow channel in these pores, only small molecules and ions can effectively diffuse through, thereby protecting the cell from larger, potentially harmful substances. Some bacterial toxins manage to destroy cells by enlarging these pores to allow the entry of large harmful molecules *(4)*. Aqueous pores may also be formed between two adjacent cells joining their cytoplasms. This type of pore is called a gap junction and is typically much wider, allowing larger molecule exchange between the cells. In either case, this route of cell entry is very fast, much more so than active transport through carrier proteins. The ion channels, which are specifically designed to transport inorganic ions across the lipid bilayer, can allow up to 100 million ions to pass per second.

One important difference between aqueous pores and ion channels is that aqueous pores remain open, but ion channels are gated and open and close in response to specific stimuli. There are several types of stimuli that can cause an ion gate to open. Voltage-gated channels open when a voltage change across the membrane occurs. Mechanical-gated channels open when a mechanical stress is applied or the membrane is stretched, and ligand-gated channels open when a specific ligand or receptor on the cell surface is bound. If the ligand is an extracellular mediator such as a neurotransmitter, it is a transmitter-gated channel. If the ligand is an intracellular mediator such as an ion or nucleotide, it is an ion-gated channel or nucleotide-gated channel, respectively. The main importance to cellular drug delivery is that there are many means to open an ion channel and these paths may be exploited to deliver drug molecules into cells. Under normal circumstances, however, ion channels allow only tiny ions to pass and may not be a feasible route to deliver any but the smallest drug. Similarly, aqueous pores may only be useful for delivering small-molecule drugs into the cell unless a novel drug carrier can act to enlarge the pores or channels, as some bacteria and neurotoxins do, permitting entry of larger substances.

2.2. Intracellular Transport

Many of the barriers encountered at the plasma membrane also occur within the cell because lipid membranes surround most of the intracellular compartments. An overall schematic of general intracellular transport is shown in Fig. 2.

As described earlier, movement of many substances through the cytosol occurs via diffusion. The rate of diffusion is inversely proportional to the size of the molecule. In addition, throughout this semi-liquid media, even the diffusion of small molecules is slow compared with that in water. Small drug molecules entering the cytosol are often metabolized and then taken up into the ER, mitochondria, or peroxisomes through transmembrane transport. Other substances, such as macromolecules or macromolecular assemblies, are transported into the nucleus through gated channels or nuclear receptors, which only allow specific molecules to pass through. In most cases, large macromolecules in the cell have sorting signals that make them selective to certain organelles. If taken up into the mitochondria or peroxisomes, the substances are handled through the metabolic pathways to generate energy. If taken up into the ER, they become part of the biosynthetic secretory pathway in which they enter the Golgi complex, and become modified or metabolized and packaged. Eventually they are sent off as secretory vesicles bound for the cell surface or are diverted to lysosomes.

3. Receptor Proteins and Active Transport

Much research has been done to determine receptor protein structure and function and how these receptors initiate uptake mechanisms within cells. The research efforts continue so that every minute detail can be known as to how cells work. This knowledge can help us better understand specifically what causes diseases and what drugs or types of therapy can be used to cure or treat them by allowing us to target specific receptors on specific cells. In this section, receptor proteins and their function will be discussed, with some specific examples given as to how proteins interact with drugs during cellular drug delivery.

3.1. Receptor Proteins

Proteins are composed of a specific sequence of amino acids. They are of great importance as enzymes in the cell, as receptors on the cell surface, and as biological motors and switches within the cell. But what makes some proteins function as receptors specific only to certain molecules and how do these certain molecules "fit" into the receptor so well? The answer lies in the numerous noncovalent interactions that play a major role in protein folding and receptor binding.

The long polypeptide chain of a protein, anywhere from 50 to 2000 amino acids long, is able to adopt many different conformations, limited mainly by steric interactions from the amino acid side chains. The sequential order of amino acids, where the

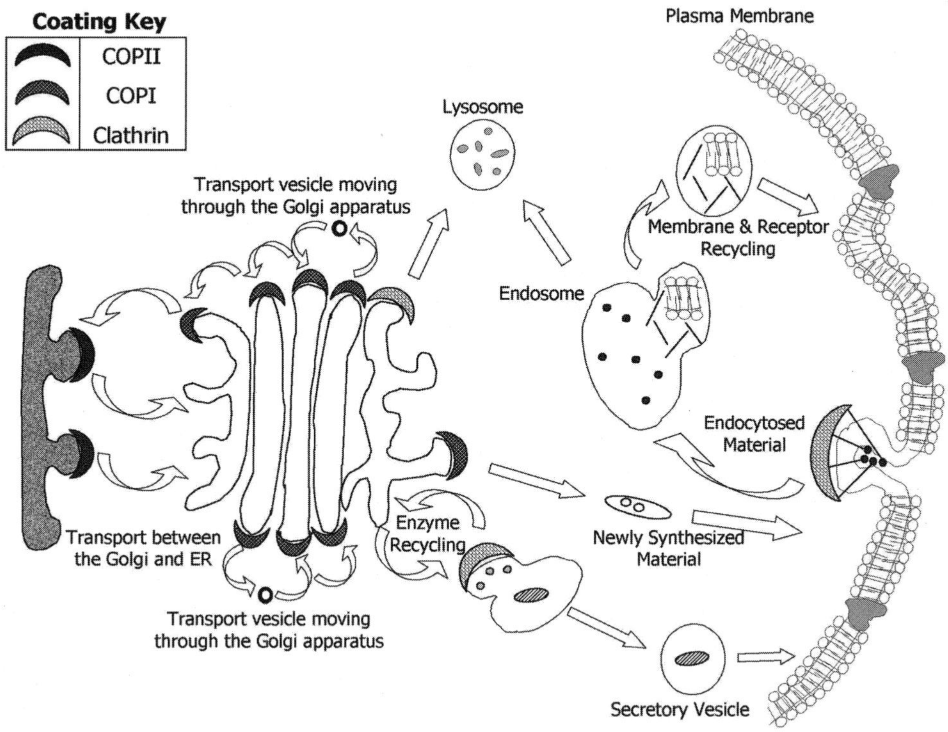

Fig. 2. General overview of the intracellular transport pathways occurring in most eukaryotic cells, including recycling of receptor proteins and membranes.

polar vs nonpolar groups are situated along the backbone, determines the folded shape of the protein as corresponding hydrogen bonds, ionic bonds, and van der Waals attractions hold specific sections together. The final conformation is usually the one in which the free energy of the system is at a minimum.

In order for a biological function to be carried out or activated, binding of a protein ligand or a drug to a protein receptor must occur. There are several different ways a ligand can bind to a particular protein receptor and most deal with only protein ligand or protein drug interactions at receptors. In the first case, the surface of one entity interacts with an extended string-like portion of another in a surface–string interaction. The second interaction, sometimes called a protein–protein interaction, occurs when two α-helices interact to form a coiled-coil structure. This type of interaction is often found in some gene-regulatory proteins and usually requires each entity to be a protein or large macromolecule with a coil structure. The third type of interaction is the most common one, generally stronger and more specific. It is the surface–surface interaction and it occurs when two protein surfaces are precisely matched up, such as the interaction between antibody and antigen. In most cases, the interaction is essential for the protein to function.

The binding of ligands or substrates to a receptor protein can initiate a protein conformational change. This plays a major role in cell signaling and enzyme regulation. For example, many antagonists or inhibitors bind to one site and prevent binding of other substances by causing the protein to shift conformations hiding other active sites.

This property of having more than one, differently shaped surface binding site is known as allostery and occurs in most receptor proteins. One site normally acts as a regulatory site for feedback mechanisms in the cell, whereas the other is the active site. When one site is bound, the protein "switches" on or off depending on which site is activated, and a new conformation with different surface contours is adopted. These types of changes not only occur to regulate pathways in the cell or signal changes, but also to move subcellular units around in the cell or move the cell itself around, as seen in motor proteins. Allosteric proteins can also use the energy from adenosine 5'-triphosphate (ATP) hydrolysis to accomplish active transport across cellular membranes. This mechanism of transport in addition to several examples will be discussed next.

3.2. Active Transport

There are several different kinds of active transport including ion transport through ion channels, electron pumps, and carrier-mediated transport through transporter proteins. In order for each of them to work, binding of one or more substrates must occur. The process involves specificity at the binding site and the potential for saturation of the binding sites owing to a high concentration of ligand. The transporters are capable of not only moving substances down a concentration or electrochemical gradient as in passive transport, but they can also pump substances against this gradient actively, which often occurs to get them inside the cell.

In general, carrier proteins bind drug molecules to bring them inside the cell without altering their chemistry. The process involves a reversible conformational change of the bilayer-embedded carrier proteins in order to initially bind to the ligand outside the cell and subsequently release it inside the cell. In most cases of active transport, energy is required to pump the ligand against the concentration gradient. There are three mechanisms in which cells achieve active transport: coupled carriers, ATP-driven pumps, and light-driven pumps.

3.2.1. Coupled Carriers

There are three kinds of carrier-mediated transporters. Those only transporting one type of substance from one side of the plasma membrane to the other are called uniporters. The other two types are symporters and antiporters, which transport either two types of substances at the same time from one side to the other or two types of substances in opposite directions across the membrane, respectively. In the latter case, the coupled carriers often use the electrochemical gradient driving force of one of the substrates to transport the other. In many cells, Na^+ is the co-substrate providing the driving force into the cell. When Na^+ comes into the cell, an ATP-driven pump embedded in the plasma membrane is activated to transport it out of the cell, thus maintaining a large electrochemical gradient for the influx of a particular substance. In other cases, H^+ ions are the co-substrate as found in the uptake of enalapril, an angiotensin-converting enzyme (ACE) inhibitor, by small intestinal brush border membrane vesicles of rabbits *(22)*. More information on transporters, their superfamilies, and prodrug approaches to cellular drug delivery utilizing membrane transporters is discussed in Chapter 7.

3.2.2. ATP-Driven Pumps and Light-Driven Pumps

Light-driven pumps are mainly found in bacterial cells and will not be discussed here in great detail. They use the energy from light to actively transport ligands against the electrochemical gradient.

ATP-driven pumps have an important role in cell regulation. The mechanisms that pump H^+ ions into lysosomes and other organelles where a low pH is required act similarly to the motor proteins. One of the best understood pumps of this sort is the Ca^{2+} pump in muscle cells, which uses the energy of ATP hydrolysis to change conformations. This allows it to bind ions, bring them inside the cell, and release them while returning to the lower energy conformation. Another critical ATP pump in cell regulation is the Na^+-K^+ pump, or Na^+-K^+ ATPase, as it is sometimes called. It is crucial not only to pH regulation, but also in the regulation of cell volume, propagation of nerve impulses, and transport of essential nutrients.

3.3 Proteins Involved in Multi-Drug Resistance

Finding a specific drug with high potency, delivered to the patient with a suitable and efficient formulation or drug carrier are not the only challenges facing drug-delivery scientists. Often, therapy is hindered by multi-drug resistance (MDR), and thus the understanding of the mechanism as well as how to overcome it becomes a tremendous task. There are many different resistance mechanisms that hinder the effective treatment of infectious and malignant diseases, but the resistance mechanism involving the ATP-binding cassette (ABC) superfamily of transporter proteins has gained the most attention over recent years owing to its role in some of the most common, widespread diseases such as malaria, leishmaniasis, and cancers *(23)*. In many cases, drug influx into the cell is not enough to overcome drug efflux out of the cell, resulting in the lack of drug accumulation inside the cell and thus unsuccessful therapy. Several transporters responsible for drug efflux are members of the ABC superfamily. These proteins act as ATP-driven pumps and use the energy from ATP hydrolysis to actively transport substances across the plasma membrane. They include P-glycoproteins (P-gp), multidrug resistance-associated proteins (MRP), and lung resistance-related proteins (LRP) *(24–26)*. Many MDR cell lines have been found to exhibit excess numbers of these transport proteins, thus causing a lack of drug accumulation within the cell owing to extensive efflux. Additionally, the intracellular pH has been linked to MDR, as found in drug-resistant mouse renal proximal tubule cells. Alkalinization of the endosomes and lysosomes by lysosomotropic agents have been found to stimulate the efflux of vinblastine, an anticancer drug *(26)*.

Cancer is one of the most well-studied illnesses in which MDR is a major problem. The reason many cancer chemotherapy regimens fail is multifactorial. Extracellular factors including bioavailability of the drug, drug metabolism, cell-cycle stage, and proliferation status, as well as intracellular factors of drug influx and efflux involving these ATP-driven efflux pump proteins, DNA replication, and cellular repair mechanisms, have all been identified as contributing significantly to MDR *(24)*. One method to address MDR is to block efflux of drugs from the cell to increase retention. Several different types of drugs have been studied as effective efflux blockers and chemosensitizing agents. They include verapamil, phenothiazine calmodulin inhibitors, cyclosporins, and tamoxifene, and have generally been shown to be effective in overcoming resistance *(24)*. The ABC superfamily and MDR is discussed in more detail in Chapters 8 and 9, and cancer and its treatment modalities are reviewed in Chapters 11 and 12.

4. Conclusion

The complexity of even a single cell presenting formidable barriers to drug delivery both intracellularly and extracellularly can be overwhelming. By studying the mechanisms employed by viruses, bacteria, and toxins in exerting their effect on cells, the drug-delivery scientist can gain important insights on how to exploit these mechanisms to deliver drugs effectively to target cells or target organelles, overcome barriers associated with uptake either passive (endocytosis) or active (transport proteins), or conquer MDR. A multidisciplinary background is necessary to understand the science of cellular drug delivery so that continuous improvements can be made in the quality of therapeutics.

Acknowledgment

The authors thank B. Neal Ware for critical reading of this manuscript.

References

1. Saltzman WM. Drug Delivery: Engineering Principles for Drug Therapy. New York, NY, Oxford University Press, 2001.
2. Alberts B, Johnson A, Lewis J, Raff M, Roberts K, Walter P. Molecular Biology of the Cell, 4th ed. New York, NY, Garland Science, 2002.
3. Tomlinson E. Pathophysiology and the temporal and spatial aspects of drug delivery. In: Site-Specific Drug Delivery. Tomlinson E, Davis SS, eds. Chichester, UK, John Wiley & Sons, 1986, pp. 1–26.
4. Voet D, Voet JG. Biochemistry, 2nd ed. New York, NY, John Wiley & Sons, 1995.
5. Campbell NA, Reese JB, Mitchell LG. Biology 5th ed. Menlo Park, CA, Benjamin/Cummings, 1999.
6. Kuchel PW, Ralston GB. Schaum's Outline of Theory and Problems of Biochemistry. New York, NY, McGraw-Hill, 1988.
7. Jensen KD, Nori A, Tijerina M, et al. Cytoplasmic delivery and nuclear targeting of synthetic macromolecules. J Control Rel 2003;87(1–3):89–105.
8. Rosenkranz AA, Lunin VG, Sergienko OV, et al. Targeted intracellular site-specific drug delivery: Photosensitizer targeting to melanoma cell nuclei. Russ J Gen 2003;39(2):198–206.
9. Dargemont C. Nuclear export of proteins: molecular mechanisms and functions. M S-Med Sci 2002;18(12):1237–1244.
10. Lackey CA, Murthy N, Press OW, et al. Hemolytic activity of pH-responsive polymer-streptavidin bioconjugates. Bioconj Chem 1999;10(3):401–405.
11. Washington N, Washington C, Wilson CG. Physiological Pharmaceutics: Barriers to Drug Absorption, 2nd ed. New York, NY, Taylor and Francis, 2001.
12. Hopkins CR. Site-specific drug delivery-cellular opportunities and challenges. In: Tomlinson E, Davis SS, eds. Site-Specific Drug Delivery. Chichester, UK, John Wiley & Sons, 1986, pp. 27–48.
13. Hillery AM. Drug delivery: the basic concepts. In: Hillery AM, Lloyd AW, Swarbrick J, eds. Drug Delivery and Targeting: For Pharmacists and Pharmaceutical Scientists. New York, NY, Taylor and Francis, 2001, pp. 1–48.
14. Gumbleton M, Hollins AJ, Omidi Y, et al. Targeting caveolae for vesicular drug transport. J Control Rel 2003;87(1–3):139–151.
15. McIntosh DP, Tan XY, Oh P, et al. Targeting endothelium and its dynamic caveolae for tissue-specific transcytosis in vivo: a pathway to overcome cell barriers to drug and gene delivery. Proc Natl Acad Sci USA 2002;99:1196–2001.
16. Sandvig K, van Deurs B. Membrane traffic exploited by protein toxins. Ann Rev Cell Dev Bio 2002;18:1–24.
17. Bertozzi C, Bednarski M. C-glycosyl compounds bind to receptors on the surface of escherichia-coli and can target proteins to the organism. Carb Res 1992;223:243–253.
18. Handel TM, Bertozzi C, Hubbell JA. Biopolymer engineering and design: beyond the genome—Editorial overview. Curr Opin Chem Biol 2001;5(6):675–676.
19. Lindau M, Almers W. Structure and function of fusion pores in exocytosis and ectoplasmic membrane-fusion. Curr Opin Cell Biol 1995;7(4):509–517.

20. Tsui MMK, Tai WCS, Banfield DK. Selective formation of Sed5p-containing SNARE complexes is mediated by combinatorial binding interactions. Mol Biol Cell 2001;12(2):521–538.
21. Lentz BR, Malinin V, Haque ME, Evans K. Protein machines and lipid assemblies: current views of cell membrane fusion. Curr Opin Struct Biol 2000;10:607–615.
22. Kitagawa S, Takeda J, Sato S. Uptake of enalapril by rabbit small intestinal brush-border membrane vesicles. Biol Pharm Bull 1999;22(7):762–764.
23. Cole SPC, Deeley RG. Multidrug resistance mediated by the ATP-binding cassette transporter protein MRP. BioEssays 1998:20(11):931–940.
24. Krishan A, Fitz CM, Andritsch I. Drug retention, efflux, and resistance in tumor cells. Cytometry 1997;29(4):279–285.
25. Sharma R, Awasthi YC, Yang Y, et al. Energy dependent transport of xenobiotics and its relevance to multidrug resistance. Curr Cancer Drug Targets 2003;3(2):89–107.
26. Ouar Z, Lacave R, Bens M, et al. Mechanisms of altered sequestration and efflux of chemotherapeutic drugs by multidrug-resistant cells. Cell Biol Toxicol 1999;15(2):91–100.

2

Cellular Growth, Development, and Defensive Response

Yan Huang, Guangliang Pan, Svein Øie, and D. Robert Lu

1. Normal and Abnormal Cell Growth

1.1. Cell Cycle

The cells of living organisms experience one of three fates: (1) live and reproduce, (2) live without reproducing, or (3) death. A cell reproduces by carrying out a series of processes in which it duplicates its contents and divides itself into two daughter cells. The cycle of duplication and division is known as the cell cycle, which involves four major events: cell growth, DNA replication, chromosome segregation into two identical sets, and cell division. The entire process can be divided into two fundamental parts: interphase (including G_1, S, G_2 phases) and mitosis (M phase) *(1–2)*, as illustrated in Fig. 1.

Interphase starts with a gap phase 1, or G_1 phase, during which the cell grows and prepares for the initiation of DNA replication. After the cell enters S phase, DNA synthesis starts and the whole chromosomes are replicated. In order to prepare for cell division, the cell enters the second gap period, or G_2 phase, and continues to grow and synthesize proteins such as mitosis-promoting factors for the next phase. The culmination of cell cycle is the triggering of the mitosis phase, during which the chromosome segregation and cell division take place. Interphase can be further subdivided into five phases: prophase, prometaphase, metaphase, anaphase, and telophase. During mitosis, chromosomes are condensed and nuclear envelope is broken down. The mitotic spindle is then formed and the chromosomes move to the opposite poles followed by cell division.

Typically, interphase lasts much longer time than the mitosis phase. For example, interphase lasts about 23 h, whereas mitosis lasts only about 1 h for human cell, with a total cycle time of 24 h. Although S phase and M phase are two major phases in a cell cycle, the two gap phases, G_1 and G_2, monitor the internal and external environment to ensure that conditions are suitable and preparations are complete for the S and M phases, respectively. The length of the G_1 phase is the most variable among all the phases, and it is greatly dependent on external conditions and extracellular signals. If extracellular conditions are unfavorable, cells delay progress through G_1 and may even enter a quiescent state known as G_0, in which they can remain metabolically active but no longer proliferate, unless called on to do so by appropriate extracellular signals. If extracellular conditions are favorable and signals for growing and dividing are present, cells in early G_1 or G_0 progress through a commitment point near the end of G_1 known

From: *Cellular Drug Delivery: Principles and Practice*
Edited by: D. R. Lu and S. Øie © Humana Press Inc., Totowa, NJ

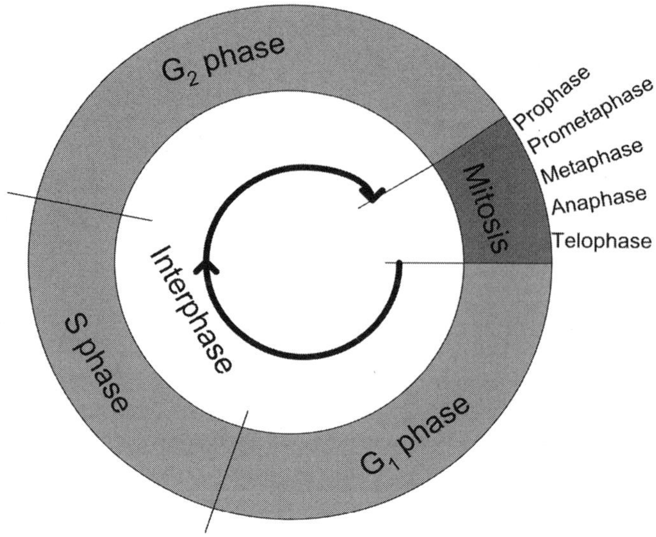

Fig 1. A typical diagram of cell cycle.

as the restriction point. After passing this point, cells are committed to DNA replication, even if the extracellular signals that stimulate cell growth and division are removed.

1.2. Intracellular and Extracellular Control of Cell Cycle

Normally, the cell cycle is a rather complicated and precisely coordinated process. In eukaryotes, it is mainly controlled by enzymes known as cyclin-dependent kinases (Cdks) that serve as the control system to coordinate the major transitions in a cell cycle. Coordination of the timing and order of these processes is achieved by a regulatory system that represents the checkpoints in the major transitions in the cell cycle. Two key checkpoints are at the G_1/S and G_2/M phase transitions. G_1 and G_2 cyclins and their associated catalytic subunits, Cdks, are responsible for controlling the transition. In most cells, the control system responds to various signals from both inside and outside the cell. Intracellularly, the control system monitors progression to ensure that the transitions are properly timed and ordered. For example, it is critically important that the cell does not begin mitosis until replication of the genome has been completed. Extracellularly, the control system analyzes its environmental conditions and regulates the progression according to the extracellular signals. Cdk activities oscillate as the cell progresses through the cell cycle. These oscillations lead directly to cyclical changes in the phosphorylation of key components of the cell-cycle process, resulting in the initiation of different cell-cycle events. For example, an increase in Cdk activity at the beginning of mitosis leads to protein phosphorylation, which controls spindle assembly, nuclear envelope structure, and chromosome condensation. The Cdk activities are in turn controlled by a combination of small inhibitory proteins know as cyclin dependent-kinase inhibitors (CKIs). It should be noted that a variety of other mechanisms also contribute to the control of Cdk activity, e.g., some Cdks exemplified by Cdk7 are able to control the activities of other Cdks, which make the control system a complex Cdk regulatory network *(3)*.

Extracellular signals contribute to the cell-cycle control by helping the cells determine whether there is need and whether it is ready to divide to ensure that the organism and its organs achieve and maintain an appropriate size. They are classified into three types based on the effects of these regulators on the progression in a cell cycle: cell division, cell growth, and cell survival *(1,2)*. By removing the restriction point that restricts the cell-cycle progression in G_1 phase, mitogens can stimulate the cell-division rate. On the other hand, growth factor can promote cell mass increase by stimulating the biosynthesis while inhibiting the biodegradation of macromolecules. Survival factors can inhibit the process of apoptosis and lead to an increase in cell numbers.

1.3. Abnormal Growth of Cell

Because the cell cycle is tightly regulated by intracellular and extracellular signals, it normally runs smoothly. However, cell mutation is a routine process in human body. Although most cell mutations are corrected efficiently by the human body, some can cause permanent disruption. Mutations, especially mutations of key regulatory genes, may cause much more rapid cell division than that of normal cells and lead to a growing mutant clone. If the mutant cells continue to grow vigorously and out of control, cancer can develop. The cancer cells do not obey the normal cell division restraints and they can invade the territories of neighboring cells. They can even transfer to other organs that are far away from them through body circulation. Cancers can be classified into two major categories based on their original cell or tissue type: carcinomas and sarcomas. The former refers to cancers arising from epithelial cells, and the latter refers to those arising from connective tissue or muscle cells. Many cancers do not fit into either of these two categories, e.g., various leukemias derive from hemopoietic cells, and others derive from the cells of nervous system.

The development of cancer can be viewed as a multistep process involving mutations and selection for cells with progressively increasing capacity of proliferation, survival, invasion, and metastasis. It starts with tumor initiation, which is the result of genetic alteration leading to abnormal proliferation of a single cell. The tumor progression turns an initially mild disorder of cell behavior gradually into a full-blown cancer. Clonal selection, a process similar to natural selection that a new clone of tumor cells has evolved on the basis of its parent by gaining increased growth rate, survival, invasion, and metastasis ability, makes the tumors become more rapidly growing and increasingly malignant.

A cell must acquire a whole range of aberrant properties as it evolves in order to turn an abnormal cell into cancer. Different cancers may require different combinations of properties. One of the common characteristics of cancer cells is that they disregard the external and internal signals that normally keep cell proliferation under tight control. A primary distinction between cancer cells and normal cells is that normal cells display density-dependent inhibition of cell proliferation. Normal cells proliferate to reach certain density and then cease proliferating, whereas cancer cells are insensitive to these signals. Cancer cells are also less stringently regulated, and insensitive to cell-contact inhibition, which makes them tend to be invasive and able to metastasize. In order to develop cancer, cancer cells tend to have longer life span as compared to their normal cell counterparts. Because cancer cells are usually blocked at an early stage of differentiation, they fail to differentiate normally, which is related to abnormal proliferation. Another important property of cancer cells is that they are genetically unstable, which makes it possible for further genetic alterations required for neoplasia and malignancy to occur.

1.4. Cell Cycle and Cancer Drug Targeting

Components of the cell-cycle machinery are frequently altered in cancer cells. Deregulated Cdk activity, combined with aberrant checkpoint control (at the G_1/S and G_2/M boundaries), leads to undesirable cell proliferation. The Cdks offer multiple mechanisms for intervention in the transformed state. In S phase, Cdk2/cyclin A is important for phosphorylation and inactivation of the E2F/DP1 transcription factor. Inhibition of Cdk2/cyclin A results in elevated E2F concentrations, which leads to S-phase arrest and apoptosis.

Among various regulatory proteins in a cell cycle, Cdks are essential for driving the cell through each stage of a cell cycle. Thus it is not surprising that Cdks are important targets for therapeutic intervention in various proliferative diseases, including cancer. By modulating Cdks, it is possible to block cell-cycle progression, induce apoptosis, promote differentiation, inhibit angiogenesis, and modulate transcription. For example, flavonoid can cause cell-cycle arrest at G_1 or G_2 and can inhibit the activation and the activity of several Cdks *(4)*. Because flavonoid affects Cdk activity directly, it is called a direct modulator. Other direct modulators include staurosporines, paullones, indirubins, roscovitine, olomoucine, and purvalanol *(3)*. On the other hand, indirect modulators, such as proteasome inhibitors, can block G_1 or G_2 transition. They have offered a promising new approach to treating cancers. The 26S proteasome regulates the turnover of proteins involved in cell-cycle control and apoptosis. It is relevant to human cancer because many intracellular proteins regulated by the ubiquitin-mediated proteasome degradative pathway govern the cell cycle, tumor growth, and survival. Lactacystin is one such inhibitor that was found to arrest umbilical vein cells at the G_1 phase of the cell cycle and to induce the nuclear accumulation of p53 in these cells *(5)*.

The variety of mechanisms controlling the major transitions in the cell cycle provide many possibilities for drug targeting along the cell cycle, such as drugs arresting at G_0/G_1, drugs arresting at G_1/S, and drugs arresting at G_2/M. For example, the anticarcinogenic potential of grape-seed polyphenols were found to be involved in the modulation of mitogenic signaling, induction of G_1 arrest, and apoptotic cell death *(6)*. Okadeic acid was found to arrest plasmacytoma cells at both G_2/M and S phases and induces vimentin expression in these cells *(7)*, thereby inducing apoptosis and causing differentiation of tumor cells. Although okadeic acid has not been tested in the clinic, it merits further development for the treatment of human proliferative diseases like cancers from the modulatory effects in the cell cycle.

2. Genetic Process and Control

2.1. DNA–RNA–Protein Process Overview

The growth and development of all cells and living organisms are dependent on the faithful transmission of genetic information from parent to offspring. Thus accurate replication of DNA is essential for all cells and organisms. DNA replication undergoes a process called DNA templating, in which the nucleotide sequence of a DNA strand is copied by complementary base-pairing (A with T, and G with C) into a complementary DNA sequence. The replication is done through a semi-conservative process, i.e., the resulting DNA contains a newly created strand and a strand from the parent DNA. DNA repair will follow to correct the mistakes in DNA replication through one of the two major DNA repair pathways: (1) direct reversal of chemical reaction responsible for the mistake, and (2) removal of the damaged bases followed by their replacement with newly synthesized DNA. The replication and repair mechanisms ensure that the daughter DNAs carry exactly the same genetic information as their parent DNA.

When a cell needs a particular protein, the nucleotide sequence of the appropriate portion in DNA molecule is first transcribed into mRNA. Once mRNA is produced, it can be used to synthesize the corresponding protein. Each group of three consecutive nucleotides in RNA is called a codon, and each codon specifies either an amino acid or a stop codon in the translation process. The translation of the nucleotide sequence of an mRNA molecule into protein takes place in the cytoplasm on a large ribonucleoprotein assembly called a ribosome, a complex catalytic machine made of the ribosomal proteins and the ribosomal RNAs (rRNAs). This process is generally divided into three stages: initiation, elongation, and termination. To initiate translation, a small ribosomal subunit binds to the mRNA molecule at a start codon (AUG) that is recognized by a unique initiator tRNA molecule. Elongation phase of protein synthesis follows, and each aminoacyl-tRNAs bearing a specific amino acid binds sequentially to the appropriate codon in mRNA by forming complementary base pairs with the tRNA anticodon. Each amino acid is added to the C-terminal end of the growing polypeptide. In the direction of 5'-to-3', the mRNA molecule progresses through the ribosome codon by codon until stop codon is reached. After the release factor binds to ribosome, translation is terminated and complete polypeptide is released. The flow of genetic information from DNA to RNA to protein is illustrated in Fig. 2.

2.2. The Control of Gene Expression and Drug Targeting

Different cell types in a multicellular organism differ dramatically in both their structures and functions, resulting from the different sets of RNA and protein molecules synthesized and accumulated.

The control of gene expression is exerted at multiple levels by changes of DNA content or position and changes in gene activity. The four main levels of gene activity control are transcriptional control, posttranscriptional control, translational control, and posttranslational control. Gene expression is a very complicated process, and many regulatory factors, including proteins and nucleic acids, are involved in this process. For example, it was recently discovered that many small RNAs play an important role in gene-expression control by affecting mRNA degradation *(8)*. In theory, all regulatory factors in this process can serve as the therapeutic targets for targeted drug development. These regulated steps and the corresponding drug targeting strategies are discussed in the following sections.

2.2.1. Transcriptional Control

In eukaryotic cells, transcription is a complex process in which many proteins work together to transcribe DNAs into mRNAs. It is the most regulated step in the genetic path from DNA to proteins. The DNA sequence to which RNA polymerase binds to initiate transcription of gene is called the promoter. Multiple proteins are required by the RNA polymerase to act as initiation factors. Some of these proteins bind to promoter sequences and direct the polymerase to the transcription start site. Transcription is initiated by eukaryotic RNA polymerase II, and it can be further stimulated by activators and inhibited by repressors. Transcription can also be regulated by some extracellular signals such as hormones.

In recent years, researchers have found that gene expression can be altered by the change of the chromatin shape, although the reason of such change is still mysterious.

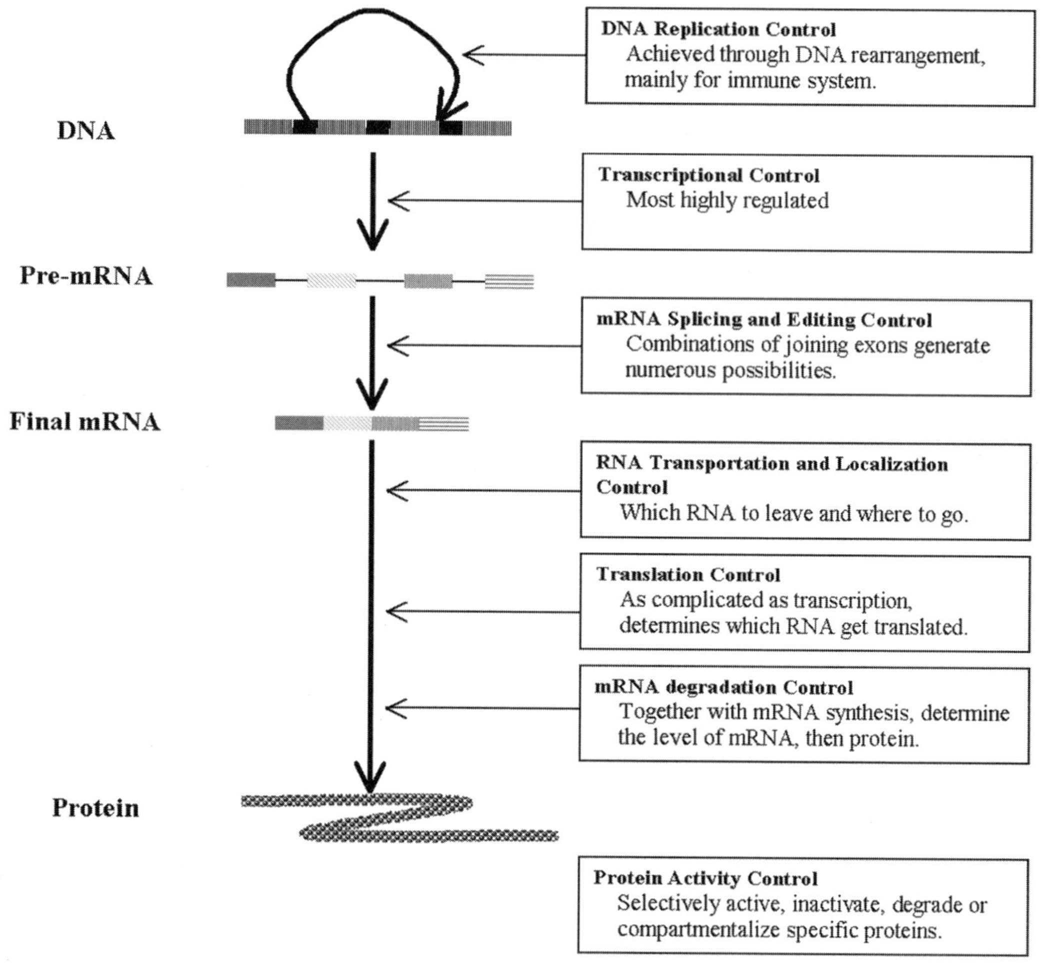

Fig 2. Diagram of genetic information flow (DNA–RNA–protein).

For example, in many chromosomes, loops or puffs of exposed DNA are sites of heavy transcription. Genes can be amplified when there is a heavy demand for a protein coded by a gene or set of genes, as the cell can make multiple copies of the genes so that more mRNAs can be simultaneously made.

The complicated control network of transcription provides numerous possibilities for drug targeting. Triple helix-forming oligonucleotides, for example, can be used to inhibit transcription initiation. The inhibition of binding of transcription activating factors by triplex formation can modulate the level of transcription of the target gene. This provides rationale for the development of new tools for cellular biology and of new therapeutic approaches to control gene expression at the transcriptional level *(9)*.

2.2.2. Posttranscriptional Control

In eukaryotes, transcription of protein-coding genes yields pre-mRNAs, which are spliced and edited into functional mature mRNAs that are used to guide protein synthesis. In eukaryotic cells, nascent pre-mRNAs are associated with a complex set of hnRNP proteins before transcription is completed. One function of these RNA-binding proteins is to help the formation of the structures recognized by RNA-processing factors.

In multicellular organisms, most pre-mRNAs are spliced to remove noncoding introns. Small nuclear ribonucleoprotein particles (snRNPs; U_1, U_2, U_4, U_5, and U_6) play a key role in this process. The snRNPs associate with splice sites to form a spliceosome (composed of many proteins and snRNPs), in which the splicing reactions occur. In this process, U_1 of snRNP is first bound to yield 5' splice site of pre-mRNA. The recognition of 5' splice sites is based on base pairing between the 5' splice site consensus sequence and a complementary sequence at the 5' end of U_1 snRNP. After U_2 snRNP binds to the branch point by a similar mechanism, a complex consisting of U_4/U_6 and U_5 snRNPs is then incorporated into the spliceosome.

Because most pre-mRNAs contain multiple introns, different mRNAs can be produced from the same gene by different combinations of 5' and 3' splice sites. The possibility of joining exons in varied combinations provides exponential ways of getting mature RNA from the same pre-mRNA. This process, called alternative splicing, increases the diversity of proteins that can be expressed from a single transcription unit. Alternative splicing plays an important role in the gene-expression process, and it is estimated that at least 35% of all human genes are alternatively spliced *(10)*.

Although the exact mechanism of splicing is unknown, it has been shown that several proteins contribute to the selection of a splice site and can affect the use of an alternative splice site in a pre-mRNA molecule. Sequence-specific RNA-binding proteins, for example, have been shown to bind near specific splice sites to either inhibit or activate splicing to the nearby site. Because transcriptional activators consist of DNA-binding domain and activation domain, which are generally encoded in separate exons, alternative splicing enables the production of many activators and repressors from the same gene. It has been shown that alternative splicing can be modified by using antisense oligonucleotides, and this provides a very promising potential chemotherapeutic target for cancer and other proliferative diseases *(10)*.

2.2.3. Control of mRNA Degradation

In mammalian cells, the levels of both rRNA and tRNA are relatively stable, whereas the stability of different mRNAs may vary greatly. These unstable mRNAs often code for regulatory proteins whose levels need to be changed rapidly to respond to various signals. For example, the hormone lymphokine produced by lymphocytes, which coordinate cell-cell interactions between the cells involved in the immune response of mammals, are synthesized and secreted in short bursts. Correspondingly, the mRNAs encoding these proteins must be synthesized and degraded in a very short period of time, and the switches in this controlling machinery must be turned on and off promptly.

The degradation of most eukaryotic mRNAs is completed by a gradual shortening of their poly-A tails. It has been demonstrated recently that some small RNAs contribute to mRNA degradation in RNA interference (RNAi). The exact mechanism about this process is still unknown, but scientists believe that enzyme complex called RISC uses the sequence information in small RNAs (miRNAs and siRNAs), which are combined to identify and degrade mRNAs with complementary sequence *(8)*.

The degradation of mRNAs can also be affected by extracellular signals. For example, mRNA encoding for transferrin receptors is regulated by the iron level within the cell. In the presence of adequate amounts of iron, transferrin receptor mRNA is degraded rapidly as a result of specific nuclease cleavage at a sequence near its 3' end.

Otherwise, the mRNA is stabilized, resulting in increased synthesis of transferrin receptor and more iron uptake by the cell. This regulation is mediated by a protein that binds to specific sequences (called the iron response element [IRE] near the 3' end of transferrin receptor mRNA, inhibiting the mRNA from cleavage.

2.2.4. Translational Control

Translation completes the flow of genetic information within the cell from DNA to protein. It is a complicated procedure wherein the information encoded in RNAs is translated into proteins. The translation process generally is divided into three stages: initiation, elongation, and termination. Ribosomes, which are composed of proteins and rRNAs, are the sites of protein synthesis for mammalian cells. During translation, the rRNA plays the dominant role in determining the overall structure of the ribosome, forming the binding sites for the tRNAs, matching the tRNAs to codons in the mRNA, and providing the peptidyl transferase enzyme to link amino acids together during translation.

Translational control can be achieved by binding repressor proteins, which block translation to specific mRNA sequences. It has been shown that malignant transformation in cancer could be caused by the increased translation of a subset of mRNAs encoding important proteins for cell growth and proliferation, which usually possess regulatory sequences that render their translation more sensitive to changes in the activity of translation initiation factors. By repressing the translation of the mRNA, that is required for cancer development, cancer cell growth and proliferation could be effectively inhibited. It has been reported that translational control in cancer cells could be an excellent target for anticancer drug development *(11)*.

2.2.5. Posttranslational Control

One important function of proteins is to serve as enzymes, and the regulation of enzyme activity plays an important role in governing cellular behavior. The catalytic activity of enzymes can be altered by changing their conformations. This can be achieved by binding small molecules, such as amino acids or nucleotides, to regulate enzyme activity. Protein activity can also be regulated by the interactions between the polypeptide chains that form proteins, and many activators and repressors function through protein–protein interactions. Proteins are targets of most traditional drugs, which use small molecules to inhibit functions of particular proteins *(12)*.

Just as with other molecules in the cells, the levels of proteins are determined by the difference between the rates of synthesis and the rates of degradation. Different rates of protein degradation affect the behavior of the cells, serving as an important factor for cell regulation. For example, regulatory molecules such as transcription factors usually have shorter half-lives to allow their levels to change quickly in response to various signals from both the inside and the outside cells. The ubiquitin/proteasome-dependent protein degradation pathway plays an essential role in both cell proliferation and cell death in human cancer cells *(13)*. The knowledge that proteasome function is required for tumor cell survival has prompted the design, synthesis, and evaluation of various pharmacological proteasome inhibitors. Both in vitro and in vivo experimental results have demonstrated the potential use of proteasome inhibitors as novel anticancer drugs.

2.2.6. Expression of Receptor Protein and Drug Targeting

It is known that many cancer cells overexpress certain cell-surface receptors, including transferrin receptor *(14)*, folate receptor *(15)*, and low-density lipoprotein (LDL) receptor *(16,17)*, to meet the increased cell proliferation and growth requirements. One strategy of developing targeted drug delivery for cancer therapy is to take advantage of these overexpressed cell-surface receptors. By incorporating the corresponding ligands for the cancer cell-surface receptors onto drug or drug carriers, an anticancer drug can be specifically delivered to the cancer site.

Tumor growth and development is related closely to tumor neovascularization, the growth of new blood vessels. The endothelial cells play a very important role in this process, and therefore, they are attractive targets for cancer therapy. More importantly, angiogenic endothelial cells overexpress certain proteins on the cell surface, which can be recognized only by a certain peptide sequences *(18,19)*. Thus, targeting angiogenic endothelial cells may become a very promising strategy in the treatment of cancers.

3. Cellular Defensive Response Systems

3.1. Human Immune System

Our body, like other living organisms, is in an environment that encounters many foreign invaders. Most of the foreign invaders that confront the human immune system are microscopic pathogens, including fungi, parasites, bacteria, and viruses. All foreign microbes in the human body display special markers, and it is these special markers that can be recognized as harmful and identified for destruction by the immune system. In order to survive in an environment with various types of pathogens, the body must have well-developed mechanisms in place to resist the infection by pathogens. These defenses can be categorized into two types: innate immune responses and adaptive immune responses *(20)*.

Innate immunity refers to antigen-nonspecific defense mechanisms and is the immunity with which one is born. It serves as the first line of defense for the body and is switched on immediately after an infection begins; it does not depend on the host's prior exposure to the pathogen. Adaptive immunity, however, operates later in an infection, is highly specific to the pathogen that has induced it, and is much more powerful. However, the adaptive immune responses are slow to develop on first exposure to a new pathogen. It may take a week or so before the responses are effective. The key difference between the two immune responses is that innate immune responses are not specific to a particular pathogen, whereas adaptive immune responses are.

The function of immune system is to remove or destroy invading pathogens and any toxic molecules they may produce. It is a crucial task for a healthy immune system to be able to determine what is foreign and what is its own. This applies to both innate and adaptive immune systems, although it is more important to the adaptive immune system, which is both more powerful and more destructive. Occasionally, the system fails to make this distinction and reacts destructively against the host's own molecules. Such autoimmune diseases can be fatal. On the other hand, tolerance can be problematic in some other cases like cancer. To the immune system, the difference between a cancer and a normal cell is so small that the immune system largely tolerates cancer cells rather than attacking them. In order to use the immune system to attack cancer cells, stimulations must be made to the immune system strongly enough to overcome this tolerance. Although the relation between the cancer and the immune system is still not clear, it is believed by most scientists that the cancer cells either have generated tolerance in the immune system or have developed ways of resisting immune recognition. In terms of cancer treatment, we need to identify ways to break the tolerance or circumvent resistance mechanisms *(21)*. Cancer vaccine is such a solution and will be discussed in later section.

3.2. Innate Immune System

Innate immunity consists primarily of a chemical response system, including complement, endocytosis, and phagocytosis. Macrophages, for example, detect and engulf extracellular molecules and materials, clearing the system of both debris and pathogens. The innate immune responses in vertebrates are also required to activate adaptive immune responses.

Innate immune responses rely on the body's ability to recognize conserved features of pathogens that are not considered as self, because the pathogen surface contains various classes of common pathogen-associated immunostimulants, such as many types of molecules on microbial surfaces and the double-stranded RNA of some viruses *(1,2)*. These molecules can be recognized by some dedicated receptors in the host, which are collectively called pattern-recognition receptors. The cell-surface receptors have two functions: (1) to initiate the phagocytosis of the pathogen, and (2) to stimulate innate immune responses.

Phagocytic cells display a variety of cell-surface receptors that enable them to recognize and engulf pathogens. These include pattern recognition receptors such as toll-like receptors (TLRs). There are two kinds of phagocytic cells: macrophages and neutrophils. Macrophages are usually long-lived cells residing in tissues throughout the body. They patrol the tissues of the body and are among the first cells to encounter invading microbes. Neutrophils, on the other hand, are short-lived cells that are abundant in blood but are not present in healthy tissues. They are rapidly recruited and dispatched to sites of infection by both activated macrophages and molecules released by the microbes themselves. Phagocytic cells trigger inflammatory responses to help fight infection and begin to activate the adaptive immune system. The adaptive immune system plays a major role after this point.

In recent years, macrophage-mediated gene delivery has been studied for cancer treatment by arming macrophages with the ability to express a therapeutic gene. In experiment, a hypoxia-regulated adenoviral vector was used to transduce human macrophages with either a reporter or a therapeutic gene encoding human cytochrome P4502B6. Infiltration of transduced macrophages into a tumor spheroid results in the induction of gene expression. They have significant tumor cell-killing ability in the presence of cyclophosphamide via activation by P4502B6. It has also been shown that this can be further targeted to tumors through hypoxia-regulated gene expression *(22)*.

It has been found that macrophages play very important roles in human immunodeficiency virus (HIV) dissemination to the whole immune systems. When they are infected by HIV, they stay alive, allowing HIV to live and replicate for a long time. Because macrophages express scavenger receptor proteins such as modified LDL receptor in the cell surface *(23)*, they have been considered to be very important therapeutic target for HIV infection. During the development of disease atherosclerosis, lipid-loaded macrophages appear in the blood vessel wall. Because of the scavenger receptors in these cells, they can serve as ideal therapeutic targets for this disease.

3.3. Adaptive Immune System

The adaptive immune system is a more advanced and powerful system than the innate immune system. The adaptive immune system is called on by the innate immune system to respond to pathogens. Unlike innate immune responses, the adaptive responses are highly specific to the particular pathogen that induced them, and the protection is also long-lasting. There are two broad classes of such responses: antibody responses and cell-mediated immune responses, which are carried out by different classes of lymphocytes, called B cells and T cells, respectively.

In the process of immune response, the pattern-recognition receptors present on the surface of various types of host cells activate intracellular signaling pathways in response to the binding of pathogen-associated immunostimulants. This leads to the production of extracellular signal molecules that promote inflammation and help activate adaptive immune responses if needed.

3.3.1. B Cells and Antibody-Mediated Drug Targeting

B cells respond to antigens by secreting antibodies. In antibody responses, the antibodies circulate in the bloodstream and permeate to other body fluids, where they bind specifically to the foreign antigen that stimulated their production. Binding of antibody inactivates viruses and microbial toxins by blocking their ability to bind to the receptors on host cells, thus blocking viruses from entering cells. Antibody binding also marks invading pathogens for destruction by making it easier for phagocytic cells of the innate immune system to ingest them.

There are two phases in the development of B cells: (1) the antigen-independent phase and (2) the antigen-dependent phase. The first phase involves the generation of diversity and the acquisition of rearranged heavy (H)- and light (L)-chain genes for synthesizing IgM antibody as a surface receptor protein. In the antigen-dependent phase, a B cell uses one of its receptors to bind to its matching antigen, which the B-cell engulfs and processes. The B cell then displays a piece of the antigen, bound to a class II major histocompatibility complex (MHC) protein on the cell surface. The whole complex binds to an activated helper T (Th) cell. This binding process stimulates the transformation of the B cell into an antibody-secreting plasma cell (Fig. 3).

Each B-cell clone makes antibody molecules with a unique antigen-binding site. Initially, during the development of B cell in the bone marrow, the antibody molecules are inserted into the plasma membrane, where they serve as receptors for antigen. In peripheral lymphoid organs, antigen binding to these receptors, together with costimulatory signals provided by Th cells, activates the B cells to proliferate and differentiate into either memory cells or antibody-secreting effector cells. The effector cells secrete antibodies with the same unique antigen-binding site as the membrane-bound antibodies. There are five classes of antibodies in mammals, namely IgM, IgD, IgG, IgA, IgE, each of which mediates a characteristic biological response following antigen binding.

Because antibodies are the product of immune response of B-cells to immunogens, antibodies can be used either as therapeutics or ligands for drug targeting. Because proteins like Her2 are abundant on surface of cancer cells, antibodies against such proteins can be armed with a toxin, or made to carry an enzyme that cleaves a harmless "prodrug" into a toxic molecule. The second case is especially useful because one molecule of enzyme can then generate a large number of toxic molecules. One virtue of this strategy is that the toxic drug generated enzymatically can then diffuse to neighboring tumor cells, increasing the odds that they can be killed, even if the antibody did not bind to them directly *(1,2)*.

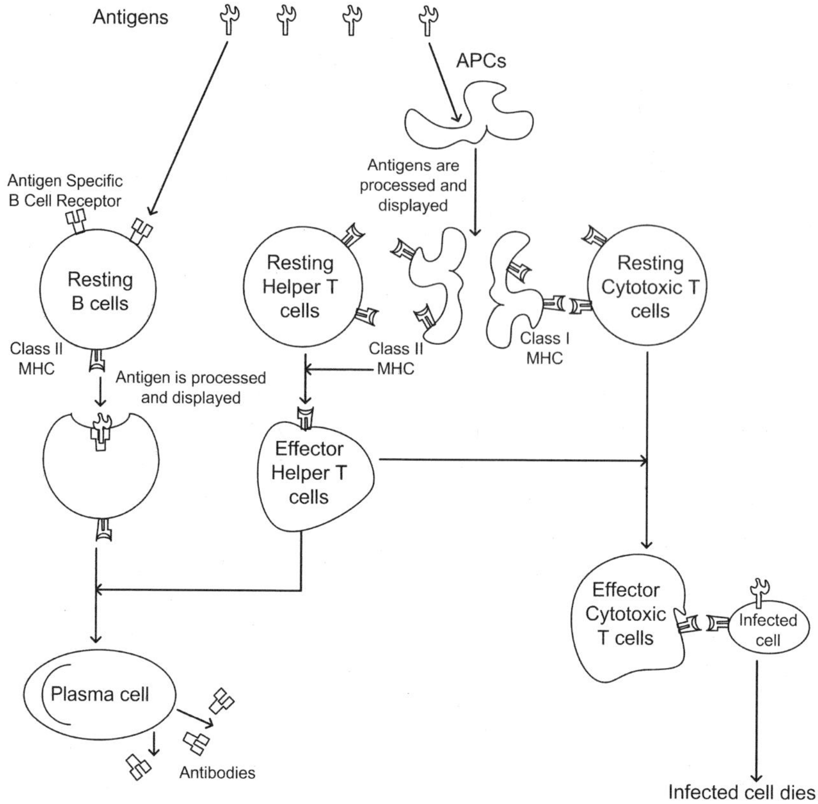

Fig 3. Activation of B cell and T cell.

3.3.2. T Cells in the Adaptive Immune System

The diverse responses of T cells are collectively called cell-mediated immune reactions. Most adaptive immune responses, including antibody responses, require Th cells for their initiation. Most importantly, unlike B cells, T cells can help eliminate pathogens that reside inside host cells. Activated T cells react directly against a foreign antigen that is presented to them on the surface of a host cell. For example, killer T cells possibly kill a virus-infected host cell, thereby eliminating the infected cell before the virus has a chance to replicate. T cells are also responsible for orchestrating, regulating, and coordinating the overall immune response. For example, Th cells alert B cells to start making antibodies (Fig. 3).

Like antibody responses, T-cell responses are exquisitely antigen-specific. But T-cell responses differ from B-cell responses in at least two crucial ways. First, they differ in their mechanism of antigen recognition. T cells depend on antigen-presenting cells (APCs) in peripheral lymphoid organs that present antigen to them. Protein antigens are partly degraded inside the APC, and are then carried to the surface of the presenting cell on special molecules called MHC proteins, which bind the fragments and present them to T cells. The second difference is the mechanism that T cells degrade the antigen. Once activated, effector T cells act only at short range, either within a secondary lymphoid organ or at the site of infection. They interact directly with other cells in the body, during which they either kill or signal in some way. Activated B cells, by contrast, secrete antibodies that can act far away.

Based on their functions in the immune system, T cells can be categorized into two classes: cytotoxic T cells and Th cells. The former kill infected cells directly by inducing them to undergo apoptosis, whereas the latter help activate B cells to make antibody responses and macrophages to destroy microorganisms. The Th cells also stimulate the activation of cytotoxic T cells. They are dependent on the unique cell-surface molecules, MHC, to help them recognize antigen fragments. They have receptors on their cell surface similar to antibodies, which recognize fragments of foreign proteins that are displayed on the surface of host cells in association with MHC proteins. Both are activated in secondary lymphoid organs by APCs, which express peptide-MHC complexes, costimulatory proteins, and various cell-cell adhesion molecules on their cell surface. Two kinds of MHC proteins, MHC-I and MHC-II, play crucial roles in presenting the antigens molecules to the cytotoxic T cells and Th cells, respectively. MHC-I are expressed in almost all vertebrate cells, whereas MHC-II are normally expressed in those cells interacting with Th cells, such as dendritic cells (DCs), macrophages, and B lymphocytes *(12)*.

MHC proteins have a single peptide-binding groove, which can bind a large and characteristic set of small peptide fragments derived from proteins. After they have formed inside the target cell, the peptide-MHC complexes are transported to the cell surface. Complexes containing a peptide derived from a foreign protein are recognized by T-cell receptors, which interact with both the peptide and the walls of the peptide-binding groove. The MHC molecules appear to be the keys to understanding and manipulating T cells, including both T-4 and T-8 cells. Based on the knowledge concerning the presentation of peptides by MHC molecules, a reliable method for predicting peptides that can bind to MHC, thus inducing T-cell immunity, can be of great value in modulating immune responses.

T cells play very important roles in human immune systems. However, they are subject to attack by foreign invaders, e.g., T-4 cells are the primary target of the HIV-1 virus. Because T-4 cells are responsible for the effective immune activity of macrophages, B lymphocytes, and other T lymphocytes, when they are depleted because of the HIV infection, the patient develops acquired immune deficiency syndrome (AIDS). The intensive study of the HIV replication has provided insight into the fundamental cellular mechanisms. One very important example is the discovery of protein TAT, an 86 amino acid transcriptional activator protein. TAT can be synthesized in one HIV-infected cell and transited into a neighboring cell without the requirement for the receptor-mediated event. This phenomenon has provided a great advantage to use TAT or fragment of TAT for efficient gene and drug delivery *(24,25)*.

3.3.3. Diversity and Specificity of the Adaptive Immune System

The most remarkable feature of the adaptive immune system is that it can respond to millions of different foreign antigens in a highly specific manner. There are two important issues related to adaptive immune system: (1) the generation of diversity, and (2) the specificity of the antibody and MHC molecules.

The immune system has the capacity to recognize and respond to a large amount of antigens. This extreme diversity can be generated in at least three possible ways: (1) multiple genes in the germ-line DNA, (2) variable recombination during the differentiation of germ-line cells into B cells, and (3) mutation during the differentiation of germ-line cells into B cells. The diversity of MHC molecules is generated in a similar process.

Once the diversified antigen-binding receptors are generated, how does the immune system pick the appropriate ones to respond to a specific antigen? This can be explained by the clonal-selection theory. The adaptive immune system is composed of millions of lymphocyte clones, with the cells in each clone sharing a unique cell-surface receptor that enables them to bind a particular antigen. With the binding of antigen on the cell-surface receptors, the lymphocytes proliferate and differentiate into functional lymphocytes. Lymphocytes have three possible fates: (1) some lymphocytes proliferate and differentiate into memory cells, which are able to respond faster and more efficiently the next time the same pathogen invades to prevent the body from the same illness in the future; (2) lymphocytes that would react against self molecules are induced to alter their receptors, induced to kill themselves, inactivated, or suppressed, ensuring self-tolerance; and (3) not to react at all, and subsequent death. In this process, both B and T cells circulate continuously between the blood and lymph until they encounter their specific foreign antigen in a peripheral lymphoid organ. After that, they proliferate and differentiate into effector cells or memory cells *(26)*.

3.4. Artificially Acquired Immunity

Based on the key properties of the immune system, we can defend our body using artificially acquired immunity. We can get the immunity artificially either passively or actively. Passive immunity is usually achieved by injecting antibodies into the body. For example, administering tetanus antitoxin to a patient is a way of conferring passive immunity *(27)*. Passive immunization is effective very quickly, but it lasts only a short time. It is mainly used to protect people when they are particularly vulnerable, such as immediately after exposure to a serious disease.

The purpose of vaccines is to stimulate the immune response like antibody formation or T-cell responses without subjecting a person to the risk of actual infection *(27)*. Most traditional vaccines rely on the vaccine's capacity to evoke antibody formation, but more attention is paid to vaccines evoking T-cell responses, especially for HIV vaccines and cancer vaccines. Recent developments in vaccine research have provided many ways to obtain active immunity, including conjugate vaccines, subunit vaccines, recombinant vector vaccines, naked DNA techniques, and vaccine presentation *(27)*. Conjugate vaccines are used to deal with certain bacteria having special outer coats that disguise antigens so that the immature immune systems are unable to recognize these harmful bacteria. Proteins or toxins from a second type of organism that is easier for an immature immune system to recognize are linked to the outer coats of the disease-causing bacteria. This enables an immature immune system to respond and defend against the disease agent. Subunit vaccines can be made by using a fragment of the microbe, because the fragment of the microbe is able to trigger the immune system while having far fewer side effects. Subunit vaccines made from *Streptococcus pneumoniae* have been used to protect against pneumonia. A recombinant subunit vaccine for hepatitis B virus infection made by inserting a tiny portion of the hepatitis B virus' genetic material into common baker's yeast is also in clinical trial.

A vaccine vector is a weakened virus or bacterium into which harmless genetic material from another disease-causing organism can be inserted. The vaccinia virus, for example, can be used to make recombinant vector vaccines because it is relatively large and has ample room to accept additional genetic fragments. A vaccinia virus with several genes from the HIV is currently being tested as a vaccine against AIDS. In addition, naked DNA technique has also been used to obtain active immunity. This is achieved by incorporating "naked DNA" encoding certain proteins from a disease-causing organism into the body's own cells. The proteins encoded by the DNA work as antigens to stimulate the immune system. In this way, the DNA will have an effect similar to that of a live, attenuated vaccine, thus produces antigens for years. The exclusion of genes critical to the disease-causing organism's survival also assures that the vaccines are safe and do not actually cause disease.

Vaccine presentation deals with presenting the vaccine to the immune system, which is closely related to drug delivery. Microspheres—tiny spheres containing bits of antigenic material—show promise in that they can release small doses of vaccine over extended periods of time as they gradually dissolve in the body. This makes two or more doses of vaccine in one administration possible.

3.4.1. HIV Vaccine

The focus of HIV vaccine research has progressed from HIV surface antigens and the role of antibodies in early stages to the importance of cytotoxic T cells. Subunit vaccines for HIV based on viral-surface proteins such as gp120 have the advantage of being safe and simple to prepare. It remains to be confirmed whether these vaccines can elicit antibodies capable of neutralizing primary HIV isolates. A vaccine candidate based on gp120 from two different HIV clades recently entered Phase III testing in the United States. Vectored vaccines employ non-HIV viruses (e.g., avian pox viruses) engineered to carry genes encoding one or more HIV epitopes. A related vector comparison study is ongoing and is evaluating three potential products to determine which vector produces the most robust immune response. By combining a canarypox-vectored product with a subunit vaccine, a hybrid vaccine has been studied to determine if a robust cellular and humoral response to HIV can be elicited. In addition, attenuated live or whole-killed HIV vaccines have shown promise in nonhuman primates.

3.4.2. Cancer Vaccines

Cancer vaccines work primarily to prevent cancer from recurring by alerting the body for certain characteristics of cancer cells and killing the remaining cancer cells. Although there are some promising results reported, the research in this area is still in a very early stage. It has been shown that the most effective anti-tumor immune responses are achieved by stimulating T-cells, which can recognize and kill tumor cells directly.

Cancer vaccine can be produced from cancer cells, parts of cells, or pure antigens *(27)*. Tumor-cell vaccines use killed cancer cells removed during surgery. The antigens on the killed tumor-cell surfaces can stimulate a specific immune-system response. As a result, cancer cells carrying these antigens are recognized and attacked. To increase the effectiveness of the vaccine, killed tumor cells may be further enhanced with nonspecific adjuvants. The advantages of using whole tumor cells is that they may expose the immune system to a large number of important cancer antigens, some of which may have not yet been identified. There are two basic kinds of tumor-cell vaccines: autologous and allogeneic. The former takes advantage of the tumor cells from the patient itself, whereas the latter utilizes cells from someone else *(28)*.

Dendritic cells are the most important APCs. When exposed to a foreign molecule, DCs can take up the foreign substance and display it, and stimulate the immune responses. This process is rather specific, as each patient's own DCs must be used as the foundation of the vaccine. It was shown recently at the Duke Comprehensive Cancer Center that RNA can be used instead of protein to make dendritic vaccines, which have dramatically expanded the scope of cancer vaccines, thus making the technology broadly applicable *(29)*.

By creating antigens that are easier for the immune system to recognize, antigen vaccines stimulate the immune system using individual antigens, rather than the whole tumor cells that contain many thousands of antigens. Although antigen vaccines may be specific for a certain type of cancer, they are not patient-specific like cell vaccines. Although a large amount of work needs to be accomplished before making significant progress, cancer vaccine holds promise in the battle with cancers.

4. Concluding Remarks

As the basic structural and functional unit of all living organisms, the cell is very complicated machinery with highly ordered structure and precisely controlled biological processes. Any deviation of the cell structure or function from the normal state can cause disease in the human body. In order to correct the cellular malfunction, drugs must be delivered efficiently to the cells to produce a pharmacological effect. Understanding the cell cycle, growth control, and cellular defensive response becomes essential for the design and development of therapeutics for targeted cellular drug delivery. The rapid development in human genomics, cell biology, and molecular biology will lead to the complete understanding of various cellular processes, and more and more targeted drug delivery systems will be developed in the near future.

References

1. Alberts B, Johnson A, Lewis J, Raff M, Roberts K, Walter P. Molecular Biology of the Cell, 4th ed., New York, Garland Science, 2002.
2. Lodish H, Baltimire D, Berk A et al. Molecular Cell Biology, 3rd ed., New York, Scientific American Boo, 1995.
3. Gali-Muhtasib H, Bakkar N. Modulating cell cycle: current applications and prospects for future drug development. Curr Cancer Drug Targets 2002;2:309–336.
4. Losiewicz MD, Carlson BA, Kaur G, Sausville EA, Worland PJ. Potent inhibition of CDC2 kinase activity by the flavonoid L86-8275. Biochem Biophys Res Commun 1994;201:589–595.
5. Kumeda SI, Deguchi A, Toi M, Omura S, Umezawa K. Induction of G_1 arrest and selective growth inhibition by lactacystin in human umbilical vein endothelial cells. Anticancer Res 1999;19:3961–3968.
6. Agarwal C, Sharma Y, Agarwal R. Anticarcinogenic effect of a polyphenolic fraction isolated from grape seeds in human prostate carcinoma DU145 cells: modulation of mitogenic signaling and cell-cycle regulators and induction of G_1 arrest and apoptosis. Mol Carcinog 2000;28:129–138.
7. Giese G, Wiegers W, Kubbies M, Scherbath A, Traub P. Okadeic acid coinduces vimentin expression and cell cycle arrest in MPC-11 mouse plastocytoma cells. J Cell Physiol 1995;163:145–154.
8. Couzin J. Breakthrough of the year: small RNAs make big splash. Science 2002;298:2296–2297.
9. Majumdar A, Khorlin A, Dyatkina N, Lin M et al. Targeted gene knockout mediated by triple helix forming oligonucleotides. Nat Genet 1998;20(2):212–214.
10. Mercatane DR, Sazani P, Kole R. Modification of alternative splicing by antisense oligonucleotides as a potential chemotherapy for cancer and other diseases. Curr Cancer Drug Targets 2001;1–3:211–230.
11. Cormier P, Pyronnet S, Salaün P, Mulner-Lorillon O, Sonenberg N. Cap-dependent translation and control of the cell cycle. Prog Cell Cycle Res 2003;5.

12. Cooper G, The Cell: A Molecule Approach, 2nd ed. Sunderland, MA, ASM Press, 2000.
13. Spence J, Gali RR, Dittmar G, Sherman F, Karin M, Finley D. Cell cycle-regulated modification of the ribosome by a variant multiubiquitin chain. Cell 2000;102:67–76.
14. Huwyler J, Wu D, Pardridge WM. Brain drug delivery of small molecules using immunoliposomes. Proc Natl Acad Sci USA 1996;93:14164.
15. Reddy JA, Low PS. Folate-mediated targeting of therapeutic and imaging agents to cancers. Crit Rev Ther Drug Carrier Syst 1998;5:587.
16. Jung-Testas I, Weintraub H, Dupuis D, Eychenne B, Baulieu D-E, Robel P. Low density lipoprotein-receptor in primary cultures of rat glial cells. J Steroid Biochem Mol Biol 1992;2:597–605.
17. Maletinska L, Blakely EA, Bjornstad KA, Deen DF, Knoff LJ, Forte TM. Human glioblastoma cell lines: levels of low-density lipoprotein receptor and low-density lipoprotein receptor-related protein. Cancer Res 2000;60:2300–2303.
18. Arap W, Pasqualini R, Ruoslahti E. Cancer treatment by targeted drug delivery to tumor vasculature in a mouse model. Science 1998;279:377–380.
19. Arap W, Kolonin MG, Trepel M, et al. Steps toward mapping the human vasculature by phage display. Nature Med 2002;8(2):121–127.
20. Paul W, Fundamental Immunology, 4th ed. Philadelphia, Lipponcott-Raven Publishers, 1999.
21. Pardoll D. Immunology: T cells and tumours. Nature 2001;411:1010–1012.
22. Griffiths L, Binley K, Iqball S, et al. The macrophage: a novel system to deliver gene therapy to pathological hypoxia. Nature 2000;7:255–262.
23. Matsumoto A, Naito M, Itakura H, et al. Human macrophage scavenger receptors: primary structure, expression, and localization in atherosclerotic lesions. Proc Natl Acad Sci USA 1990;87:9133–9137.
24. Torchilin VP. TAT peptide-modified liposomes for intracellular delivery of drugs and DNA. Cell Mol Biol Lett 2002;7(2):265–267.
25. Fawell S, Seery J, Daikh Y, et al. Tat-mediated delivery of heterologous proteins into cells. Proc Natl Acad Sci USA 1994;91:664–648.
26. Wolfe S. Molecular and Cellular Biology. Belmont, CA, Wadworth Publishing Company, 1993.
27. http://www.niaid.nih.gov/publications/vaccine/
28. http://www.thecancergroup.org/
29. Nair S, Hull S, Coleman D, Gilboa E, Lyerly HK, Morse M. Induction of carcinoembryonic antigen (CEA)-specific CTL responses in vitro using autologous dendritic cells loaded with CEA peptide or CEA RNA in patients with metastatic malignancies expressing CE. Int J Cancer 1999;82:121–124.

II

Principles of Cellular Drug Delivery

3

Augmentation of Cell-Mediated Immunity to Virus

Erica L. Woodahl and Rodney J. Y. Ho

1. Immune Responses to Viral Infection

Although most infections due to self-replicating pathogens such as bacteria and fungi can be treated by drug therapy and cleared by subsequent host antibody responses, viral infection may require additional defense mechanisms known as cell-mediated immunity. Unlike bacteria and fungi, virus replication requires host-cell machinery, including mechanisms of DNA, RNA, and protein synthesis. Therefore, after initial viral infection of host cells and tissues, viral RNA or DNA may persist in infected but viable cells. At an opportune time, viral replication is initiated, and the infected host cells are lysed and viral progeny released. These processes are observed clinically as an active, or recurring, viral infection and pathogenesis in the host. Although antibody, or humoral, responses elicited during the first exposure (primary infection) may help clear the reactivated virus, such mechanisms may develop well after severe consequences to the host have occurred. Therefore, it is important to control viral reactivation by selectively eliminating virus-infected cells, which is a specialized function of cell-mediated immunity. In order to elicit a cell-mediated response, select fragments of viral antigens must be presented on the surface of infected cells, along with major histocompatibility complex (MHC) molecules. MHC molecules are polymorphic glycoproteins that bind antigen and mediate migration of the antigen to the cell surface to be presented to immune cells. The induction of cell-mediated responses, which can directly or indirectly kill virus-infected cells, is critical in clearing viral infection from the body.

The unpredictable reoccurrence of herpes simplex virus (HSV) is an example in which the presence of humoral responses is not sufficient to clear infection, highlighting the role of cell-mediated immunity. The two major strains of HSV are HSV-1, which primarily infects the lips and manifests in cold sores, and HSV-2, which produces genital lesions. Primary HSV-1 infection is followed by a stage of latency, or a dormant state, where virus sequences can be found in neurons of the trigeminal ganglion *(1)*. HSV-1 infection starts in epithelial cells, causing a cold sore, and spreads to sensory neurons, whose axons innervate the lips. The initial infection is cleared by immune responses but HSV-1 then becomes latent in these neurons. In this latent state, HSV-1 does not replicate or synthesize protein and, therefore, exists in infected cells with low antigen density that is insufficient to trigger a cell-mediated response. In

From: *Cellular Drug Delivery: Principles and Practice*
Edited by: D. R. Lu and S. Øie © Humana Press Inc., Totowa, NJ

addition, HSV-1 and HSV-2 have been shown to avoid recognition by cell-mediated responses by expressing a viral protein called ICP47, which inhibits viral antigen processing and presentation through class I MHC (MHC-I) molecules on the surface of infected cells *(2)*. Although HSV-1 and HSV-2-infected individuals have circulating antibodies against HSV-1, they are unprotected from recurrences *(3)*. All of these data point to the requirement of functional cell-mediated immunity to control HSV viral replication and reactivation.

The immune system consists of distinct cell types with specialized roles at different stages of the immune response *(4)*. The differentiated immune cells are derived from hematopoietic stem cells, which are localized primarily in bone marrow. Immune system cells are found in the spleen, lymph nodes, tonsils, bone marrow, blood, thymus, and Peyer's patches in the intestine. White blood cells, including macrophages, dendritic cells (DCs), Langerhans cells, natural killer (NK) cells, mast cells, and basophils may interact with lymphocytes in the presentation of an antigen and thereby mediate the immune response. Two classes of lymphocytes, central to immunity against pathogenic challenge, are T lymphocytes (T cells), which mediate cell-mediated responses and B lymphocytes (B cells), which are important primarily in mediating humoral responses. Some of the key functions of T lymphocytes are regulation of the development of certain types of immune responses, such as autoimmune response, graft-v-host reactions, facilitation of antibody production, enhancement of the microbicidal activity of macrophages, and lysis of virus-infected cells and certain cancer cells. The B lymphocytes are precursors of antibody-secreting cells and may be involved in antigen presentation to T lymphocytes. When a naive B cell binds to an antigen, the cell rapidly divides and differentiates into mature B cells, called plasma cells. Plasma cells secrete immunoglobulins, which are antibody molecules that exhibit antitoxin activity and thereby neutralize pathogens. Immunoglobulins are found at high concentrations in blood or plasma. When viral particle levels subside in the host, the plasma cells stop secreting antibodies, but a small fraction of cells specific for that particular antigen, called memory B-cells, continue to exist. Additional details on the biology of antibodies and immunoglobulins and their therapeutic actions have been recently reviewed *(5)*.

Although the humoral response, mediated by antibody-secreting B cells, is an important component to combating viral infection, its effectiveness may be limited to clearance of extracellular viruses and to some extent preventing viral entry by processes of neutralization, complement activation, and opsonization *(6)*. Intracellular virus, either as residual or latent infection in host cells, must be controlled or eliminated selectively in order to completely recover from viral infection. Our immune system has developed antigen-specific cell-mediated mechanisms to seek out and clear infected cells. One of the key responses is the ability of cytotoxic T lymphocytes (CTLs) to recognize and produce contact-dependent cell-mediated lysis of virus-infected host cells. Because CTLs exhibit exquisite recognition of specific antigens and safeguards against autoimmune reactions, their mechanisms of action and capacity for clonal expansion have been well-studied over the past three decades. Figure 1 displays a simplified schematic of the interrelationships between antigen-presenting cells (APCs), and T and B lymphocytes, leading to humoral and cell-mediated responses.

In this chapter, we will briefly review the biological processes of a cell-mediated response and discuss strategies aimed at promoting cell-mediated responses against viral infection. We will also discuss how vaccine design can be improved to elicit cell-mediated immune responses through selective tissue and cell-delivery strategies.

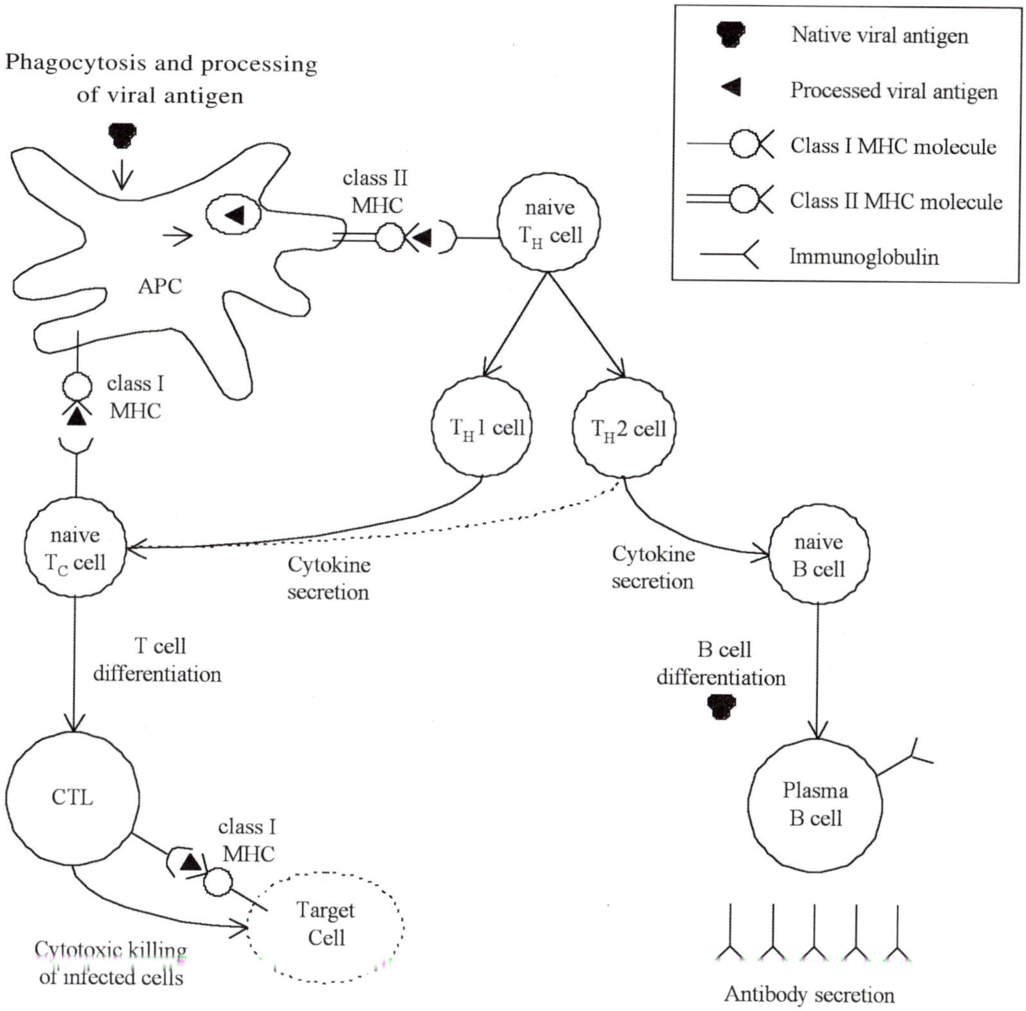

Fig. 1. Schematic representation of the relationships between humoral and cell-mediated responses to viral infection. Viral antigen is phagocytosed by APCs, and processed and presented with MHC-II molecules on the cell membrane. MHC-II molecules trigger differentiation of naive Th (or T_H) cells into two subpopulations, Th1 and Th2, which secrete a variety of cytokines. Th1 cells secrete cytokines that primarily induce the differentiation of naïve T_C cells into cytotoxic T lymphocytes, which seek out and lyse virus-infected cells expressing MHC-I molecules. Cytokines secreted by Th2 primarily promote naive B cells to differentiate into plasma cells that secrete large amounts of antibody specific for the viral pathogen. Live-attenuated vaccines generate an immune response similar to natural infection, initiating both humoral and cell-mediated responses. Without the enhancement of antigen-delivery systems and adjuvants, inactivated and subunit antigens are not optimally processed and presented in APCs and, therefore, primarily generate an antibody response through stimulation of B cells.

Advances in recombinant vaccine technologies have produced virus-free vaccines that are safe, but in need of enhanced cell-mediated responses. We will discuss some delivery systems and adjuvants that have been developed to augment the capacity and magnitude of cell-mediated responses, thereby influencing vaccine safety and efficacy.

2. Role and Mechanisms of Cell-Mediated Immunity

The cellular response to viral infection can be roughly divided into a nonspecific component, called innate immunity, and a specific component, called adaptive immunity. The innate immune response relies on a rather nonspecific recognition of most pathogens and is quickly activated, but it is often not sufficient to eliminate an infection. The innate system provides some control of the infection while the cells of the adaptive immunity are being activated. Typically, there is a 5- to 6-d delay between infection and the initiation of the specific adaptive, or acquired, immune response. Mounting an effective immune response against infection often requires selective recognition of a pathogen. Pathogen-specific immunity is provided by adaptive cell-mediated responses including CTL cell-dependent clearance of virus-infected cells. These effector cells are also maintained as memory cells specific for the particular pathogen. Viruses often have an ability to evade certain aspects of the immune responses. Therefore, duration and extent of viral infection would depend on the ability of the host to overcome the dynamics of virus–host interactions.

Mechanisms of both innate and adaptive immune systems work in concert to provide optimal immune responses in clearing extracellular virus and virus-infected cells. Individuals lacking effective innate immunity are not able to contain the initial stages of viral infection and, therefore, the virus replicates uncontrollably. Individuals lacking adaptive immunity are able to limit initial viral replication through processes of innate immunity, but ultimately are not able to clear infection without a specific response to the pathogen. Therefore, both mechanisms are essential for clearing viral infection from the body. The ability of cell-mediated and humoral responses to eliminate viral infection is discussed further in Box 1 using influenza infection as an example.

2.1. Innate Immunity Contributes to Cell-Mediated Immunity

Instead of specific recognition of a pathogen, mechanisms of the innate immune response rely on recognition of common features of a wide spectrum of pathogens or components of the immune system bound to pathogens. Innate immunity, unlike adaptive immunity, is not increased by exposure to a pathogen. An important element of the innate immunity is phagocytosis, which involves internalization, degradation, and presentation of antigen fragments on the cell surface. Phagocytic cells of the immune system include macrophages, dendritic cells, and neutrophils. Macrophages, dendritic cells, and B cells can act as professional antigen-presenting cells (APCs), which bind antigenic peptides from the ingested pathogen to MHC molecules and present them on the cell membrane. MHC molecules consist of two major types: MHC-I molecules, expressed on nearly all nucleated cells, and class II MHC (MHC-II) molecules, expressed only by APCs. APCs are important in communicating with the adaptive immune system by eliciting or recruiting T helper (Th) cells and CTLs, and play a major role in both humoral and cell-mediated immune responses.

The major antiviral activities of the innate response are the induction of cytokines and the activation of NK cells. A human cell infected with virus will secrete interferon-α (IFN-α) and interferon-β (IFN-β) *(7)*. The expression of IFN-α and IFN-β is triggered by double-stranded RNA, which is required for virus replication and not normally found in human cells. The antiviral response generated by IFN-α and IFN-β is mediated by binding to IFN-α/β receptors, which induces the expression of a ribonuclease that degrades viral RNA, and the expression of a protein kinase that inactivates viral protein synthesis. IFN-α and IFN-β also activate NK cells, important for combating viral infection. Interleukins 2 (IL-2) and 12 (IL-12) are also important in antiviral activity: IL-2, secreted by Th cells, is important in recruitment of CTLs and IL-12 is important in activation of NK cells.

NK cells are another important component of innate immunity, and are particularly crucial in containing viral infections through cell-mediated lysis of infected cells. NK cells are nonspecific lymphocytes recruited to the site of infection by inflammatory cytokines. The primary role of NK cells is to control viral replication through cytotoxic activity and the secretion of cytokines until cells of the adaptive immunity have been activated. NK cells also secrete tumor necrosis factor (TNF), which can further enhance their cytotoxic effect. NK cells can also be activated by the humoral response causing apoptosis in virus-infected cells in a process known as antibody-dependent cell-mediated cytotoxicity (ADCC). Target-cell killing by ADCC involves the binding of Fc receptors expressed on NK cells to the Fc region of IgG antibody molecules bound to the surface of infected cells. Macrophages, monocytes, neutrophils, and eosinophils can also mediate ADCC by similar mechanisms. NK cells are an important component in combating viral infection, especially in the early phase of infection before CTLs are developed.

2.2. Antigen Specific Cell-Mediated Immunity

Once an infection is established, cell-mediated responses are crucial in clearing virus-infected cells from the body and eliminating the infection. Cell-mediated responses are generated by various subpopulations of T cells. When a naïve T cell encounters an antigen presented with MHC, it differentiates into memory and various effector T cells. Effector T cells are the most important component of cell-mediated immunity and express a T-cell receptor that recognizes specific antigen epitopes bound to MHC molecules. Antigen-specific T cells are selected and expanded based on the particular viral antigen. The two main subpopulations of T cells, T helper (Th), and T cytotoxic (Tc) cells, work together to produce cell-mediated responses. Th cells express the membrane glycoprotein CD4 and are therefore also known as CD4⁺ T cells. Th cells recognize antigen displayed with MHC-II molecules on the surface of APCs, becoming activated to cytokine-secreting effector cells. Cytokines play an important role in activating and expanding cell-mediated immunity by inducing phagocytosis and activating B and Tc cells. Th cells are further subdivided into Th1 and Th2 according to their cytokine secretion profile. Th1 cells secrete IL-2, IFN-γ, and TNF-β, and are important in the activation of many cell-mediated responses including hypersensitivity reactions and CTL stimulation. Th2 cells secrete IL-4, IL-5, IL-6, and IL-10, and function as activators of B cells. Both IFN-γ and IL-2 activate NK cells; their importance in controlling viral infection was discussed in a previous section.

The second class of T lymphocytes, Tc cells, express the glycoprotein CD8 and are often referred to as $CD8^+$ T cells. Tc cells are activated by the recognition of MHC-I-presented antigens and by cytokines secreted by Th cells, and differentiate into CTL effector cells. CTLs secrete few cytokines; their primary function is to exhibit cytotoxic activity. Because nearly all nucleated cells express MHC-I-molecules, CTLs can recognize many types of virus-infected cells virtually anywhere in the body. The major function of CTLs in viral infection is to mediate contact-dependent cell-mediated lysis of infected cells, thereby eliminating the source of viral replication and reactivation.

CTL lysis is one of the most potent immune mechanisms for clearing a viral infection from the body *(8,9)*.

Box 1. The Role of Cell-Mediated Immunity in Influenza Infection

A classic example of viral infection that can be completely cleared by the body's humoral and cell-mediated immune systems is influenza infection, which has been extensively reviewed in the literature *(10,11)*. Influenza is a rapidly replicating virus; therefore, it is an effective stimulator of cell-mediated responses and is also efficiently cleared by CTLs and antibodies. Influenza viral particles are surrounded by an outer envelope, which is covered with two glycoproteins, hemagglutinin (HA) and neuraminidase (NA). Both of these glycoproteins have unique functions that allow them to infect and replicate in host cells. HA binds to sialic acid groups on the plasma membrane of host cells facilitating cell entry. NA cleaves *N*-acetylneuraminic (sialic) acid from viral and host-cell glycoproteins to facilitate viral budding from the host cell. A distinctive feature of influenza virus is the antigenic variation observed between different subtypes. The influenza virus has the ability to genetically modify the genes encoding HA and NA glycoproteins expressed on its surface so that the immune responses to different subtypes are completely altered.

Clearance of the influenza virus from the body involves both the innate and adaptive arms of the immune system. Innate mechanisms include the initiation of fever, production of interferon-α (IFN-α) and IFN-β, macrophages, NK cells, and complement. Influenza virus titer does decline in the early phase of infection; however, there is persistence of the virus, indicating that innate immunity is not sufficient for recovery. The humoral response to influenza is generated against the HA glycoprotein, but this specificity is dependent on the individual's prior antigenic exposure. However, adults who have been previously exposed to a different subtype produce cross-reactive antibodies, as well as strain-specific antibodies. The humoral response is capable of neutralizing the virus before infection of host cells and mediating the lytic action of ADCC in cells that have already become infected. However, individuals lacking antibodies to HA are still able to recover from influenza infection, indicating that cell-mediated responses are also important. Memory T cells, both species-specific and cross-reactive, correlate with rapid clearing of the virus. Effector T cells, including CTLs, are critical in the removal of infected cells and in the recovery from infection.

3. Immune Responses Elicited by Viral Vaccines

After an infection subsides, most effector B and T lymphocytes die; however, memory cells continue circulating and form the basis for protection against the same pathogen in the future. Vaccines often mimic natural infection by establishing pathogen-specific immune memory in vaccinated individuals. When a successfully vaccinated individual is exposed to the live pathogen, their immune system is armed and ready, and more likely to mount a rapid response. Often conventional vaccines are empirically designed to produce sufficient immunity. Vaccine antigens could be derived from live-attenuated viruses, inactivated (or killed) viruses, or recombinant sub-

Table 1
Current Viral Vaccines

Live-attenuated	Inactivated or killed	Subcellular/subunit
Adenovirus	Hepatitis A	Hepatitis A
Influenza (cold-adapted)	Influenza	Hepatitis B
Measles/Mumps/Rubella	Japanese encephalitis	HIV-1[a]
Polio	Polio	HSV-2[a]
Smallpox	Rabies	Influenza
Varicella zoster		
West Nile[a]		
Yellow fever		

[a]Vaccines in clinical trials.

unit fractions to elicit protective immune responses. Vaccines approved for human use in the United States, and those in development, are listed in Table 1.

Many US Food and Drug Administration (FDA)-approved viral vaccines provide sufficient antibody responses to abort the primary infection; for example, the influenza vaccine composed of inactivated or killed antigen and the alum adjuvant. Even with the help of the only adjuvant approved for human use, alum-induced inactivated viral antigens are presented to APC extracellularly, as opposed to intracellular presentation by virus-infected cells, leading to a bias towards antibody-mediated responses. However, many inactivated vaccines directed against viruses, such as human immunodeficiency virus-1 (HIV-1), may not provide a sufficient margin of safety; that is, vaccine candidates free of genetic material. Recombinant technologies have allowed for the production of vehicles with which to deliver vaccines intracellularly (i.e., particle systems, bacterial vectors, and viral-fusion proteins); however, the ability to produce safer avirulent or apathogenic vehicles remains elusive.

3.1. Traditional Vaccine Approaches

Traditional vaccine approaches have focused on immunogens derived from inactivated whole pathogens or components of those pathogens. The rationale is that inactivated pathogens will retain the ability to initiate a specific immune response, but will be unable to elicit a disease response. The majority of vaccines in use are live-attenuated and inactivated or killed vaccines.

Live-attenuated vaccines have the ability to transiently infect a vaccinated individual but are designed to prevent significant adverse effects. Attenuation is achieved by growing the pathogen in non-natural culture conditions and selecting for variants that show reduced pathogenicity. Therefore, these attenuated pathogens will be less capable of growing in a natural environment. Because of the pathogen's limited ability for transient infection in a vaccinated individual, there will be a prolonged exposure to the immune system, allowing for stronger immunity. A disadvantage of live-attenuated vaccines is that there may be a low frequency of reversion of the pathogen to a more virulent stage. Nevertheless, live-attenuated vaccines are good at initiating both humoral and cell-mediated immunity.

The second type of vaccines in use is inactivated or killed vaccines. A whole pathogen is treated either chemically (formaldehyde or various alkylating agents) or physically (heat, irradiation, ultraviolet [UV]) to inactive the pathogen. The major benefit with inactivated pathogens is that they can no longer replicate in a vaccinated host. However, many of the methods used to inactivate pathogens could also destroy the epitopes essential for immune-system recognition. Inactivated vaccines are less potent than live-attenuated vaccines because killed viruses can no longer infect cells. Consequently, inactivated vaccines are better at eliciting a humoral response and less effective at generating a cell-mediated response. Under stringent inactivation processes, inactivated pathogens could no longer produce a transient infection in their host, and repeated booster immunizations are necessary to achieve the desired immunity.

3.2. Recombinant Subunit Vaccines

Instead of using the whole pathogen to formulate vaccines, scientists have also made vaccines using purified subcellular macromolecules of a pathogen, usually recombinant proteins. Only one or a limited number of the most important antigens of a pathogen are used in formulating subunit vaccines. Subunit vaccines, free of pathogen genetic material, provide significantly reduced risk of adverse effects. The primary advantage of subunit vaccines is an improved margin of safety because the vaccine is manufactured from recombinant DNA, eliminating the risk of accidental infection owing to viral DNA or RNA. Recombinant proteins can be produced in bacterial or mammalian hosts, allowing for greater yield and purity, better quality control, and lower costs. However, to engineer appropriate subunit vaccines, the key antigenic target of the pathogen must be identified. This can prove to be difficult if a particular pathogen contains a wide spectrum of subspecies or highly variable strains.

Like inactivated or killed vaccines, subunit vaccines are nonreplicative in cells of a vaccinated individual. Therefore, subunit vaccines are also much better at eliciting a humoral response than they are at activating a cell-mediated response. The potential benefits of the improved safety have prompted efforts to develop or improve vaccines using delivery systems and adjuvants that will augment, or enhance, the cell-mediated response to subunit vaccines.

4. Rational Augmentation Strategies

Although a wide spectrum of adjuvants have been developed and demonstrated to enhance the immunogenicity of experimental vaccine antigens, the only adjuvant currently approved by the FDA in the United States for human administration is alum (aluminum salts). Although alum is sufficient to increase antibody response, it is not effective in enhancing cell-mediated immunity. Recent advances in immune recognition mechanisms have allowed for the development of adjuvant designs that can augment cell-mediated responses of antigen. Viral vaccine strategies may benefit from formulation designs that consider two key steps in delivery of antigen to tissues and cells of the immune system. First, antigen must be delivered to target immune tissues where antigens are processed and presented. Second, after antigen has been delivered to these tissues, some degree of success in intracellular delivery to APCs must be achieved to elicit Th cell and effector CTL responses, mimicking natural viral infection in host cells.

The advent and maturation of recombinant DNA technology have provided the ability to readily produce recombinant viral antigens as vaccine candidates. Therefore, if recombinant viral proteins can provide sufficient cellular and humoral responses, their

Table 2
Targeted Delivery of Antigen to Lymphoid Tissues

Carrier	Route of administration	Lymphoid tissue target	References
Liposomes	Intravenous	Spleen/liver	*(13–16,47,49,55,56)*
	Subcutaneous	Lymph nodes/ Lymphatic tissues	
Saponins; ISCOMs	Intravenous	Spleen/liver	*(47,58–60)*
	Subcutaneous	Draining lymph nodes	
Emulsions	Intramuscular or Subcutaneous	Draining lymph nodes	*(12,61–65,67–70)*
Gene gun	Transcutaneous	Skin	*(25,26)*
Imiquimod	Transcutaneous	Skin	*(21–24)*
Toxin-mediated	Transcutaneous	Skin	*(17,19,27)*
Edible vaccines	Oral	GALT	*(17,18,78,79)*
Immunoglobulin complexes	Oral/Inhalation	MALT	*(35–37)*
Microparticles (e.g., PLG)	Oral/Inhalation	MALT	*(20,80,81)*

potential is tremendous. In addition, adverse vaccine reaction owing to pathogenic reversions encountered in using avirulent virus or bacterial vaccine delivery systems could be avoided. As it is more challenging to elicit cell-mediated responses than humoral responses with recombinant proteins, we will focus our discussion on enhancing cell-mediated responses to viral protein antigens with some experimental delivery systems and adjuvant approaches.

4.1. Strategies to Deliver to Lymphoid Tissues

The first step in delivery of viral vaccine is targeting the vaccine to lymphoid tissues where the antigen can be exposed to cells of the immune system. Diverse organs and tissue are involved in immune presentation throughout the body. The lymph nodes, spleen, cutaneous tissues, and various mucosal-associated lymphoid tissues (MALT) are tissues that trap antigen and allow the antigen to interact with lymphocytes and APCs. Mucosal sites include tonsils, appendix, Peyer's patches in gut-associated lymphoid tissue (GALT), and mucous membranes lining the digestive, respiratory, and urogenital tracts. Targeting or concentrating antigen in these lymphoid tissues will enhance the ability of the immune system to develop cell-mediated responses. Alum-precipitated antigen typically is administered either intramuscularly or subcutaneously and distributes nonspecifically to muscle and other cells in the vicinity of administration and therefore is not very efficient at delivery of antigen to specific tissues of the immune system. Various other delivery systems have been used to increase delivery of antigen to tissues important in immune presentation (Table 2).

Delivery of viral antigen to lymph nodes has been well-studied with several different delivery and adjuvant systems and provides an extremely efficient mode of immune presentation. Emulsions have been shown to concentrate antigen in draining lymph nodes where antigen encounters phagocytic cells of the immune system *(12)*. Liposomes are also extremely efficient in delivering antigen to lymphatic tissues, and the route of administration influences the ultimate localization of antigen in specific tissues *(13–16)*. After intravenous administration, it is clear that large liposomes are taken up efficiently by macrophages and DCs of the reticuloendothelial system in blood and tissues, including the liver and spleen. Subcutaneously and intramuscularly administered large liposomes may be trapped at injection sites and serve as a drug depot. Small liposomes (50–80 nm) administered subcutaneously will be retained in draining lymph nodes and eventually distribute throughout the lymphatic system and lymphoid tissues *(16)*. Saponins and immunostimulating complexes (ISCOMs) also concentrate antigen in the lymphatic system, perhaps by mechanisms similar to liposomes.

Mucosal immunity, especially in the lungs, intestine, and urogenital sites, is an important first line of defense against infection. Induction of systemic immunity often cannot induce sufficient mucosal immunity; therefore, many researchers have investigated approaches to deliver antigen directly to MALT, which are rich in APCs. Edible vaccines provide a method for oral delivery of viral antigens into Peyer's patches in GALT without the use of additional adjuvants *(17,18)*. Ingestion of transgenic plants expressing viral antigen can provide antigen-specific immunity, and could provide an economical alternative to purified recombinant protein. Genetically modified bacterial enterotoxins, including cholera toxin, heat labile toxin, and pertussis toxin, can induce mucosal immunity, although there are still concerns about safety *(17,19)*. Liposomes, saponins, and microparticles may also be important adjuvants in enhancing mucosal immunity either by oral administration or inhalation into the respiratory tract *(17,20)*. Typically, mucosal vaccines display low immunogenicity and therefore may require higher doses to elicit a response.

Delivery of viral antigens through the skin is an efficient route to eliciting cell-mediated responses. The epidermal layer of skin contains a type of APCs called Langerhans cells, which internalize and present antigen with MHC-II molecules and migrate to lymph nodes were they encounter and activate Th cells. Langerhans cells also stimulate the secretion of IFN-α, interleukin (IL)-12, and IFN-γ, further enhancing cell-mediated responses. A novel method of targeting antigen to cutaneous tissues is with Imiquimod, a topically applied immune response modifier, which activates Langerhans cells. Increases in Th1 and CTL cell-mediated responses have been observed against HSV and human papillomavirus (HPV) when formulated with Imiquimod *(21–24)*. Cell-mediated immune responses have been augmented with other transcutaneous delivery mechanisms such as gene-gun antigen-delivery approaches, which enhance DNA-particle complex penetration across skin cells *(25,26)*, and delivery of antigen and enterotoxin via an adhesive patch *(27)*.

Increased understanding of mechanisms involved in immune responses has brought about new ways for researchers to think about designing and targeting more specific and effective vaccines, so called "virtual pathogens" *(28)*. By combining various elements that are known to induce or regulate cell-mediated responses, researchers could generate prototypic viral vaccines that will mimic natural infection. These viral vaccines potentially could deliver subunit vaccines containing viral antigen epitopes, known to initiate immune responses, and cytokines, known to regulate the responses, in a lipid bilayer. Other molecules can be added to further enhance immune responses, including mannose receptors targeting APC receptors important for antigen uptake and processing, and prokaryotic signals such as CpG motifs, which will also activate APCs. Virtual pathogen vaccines may hold promise for incorporating the immunogenic properties of a virus without causing pathogenicity.

Table 3
Delivery Vehicles for Intracellular Targeting

Delivery vehicles	Examples	Viral protein antigen	References
Particles	Liposomes	HIV; HSV; influenza; hepatitis A; hepatitis B; adenovirus; rabies; measles; rubella	*(14,15,48,50–54)*
	Saponins; ISCOMs	HIV; SIV; influenza; HSV; hepatitis C; hepatitis B	*(82–93)*
	Emulsions	HSV; influenza; HIV; CMV; hepatitis B; SIV	*(63–65,67–70,94)*
	Microparticles	HIV; SIV; hepatitis B; Influenza	*(19,20,80,81,95)*
Viral vectors	Vaccinia virus	HIV; HSV; rabies; Newcastle disease; Japanese encephalitis	*(96,97)*
	Adenovirus	hepatitis B; rabies	*(98,99)*
	Alphavirus	HIV; SIV; influenza	*(100,101)*
	Adeno-associated virus	HSV	*(102–104)*
	Poliovirus	SIV	*(105)*
	Yellow Fever virus	West Nile virus	*(106)*
Bacterial vectors	*Salmonella*	HSV; influenza; lymphocytic choriomeningitis virus	*(107,108)*
	Listeria monocytogenes	HIV; influenza; lymphocytic choriomeningitis virus	*(109,110)*
	Vibrio cholera	HIV; rotavirus	*(108,111)*

4.2. Delivery Strategies for Intracellular Targeting

After delivery of viral antigen to lymphoid tissues, the next issue in augmenting cell-mediated immunity is intracellular delivery. Several delivery systems can target viral antigen to intracellular compartments of cells, necessary for MHC-I processing and presentation and CTL responses. Some of the approaches, including viral and bacterial vectors, designed for intracellular delivery are summarized in Table 3. Live vectors are quite efficient at mediating intracellular delivery of antigen; this issue has been extensively reviewed *(29–31)*. Hence, we will focus on nonviral approaches, free of genetic material, to target viral antigens intracellularly.

Particulate carriers such as liposomes, ISCOMs, emulsions, and microparticles efficiently deliver viral antigen into the cytosol of cells, mimicking natural infection, to elicit a cell-mediated response *(19,32)*. The dimensions of these particles, similar to that of infecting microorganisms, allow for phagocytosis and some degree of success in delivery of antigen of APCs for presentation by both MHC-I and MHC-II molecules. These vaccine carriers may also induce cytokine secretion, further augmenting cell-mediated responses.

Entrapment of antigens in biodegradable microparticles may enhance intracellular delivery as well as lymphoid tissues targeting. Poly(lactide-co-glycolide) (PLG) polymers and other polymeric materials are especially efficient in delivery of antigen to mucosal tissues *(17,19,20)*. When delivered orally, PLG polymers are taken up, with some degree of preference, by cells in Peyer's patches of the lower intestine where enriched populations of APCs are located. Intranasal administration of PLG polymers leads to concentration of antigen in draining lymph nodes. In addition, PLG polymers fabricated as microparticles may enter the cytosol of epithelial cells on the mucosal surface, allowing for presentation by MHC-I molecules eliciting a cell-mediated response.

Several other techniques are also under investigation to enhance intracellular delivery. Modified liposomes, called cochleate delivery vehicles, are multilayered structures composed of a lipid bilayer sheet rolled up or stacked in sheets with no internal aqueous space *(33)*. Antigens incorporated within the interior of the cochleate structure are protected from degradation. Cochleates fuse with cell membranes and deliver antigens into the cell. The use of lipopeptides is another vaccine strategy that may provide enhanced intracellular delivery of antigens without the need for additional adjuvants *(34)*. Lipopeptides are composed of antigen covalently linked to lipid moieties, which may help penetrate the cell membrane of APCs, allowing rapid intracellular delivery of the antigen. Immunotargeting of antigen, fused to an Fc fragment, may also enhance antigen uptake into APCs *(35–37)*. The antigen-Fc complex binds Fc receptors on dendritic cells, is internalized and processed, and is subject to MHC-I presentation.

Enhanced cellular uptake of antigen could be achieved with immunostimulatory oligonucleotides containing cytosine-phosphate-guanosine (CpG) motifs. These motifs are found in bacterial DNA and have been shown to induce multiple immune responses. CpG motifs are composed of approx 20 unmethylated cytosine and guanine dinucleotides, usually with two 5' purines and two 3' pyrimidines at either end *(38)*. CpG motifs are recognized as common features of an infectious agent by cells of the innate immune system through the Toll-like receptor (TLR) 9 *(39)*. Binding of the CpG DNA to the TLR initiates endocytosis and induces macrophages and DCs to secrete cytokines, which induces NK cells and aids in Th1 differentiation and enhancement of CTL levels *(40–42)*. CpG motifs may also induce secretion of a broad range of cytokines that may further enhance immune responses. The cell-mediated responses elicited by CpG motifs appear to be enhanced by attachment to protein antigens and formulation with other delivery systems such as alum, liposomes, and emulsions *(19,43)*. CpG oligonucleotides have been shown to induce humoral and cell-mediated responses, either alone or in combination with alum or incomplete Freund's adjuvant, to a variety of viral pathogens, including hepatitis B *(44,45)*, HSV-1 and -2 *(46)*, and HIV-1 *(41)*.

In the following, we will discuss in some detail some the mechanisms of vaccine delivery systems for protein antigens designed to augment cell-mediated responses. We will also discuss the role administered cytokines play in enhancement and regulation of immune responses.

4.2.1. Liposomes Mediate Tissue and Intracellular Delivery

Liposomes are colloidal particles composed of phospholipid molecules in the formation of an enclosed lipid bilayer. Soluble antigens can be enclosed in the internal aqueous space or amphipathic antigens can incorporate in the lipid bilayer. Liposomes have been used to enhance antigen-specific immune responses for various vaccines. The influence of physiochemical properties of liposomes on the antigen immune response, such as size, charge, membrane fluidity, and epitope density, has been studied extensively *(47)*. Liposomes deliver encapsulated antigen both to lymphoid tissues for antigen presentation and intracellularly to elicit cell-mediated responses. As discussed earlier, the route of administration of liposomes is important in delivering antigen to appropriate tissues. Once in lymphatic tissues, liposomes will be phagocytosed and liposome encapsulated antigen will be processed and presented by APCs. In addition to macrophages that predominantly present antigen with MHC-II molecules, antigen delivered to some endothelial, Langerhans, and dendritic cells is more likely to be presented by MHC-I molecules, therefore mediating cellular immune responses, including CTL activation *(48)*.

Liposomes also can deliver antigen intracellularly by fusing with the plasma membrane to deliver antigen into the cytoplasm of APCs to induce a cell-mediated response. Therefore, liposome formulation may provide an excellent way to enhance antigen delivery and presentation to stimulate both humoral and cell-mediated immune response to vaccines. The efficiency of intracellular delivery will depend on the composition of liposome membranes and the inclusion of bio-sensors to amplify delivery. For example, listeriolysin O (LLO) is a purified protein from the cytosol-invading bacterium, *Listeria monocytogenes*, that enhances intracellular delivery of antigens when encapsulated in liposomal carriers. LLO has been shown to deliver ovalbumin into the cytosol of APCs; however, the technology will allow for other antigens, such as viral peptides, to be introduced in this manner *(49)*. Researches have shown that subcutaneous immunizations of LLO-containing liposomes generated stronger CTL responses than intravenous immunizations, consistent with our hypothesis that subcutaneous injection directs liposomes to draining lymph nodes, and eventually the entire lymphatic system, for antigen presentation. Immune stimulators have also been incorporated into liposomes to increase their stimulation of the immune system. When the lipophilic adjuvant muramyl tripeptidyl-phosphatidylethanolamine (MTP-PE) is in liposome formulation with HIV-envelope protein, a dramatic increase is observed in humoral and cell-mediated responses *(50)* or HSV-1 and -2 antigens *(51–54)*. Liposomes can also be actively targeted to specific cells within tissues to enhance the immune response *(16)*. A variety of ligands or receptors unique to the target cells have been formulated in the liposome membrane, including antibodies, growth factors, cytokines, hormones, and toxins *(16)*.

Another method to increase intracellular delivery of viral antigens is to use specialized liposomes designed with viral [illegible] proteins, called virosomes. Virosomes enhance intracellular delivery by utilizing targeting and fusogenic properties of viruses *(19)*. Two new virosome-based recombinant protein vaccines currently on the market incorporate influenza virus surface glycoproteins into the lipid bilayer to mediate intracellular delivery of viral antigens. The first virosome vaccine approved for human use was a hepatitis A vaccine (Epaxal®) containing liposome-encapsulated formalin-inactivated hepatitis A virus particles *(55)*. The advantage of Epaxal is that the immune response is enhanced for both hepatitis A antigen and influenza virus antigen. A second virosome-based influenza vaccine is Inflexal V®, a trivalent product that generates enhanced immunogenicity compared with conventional influenza vaccines *(56)*.

4.2.2. Saponins Provide Intracellular Delivery of Antigen

Saponins are compounds with potent immunological activity purified from *Quillaja saponaria*, a South American soap-bark tree. Saponins, which form micellar suspensions, have been shown to have a strong adjuvant effect on antibody and cell-mediated responses. The commonly used saponin Quil A is actually a partially purified mixture of various saponins from crude extracts of *Quillaja saponaria*. Further purifications of Quil A have provided purified saponins with potent adjuvant effects such as QS-21 and QS-7, which may also produce less toxicity than Quil A. QS-21 has been shown to enhance immune responses to gp120 HIV-1_{MN} envelope protein in human immunizations, and may allow for the administration of a lower dose of antigen compared to alum *(57)*.

A unique feature of saponins is their strong affinity for cholesterol; this ability is exploited in the formulation of ISCOMs, which are spherical, colloidal particles composed of saponins (typically Quil A), cholesterol, phospholipids, and antigen *(47,58)*. ISCOMs are composed of regularly ordered micellar subunits that form the shape of a ring. ISCOM formulation with antigenic proteins is more restricted than formulation of antigens with liposomes. Although amphipathic proteins are usually easily incorporated into ISCOMs, extremely hydrophobic, large, and soluble proteins may be harder to formulate. The adjuvant effect of ISCOMs formulated with Quil A has been shown to be greater than the effect of Quil A alone. The mechanisms by which Quil A and ISCOMs elicit an enhanced immune response are not entirely understood; however, they appear to work in a similar fashion. Quil A and ISCOMs cause a local inflammatory response, which recruits lymphocytes and macrophages important in antigen presentation. Various saponins may stimulate T and B cells differently and to various degrees *(59)*. Like liposomes, ISCOMs and saponins strongly enhance CTL responses, perhaps by fusion with the cell membrane of APCs *(47)*. ISCOMs are able to generate a number of immune responses, including APC activation, secretion of IL-2 and IFN-γ, and the induction of Th cell and CTL responses *(60)*.

4.2.3. Emulsions Target Antigen to Lymph Nodes

Oil-in-water emulsions have been shown to induce both humoral and cellular immune responses to subcellular or subunit vaccines. Emulsions are made through a microfluidization process that creates small oil droplets formulated with antigen in an aqueous solution. A mechanism of action for emulsions may be the depot effect, in which emulsions facilitate the transport of antigen to the lymph nodes *(12)*. Emulsions may also interact with cell membranes in a manner similar to liposomes. Two prototypical emulsion adjuvants that have been studied to enhance the immune response are complete Freund's adjuvant (CFA) and incomplete Freund's adjuvant (IFA); however, both have been found to exhibit toxicity and have not been approved for human use. Freund's adjuvants stimulate a local inflammatory response at the site of injection, which attracts both macrophages and lymphocytes. IFA, composed of antigen in an aqueous solution, mineral oil, and an emulsifying agent, allows for the antigen to be released slowly from the injection site. CFA is similar to IFA, except that it contains immunostimulatory mycobacterial cell-wall components, such as muramyl dipeptide in the emulsion that makes CFA more potent that the incomplete form. However, the inclusion of bacterial elements in CFA increases the potential for toxicity. A less toxic oil-in-water emulsion, Adjuvant 65, has been used as an adjuvant but is also not approved for human administration *(61,62)*.

Two of the new safer approaches to using emulsions as a modifier of immune responses are MF59 and Syntex Adjuvant Formulation-m (SAF-m). MF59, a squalene-in-water emulsion, has been shown to augment cell-mediated responses to HSV *(63)*, influenza *(63)*, HIV-1 *(64)*, and cytomegalovirus (CMV) *(65)* subunit vaccines, and can also be formulated with the immunostimulator MTP-PE, which provides further enhancement. The release rate of MTP-PE from MF59 is higher than from liposomes, therefore increasing the risk of severe local and systemic reactogenicity *(66)*. Regardless, MF59 appears to be internalized by DCs, which function as APCs and activate mechanisms of cell-mediated responses *(12)*. SAF-m is another squalene-in-water emulsion used as an adjuvant to enhance cell-mediated immunity to a variety of viruses, including hepatitis B *(67)*, influenza *(68)*, simian immunodeficiency (SIV) *(69)*, and HSV *(70)*. However, some of the newer emulsions may also have significant adverse effects.

4.2.4. Cytokines Play a Role in Regulating and Enhancing Immune Responses

Cytokines are an important additional consideration in enhancing cell-mediated responses to subunit vaccines. Cytokines are a group of small proteins important in cell-to-cell communications that influence immune interactions throughout the body. The biological activities of cytokines are often mediated by multiple receptors expressed on cell surfaces. Cytokines can generally be divided into two categories: colony-stimulating factors (CSF) and lymphokines. Colony-stimulating factors, granulocyte colony-stimulating factor (G-CSF), granulocyte-macrophage (GM-CSF), and monocyte CSF (M-CSF), are large glycoproteins produced primarily by APCs to stimulate the growth of immune progenitor cells. Lymphokines, such as ILs, are produced by leukocytes and stimulate leukocyte development and the production of IFNs *(71)*.

Cytokines regulate the immune system by stimulating or inhibiting the activation and differentiation of various cells and by regulating the secretion of antibodies and other cytokines. Cytokines may modulate antigen presentation by activating APCs and increasing the number of activated APCs. APCs express numerous cytokine receptors on their cell surface, and binding of a cytokine to its receptor results in activation, migration, and maturation *(28)*. Cytokines also regulate B cell proliferation and differentiation into antibody-secreting cells. Once antigen is presented to T cells by APCs, cytokines are important in enhancing cell-mediated immunity by regulating the clonal expansion of antigen-specific T cells into Th cells and CTLs.

A number of laboratories have investigated the ability of interleukins to enhance cell-mediated immunity. IL-2 has been shown to influence multiple immune functions in addition to its ability to induce T-cell proliferation. IL-2 stimulates the growth of both cytotoxic and Th cells as well as B cells, macrophages, and NK cells *(72)*. Mice vaccinated with a fusion protein composed of HSV glycoprotein D (gD) and human IL-2, displayed complete immunity against HSV challenge, whereas those vaccinated with gD formulated in alum showed no such protection *(73)*. IL-2 and IL-7 also produced similar protection against HSV infection through enhancement of CTL responses *(74)*. IL-2 may also have clinical relevance in HIV infection. Studies have shown that IL-2 therapy in HIV patients, in combination with highly active antiretroviral therapy (HAART), increases CD4 and CD8 cell counts, decreases HIV-1 latently infected CD4 cells, and decreases viral load in peripheral blood mononuclear cells (PBMCs) compared to HAART alone *(75)*.

A second interleukin, IL-12, may also be important in enhancing cell-mediated immune responses to viruses. The primary functions of IL-12 are promoting Th1 and NK cell growth and secretion of IFN-γ, and activation of CTLs. Cell-mediated immunity enhancement by IL-12 may also be important in HIV vaccine and therapy; some approaches using IL-12 have been reviewed *(76)*. IL-12 has also been shown to work synergistically with GM-CSF and TNF-α to increase CTL response to HIV-1 antigen. An HIV vaccine trial in nonhuman primates found that IL-12 also stimulates both humoral and cell-mediated immune responses to HIV-1 envelope protein gp120 *(77)*.

Cytokines are the messengers of the immune system and generally act locally on cells in close proximity to the producer cell. Consequently, cytokines have short half-lives and systemic concentrations are low. The high systemic levels of cytokines often administered to induce cell-mediated immunity to virus are frequently associated with adverse effects. Therefore, the toxicity and short half-life of cytokines in vivo has limited their effectiveness as an adjuvant *(28)*. Targeted delivery of cytokines could provide sufficient concentrations at sites of antigen presentation and expansion of B and T cells while limiting systemic cytokine levels. When successfully delivered at the site of action, cytokines may provide cell-mediated immune enhancement not achieved by increased delivery of antigen alone. Therefore, incorporating cytokines into other targeted delivery systems and adjuvants may enhance their ability to elicit and regulate cell-mediated responses.

5. Summary

Cell-mediated responses that can eliminate virus-infected cells are important for combating viral infections. Mechanisms of both innate and adaptive immunity play a role in cell-mediated responses to viral infection, but the activation of antigen-specific CTL responses is critical for complete viral clearance from the body. Advances in recombinant protein technologies have allowed for the development of subunit vaccines. These recombinant vaccines are safer than traditional vaccines but tend to elicit humoral rather than cell-mediated immunity. Several vaccine delivery and adjuvant strategies discussed in this chapter could provide significant augmentation of cell-mediated immune responses to viral antigens, thus providing a greater degree of protection. The two important issues in enhancing cell-mediated responses to viruses are targeting viral antigens and cytokines to immune tissues central to proper antigen presentation, and efficient intracellular delivery of antigen to APCs. It is likely that one or more of these vaccine and cytokine delivery systems and adjuvants will provide breakthroughs in recombinant vaccines capable of inducing both humoral and cell-mediated responses.

Acknowledgments

The authors wish to thank Tami Daley for her editorial assistance. Supported in part by NIH Grants GM62883, AI 31854, AI 52663, HL 56548, and ES07033, NS 39178. ELW was the recipient of the William E. Bradley Fellowship in Pharmaceutics and the NIH Pharmacological Sciences Training Grant GM 07750.

References

1. Simmons A. Clinical manifestations and treatment considerations of herpes simplex virus infection. J Infect Dis 2002;186 (Suppl)1:S71–S77.
2. Whitley RJ, Roizman B. Herpes simplex viruses: is a vaccine tenable? J Clin Invest 2002;110:145–151.
3. Morrison LA. Vaccines against genital herpes: progress and limitations. Drugs 2002;62:1119–1129.
4. Paul WE. Fundamental immunology. New York, Raven Press, 1993:xvii;1490.
5. Ho RJ, Gibaldi M. Antibodies and derivatives. In: Ho RJ, Gibaldi M, eds. Biotechnology and Biopharmaceuticals. Hoboken, NJ, Wiley-Liss, 2003.
6. Lachmann PJ, Davies A. Complement and immunity to viruses. Immunol Rev 1997;159:69–77.

7. Welsh RM, Lin MY, Lohman BL, Varga SM, Zarozinski CC, Selin LK. Alpha beta and gamma delta T-cell networks and their roles in natural resistance to viral infections. Immunol Rev 1997;159:79–93.
8. Ramshaw IA, Ramsay AJ, Karupiah G, et al. Cytokines and immunity to viral infections. Immunol Rev 1997;159:119–35.
9. Doherty PC, Topham DJ, Tripp RA, et al. Effector CD4+ and CD8+ T-cell mechanisms in the control of respiratory virus infections. Immunol Rev 1997;159:105–117.
10. Ada GL, Jones PD. The immune response to influenza infection. Curr Top Microbiol Immunol 1986;128:1–54.
11. Ghendon Y. The immune response to influenza vaccines. Acta Virol 1990;34:295–304.
12. Vogel FR. Improving vaccine performance with adjuvants. Clin Infect Dis 2000;30(Suppl)3:S266–S270.
13. Rao M, Alving CR. Delivery of lipids and liposomal proteins to the cytoplasm and Golgi of antigen-presenting cells. mangala.rao@na.amedd.army.mil. Adv Drug Deliv Rev 2000;41:171–188.
14. Zheng L, Huang XL, Fan Z, et al. Delivery of liposome-encapsulated HIV type 1 proteins to human dendritic cells for stimulation of HIV type 1-specific memory cytotoxic T lymphocyte responses. AIDS Res Hum Retroviruses 1999;15:1011–1020.
15. Ahmad N, Khan MA, Owais M. Liposome mediated antigen delivery leads to induction of CD8+ T lymphocyte and antibody responses against the V3 loop region of HIV gp120. Cell Immunol 2001;210:49–55.
16. Lian T, Ho RJ. Trends and developments in liposome drug delivery systems. J Pharm Sci 2001;90:667–680.
17. Ryan EJ, Daly LM, Mills KH. Immunomodulators and delivery systems for vaccination by mucosal routes. Trends Biotechnol 2001;19:293–304.
18. Mason HS, Warzecha H, Mor T, Arntzen CJ. Edible plant vaccines: applications for prophylactic and therapeutic molecular medicine. Trends Mol Med 2002;8:324–329.
19. Singh M, O'Hagan DT. Recent advances in vaccine adjuvants. Pharm Res 2002;19:715–728.
20. Brayden DJ, Baird AW. Microparticle vaccine approaches to stimulate mucosal immunisation. Microbes Infect 2001;3:867–876.
21. Miller RL, Gerster JF, Owens ML, et al. Imiquimod applied topically: a novel immune response modifier and new class of drug. Int J Immunopharmacol 1999;21:1–14.
22. Hengge UR, Benninghoff B, Ruzicka T, Goos M. Topical immunomodulators—progress towards treating inflammation, infection, and cancer. Lancet Infect Dis 2001;1:189–198.
23. Miller RL, Tomai MA, Harrison CJ, Bernstein DI. Immunomodulation as a treatment strategy for genital herpes: review of the evidence. Int Immunopharmacol 2002;2:443–451.
24. Rudy SJ. Imiquimod (Aldara): modifying the immune response. Dermatol Nurs 2002;14:268–270.
25. Fynan EF, Webster RG, Fuller DH, et al. DNA vaccines: protective immunizations by parenteral, mucosal, and gene-gun inoculations. Proc Natl Acad Sci USA 1993;90:11478–11482.
26. Ellis RW. Technologies for the design, discovery, formulation and administration of vaccines. Vaccine 2001;19:2681–2687.
27. Glenn GM, Taylor DN, Li X, Frankel S, Montemarano A, Alving CR. Transcutaneous immunization: a human vaccine delivery strategy using a patch. Nat Med 2000;6:1403–1406.
28. Buchan GS, Young SL, Lockhart EA, et al. Targeting early events in T cell activation to construct improved vaccines. Mol Immunol 2000;37:545–552.
29. Monahan SJ, Salgaller ML. Viral vectors for gene transfer into antigen presenting cells. Curr Opin Mol Ther 1999;1:558–564.
30. Morrow CD, Novak MJ, Ansardi DC, et al. Recombinant viruses as vectors for mucosal immunity. Curr Top Microbiol Immunol 1999;236:255–273.
31. Bramson JL, Wan YH. The efficacy of genetic vaccination is dependent upon the nature of the vector system and antigen. Expert Opin Biol Ther 2002;2:75–85.
32. Foged C, Sundblad A, Hovgaard L. Targeting vaccines to dendritic cells. Pharm Res 2002;19:229–238.

33. Gould-Fogerite S, Mannino RJ, Margolis D. Cochleate delivery vehicles: applications to gene therapy. Drug Deliv Tech 2003;3:40–47.
34. BenMohamed L, Wechsler SL, Nesburn AB. Lipopeptide vaccines: yesterday, today, and tomorrow. Lancet Infect Dis 2002;2:425–431.
35. Barber BH. The immunotargeting approach to adjuvant-independent subunit vaccine design. Semin Immunol 1997;9:293–301.
36. Guyre PM, Graziano RF, Goldstein J, et al. Increased potency of Fc-receptor-targeted antigens. Cancer Immunol Immunother 1997;45:146–148.
37. Bot AI, Smith DJ, Bot S, et al. Receptor-mediated targeting of spray-dried lipid particles coformulated with immunoglobulin and loaded with a prototype vaccine. Pharm Res 2001;18:971–979.
38. Moingeon P, de Taisne C, Almond J. Delivery technologies for human vaccines. Br Med Bull 2002;62:29–44.
39. Eriksson K, Holmgren J. Recent advances in mucosal vaccines and adjuvants. Curr Opin Immunol 2002;14:666–672.
40. Chu RS, Targoni OS, Krieg AM, et al. CpG oligodeoxynucleotides act as adjuvants that switch on T helper 1 (Th1) immunity. J Exp Med 1997;186:1623–1631.
41. Krieg AM, Davis HL. Enhancing vaccines with immune stimulatory CpG DNA. Curr Opin Mol Ther 2001;3:15–24.
42. Stern BV, Boehm BO, Tary-Lehmann M. Vaccination with tumor peptide in CpG adjuvant protects via IFN-gamma-dependent CD4 cell immunity. J Immunol 2002;168:6099–6105.
43. Baca-Estrada ME, Foldvari M, Babiuk SL, Babiuk LA. Vaccine delivery: lipid-based delivery systems. J Biotechnol 2000;83:91–104.
44. Brazolot Millan CL, Weeratna R, Krieg AM, et al. CpG DNA can induce strong Th1 humoral and cell-mediated immune responses against hepatitis B surface antigen in young mice. Proc Natl Acad Sci USA 1998;95:15553–15558.
45. Weeratna RD, Brazolot Millan CL, McCluskie MJ, Davis HL. CpG ODN can re-direct the Th bias of established Th2 immune responses in adult and young mice. FEMS Immunol Med Microbiol 2001;32:65–71.
46. Gallichan WS, Woolstencroft RN, Guarasci T, et al. Intranasal immunization with CpG oligodeoxynucleotides as an adjuvant dramatically increases IgA and protection against herpes simplex virus-2 in the genital tract. J Immunol 2001;166:3451–3457.
47. Kersten GF, Crommelin DJ. Liposomes and ISCOMS as vaccine formulations. Biochim Biophys Acta 1995;1241:117–138.
48. Lian T, Bui T, Ho RJ. Formulation of HIV-envelope protein with lipid vesicles expressing ganglioside GM1 associated to cholera toxin B enhances mucosal immune responses. Vaccine 1999;18:604–611.
49. Mandal M, Lee KD. Listeriolysin O-liposome-mediated cytosolic delivery of macromolecule antigen in vivo: enhancement of antigen-specific cytotoxic T lymphocyte frequency, activity, and tumor protection. Biochim Biophys Acta 2002;1563:7–17.
50. Bui T, Dykers T, Hu SL, Faltynek CR, Ho RJ. Effect of MTP-PE liposomes and interleukin-7 on induction of antibody and cell-mediated immune responses to a recombinant HIV-envelope protein. J Acquir Immune Defic Syndr 1994;7:799–806.
51. Ho RJ, Burke RL, Merigan TC. Antigen-presenting liposomes are effective in treatment of recurrent herpes simplex virus genitalis in guinea pigs. J Virol 1989;63:2951–2958.
52. Ho RJ, Burke RL, Merigan TC. Physical and biological characterization of antigen presenting liposome formulations: relative efficacy for the treatment of recurrent genital HSV-2 in guinea pigs. Antiviral Res 1990;13:187–199.
53. Ho RJ, Burke RL, Merigan TC. Liposome-formulated interleukin-2 as an adjuvant of recombinant HSV glycoprotein gD for the treatment of recurrent genital HSV-2 in guinea- pigs. Vaccine 1992;10:209–213.
54. Ho RJ, Burke RL, Merigan TC. Disposition of antigen-presenting liposomes in vivo: effect on presentation of herpes simplex virus antigen rgD. Vaccine 1994;12:235–242.
55. Ambrosch F, Wiedermann G, Jonas S, et al. Immunogenicity and protectivity of a new liposomal hepatitis A vaccine. Vaccine 1997;15:1209–1213.

56. Gluck R, Metcalfe IC. New technology platforms in the development of vaccines for the future. Vaccine 2002;20(Suppl 5):B10–B16.
57. Evans TG, McElrath MJ, Matthews T, et al. QS-21 promotes an adjuvant effect allowing for reduced antigen dose during HIV-1 envelope subunit immunization in humans. Vaccine 2001;19:2080–2091.
58. Kersten GF, Crommelin DJ. Liposomes and ISCOMs. Vaccine 2003;21:915–920.
59. Behboudi S, Morein B, Villacres-Eriksson MC. Quillaja saponin formulations that stimulate proinflammatory cytokines elicit a potent acquired cell-mediated immunity. Scand J Immunol 1999;50:371–377.
60. Morein B, Villacres-Eriksson M, Lovgren-Bengtsson K. Iscom, a delivery system for parenteral and mucosal vaccination. Dev Biol Stand 1998;92:33–39.
61. Hilleman MR, Woodhour AF, Friedman A, Phelps AH. Studies for safety of adjuvant 65. Ann Allergy 1972;30:477–483.
62. Hilleman MR, Woodhour A, Friedman A, Weibel RE, Stokes J, Jr. The clinical application of adjuvant 65. Ann Allergy 1972;30:152–158.
63. Ott G, Barchfeld GL, Chernoff D, Radhakrishnan R, van Hoogevest P, Van Nest G. MF59. Design and evaluation of a safe and potent adjuvant for human vaccines. Pharm Biotechnol 1995;6:277–296.
64. Graham BS, Keefer MC, McElrath MJ, et al. Safety and immunogenicity of a candidate HIV-1 vaccine in healthy adults: recombinant glycoprotein (rgp) 120. A randomized, double-blind trial. NIAID AIDS Vaccine Evaluation Group. Ann Intern Med 1996;125:270–279.
65. Bernstein DI, Schleiss MR, Berencsi K, et al. Effect of previous or simultaneous immunization with canarypox expressing cytomegalovirus (CMV) glycoprotein B (gB) on response to subunit gB vaccine plus MF59 in healthy CMV-seronegative adults. J Infect Dis 2002;185:686–690.
66. Keefer MC, Graham BS, McElrath MJ, et al. Safety and immunogenicity of Env 2-3, a human immunodeficiency virus type 1 candidate vaccine, in combination with a novel adjuvant, MTP- PE/MF59. NIAID AIDS Vaccine Evaluation Group. AIDS Res Hum Retroviruses 1996;12:683–693.
67. Byars NE, Nakano G, Welch M, et al. Improvement of hepatitis B vaccine by the use of a new adjuvant. Vaccine 1991;9:309–318.
68. Byars NE, Allison AC, Harmon MW, Kendal AP. Enhancement of antibody responses to influenza B virus haemagglutinin by use of a new adjuvant formulation. Vaccine [illegible]
69. Murphey-Corb M, Martin LN, Davison-Fairburn B, et al. A formalin-inactivated whole SIV vaccine confers protection in macaques. Science 1989;246:1293–1297.
70. Byars NE, Fraser-Smith EB, Pecyk RA, et al. Vaccinating guinea pigs with recombinant glycoprotein D of herpes simplex virus in an efficacious adjuvant formulation elicits protection against vaginal infection. Vaccine 1994;12:200–209.
71. Dong P, Brunn C, Ho RJ. Cytokines as vaccine adjuvants. Current status and potential applications. Pharm Biotechnol 1995;6:625–643.
72. Ho RJ, Merigan TC. Interleukin-2: A potentially useful vaccine adjuvant. In: Spriggs DR, Koff WC, eds. Topics in Vaccine Ajuvant Research. Boca Raton, FL, CRC Press, 1991.
73. Hazama M, Mayumi-Aono A, Asakawa N, et al. Adjuvant-independent enhanced immune responses to recombinant herpes simplex virus type 1 glycoprotein D by fusion with biologically active interleukin-2. Vaccine 1993;11:629–636.
74. Wiryana P, Bui T, Faltynek CR, Ho RJ. Augmentation of cell-mediated immunotherapy against herpes simplex virus by interleukins: comparison of in vivo effects of IL-2 and IL-7 on adoptively transferred T cells. Vaccine 1997;15:561–563.
75. Mitsuyasu R. Immune therapy: non-highly active antiretroviral therapy management of human immunodeficiency virus-infected patients. J Infect Dis 2002;185(Suppl 2):S115–S122.
76. Gherardi MM, Ramirez JC, Esteban M. Towards a new generation of vaccines: the cytokine IL-12 as an adjuvant to enhance cellular immune responses to pathogens during prime-booster vaccination regimens. Histol Histopathol 2001;16:655–667.

77. van der Meide PH, Villinger F, Ansari AA, et al. Stimulation of both humoral and cellular immune responses to HIV-1 gp120 by interleukin-12 in Rhesus macaques. Vaccine 2002;20:2296–2302.
78. Kong Q, Richter L, Yang YF, Arntzen CJ, Mason HS, Thanavala Y. Oral immunization with hepatitis B surface antigen expressed in transgenic plants. Proc Natl Acad Sci USA 2001;98:11539–11544.
79. Lauterslager TG, Florack DE, van der Wal TJ, et al. Oral immunisation of naive and primed animals with transgenic potato tubers expressing LT-B. Vaccine 2001;19:2749–2755.
80. Jones DH, Clegg JC, Farrar GH. Oral delivery of micro-encapsulated DNA vaccines. Dev Biol Stand 1998;92:149–155.
81. Vajdy M, O'Hagan DT. Microparticles for intranasal immunization. Adv Drug Deliv Rev 2001;51:127–141.
82. Sasaki S, Sumino K, Hamajima K, et al. Induction of systemic and mucosal immune responses to human immunodeficiency virus type 1 by a DNA vaccine formulated with QS-21 saponin adjuvant via intramuscular and intranasal routes. J Virol 1998;72:4931–4939.
83. Newman MJ, Wu JY, Gardner BH, et al. Saponin adjuvant induction of ovalbumin-specific CD8+ cytotoxic T lymphocyte responses. J Immunol 1992;148:2357–2362.
84. Wu JY, Gardner BH, Murphy CI, et al. Saponin adjuvant enhancement of antigen-specific immune responses to an experimental HIV-1 vaccine. J Immunol 1992;148:1519–1525.
85. Newman MJ, Munroe KJ, Anderson CA, et al. Induction of antigen-specific killer T lymphocyte responses using subunit SIVmac251 gag and env vaccines containing QS-21 saponin adjuvant. AIDS Res Hum Retroviruses 1994;10:853–861.
86. Yamamoto H, Miller MD, Tsubota H, et al. Studies of cloned simian immunodeficiency virus-specific T lymphocytes. gag-specific cytotoxic T lymphocytes exhibit a restricted epitope specificity. J Immunol 1990;144:3385–3391.
87. Windon RG, Chaplin PJ, McWaters P, et al. Local immune responses to influenza antigen are synergistically enhanced by the adjuvant ISCOMATRIX. Vaccine 2001;20:490–497.
88. Sjolander S, Drane D, Davis R, Beezum L, Pearse M, Cox J. Intranasal immunisation with influenza-ISCOM induces strong mucosal as well as systemic antibody and cytotoxic T-lymphocyte responses. Vaccine 2001;19:4072–4080.
89. Mohamedi SA, Heath AW, Jennings R. A comparison of oral and parenteral routes for therapeutic vaccination with HSV-2 ISCOMs in mice; cytokine profiles, antibody responses and protection. Antiviral Res 2001;49:83–99.
90. Simms JR, Heath AW, Jennings R. Use of herpes simplex virus (HSV) type 1 ISCOMS 703 vaccine for prophylactic and therapeutic treatment of primary and recurrent HSV-2 infection in guinea pigs. J Infect Dis 2000;181:1240–1248.
91. Polakos NK, Drane D, Cox J, et al. Characterization of hepatitis C virus core-specific immune responses primed in rhesus macaques by a nonclassical ISCOM vaccine. J Immunol 2001;166:3589–3598.
92. Qiao M, Murata K, Davis AR, et al. Hepatitis C virus-like particles combined with novel adjuvant systems enhance virus-specific immune responses. Hepatology 2003;37:52–59.
93. Guan XJ, Wu YZ, Jia ZC, et al. Construction and characterization of an experimental ISCOMS-based hepatitis B polypeptide vaccine. World J Gastroenterol 2002;8:294–297.
94. Marx PA, Pedersen NC, Lerche NW, et al. Prevention of simian acquired immune deficiency syndrome with a formalin-inactivated type D retrovirus vaccine. J Virol 1986;60:431–435.
95. Zhou S, Liao X, Li X, et al. Poly-D,L-lactide-co-poly(ethylene glycol) microspheres as potential vaccine delivery systems. J Control Release 2003;86:195–205.
96. Moss B. Genetically engineered poxviruses for recombinant gene expression, vaccination, and safety. Proc Natl Acad Sci USA 1996;93:11341–11348.
97. Paoletti E. Applications of pox virus vectors to vaccination: an update. Proc Natl Acad Sci USA 1996;93:11349–11353.
98. Davis AR, Kostek B, Mason BB, et al. Expression of hepatitis B surface antigen with a recombinant adenovirus. Proc Natl Acad Sci USA 1985;82:7560–7564.
99. Xiang ZQ, Yang Y, Wilson JM, Ertl HC. A replication-defective human adenovirus recombinant serves as a highly efficacious vaccine carrier. Virology 1996;219:220–227.

100. Tubulekas I, Berglund P, Fleeton M, Liljestrom P. Alphavirus expression vectors and their use as recombinant vaccines: a minireview. Gene 1997;190:191–195.
101. Lundstrom K. Alphavirus vectors for gene therapy applications. Curr Gene Ther 2001;1:19–29.
102. Manning WC, Paliard X, Zhou S, et al. Genetic immunization with adeno-associated virus vectors expressing herpes simplex virus type 2 glycoproteins B and D. J Virol 1997;71:7960–7962.
103. Romano G, Michell P, Pacilio C, Giordano A. Latest developments in gene transfer technology: achievements, perspectives, and controversies over therapeutic applications. Stem Cells 2000;18:19–39.
104. Clark KR, Johnson PR. Gene delivery of vaccines for infectious disease. Curr Opin Mol Ther 2001;3:375–384.
105. Crotty S, Miller CJ, Lohman BL, et al. Protection against simian immunodeficiency virus vaginal challenge by using Sabin poliovirus vectors. J Virol 2001;75:7435–7452.
106. Monath TP. Prospects for development of a vaccine against the West Nile virus. Ann NY Acad Sci 2001;951:1–12.
107. Russmann H, Shams H, Poblete F, Fu Y, Galan JE, Donis RO. Delivery of epitopes by the Salmonella type III secretion system for vaccine development. Science 1998;281:565–568.
108. Killeen K, Spriggs D, Mekalanos J. Bacterial mucosal vaccines: Vibrio cholerae as a live attenuated vaccine/vector paradigm. Curr Top Microbiol Immunol 1999;236:237–254.
109. Medina E, Guzman CA. Use of live bacterial vaccine vectors for antigen delivery: potential and limitations. Vaccine 2001;19:1573–1580.
110. Lieberman J, Frankel FR. Engineered Listeria monocytogenes as an AIDS vaccine. Vaccine 2002;20:2007–2010.
111. Lindberg AA. The history of live bacterial vaccines. Dev Biol Stand 1995;84:211–219.

4

Development of Nonviral DNA Delivery Systems

Hans E. J. Hofland, Frank L. Sorgi, and Edward G. Spack

1. Introduction

DNA delivery holds great therapeutic potential, but several barriers have frustrated many creative approaches over the last decade. The presence of an established antiviral immunity in many patients and the rapid induction of an adaptive immune response in naïve patients continues to block many attempts to introduce DNA by viral vectors. Nonviral delivery strategies avoid the problems of viral-coat proteins, but encountered new challenges of low and transient expression related to physiological and innate immune barriers. Naked DNA must overcome serum nucleases, conserved immune receptors, nonspecific clearance, cellular membrane barriers, endosomal degradation, and intracellular trafficking to ensure optimal localization and expression (Fig. 1). This chapter reviews the obstacles to nonviral DNA delivery and highlights current formulation strategies designed to ensure efficient localization and expression of therapeutic genes.

2. Overcoming Barriers to DNA Delivery

2.1. Physiological Barriers

2.1.1. DNA Protection

Systemic gene delivery faces formidable biological barriers. Following injection, DNA is subjected to immediate degradation by nucleases in the blood stream. Approximately 70% of the DNA is cleared from the circulation within the first minute after injection, thereby completely eliminating any chance for gene expression *(1)*. In order to avoid this rapid degradation, the DNA must be protected from nuclease attack. Nonviral gene-delivery systems, including cationic lipids and polymers, have been shown to protect DNA effectively *(2–4)*. These complexes are formed by the charge–charge interaction between the cationic lipid/polymer and the anionic DNA. This electrostatic interaction produces an uncontrolled aggregation and subsequent particle formation *(5)*. The resulting lipid/DNA complexes or polymer/DNA complexes must meet stringent size criteria, such that the particles are sufficiently small for filter-sterilization and intravenous injection. This limits the final particle size to less than 0.2 μm. Therefore, in addition to shielding, the DNA packaging process must include DNA compaction. Compounds that are very effective in both protection and compaction include

From: *Cellular Drug Delivery: Principles and Practice*
Edited by: D. R. Lu and S. Øie © Humana Press Inc., Totowa, NJ

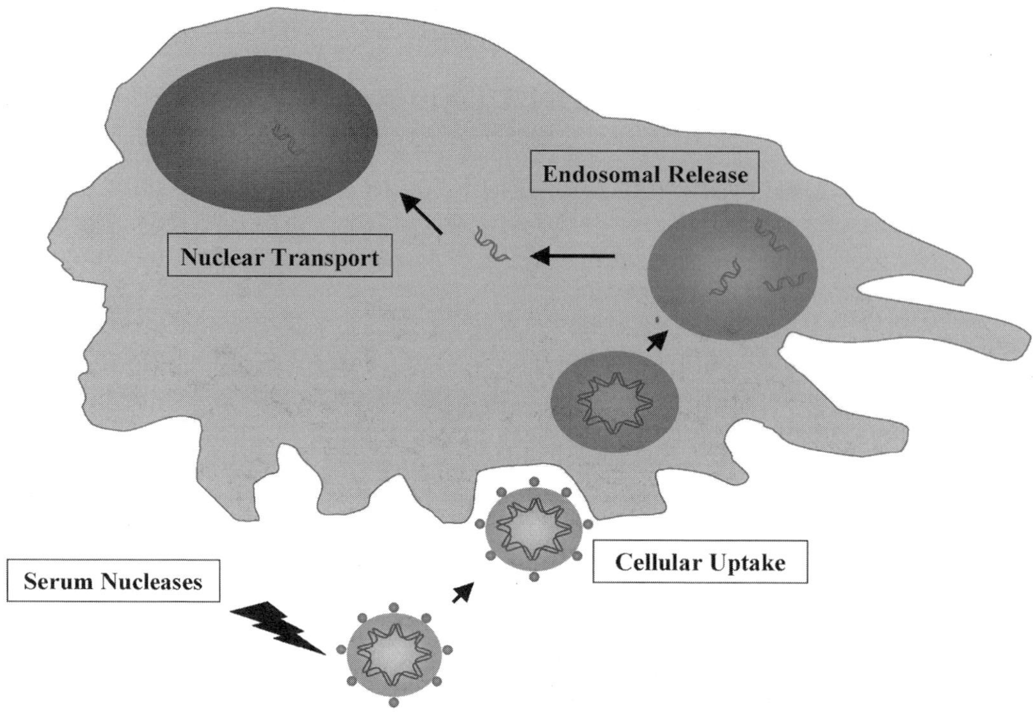

Fig. 1. Potential barriers to nonviral DNA delivery. Plasmids are vulnerable to serum nucleases and endosomal degradation. Various strategies are designed to protect DNA, improve targeting to specific cells or regions, and increase the efficiency of cellular uptake and nuclear transport.

polycationic lipids *(6–10)*, polymers (e.g., starburst polymers, polyethyleneimine REF), peptides *(11,12)*, or a combination of these components *(13,14)*.

2.1.2. Biodistribution (Delivery to Target Cells)

Compacted DNA particles injected into the circulation face the daunting task of avoiding immediate phagocytosis by mature macrophages residing in the tissues, i.e., the reticulo- endothelial system (RES). Both the size and charge of these particles determine how rapidly they are scavenged from the circulation *(15–17)*. Changing the zeta-potential of the polycation/DNA complex from negative to positive results in a sharp decrease in lung accumulation, accompanied by an increase in spleen accumulation *(18)*. However, the effect of charge on biodistribution can be clouded by the effect of size of the complex. Large particles (>400 nm) will simply embolize in the capillary beds, which is the probable mechanism for complex accumulation in the lung *(19,20)*.

Analogous to what has been described in the liposome literature, coating of DNA complexes with polyethylene glycol (PEG) may help polycation/DNA complexes avoid RES uptake and prolong their circulation time *(21)*. Tumors are known to have ill-defined, leaky vasculature *(22)*. This allows particles to extravasate into the tumor tissue, where they are retained owing to a lack of lymphatic drainage, i.e., enhanced permeation and retention (EPR). Several groups have tried to exploit this phenomenon in order to achieve systemic tumor targeting *(23–26)*. However, targeting to the tumor site does not necessarily result in subsequent expression within the tumor cell. Often, the pegylation (of DNA or the lipid/polymer) provides a substantial hindrance to cellular uptake.

The systemic biodistribution of gene expression is also dependent on the route of administration. Following intravenous administration of lipid/DNA complexes, the highest levels of gene expression can be found in the lungs *(20)*, possibly resulting from a physical embolism mechanism. Relative to this expression level, 78% was found in the spleen and 63% in the liver. Following intraperitoneal (ip) injection, however, the highest expression was found in the spleen, and only 25% of this level was found in the liver and 18% in the lungs. Expression in the spleen was also the highest following subcutaneous (sc) injection. Respective liver and lung expression were 28% and 72%.

The most direct means to avoid biodistribution problems and assist the DNA in reaching its target cells is by local administration: simply placing the formulation in the proximity of the target cells. Thus far, most clinical applications in nonviral gene therapy have been limited to local therapies: intratracheal administration for the treatment of cystic fibrosis (CF) *(27–29)*; direct intratumor injections of suicide genes *(30,31)*, apoptosis pathway genes *(32)*, or immune-stimulating genes *(33)*; stereotactic/intracranial injections aimed to treat localized disease in the brain such as Parkinson's disease (PD) *(34)* and Canavan's disease *(35)*; intramuscular injections *(36)* or vaccination *(37)* to induce angiogenesis in peripheral artery *(38)* or coronary artery diseases *(39)*; and direct intra-arterial instillation to treat restenosis *(40)*.

Although local administration has obvious advantages based on reduced exposure to potential toxic materials and the elimination of biodistribution problems, its use will likely be limited. A "next generation" of gene-delivery vehicles is needed that is designed for systemic administration and targeting to specific tissues.

2.2. Cellular Barriers

2.2.1. Cell Entry

Cellular uptake of nonviral DNA-delivery systems is a nonspecific process. Because of the excess positive charge of the complex, it will bind nonspecifically to a negatively charged cell membrane and is subsequently taken up by endocytosis or membrane fusion *(41)*. Electron micrographs have shown that both lipid/DNA complexes *(42)* and lipopoly(L-lysine)/DNA complexes *(43)* are taken up predominantly by endocytosis. These studies contradict earlier suggestions that the main mechanism of cell entry is by membrane fusion *(44,45)*. Other studies have demonstrated that the effect of lysosomotropic agents is dependent on both the formulation and cell type, suggesting that either mechanism may occur and is dictated by the cell type and formulation used *(46,47)*.

Although the mechanisms behind cellular uptake of DNA (with or without a carrier system) are still unclear, the current belief is that, at least in an in vivo situation, DNA enters the cell by endocytosis *(47,42,43)*. The specificity of gene expression can be increased and potential toxicity can be reduced by coating the particles with PEG. Unfortunately, although PEG coating renders these particles invisible to the scavenging RES, it also reduces uptake by target cells. Therefore, cell-specific uptake mechanisms must be incorporated into the formulation.

Current targeting ligands under study for oncology applications include the use of the folate receptor *(23,48)*, transferrin *(49)*, endothelial growth factor (EGF) receptor *(50)*, c-erbB-2 receptor *(51)*, and angiogenic-specific receptors *(52)*. Coating lipid/DNA complexes with PEG dramatically reduced gene-transfer to all tissues. However, the attachment of folic acid at the distal end of the PEG-lipid restored gene-transfer activity in specific tumor tissue only. The gene transfer activity in lungs was reduced by 50–100-fold compared with nontargeted lipid/DNA complexes without PEG. This approach demonstrates in vivo proof-of-concept necessary to achieve targeted tumor gene delivery *(23)*.

Strategies for other disease applications include targeting to inflamed tissues by using selectins such as asialo-LewisX. Targeting to the lungs has been pursued by using the polymeric immunoglobulin receptor *(53)* or specific antibodies *(54)*. Liver hepatocyte targeting has been demonstrated by galactosylated delivery systems *(55,56)*. Using proteins, such as asialoglycoprotein or antibodies as targeting ligands, is inherently associated with the possibility of evoking an immune response *(57)*. However, reducing the size of antibodies to Fab fragments or smaller may prevent an immune response against these molecules *(58)*.

Next-generation gene-delivery systems are becoming increasingly complex. Recent innovations include the combination of both nonviral DNA and viral elements into a single gene-delivery system *(59)*. Ultimately, the success of nonviral gene therapy will greatly depend on the ability to design systems with reduced toxicity, increased stability in serum, and specific tissue targeting.

2.2.2. Endosomal Release

Once the DNA reaches the cells, the degree of cellular uptake is most likely not the cause of the low transfection efficiency of nonviral vectors *(42,46,60)*. In vitro studies demonstrated that 6 h after adding lipid/DNA complex at a dose of 2 μg DNA to 2 × 10^5 COS cells, each cell contained an average of 3 × 10^5 copies of the plasmid. However, only less than 50% of these cells actually expressed the transgene. This suggests that the success of nonviral delivery of DNA could be owing to a mass action effect.

Zabner *(42)* showed that following endocytosis, the DNA containing particles are retained in perinuclear vesicles, or in the lysosomes, where it will be degraded. Escape from these vesicles is thought to be a major barrier for transfection. Therefore, two distinctly different strategies are followed to enhance endosomal release. First, fusogenic lipids or peptides are used to disrupt membranes *(61–63)*. Fusogenic lipids form hexagonal structures or pores in membranes. Enhanced expression of genes transferred by pH-sensitive liposomes is attributed to the instability of these formulations to environments of lower pH *(64)*. The pH-sensitive liposomes will destabilize the endosome, allowing the plasmid to escape into the cytoplasm, before lysosomal degradation can take place. Others have also described the use of fusogenic delivery systems. Plasmid DNA was encapsulated into liposomes containing the nuclear protein high mobility group I, and fused with inactivated Hemagglutinating Virus of Japan (HVJ). Expression of various genes was achieved with this fusion hybrid of liposome and virus *(65–69)*. In vitro levels of transfection found with this system were similar or slightly better than Lipofectin®-mediated transfection, depending on the cell type *(70)*. Unfortunately, these fusogenic liposome formulations are not applicable for iv delivery because their particle size is well over 1 μm *(71)*. These large liposomes are readily taken up by the reticuloendothelial system. The circulation half-life could be improved by size reduction, but this would result in a substantial reduction in the encapsulation efficiency of the DNA. Cationic lipid/DNA complexes generally contain the fusogenic helper lipid DOPE *(72,73)*. Upon adhesion of the cationic lipids to the negatively charged lipids in the cellular membrane, it is believed that phase separation occurs. This results in the presence of DOPE-rich regions, which induce hexagonal phases and therefore, membrane destabilization.

A second strategy involves the use of DNA delivery systems that possesses a high buffering capacity. This will prevent acidification of the endosome and subsequently leads to the rupture of the endosomal membrane *(74)*. The rationale that polymer/DNA complexes are taken up by cells via endocytosis, and that endosomes can be ruptured if the pH drop of the late endosomes is inhibited by the buffering capacity of the formulation, led to the use of polyethylenimine as a DNA-delivery polymer *(75)*. However, concerns of cellular toxicity may become an issue as a result of the emptying of the endosomal contents into the cell following the rupture of the endosome.

2.2.3. Nuclear Localization

Following endosome escape, transport of DNA to the nucleus is required for expression. Some reports indicate that it is not the release from the endosome, nor the uncoating of the DNA, but the transport of the DNA from the cytoplasm into the nucleus *(76)* that is the rate-limiting step for expression to occur *(77,78)*. The DNA in a cationic liposome/DNA complex can be displaced from the complex upon the addition of anionic liposomes. This DNA release may be owing to the multivalent nature of the anionic lipid surface and the collaborative effects of electrostatic interactions and hydrophilic-hydrophobic interactions of the lipids. Lipid mixing results in neutralization of the charges, which allows for diffusion of the cationic lipids away from the DNA. Thus, the anionic lipids in the liposomes will compete with the DNA for binding to cationic lipids. Likewise, it is suggested that the anionic lipids present in the endosome membrane can displace the DNA from cationic lipid/DNA complex. This study shows that DNA release by competition is a very efficient process. Equal moles of anionic lipid added to the cationic lipid/DNA complex leads to 80% release of the DNA.

Micro-injection studies also revealed that nuclear localization is an inefficient process *(79)*. DNA injected into the nucleus of a cell led to expression of the transgene, whereas micro-injection of the same DNA into the cytoplasm did not. Micro-injection studies also revealed that the DNA must be uncoated before it enters the nucleus, because lipid/DNA complexes that were directly injected into the nucleus did not lead to gene expression. This is in agreement with the studies of Zabner et al. *(76)* who observed that even at a 100-fold charge excess of DNA, neither tRNA, ATP, poly(glutamic acid), spermine, spermidine, nor histones were able to displace DNA from the cationic lipid/DNA complex.

One strategy to circumvent the requirement for nuclear transport employs a powerful cytoplasmic expression system utilizing the T7 promoter. Purified T7 RNA polymerase could be co-delivered with the DNA using DC-Chol liposomes *(80,81)*. This system is particularly suitable for expression of short RNA molecules, such as antisense oligonucleotides or ribozymes.

2.2.4. Gene Expression

After conquering all the hurdles described earlier, the DNA finally reaches the nucleus where the transgene can be expressed. Zabner et al. *(42)* showed that upon micro-injection of the monocationic lipid DMRIE/DNA complex into the nucleus, the transfection efficiency is dependent on the lipid to DNA ratio. At ratios where the lipid/DNA has been optimized for in vitro cellular transfection, the expression of the "optimized" complexes were less than that of naked DNA following micro-injection into the nucleus. At suboptimal lipid/DNA ratios however, the expression was increased. This suggests that gene expression may be inhibited owing to lack of dissociation of the lipid from the DNA. Remy et al. *(74)* proposed an explanation for the increased transfection efficiencies of polyamine containing delivery systems lies in the high affinity of the polyamines for DNA that leads to competitive uncoating of the plasmid DNA by the chromosomal DNA. This facilitation in lipid uncoating may be the cause of the increased levels of expression.

When gene delivery is no longer the limiting factor to reaching therapeutic levels of the transgene, the problem of how to regulate gene expression will have to be addressed. Tissue-specific promoters, or even inducable promoters, would be an elegant means to keep the expression of the desired transgene within the therapeutic window. In one example, pH-sensitive anionic immunoliposomes delivered the Herpes Tyrosine Kinase gene under the rat promoter for the phosphoenolpyruvate carboxykinase gene *(82)*. This promoter can be regulated by cAMP drugs *(83)*. Another example of inducible gene expression is via the tetracycline (tet)-regulated transactivation system *(84)*. This system is based on the transcriptional transactivators that interact specifically with bacterial *cis*-regulatory elements, and tet that modulate the binding of the transactivators. The gene will be switched on by giving the patient low, nontoxic doses of the antibiotic.

2.3. Immunological Barriers

Maintaining genomic integrity is the biological prime directive, and strategies to prevent the insertion of foreign genes by pathogens such as viruses date far back in the timescale of evolution. Within the past decade, immunologists have realized that vertebrate immune defense is a balance of innate and adaptive systems. The adaptive system, controlled by T and B lymphocytes, is characterized by receptors (T-cell receptors and immunoglobulins, respectively) whose sequence diversity is generated by a recombination of germline sequences and transcripts. This system provides a broad set of receptors for surveillance against pathogens. If one of these receptors binds to a pathogen, the cell bearing the receptor is activated to expand clonally and produce effector T cells and antibodies that will neutralize the pathogen and prevent replication or incorporation of pathogen genes. Anti-viral antibodies and cytotoxic T lymphocytes that react to viral coat proteins remain a major barrier to viral-delivery systems.

The innate immune system is a more ancient defense. Ironically, whereas the discovery of adaptive immunity dates back over a century to the work of Pasteur and Ehrlich and many others, the role of innate immunity gained general recognition within the last decade. As a result, several barriers to nonviral DNA delivery were encountered before innate immunity was recognized. In contrast to adaptive immunity, the receptors that mediate innate immunity are invariant, encoded in the germline, and bind to conserved structures common to many pathogens, referred to generally as pathogen-associated molecular patterns (PAMPs) *(84a)*. PAMPs are conserved, essential microbial structures including mannose (yeast), lipopolysaccharides (gram-negative bacteria), bacterial lipoprotein, lipoteichoic acid, and peptidoglycan *(85)*. The first PAMP receptor, Toll, was discovered in Drospohila as a key receptor in antifungal responses. Homologs termed Toll-like receptors (TLRs) were subsequently identified in mammals *(86)*.

Work in several labs demonstrated that bacterial DNA also contains PAMPs. Microbial DNA, but not vertebrate DNA, activates the innate immune system *(87)*. The dinuleotide combination CpG is suppressed in mammalian DNA, and in 75% of mammalian CpG pairs the cytosine is methylated to 5-methylcytosine. In contrast, bacterial DNA sequences contain abundant, unmethylated CpG pairs *(88)*. These unmethylated DNA sequences are recognized as PAMPs and trigger the mammalian innate immune system. Synthetic, unmethylated oligodeoxynucleotides (ODNs) containing CpG motifs are also immunostimulatory *(89)*, enabling the identification of PAMP sequences. Different CpG sequences are immunostimulatory in mice and man. In mice, the core PAMP sequence consists of an unmethylated CpG flanked by two 5' purines and two 3' pyrimidines, e.g., GACGTT *(89,90)*. These sequences are relatively inactive when added to human peripheral blood mononuclear cells (PBMCs) *(91)*. Instead, two distinct subclasses of CpG sequences stimulate different sets of human immune cells *(92,93)*. Sequences classified as "CpG-A" or "CpG-D" by different laboratories activate natural killer (NK) cells and plasmacytoid dendritic cells (PDCs), and induce monocytes to differentiate into DCs. CpG-A/CpG-D sequences contain a purine/pyrimidine/CpGpurine/pyrimidine motif flanked by three to four self-complementary sequences (e.g., GGTGCATCGATGCAGGGGG) *(94)*. A different sequence, variously dubbed "CpG-B" or "CpG-K," activates B cells and induces monocyte proliferation. These sequences contain one or more unmethylated CpG dinucleotide with a thymidine immediately 5' of the pair and a TpT or ApT on the 3' side (e.g., TCGTTCGTTCTC) *(94)*.

Several lines of evidence recently demonstrated that TLR9 is the cellular receptor for unmethylated CpG PAMPs. TLR9-deficient mice do not respond to CpG DNA *(95)*. Conversely, transfection of human cells with TLR9 renders them responsive to CpG DNA *(96)*. TLR9 is expressed by all CpG DNA-responsive cells, including the PDC subset, whereas myeloid dendritic cells (MDC) express TLR4 and respond to lipopolysaccharide (LPS) *(97)*. All 10 members of the TLR family contain extracellular leucine-rich repeats and a cytoplasmic Toll/IL-1R homology domain. Binding of a PAMP to a TLR (e.g., CpG DNA:TLR9; LPS:TLR4) initiates signaling by recruitment of the adaptor molecule MyD88, which in turn engages the IL-1R associated kinases (IRAK) and oligomerizes the adaptor TRAF (Fig. 2). This leads to activation of a trio of signal transducers commonly involved in inflammatory stimuli: Jun N-terminal kinase (JNK), p38 mitogen-activated protein kinase (MAPK), and nuclear factor κB (NF-κB) *(98)*. Most TLRs are expressed on the cell surface, but TLR9 is located in the lysosomal compartment. Internalization of CpG DNA and endosomal maturation are required for TLR9 activation *(99)*. This makes evolutionary sense, because ordinarily a host cell would not encounter bacterial DNA until the bacteria was phagocytosed and the cell wall was digested. As discussed earlier, a significant proportion of plasmid DNA delivered in gene-therapy applications is also routed to endosomes, where CpG sequences within the plasmid vector and encoded gene(s) trigger TLR9. The result of this signal transduction through TLR9 is the activation of genes encoding inflammatory cytokines (e.g., interleukin-12 [IL-12], IL-18, interferon-γ [IFN-γ], tumor necrosis factor-α [TNF-α]).

The strong induction of inflammatory cytokines by CpG DNA can in extreme cases induce systemic toxicity. Because TLRs signal by a common pathway, described earlier, it is not surprising that bacterial DNA can induce septic shock comparable to that caused by the binding of LPS to TLR4 *(100)*. Complexing of bacterial plasmid DNA with cationic lipids and other agents may in some cases exacerbate these toxic effects. Furthermore, inflammatory cytokines inhibit the expression of plasmid DNA that is successfully delivered to target cells. Most plasmid expression vectors designed for

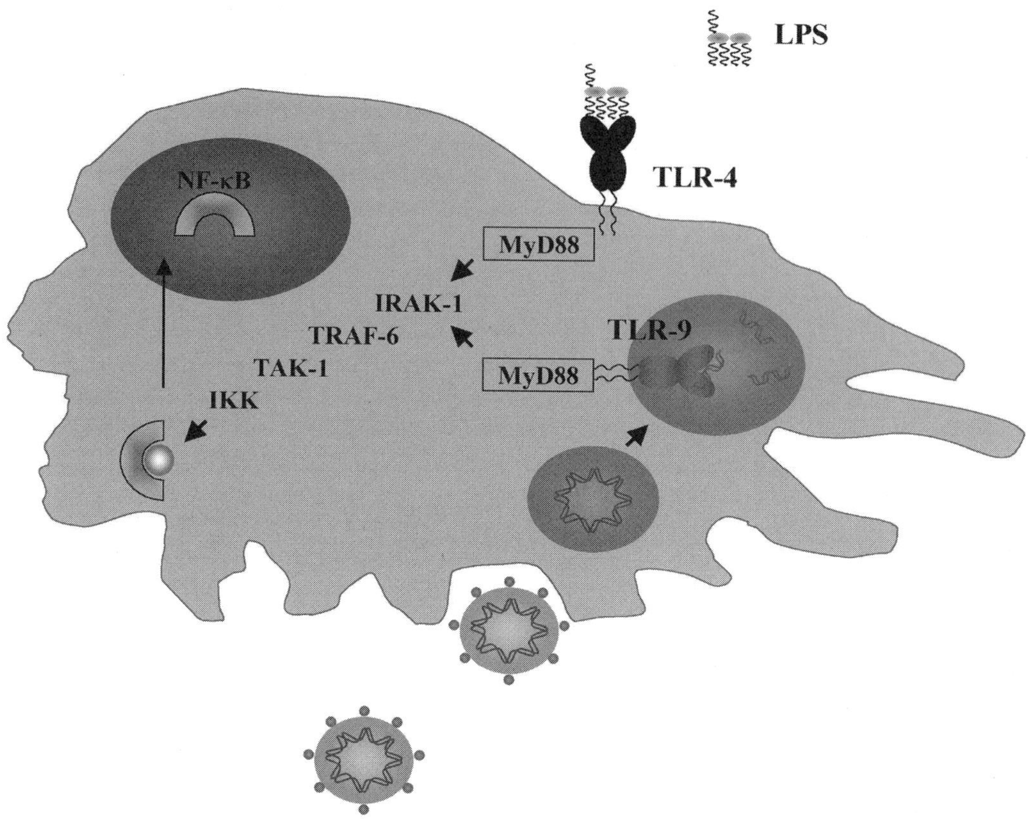

Fig. 2. Plasmid DNA activates the innate immune system. Unmethylated CpG sequences in bacterial DNA bind to the Toll-like receptor 9 (TLR9) in the endosomal compartment. This recruits the adaptor molecule MyD88, which in turn triggers an intracellular signaling cascade (e.g., IRAK-1, TRAF-1, TAK-1), culminating in the activation of transcription factors (e.g, NF-κB, JNK). TLR9 activates a subset of the genes activated by TLR4 (the cell surface receptor for lipopolysaccharide, LPS), including inflammatory cytokines (e.g., TNF-α, IFN-γ).

gene delivery use viral promoters derived from cytomegalovirus (CMV), Simian virus 40 (SV40), Moloney murine leukemia virus (MMLV), or Rous sarcoma virus (RSV) *(101)*. These promoters drive high transcription rates in vitro, but CpG sequences in the plasmid trigger inflammatory cytokines in vivo that repress the transcription of viral promoters *(102,103)*. Viral promoters are particularly susceptible to suppression by IFN-α, IFN-γ, and TNF-α, inflammatory cytokines produced by the triggering of TLR9 in PDCs and NK cells. Rising levels of inflammatory cytokines limit the level and duration of expression by genes encoded in naked or formulated DNA plasmids. Furthermore, the persistence of these cytokines blocks attempts to re-dose with plasmid DNA for several days until their level drops below inhibitory levels. In summary, the innate immune response induced by binding of unmethylated CpG sequences to TLR9 in the endosomes of phagocytic cells severely limits the expression of genes encoded within these plasmids.

3. Conclusion

DNA delivery to target cells faces several formidable barriers. Viral vectors, despite substantial elimination of viral components and the use of helper cells, trigger antibodies and cytotoxic T lymphocytes that limit transfection and expression. Nonviral delivery systems also encounter immunological barriers. Plasmid DNA is synthesized in bacterial hosts, and therefore contains unmethylated CpG sequences that bind to TLR9, a receptor in the endosomes of NK and DCs. The binding of DNA with TLR9 triggers a signaling pathway leading to the secretion of inflammatory cytokines that suppress viral promoters. To overcome this innate immune response, new strategies may be required, including antagonists of TLR9 and suppression of stimulatory CpG sequences through deletion or methylation. A better understanding of the role different carrier lipids play in the stimulation of TLR9 may also aid better expression and persistence. Although the cytokine storm induced by nonviral DNA delivery limits many applications, it may prove an ally in vaccines and tumor therapy. In these cases, the lipid component of lipid:DNA formulations may be optimized to provide improved adjuvants.

The physiological and cellular barriers to successful DNA delivery and expression are also formidable. There are several options for route of delivery, each with its challenges. Intramuscular delivery often produces transfection limited to the injection track. Subcutaneous injections by syringe or ballistic microspheres can improve local expression, but can also transfect Langerhans cells in the skin, triggering migration to regional lymph nodes and initiating a systemic immune response. Many cationic lipids are toxic when administered intravenously, and some particles may cause pulmonary embolisms when trapped in lung capillaries. The enzymes found in serum and lung fluid also pose a danger to the integrity of DNA enroute to the target cells. When lipid:DNA complexes successfully reach target cells, they must negotiate an equally perilous pathway through uptake and nuclear translocation. Many new strategies are evolving for efficient cellular and intracellular targeting, often including ligands and protective lipids. In some cases, the complexities of these solutions may pose practical challenges for manufacturing cost and reproducibility.

The development of monoclonal antibody (MAb) therapy may provide a useful parable for the development of effective nonviral DNA delivery. Initial attempts to treat human diseases with mouse MAbs met with failure owing to the development of immune responses to the foreign mouse sequences. Following this initial disappointment came a new wave of strategies to construct chimeric antibodies, to predict and remove immunostimulatory sequences, and to induce fully human antibodies in immunoglobulin-transgenic mice. Several MAbs now receive FDA approval each year, and the pipeline of these molecules is among the fastest growing sectors of biotechnology. Physiological and immunological barriers to nonviral DNA delivery have dampened initial enthusiasm for long-term gene replacement and required a reassessment that is reminiscent of the early days of MAb development. As the limits and opportunities of DNA delivery are better recognized, there is reason to hope that some of the approaches discussed in this review will survive the perilous journey from the bench to the clinic.

References

1. Mahato RI, Kawabata K, Takakura Y, and Hashida M. In vivo disposition characteristics of plasmid DNA complexed with cationic liposomes. J Drug Targeting 1995b;3:149–157.

2. Borchard, G. Chitosans for gene delivery. Adv Drug Deliv Rev 2001;52:145–150.
3. Wang J, Zhang PC, Lu HF, et al. New polyphosphoramidate with a spermidine side chain as a gene carrier. J Control Release 2002;83:157–168.
4. Yi SW, Yune TY, Kim TW, et al. A cationic lipid emulsion/DNA complex as a physically stable and serum- resistant gene delivery system. Pharm Res 2000;17:314–320.
5. Sternberg B, Sorgi F, Huang L. New structures in complex formation between DNA and cationic liposomes visualized by freeze-fracture electron microscopy. FEBS Lett 1994;356:361–366.
6. Barthel F, Remy JS, Loeffler JP, Behr JP. Gene transfer optimization with lipospermine-coated DNA. DNA Cell Biol 1993;12:553–560.
7. Blessing T, Remy JS, Behr JP. Monomolecular collapse of plasmid DNA into stable virus-like particles. Proc Natl Acad Sci USA 1998;95:1427–1431.
8. Boukhnikachvili T, Aguerre-Chariol, O, Airiau M, et al. Structure of in-serum transfecting DNA-cationic lipid complexes. FEBS Lett 1997;409:188–194.
9. Pitard B, Oudrhiri N, Vigneron JP, et al. Structural characteristics of supramolecular assemblies formed by guanidinium-cholesterol reagents for gene transfection. Proc Natl Acad Sci USA 1999;96:2621–2626.
10. Zou SM, Erbacher P, Remy JS, Behr JP. Systemic linear polyethylenimine (L-PEI)-mediated gene delivery in the mouse. J Gene Med 2000;2:128–134.
11. Brewer LR, Corzett M, Balhorn R. Protamine-induced condensation and decondensation of the same DNA molecule. Science 1999;286:120–123.
12. Mahat RI, Monera OD, Smith LC, Rolland A. Peptide-based gene delivery. Curr Opin Mol Ther 1999;1:226–243.
13. Murphy EA, Waring AJ, Murphy JC, Willson RC, Longmuir KJ. Development of an effective gene delivery system: a study of complexes composed of a peptide-based amphiphilic DNA compaction agent and phospholipid. Nucleic Acids Res 2001;29:3694–3704.
14. Sorgi FL, Bhattacharya S, Huang L. Protamine sulfate enhances lipid-mediated gene transfer. Gene Ther 1997;4:961–968.
15. Lasic D and Martin F, eds. Stealth Liposomes. Boca Raton, FL, CRC Press, Inc., 1995.
16. Litzinger DC, Brown JM, Wala I, et al. Fate of cationic liposomes and their complex with oligonucleotide in vivo. Biochim Biophys Acta 1996;1281:139–149.
17. Schiffelers RM, Bakker-Woudenberg IA, Snijders SV, Storm G. Localization of sterically stabilized liposomes in Klebsiella pneumoniae-infected rat lung tissue: influence of liposome characteristics. Biochim Biophys Acta 1999;1421:329–339.
18. Mahato RI, Kawabata K, Nomura T, et al. Physicochemical and pharmacokinetic characteristics of plasmid DNA cationic liposome complexes. J Pharm Sci 1995a;84:1267–1271.
19. Bragonzi A, Dina G, Villa A, et al. Biodistribution and transgene expression with nonviral cationic vector/DNA complexes in the lungs. Gene Ther 2000;7:1753–1760.
20. Thierry AR, Lunardi-Isakandar Y, Bryant JL, et al. Systemic gene therapy: Biodistribution and long-term expression of a transgene in mice. Proc Natl Acad Sci USA 1995;92:9742–9746.
21. Woodle MC, Martin FJ, Yau-Young A, Redemann CT. Liposomes with enhanced circulation time. US5013556, 2000.
22. Lichtenbeld HC, Yuan F, Michel CC, Jain RK. Perfusion of single tumor microvessels: application to vascular permeability measurement. Microcirculation 1996;3:349–357.
23. Hofland HE, Masson C, Iginla S, et al. Folate-targeted gene transfer in vivo. Mol Ther. 2002;5:739–744.
24. Ogris M, Wagner E. Targeting tumors with non-viral gene delivery systems. Drug Discov Today 2002;7:479–485.
25. Song LY, Ahkong QF, Rong Q, et al. Characterization of the inhibitory effect of PEG-lipid conjugates on the intracellular delivery of plasmid and antisense DNA mediated by cationic lipid liposomes. Biochim Biophys Acta 2002;1558:1–13.
26. Wightman L, Patzelt E, Wagner E, Kircheis R. Development of transferrin-polycation/DNA based vectors for gene delivery to melanoma cells. J Drug Target 1999;7:293–303.
27. Chadwick SL, Kingston HD, Stern M, et al. Safety of a single aerosol administration of escalating doses of the cationic lipid GL-67/DOPE/DMPE-PEG5000 formulation to the lungs of

normal volunteers [published erratum appears in Gene Ther 1998 Apr;5(4):569]. Gene Ther 1997;4:937–942.
28. Driskell RA, Engelhardt JF. Current status of gene therapy for inherited lung diseases. Annu Rev Physiol 2003;65:585–612.
29. Rose AC, Goddard CA, Colledge WH, et al. Optimisation of real-time quantitative RT-PCR for the evaluation of non- viral mediated gene transfer to the airways. Gene Ther 2002;9:1312-1320.
30. Kan O, Kingsman S, Naylor S. Cytochrome P450-based cancer gene therapy: current status. Expert Opin Biol Ther 2002;2:857–868.
31. Tamura T, Sakata T. Application of in vivo electroporation to cancer gene therapy. Curr Gene Ther 2003;3:59–66.
32. Xu L, Pirollo KF, Chang EH. Tumor-targeted p53-gene therapy enhances the efficacy of conventional chemo/radiotherapy. J Control Release 2001;74:115–128.
33. Stopeck AT, Jones A, Hersh EM, et al. Phase II study of direct intralesional gene transfer of allovectin-7, an HLA-B7/beta2-microglobulin DNA-liposome complex, in patients with metastatic melanoma. Clin Cancer Res 2001;7:2285–2291.
34. Tenenbaum L, Chtarto A, Lehtonen E, et al. Neuroprotective gene therapy for Parkinson's disease. Curr Gene Ther 2002;2:451–483.
35. Leone P, Janson CG, Bilianuk L, et al. Aspartoacylace gene transfer to the mammalian central nervous system with therapeutic implications for Canavan disease. Ann Neurol 2000;48:27–38.
36. Pradat PF, Kennel P, Naimi-Sadaoui S, et al. Viral and non-viral gene therapy partially prevents experimental cisplatin-induced neuropathy. Gene Ther 2002;9:1333–1337.
37. Robertson JS, Cichutek K. European Union guidance on the quality, safety and efficacy of DNA vaccines and regulatory requirements. Dev Biol (Basel) 2000;104:53–56.
38. Comerota AJ, Throm RC, Miller KA, et al. Naked plasmid DNA encoding fibroblast growth factor type 1 for the treatment of end-stage unreconstructible lower extremity ischemia: preliminary results of a phase I trial. J Vasc Surg 2002;35:930–936.
39. Schumacher B, Hannekum A, Pecher P. [Neoangiogenesis by local gene therapy: a new therapeutic concept in the treatment of coronary disease]. Z Kardiol 2000;89(Suppl 7):23–30.
40. Rutanen J, Markkanen J, Yla-Herttuala S. Gene therapy for restenosis: current status. Drugs 2002;62:1575–158 .
41. Uyechi LS, Gagne L, Thurston G, Szoka FC Jr. Mechanism of lipoplex gene delivery in mouse lung: binding and internalization of fluorescent lipid and DNA components. Gene Ther 2001;8:828–836.
42. Zabner J, Fasbender AJ, Moninger T, et al. Cellular and Molecular Barriers to Gene Transfer by a Cationic lipid. J Biol Chem 1995;270:18997–19007.
43. Zhou X, Huang L. DNA transfection mediated by cationic liposomes containing lipopolylysine: characterization and mechanism of action. Biochim Biophys Acta 1994;1189:195–203.
44. Felgner PL, Ringold GM. Cationic liposome-mediated transfection. Nature 1989;337:387–388.
45. Legendre J-Y, Szoka F. Cyclic amphipathic peptide-DNA complexes mediate high-efficiency transfection of adherent mammalian cells. Proc Natl Acad Sci USA 1993;90:893–897.
46. Lechardeur D, Lukacs GL. Intracellular barriers to non-viral gene transfer. Curr Gene Ther 2002;2:183–194.
47. Legendre J-Y, Szoka FC. Delivery of plasmid DNA into mammalian cell lines using pH-sensitive liposomes: comparison with cationic liposomes. Pharm Res 1992;9:1235–1242.
48. Reddy JA, Abburi C, Hofland H, et al. Folate-targeted, cationic liposome-mediated gene transfer into disseminated peritoneal tumors. Gene Ther 2002;9:1542–1550.
49. Kircheis R, Wightman L, Schreiber A, et al. Polyethylenimine/DNA complexes shielded by transferrin target gene expression to tumors after systemic application. Gene Ther 2001;8:28–40.
50. Wolschek MF, Thallinger C, Kursa M, et al. Specific systemic nonviral gene delivery to human hepatocellular carcinoma xenografts in SCID mice. Hepatology 2002;36:1106-1114.
52. Aoki Y, Hosaka S, Kawa S, Kiyosawa K. Potential tumor-targeting peptide vector of histidylated oligolysine conjugated to a tumor-homing RGD motif. Cancer Gene Ther 2001;8:783–787.
53. Ferkol T, Perales JC, Eckman E, et al. Gene transfer into the airway epithelium of animals by targeting the polymeric immunoglobulin receptor. J Clin Invest 1995;95:493–502.

54. Balyasnikova IV, Yeomans DC, McDonald TB, Danilov SM. Antibody-mediated lung endothelium targeting: in vivo model on primates. Gene Ther 2002;9:282–290.
55. Hashida M, Nishikawa M, Yamashita F, Takakura Y. Cell-specific delivery of genes with glycosylated carriers. Adv Drug Deliv Rev 2001;52:187–196.
56. Wu J, Nantz MH, Zern MA. Targeting hepatocytes for drug and gene delivery: emerging novel approaches and applications. Front Biosci 2002;7:d717–d725.
57. Stankovics J, Crane A M, Andrews E, et al. Overexpression of human methylmalonyl CoA mutase in mice after in vivo gene transfer with asialoglycoprotein/polylysine/DNA complexes. Human Gene Ther 1994;5:1095–1104.
58. Harding JA, Engbers CM, Newman MS, et al. Immunogenicity and pharmacokinetic attributes of poly(ethylene glycol)- grafted immunoliposomes. Biochim Biophys Acta 1997;1327:181–192.
59. Tagawa T, Manvell M, Brown N, et al. Characterisation of LMD virus-like nanoparticles self-assembled from cationic liposomes, adenovirus core peptide mu and plasmid DNA. Gene Ther 2002;9:564–576.
60. Harashima H, Shinohara Y, Kiwada H. Intracellular control of gene trafficking using liposomes as drug carriers. Eur J Pharm Sci 2001;13:85–89.
61. de Lima MC, Simoes S, Pires P, et al. Gene delivery mediated by cationic liposomes: from biophysical aspects to enhancement of transfection. Mol Membr Biol 1999;16:103–109.
62. Pereira FB, Goni FM, Nieva JL. Liposome destabilization induced by the HIV-1 fusion peptide effect of a single amino acid substitution. FEBS Lett 1995;362:243–246.
63. Saravolac EG, Ludkovski O, Skirrow R, et al. Encapsulation of plasmid DNA in stabilized plasmid-lipid particles composed of different cationic lipid concentration for optimal transfection activity. J Drug Target 2000;7:423–437.
64. Zhou X, Klibanov A, Huang L. Improved encapsulation of DNA in pH-sensitive liposomes for transfection. J Liposome Res 1992;2:125–139.
65. Kaneda Y, Iwai K, Uchida T. Introduction and expression of the human insulin gene in adult rat liver. J Biol Chem 1989;264:12126–12129.
66. Kato K, Dohi Y, Yoneda Y, et al. Use of the hemagglutinating virus of Japan (HVJ)-liposome method for analysis of infiltrating lymphocytes induced by hepatitis B virus gene expression in liver tissue. Biochim Biophys Acta 1993;1182:283–290.
67. Sawa Y, Suzuki K, Bai HZ, et al. Efficiency of in vivo gene transfection into transplanted rat heart by coronary infusion of HVJ liposome. Circulation 1995;92:479–482.
68. Tomita N, Higaki J, Kaneda Y, et al. Hypertensive rats produced by in vivo introduction of the human renin gene. Circ Res 1993;73:898–905.
69. Tomita N, Higaki J, Morishita R, et al. Direct in vivo gene introduction into rat kidney. Biochem Biophys Res Commun 1992;186:129–134.
70. Morishita R, Gibbons GH, Kaneda Y, et al. Novel in vitro gene transfer method for study of local modulators in vascular smooth muscle cells. Hypertension 1993;21:894-899.
71. Hung MC, Hortobagyi GN, Ueno NT. Development of clinical trial of E1A gene therapy targeting HER-2/neu-overexpressing breast and ovarian cancer. Adv Exp Med Biol 2000;465:171-180.
71. Kaneda Y, Uchida T, Kim J, et al. The improved efficient method for introducing macromolecules into cells using HVJ (Sendai virus) liposomes with gangliosides. Exp Cell Res 1987;173:56–69.
72. Farhood H, Serbina N, Huang L. The role of dioleoyl phosphatidylethanolamine in cationic liposome mediated gene transfer. Bba-Biomembranes 1995;1235:289–295.
73. Litzinger DC, Huang L. Phosphatidylethanolamine liposomes: drug delivery, gene transfer and immunodiagnostic applications. Biochim Biophys Acta 1992;1113:201–227.
74. Remy J-S, Sirlin C, Vierling P, Behr JP. Gene transfer with a series of lipophilic DNA-binding molecules. Bioconjugate Chem 1994;5:647–654,.
75. Boussif O, Lezoualc'h F, Zanta MA, et al. A versatile vector for gene and oligonucleotide transfer into cells in culture and in vivo: Polyethylenimine. Proc Natl Acad Sci USA 1995;92:7297–7301.

76. Xu Y, Szoka FC. Mechanism of DNA release from cationic liposome/DNA complexes used in cell transfection. Biochemistry 1996;35:5616–5623.
77. Chan CK, Jans DA. Using nuclear targeting signals to enhance non-viral gene transfer. Immunol Cell Biol 2002;80:119–130.
78. Johnson-Saliba M, Jans DA. Gene therapy: optimising DNA delivery to the nucleus. Curr Drug Targets 2001;2:371–399.
79. Capecchi MR. High efficiency transformation by direct microinjection of DNA into cultured mammalian cells. Cell 1980;22:479–488.
80. Gao X, Huang L. Cytoplasmic expression of a reporter gene by co-delivery of T7 RNA polymerase and T7 promoter sequence with cationic liposomes. Nucleic Acids Res 1993;21:2867–2872.
81. Gao X, Jaffurs D, Robbins PD, Huang L. A sustained, cytoplasmic transgene expression system delivered by cationic liposomes. Biochem Biophys Res Commun 1994;200:1201–1206.
82. Wang C-Y, Huang L. Highly efficient DNA delivery mediated by pH-sensitive immunoliposomes. Biochemistry 1989;28:9508–9514.
83. Wynshaw-Boris A, Lugo TG, Short JM, et al. Identification of a cAMP regulatory region in the gene for rat cytosolic phosphoenolpyruvate carboxykinase (GTP). Use of chimeric genes transfected into hepatoma cells. J Biol Chem 1984;259:12161–12169.
84. Shockett PE, Schatz DG. diverse strategies for tetracyline-regulated inducivle gene expression. Proc Natl Acad Sci USA 1996;93;5173–5176.
84a. Medzhitov R, Janeway CA Jr. Innate immune recognition: mechanisms and pathways. Immunol Rev 2000;173:89–97.
85. Lien E, Ingalls RR. Toll-like receptors. Crit Care Med 2002;30:S1–S11.
86. Medzhitov R, Preston-Hurlburt P, Janeway CA Jr. A human homologue of th Drosophila Toll protein signals activation of adaptive immunity. Nature 1997;388:394–397.
87. Tokunaga T H, Yammamoto S, Shimada H, et al. Antitumor activity of deoxyribonucleic acid fraction from mycobacterium bovis BCG. I. Isolation, physicochemical characterization, and antitumor activity. J Natl Cancer Inst 1984;72:955–962.
88. Bird AP. CpG islands as gene markers in the vertebrate nucleus. Trends Genet 1987;3:342–346.
89. Klinman DM, Yi A, Beaucage SL, et al. CpG motifs expressed by bacterial DNA rapidly induce lymphocytes to secrete IL-6, IL-12, and IFN-γ. Proc Natl Acad Sci USA 1996;93:2879–2883.
90. Krieg AM, Yi AK, Matson S, et al. CpG motifs in bacterial DNA trigger direct B cell activation. Nature 1995;374:546–549.
91. Hartmann G, Weeratna RD, Ballas ZK, et al. Delineation of a CpG phosphorothioate oligonucleotide for activating primate immune responses *in vitro* and *in vivo*. J Immunol 2000;164:1617–1624.
92. Klinman DM, Takeshita F, Gursel I, et al. CpG DNA: recognition by and activation of monocytes. Microbes Infect 2002;4:897–901.
93. Krieg AM. From A to Z on CpG. Trends Immunol 2002;23:64–65.
94. Verthelyi D, Ishii KJ, Gursel M, et al. Human peripheral blood cells differentially recognize and respond to two distinct CpG motifs. J. Immunol. 2001;166:2372–2377.
95. Hemmi H, Takeuchi O, Kawai T, et al. A Toll-like receptor recognizes bacterial DNA. Nature 2000;408:740–745.
96. Bauer S, Kirschning CJ, Hacker H, et al. Human TLR9 confers responsiveness to bacterial DNA via species-specific CpG motif recognition. Proc Natl Acad Sci USA 2001;98:9237–9242.
97. Krug A, Towarowski A, Britsch S, et al. Toll-like receptor expression reveals CpG DNA as a unique microbial stimulus for plasmacytoid dendritic cells which synergizes with CD40 ligand to induce high amounts of IL-12. Eur J Immunol 2001;31:3026–3037.
99. Ahmad-Nejad P, Hacker H, Rutz M, et al. Bacterial CpG-DNA and lipopolysaccharides activate Toll-like receptors at distinct cellular compartments. Eur J Immunol 2002;32:1958–1968.
100. Sparwasser T, Miethke T, Lipford G, et al. Bacterial DNA causes septic shock. Nature 1997;386:336–337.

101. Paillard F. Promoter attenuation in gene therapy: causes and remedies. Hum Gene Ther 1997;8:2009–2010.
102. Harms JS, Splitter GA. Interferon-γ inhibits transgene expression driven by SV40 or CMV promoters but augments expression driven by the mammalian MHC I promoter. Hum Gene Ther 1995;6:1291–1297.
103. Qin L, Ding Y, Pahud DR, et al. Promoter attenuation in gene therapy: Interferon-γ and tumor necrosis factor-α inhibit transgene expression. Hum Gene Ther 1997;8:2019–2029.

5

Nucleic Acid Cellular Delivery

Wu Xiao, Chun Zhang, Preeti Yadava, and Jeffrey Hughes

1. Introduction

The potential of using nucleic acids to correct a malfunctioning, nonexpressed gene or to supplement the production of a natural occurring protein has generated much enthusiasm from researchers and the lay public. Even though the potential for medical advances exists, there are still obstacles that must be overcome before this gene-therapy paradigm becomes a viable option. One of the more substantial obstacles is the delivery vector for the nucleic acids (e.g., plasmid DNA, oligonucleotides, siRNA, ribozymes, etc.). As a whole, nucleic acids offer unique challenges in the design and development of drug-delivery systems because of their poly anionic nature and their lack of resistance to metabolic degradation. Historically, the major theme in the design of modern nucleic acid delivery systems is to mimic the viral method of nucleic acid transfer.

In this chapter, the adeno-associated virus (AAV) will be presented as a model system for understanding efficient particle based delivery of nucleic acids. This virus was chosen because it is one of the simples ones that is currently being evaluated in gene-therapy paradigms. In light of the mechanism AAV uses to gain intracellular access and movement, a review of the similarities and differences between AAV and nonviral systems will be presented. In closing, an outlook will be provided for what the future might be for nonviral gene systems.

2. AAV-2 As a Model for Nucleic Acid Transfer

AAV has demonstrated a broad tropism in terms of infected cell types *(1)* and is a very effective gene transfer cassette for the potential treatment of a variety of diseases *(1–3)*. AAV, composed of a single strand of DNA encapsulated with a protein shell, belongs to the parvovirus family. It is an extremely simple and small virus. Its viral protein coat is composed of three proteins and the viral particle size is in the range of 20–30 nm. It has been shown that the AAV enters cells via clathrin-coated pits, and that this process requires dynamin, a 100-kDa cytosolic GTPase *(4,5)*. Three cellular receptors have been identified for AAV to enter its host cells. Heparin sulfate proteoglycan (HSPG) is necessary for a stable contact with the cell surface *(6,7)*. Efficient cell internalization of AAV, however, requires another cellular receptor, such as integrin $\alpha_v\beta_5$ receptor *(8,9)* or fibroblast growth factor receptor 1 (FGFR1) *(10)*. It is

From: *Cellular Drug Delivery: Principles and Practice*
Edited by: D. R. Lu and S. Øie © Humana Press Inc., Totowa, NJ

likely that there may be other co-receptors. The AAV cytoplasmic trafficking events, viral uncoating, and nuclear translocation, although under extensive studies *(9,11–14)*, are still not fully understood. In order to gain a greater understanding of AAV-mediated nucleic-acid transfer, we studied the infectious entry process of AAV-2 in HeLa cells by tracing AAV proteins, viral DNA, and intact viral particles using fluorescent-dye labeled AAV, A20 antibody immunocytochemistry, and subcellular fractionation techniques. The focus of these studies was to obtain a better insight into how this extremely simple virus deposits its DNA into the nucleus, in order to design more efficient nonviral delivery systems

The first step in AAV cellular entry is the virus' interaction with the extracellular matrix (ECM). AAV can only transduce particular cell types and has tropism for particular cell types depending on the composition of this matrix. Previously, investigators *(15,16)* have used fluorescent-labeled viruses to study the early stages of the viral-infection pathway, AAV in particular. In our experiments, a green fluorescent dye, Alexa Fluor™ 488 (Molecular Probes), was used to covalently label AAV serotype 2. Before incubation, cells were pulsed by labeled AAV for 37°C for 10 min to allow internalization. Cell samples were counter-stained with red SYTO 64® dye, which provides the shape of the whole cell (Fig. 1A,B), and nuclei position was detected by DAPI staining (data not shown). In agreement with a previous report *(4)*, fluorescent virus (yellow) demonstrated a gradual perinuclear accumulation during the first 4-h infection period (Fig. 1A). Furthermore, the majority of the viral signal maintained this perinuclear pattern up to 12 h postinfection before diffusing into the nuclear area at 16 h postinfection, as shown in Fig. 1A. A large amount of fluorescence could be detected inside the nuclei, 24 h postinfection, and after 48 h postinfection, fluorescent signals could be detected solely in the nuclear region. We also performed a free-dye control experiment to eliminate the possibility of free-dye artifact (Fig. 1B). In contrast to AAV infection, the majority of free Alexa Fluro™ 488 dye rapidly entered cell nuclei from 2 h to 8 h postinfection when labeled virus still resided outside the nuclei. This distinguished distribution pattern of free dye from labeled AAV suggests that the nonconjugated dye molecules did not bring contamination to the AAV results.

Fluorescent-dye-labeled virus can only provide information about viral proteins. Whether those fluorescent signals represent intact viral particles or dissociated capsid proteins, however, is unknown. To address this question, we performed an immunocytochemistry assay using A20 monoclonal antibody (MAb) (Fig. 2A,B). A20 antibody has been demonstrated to be able to specifically recognize the intact AAV particles with a defined three-dimensional (3D) structure and is used to detect intact AAV particles in immunocytochemistry *(17,18)*. In agreement to fluorescent virus data (Fig. 1), AAV showed a gradual perinuclear accumulation pattern during its early infection (up to 12 h postinfection). Figure 2B shows a DAPI-stained nucleus with the co-stained A20 viral particles. In contrast to fluorescent virus results, however, intact AAV particles continuously remain outside the cell nuclei throughout the 48-h experimental period, although some small clusters of intact AAV particles might be observed inside nuclei after 24 h postinfection (Fig. 2A). Whether or not this small amount of intact AAV particles found in nuclei was caused by AAV nuclear translocation, cell division, or capsid reassembling is unknown. However, the seemingly co-localization of AAV signals with nucleoli suggests the last possibility, because empty AAV capsids are first assembled in nucleoli before spreading to the whole cell area. Collectively, these data and fluorescent virus data indicate that the nuclear entry of AAV follows a slow and

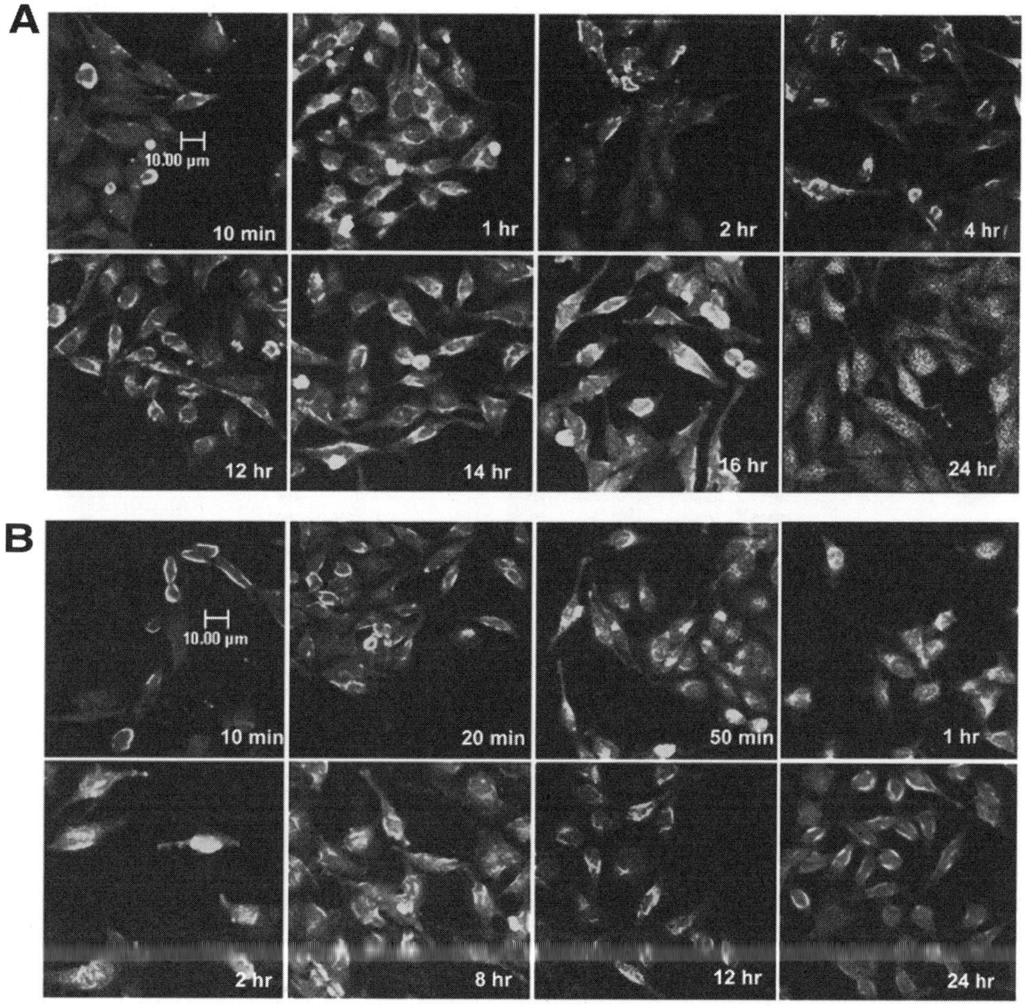

Fig. 1. Fluorescent-dye-conjugated AAV infection. **(A)** Infection of HeLa cells using fluorescent AAV. **(B)** Incubation of HeLa cells using free fluorescent dye.

regulated mode. A control experiment was performed to show only when both AAV and A20 are present can we observe the signal (Fig. 2C).

With the knowledge of the migration of viral proteins (Fig. 1) and intact viral particles, we further studied the viral infection by tracing the viral DNA using a nucleic acid slot-blot method. Cells were fractionated into cytoplasmic and nuclear parts, and no cross-contamination was confirmed by the fact that no histone H3 was shown in the cytoplasmic part (Fig. 3C) and less than 0.5% acid phosphatase activity was in the nuclear fraction (data not shown). As supplementary control experiments, the presence of proliferating cell nuclear antigen (PCNA) and heterogeneous nuclear ribonucleoprotein complex (hnRNPC) in each fraction was also examined by Western blotting (Fig. 3C). PCNA is normally found in cell cytoplasm, and is only present in the nuclei of dividing cells, such as HeLa cells. hnRNPC shuttles between the cytoplasm and nucleus, but predominantly resides in the nucleus. Typical distributions of PCNA and

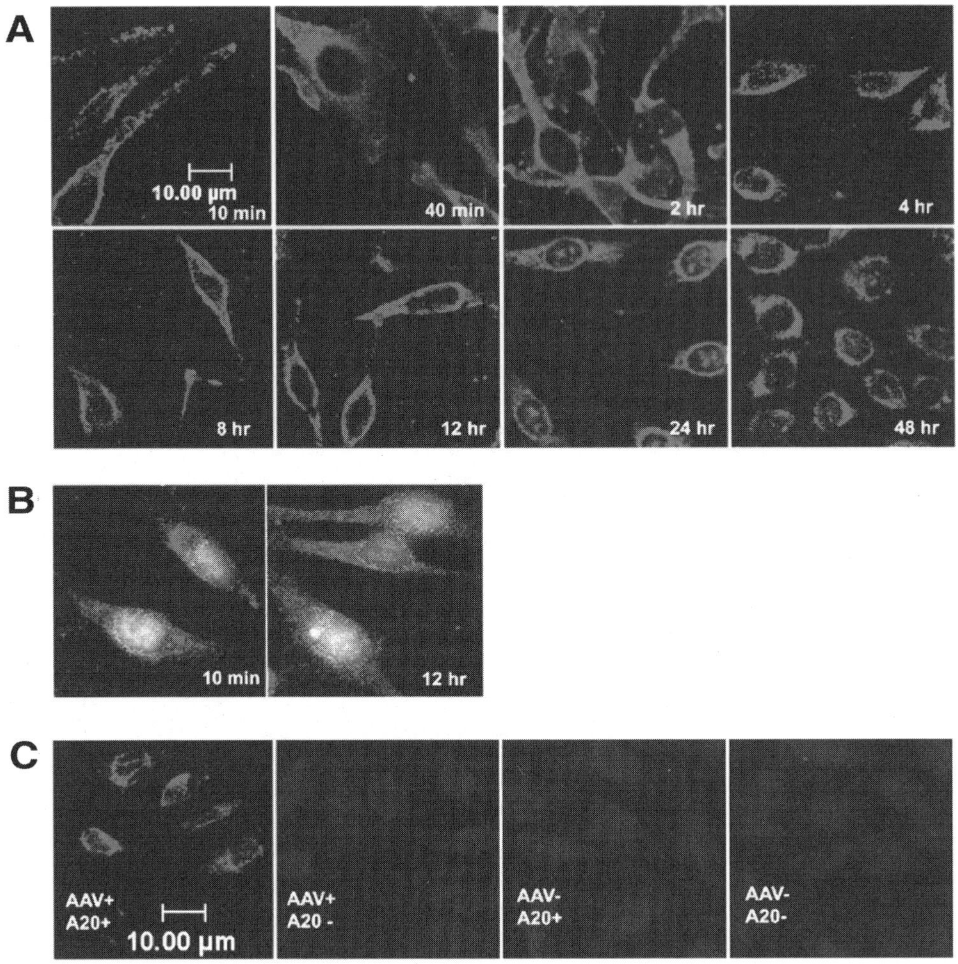

Fig. 2. A20 immunocytochemistry detection of intact AAV particles. **(A)** Immunocytochemistry detection of AAV intact particles at different time points. **(B)** Nuclear staining of 10-min and 12-h time points. **(C)** Control experiments showing the specificity of A20 antibody.

hnRNPC further prove the integrity of the cytoplasmic and nuclear fractions. As for the slot-blotting results, no observable amount of viral DNA can be seen in the nuclear fractions until 16 h postinfection, and the amount of DNA inside cell nuclei then gradually increased to a significant level at 48 h postinfection (Fig. 3A). The nuclear translocation of viral DNA coincides with that of the viral proteins (Fig. 3A), while intact AAV particles remain outside nuclei. Figure 3B shows the ratio of nuclear viral DNA to input DNA over time. These data suggest that the AAV, after accumulating perinuclearly, uncoats prior to or during the entry into cell nuclei.

AAV enters cells through a receptor-mediated endocytosis pathway via clathrin-coated pits. In order to evaluate the intracellular disposition of the viral DNA the cells were fractionated after infection and the cytoplasm separated from the nucleus. The cell postnuclear supernatants (PNS) contents were fractionated by a continuous iodixanol gradient centrifugation method *(19)*. After an initial 10-min incubation with

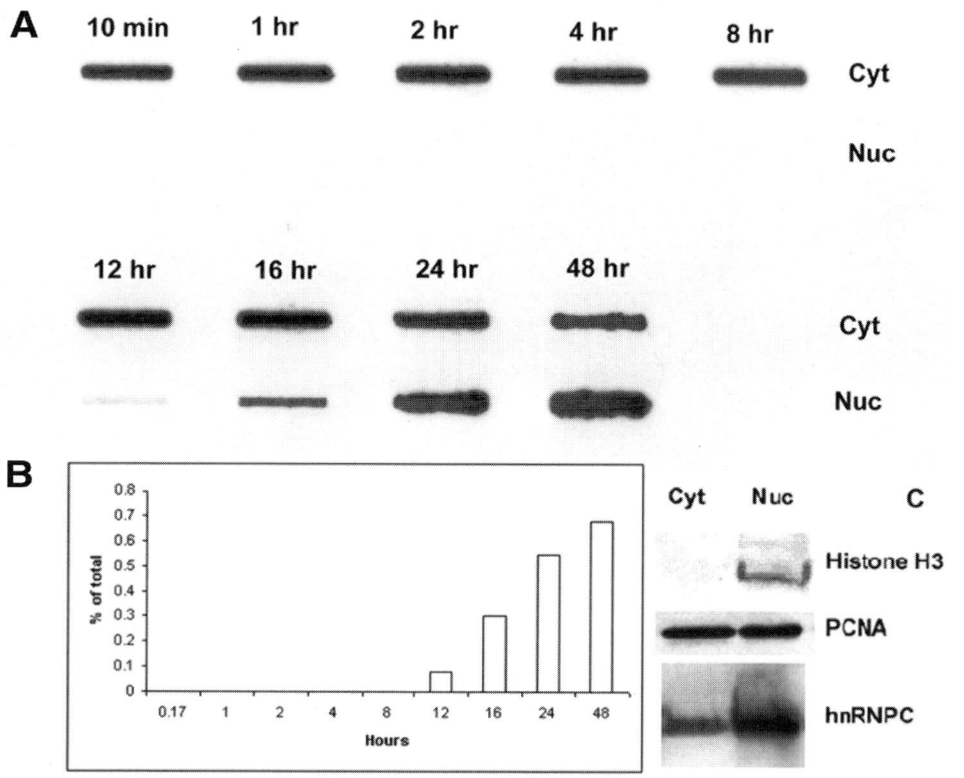

Fig. 3. Slot-blot detection of AAV DNA. **(A)** Slot-blot results of AAV DNA extracted from cytoplasmic and nuclear fractions after indicated periods. **(B)** Quantification of nuclear DNA vs total DNA. **(C)** Western control experiments showing the purity of cytoplasmic (cyt) or nuclear (nuc) fractions.

HeLa cells, AAV was readily detected in the cytoplasm in both early-endosome-trapped and endosome-escaped forms (Fig. 4A, "37°C 10 min" to "37°C 8h"). As the infection proceeded, increasing amounts of virus left the early endosomes as shown by an elevated ratio of escaped vs endosome-associated AAV (Fig. 4B). Incubation of virus with cells at 4°C for 1 h showed no free virus in this gradient (Fig. 4A, "4°C entry"), indicating the virus, attached to the cell surface, was efficiently removed by trypsinization and the free virus signals mentioned earlier truly belong to the virus escaped from endosomes. However, virus in this treatment was found in early endosomes (Fig. 4A, "4°C entry"), suggesting a rapid viral endocytosis between the addition of virus to cells and transfer of cells to 4°C. When both cells and virus were precooled to 4°C before incubation at 4°C for 1 h, no signal was identified (Fig. 4A, "4°C no entry"), indicating no viral entry occurred. No significant amount of AAV was detected in dense-endocytic-vesicle region throughout the research courses. Our data, similar to the results of a study that demonstrated canine parvovirus co localizes with transferrin—an early endosome marker *(5)*—agree with the early-endosome-escaping model, and more than half of the entered AAV could be found in the cytosol as early as 10 min after the incubation (Fig. 4A,B).

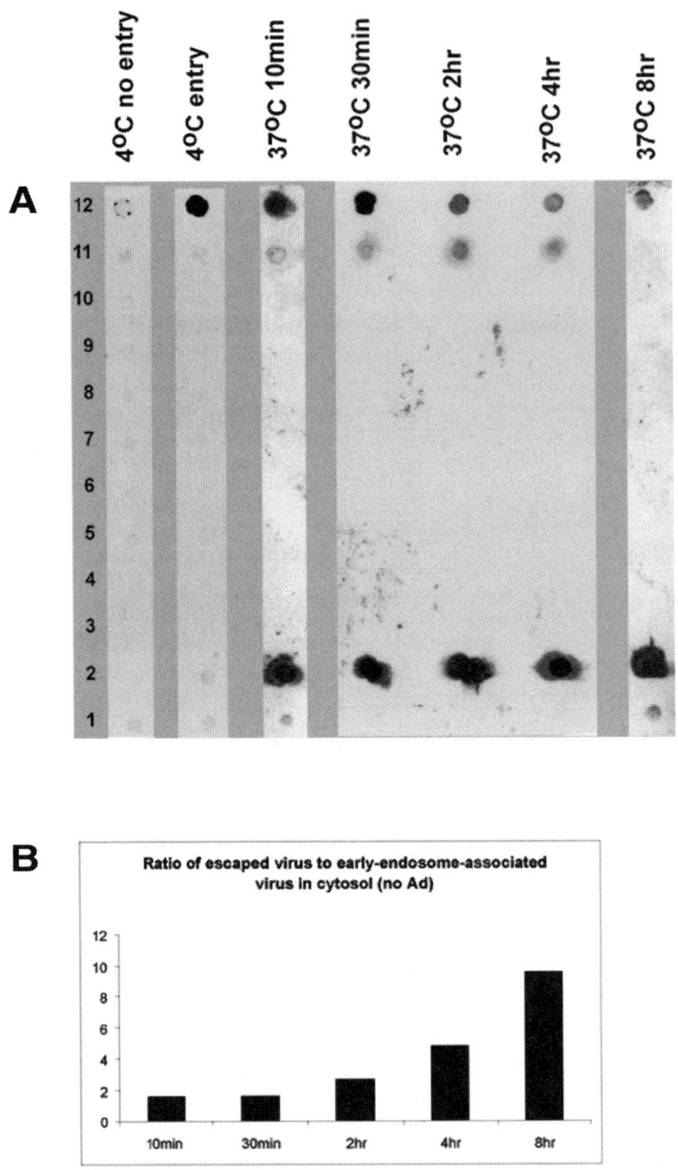

Fig. 4. Subcellular fractionation of AAV. **(A)** Dot-blot detection of AAV DNA from the fractionated PNS of AAV infected HeLa cells at the indicated conditions. **(B)** Intensities of escaped virus signals (fraction 2) and early-endosome-associated virus signals (fraction 11 and 12) were plotted as the ratio of escaped to endosome-associated virus.

From prior studies it is clear that AAV-2 has a highly regulated method of cellular entry and subcellular distribution within HeLa cells. The first step is a somewhat selective interaction with cells expressing particular extracellular ligands. This is followed by a signaling event that facilitates the endocytosis of the AAV-2 particles within the cell. Endocytosis is followed by a rapid viral particle transverse from the endosome into the cytoplasm by a poorly defined mechanism. Once the AAV-2 particles are in the cytoplasm, there is gradual accumulation around the nuclear membrane.

Based on our experiments and the results of other researchers, it appears the AAV protein coat has several important functions, because its unlikely that the single-stranded DNA of the virus impacts on these steps: (1) it protects the viral nucleic acid from metabolic degradation; (2) it allows for the interaction of the viral particle with the cellular membrane allowing attachment and entry; (3), it may facilitate endosomal escape of the viral particle; (4) the viral proteins likely have roles in shuttling the particle to the nuclear membrane. The viral coat of AAV-2 is comprised of three proteins and it appears these proteins work in harmony to facilitate viral particle internalization and intracellular disposition. AAV is one of the simplest and smallest of the viral vectors being used in gene-delivery studies. It is reasonable to assume that this virus is an excellent model for nonviral gene-delivery scientists to model in the crafting of their delivery vectors.

3. Nonviral Mediated Gene Transfer

Nonviral systems for nucleic-acid delivery have not been viewed as effective as viral delivery for numerous reasons, including poor cellular uptake, inefficient nucleic acid transfer from endosomes, metabolic degradation, and limited nucleic acid entry into the nucleus. Recently, advancements have been made in these areas to improve transfection reagents. Researchers are utilizing a variety of approaches for nucleic acid transfer, from using naked plasmid DNA to the crafting of highly advanced supramolecular complexes. In this chapter, we will focus on two of the more studied cationic vector systems, looking for similarities and differences with AAV.

Cationic molecules, either liposomes or polymers, demonstrate several viral-like properties when combined with nucleic acids. After the nucleic acid interacts with the cationic material, a particle is formed that is resistant to nuclease degradation *(2)*. Often an excess of cationic charges is required for efficient protection. These transfection particles are thought to be able to interact with extracellular proteoglycans, thus serving in the initial step of cellular entry *(21,22)* leading to transfer across the cell membrane. Currently, the exact mechanism of cellular entry of the nonviral particles is not clear, and it is unknown whether or not entry differs between cell type and cell cycle.

After entry into endosomes, the gene-transfer complex needs to exit the endosome and gain entrance to the cytoplasm. As described earlier, AAV has an effective, yet unclear mechanism: to accomplish this feat. Other virus, such as andenovirus, can use the gradual drop in pH of endosome as a trigger mechanism for part of its viral protein coat to facilitate endosomal escape *(23,24)*.

Different mechanisms have been proposed for nucleic acid escape from endosomes with cationic macromolecules. Cationic lipids can destabilize lipid bilayers by promoting the formation of nonbilayer lipid structures. Mixtures of cationic lipids and anionic phospholipids preferentially adopt the inverted hexagonal (H_{II}) phase. Furthermore, the presence of "helper" lipids such as dioleoylphosphatidylethanolamine or cholesterol—lipids that enhance cationic lipid-mediated transfection of cells—also facilitate the formation of the H_{II} phase *(25–27)*. It is suggested that the ability of cationic lipids to promote nonbilayer structures in combination with anionic phospholipids leads to disruption of the endosomal membrane following uptake of nucleic acid-cationic lipid complexes. Szoka and colleagues have proposed a more extensive membrane model in which the anionic lipids on the external leaflet of the endosome membrane interact with the cationic molecules of the transfection complex, resulting in movement of the nucleic acids into the cytoplasm *(28,29)*. Or course, the lipid composition of the target cell membrane and fusogenicity of the lipoplex also impacts on transfection efficiency, the latter in turn is influenced by the structure of the lipid and by the structure and relative amount of the neutral helper lipid used in lipoplex formulation.

For example, mixtures of the cationic lipid *N,N*-dioleyl-*N,N*-dimethylammonium chloride (DODAC) with the anionic lipid cholesteryl hemisuccinate (CHEMS) can form nonbilayer structures such as the hexagonal H_{II} phase under conditions of neutral surface charge *(30)*. Similarly, mixtures of the cationic lipid 3α-[*N*-(*N'*,*N'*-dimethylaminoethane)-carbamoyl] cholesterol hydrochloride (DC-Chol) and dioleoylphosphatidic acid (DOPA) also exhibit H_{II}-phase preferences *(30)*. The addition of these helper lipids alters poor transfecting cationic lipids to highly effective transfection complexes. These studies demonstrate the importance of membrane fusion events on nucleic-acid transfer.

As with the AAV particles, at some point the transfer vector is required to release its nucleic acids. It appears for AAV that this occurs at the nuclear membrane or within the nucleus. It is unclear where release of the DNA occurs for the two types of cationic transfer vectors or if there is an advantage of release site. Several nucleases are in the cytoplasm that could degrade noncomplexed plasmid DNA *(31)*, however, so there is an advantage to maintaining the complex. After vesicles escape from the endosome, the nucleic acid must traffic to and enter the nucleus for gene expression. The mechanism by which polyethlenimine (PEI) and DNA polyplex arrives in cell nuclei has yet to be elucidated. Lipoplexes, polyplexes, and plasmid DNA are too large to enter the nucleus through the nuclear-pore complex. Furthermore, it is well-known that the expression of plasmids microinjected into the cytoplasm of cultured cells is poor, whereas the expression of those microinjected into the nucleus is high. These results and others raise the issues with regard to plasmid trafficking in cells that are not dividing. During mitosis, the loss of the nuclear membrane could eliminate these trafficking barriers. Consistent with this postulate, gene transfer in cultured cells is greatly enhanced by mitotic activity for both lipoplexes and polyplexes. For the polymer system (PEI), there is evidence that the entire particle may be transferred to the nucleus *(32)* via specific mechanisms, which may be one of the reasons this vector is so effective.

4. Specific Examples on Transfection Complexes

Nonviral vectors are composed of nucleic acid, usually a plasmid and the transfer vector. Plasmid DNA is extrachromosomal segments of nucleic acids that bacteria use for genetic information storage. Plasmids can easily be manipulated with current molecular-cloning methods into useful mammalian expression vectors. The second component is the DNA carrier. Commonly used systems include cationic lipids; and cationic polymers such as poly-lysine, and synthetic cationic polymers. Cationic liposomes employ specific types of cationic lipids *(33–35)* as the functional component to transport DNA. These vesicles are usually not in the classical bilayer structure of liposomes. This is particularly true after cationic lipids have interacted with nucleic acids *(36)* resulting in a variety of lipid/DNA shapes. Nevertheless, the term cationic liposome is used because they are based on lipids. Synthesized cationic lipids show activity in delivering genes into the lung, liver, and other tissues *(33)*. Although these compounds are diverse in chemical structure, they do have common features. In most cases, there is a cationic head group composed of primary, secondary, tertiary, or quaternary amines, which is attached to a hydrophobic tail group. In plasmid-mediated delivery studies, there is often a relationship between the net charge of the plasmid-DNA/cationic liposome and its effectiveness in transgene expression. Generally, a small net positive charge is required for transfection. Charge ranges from 1:1 to 1:20 (charge ratio) have been reported in the literature *(36–39)*. A slight positive charge of the complex enhances the interaction with the net negative charge of cellular membranes with anionic lipids and carbohydrates, especially in tissue-culture studies.

Cationic liposomes remain promising nonviral systems for use in gene delivery. Although the exact biochemical and biophysical mechanisms of cationic liposome-assisted gene transfection and expression are not thoroughly understood, the barriers involved in the transfection process in vitro generally include the following events: (1) formation of the liposome/DNA complex; (2) entry of complex into the cell; (3) escape of DNA from the endosome; (4) dissociation of DNA from the liposome; (5) entry of DNA into the nucleus; and (6) DNA transcription. Some of the barriers are approached by rational design of the delivery systems (1–5) while others (6) will relate to the innate properties of the plasmid itself (e.g., promoters and enhancers). The cationic lipid assists in protection from nucleases, increases cellular uptake, and escape from the endosome. In a majority of reported studies, the cationic liposomes function most efficiently when the cationic lipid is combined with a second lipid knows as a helper lipid.

Several cationic polymers, such as poly(L-lysine), PEI, and polyamidoamine dendrimers, have been shown to form complexes with DNA and thus facilitate gene transfer. These cationic polymers bear groups that are protonated at physiological pH (Fig. 5). Figure 6 is a scheme demonstrating the transfer of nucleic acid with PEI. The transfecting unit-polyplex is formed by electrostatic attraction between the cationic charge on the polymer and the negatively charged DNA.

Polylysine was used to condense DNA as early as 1969 and was initially used merely as a model for the interaction of biopolymers such as DNA and histone proteins *(40)*. Poly-L-lysine conjugated with asialoorosomucoid was the first polycation employed for gene delivery *(41)*. The gene-transfer efficiency of these early poly-L-lysine polyplexes was very low *(42)*; however, with the conjugation of targeting ligands, the gene-transfer activity of poly-L-lysine polyplexes was enhanced *(43)*. Copolymers of poly-L-lysine and other amino acids have also been shown to transfer genes into mammalian cells. The conjugation of histidine to poly-L-lysine produced a transfecting polyplex that was more efficient than a poly-L-lysine–chloroquine mixture *(44,45)*. The toxicity of poly L-lysine was reduced with poly(lactic-co-glycolic acid) (PLGA)-grafted poly(L-lysine) (PLL) (PLL-g-PLGA) and the cationic PLL-g-PLGA polymeric micelles exhibited better gene-transfer efficiency than PLL *(46)*.

Of the cationic polymers, PEI is the most popular for gene delivery, because DNA/PEI complexes display high transfection efficiency in cell culture and have potential for gene delivery in vivo *(47,48)*. PEI shows efficient gene transfer without the need for endosomolytic or lysosomotropic agents. The molecular weight of PEI can affect its gene-delivery efficiency. Some reports revealed an increase in gene-transfer activity with a decrease in molecular weight (from 100 to 11.9 kDa) *(49)*, whereas others demonstrated a decrease in activity on decreasing the molecular weight (from 70 to 1.8 kDa) *(50)*. It appears that an optimum transfection result can be obtained with molecu-

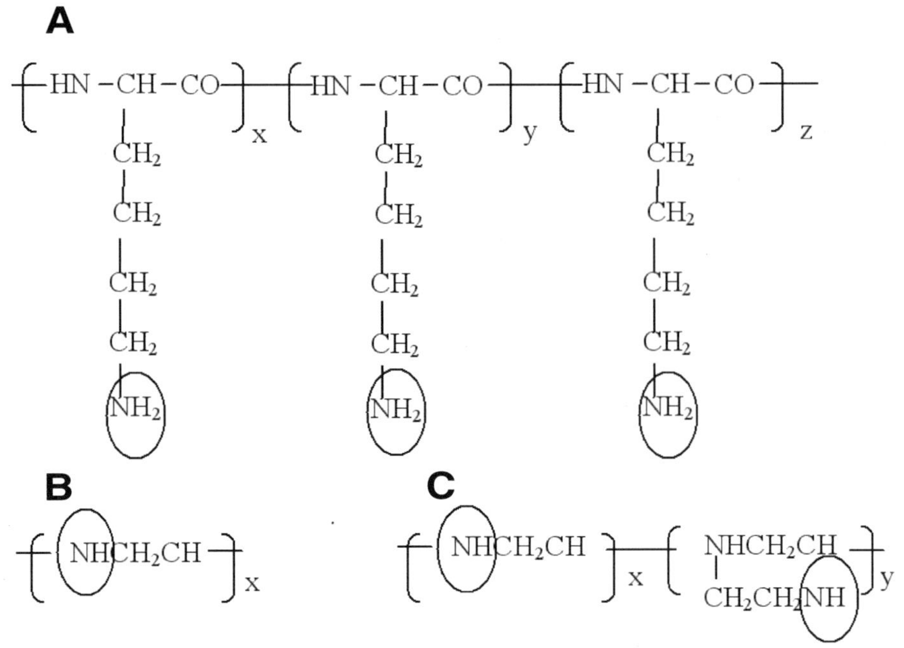

Fig. 5. Examples of cationic polymers used for gene delivery. **(A)** Poly L-lysine. **(B)** Linear polyethylenimine. **(C)** branched polyethylenimine.

lar weight between 11.9 and 70 kDa. High efficiency gene transfer has been obtained without the help of nuclear localization signals with linear PEI as a DNA-complexing agent. However, the coupling of targeting ligands to PEI enhances gene transfer in some cell lines *(51)*. Recently, complexes were generated by the mixing of plasmid DNA, linear polyethylenimine (PEI22, 22 kDa) as the main DNA condensing agent, PEG-PEI [poly(ethyleneglycol)-conjugated PEI] for surface shielding, and Tf-PEG-PEI (transferrin-PEG-PEI) to provide a ligand for receptor-mediated cell uptake *(52)*. These new DNA complexes offer simplicity and convenience, with tumor targeting activity in vivo after freeze-thawing.

Encouraging in vivo results of PEI polyplexes mediated gene expression were obtained in experimental animals. The examined animal anatomical sites included mouse lung *(53)*, rat kidneys *(54)*, and mouse brains *(55)* by intratracheal administration, intraterial injection, and with direct injection. For intravenous injection of PEI polyplexes, transfection is found predominantly in the lung endothelium *(56)*. However, with PEG-coated PEI, gene expression is observed in spleen, heart, lungs, and liver 24 h after intravenous injection of this complex in mice *(57)*. Furthermore, the combination of a PEG-coated PEI and the transferrin targeting ligand resulted in reporter gene expressed in tumor without significant toxicity *(58)*. Thus it appears that a simple polymer has some of the attractive attributes associated with AAV particles.

5. Conclusion

In summary, we have presented information describing gene-transfer vectors. This was accomplished by first addressing how one of the simplest viral vectors (AAV)

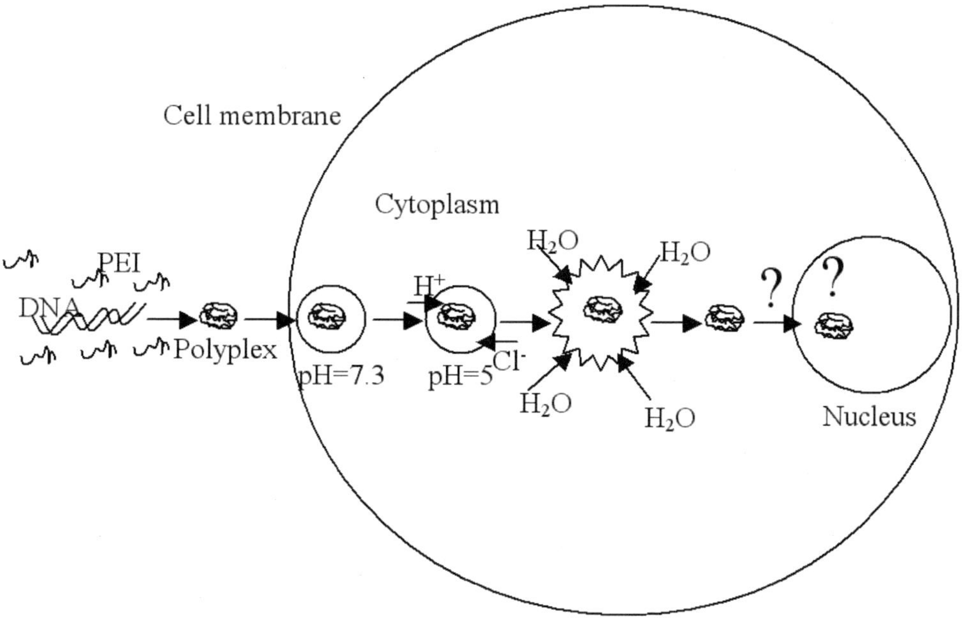

Fig. 6. Schematic representation of PEI mediated DNA uptake by mammalian cells. DNA is compacted in the presence of polycations into a transfection particle. This particle interacts with the cell membrane and is internalized within endosomes. The transfection particle is then released via a proton sponge effect resulting in an osmotic opening of the endosome. The transfection particle than progress to the nucleus.

facilitates nucleic acid movement into the cell. Several distinct stages of this transport were identified, including membrane binding, viral-particle internalization, viral-particle translocation from endosomes, and finally viral accumulation within the nucleus. The only factor that governs these processes in AAV-2 would be the three viral proteins occurring in the viral coat. At this time it is not clear how these proteins are operating, but through site-directed mutagenesis studies *(59)* it is hoped that these protein roles can be teased out. The current similarities between most nonviral vectors systems and how viruses function are amazing. The main difference is that most current nonviral vectors have not been designed with the entire transportation process in mind. In the early stages of their utility, the biggest concern was creating a system that could protect DNA, could transfer it to the cell demonstrating gene expression, and would be nontoxic. Currently efforts are underway to create supramolecular transfection systems that can mimic the viral pathway in several manners. These include systems that have select tropism for specific cells, nonviral vectors with built in enhanced endosome escape mechanisms, and systems that can target the nuclear membrane. Its clear we still have several lessons to learn from this simple virus.

Acknowledgments

The authors would like to thank the National Institutes of Health (PO1 AG 10485) for help in supporting this work.

References

1. Lai CM, Lai YK, Rakoczy PE. Adenovirus and adeno-associated virus vectors. DNA Cell Biol 2002;21:895–913.
2. Martin KR, Klein RL, Quigley HA. Gene delivery to the eye using adeno-associated viral vectors. Methods 2002;28:267–275.
3. Kapturczak MH, Flotte T, Atkinson MA. Adeno-associated virus (AAV) as a vehicle for therapeutic gene delivery: improvements in vector design and viral production enhance potential to prolong graft survival in pancreatic islet cell transplantation for the reversal of type 1 diabetes. Curr Mol Med 2001;1:245–258.
4. Bartlett JS, Wilcher R, Samulski RJ. Infectious entry pathway of adeno-associated virus and adeno-associated virus vectors. J Virol 2000;74:2777–2785.
5. Duan D, Li Q, Kao AW, et al. Dynamin is required for recombinant adeno-associated virus type 2 infection. J Virol 1999;73:10371–10376.
6. Summerford C, Samulski RJ. Membrane-associated heparan sulfate proteoglycan is a receptor for adeno-associated virus type 2 virions. J Virol 1998;72:1438–1445.
7. Hauck B, Xiao W. Characterization of tissue tropism determinants of adeno-associated virus type 1. J Virol 2003;77:2768–2774.
8. Summerford C, Bartlett JS, Samulski RJ. AlphaVbeta5 integrin: a co-receptor for adeno-associated virus type 2 infection. Nat Med 1999;5:78–82.
9. Sanlioglu S, Benson PK, Yang J, et al. Endocytosis and nuclear trafficking of adeno-associated virus type 2 are controlled by rac1 and phosphatidylinositol-3 kinase activation. J Virol 2000;74:9184–9196.
10. Qing K, Mah C, Hansen J, et al. Human fibroblast growth factor receptor 1 is a co-receptor for infection by adeno-associated virus 2. Nat Med 1999;5:71–77.
11. Hansen J, Qing K, Kwon HJ, et al. Impaired intracellular trafficking of adeno-associated virus type 2 vectors limits efficient transduction of murine fibroblasts. J Virol 2000;74:992–996.
12. Duan D, Yue Y, Yan Z, et al. Endosomal processing limits gene transfer to polarized airway epithelia by adeno-associated virus. J Clin Invest 2000;105:1573–1587.
13. Nicklin SA, Buening H, Dishart KL, et al. Efficient and selective AAV2-mediated gene transfer directed to human vascular endothelial cells. Mol Ther 2001;4:174–181.
14. Hansen J, Qing K, Srivastava A. Adeno-associated virus type 2-mediated gene transfer: altered endocytic processing enhances transduction efficiency in murine fibroblasts. J Virol 2001;75:4080–4090.
15. Bantel-Schaal U, Hub B, Kartenbeck J. Endocytosis of adeno-associated virus type 5 leads to accumulation of virus particles in the Golgi compartment. J Virol 2002;76:2340–2349.
16. Xiao W, Warrington KH Jr, Hearing P, et al. Adenovirus-facilitated nuclear translocation of adeno-associated virus type 2. J Virol 2002;76:11505–11517.
17. Wobus CE, Hugle-Dorr B, Girod A, et al. Monoclonal antibodies against the adeno-associated virus type 2 (AAV-2) capsid: epitope mapping and identification of capsid domains involved in AAV-2-cell interaction and neutralization of AAV-2 infection. J Virol 2000;74:9281–9293.
18. Grimm D, Kern A, Rittner K, Kleinschmidt JA. Novel tools for production and purification of recombinant adenoassociated virus vectors. Hum Gene Ther 1998;9:2745–2760.
19. Auricchio A, Hildinger M, O'Connor E, et al. Isolation of highly infectious and pure adeno-associated virus type 2 vectors with a single-step gravity-flow column. Hum Gene Ther 2001;12:71–76.
20. Houk BE, Hochhaus G, Hughes JA. Kinetic modeling of plasmid DNA degradation in rat plasma. AAPS PharmSci 1999;1:E9.
21. Mounkes LC, Zhong W, Cipres-Palacin G, et al. Proteoglycans mediate cationic liposome-DNA complex-based gene delivery in vitro and in vivo. J Biol Chem 1998; 273:26164–26170.
22. Wiethoff CM, Smith JG, Koe GS, Middaugh CR. The potential role of proteoglycans in cationic lipid-mediated gene delivery. Studies of the interaction of cationic lipid-DNA complexes with model glycosaminoglycans. J Biol Chem 2001;276:32806–32813.

23. Wang K, Guan T, Cheresh DA, Nemerow GR. Regulation of adenovirus membrane penetration by the cytoplasmic tail of integrin beta5. J Virol 2000;74:2731–2739.
24. Kukowska-Latallo JF, Chen C, Eichman J, et al. Enhancement of dendrimer-mediated transfection using synthetic lung surfactant exosurf neonatal in vitro. Biochem Biophys Res Commun 1999;264:253–261.
25. Molina MC, Allison SD, Anchordoquy TJ. Maintenance of nonviral vector particle size during the freezing step of the lyophilization process is insufficient for preservation of activity: insight from other structural indicators. J Pharm Sci 2001;90:1445–1455.
26. Simberg D, Danino D, Talmon Y, et al. Phase behavior, DNA ordering, and size instability of cationic lipoplexes. Relevance to optimal transfection activity. J Biol Chem 2001;276:47453–47459.
27. Zuhorn IS, Oberle V, Visser WH, et al. Phase behavior of cationic amphiphiles and their mixtures with helper lipid influences lipoplex shape, DNA translocation, and transfection efficiency. Biophys J 2002;83:2096–2108.
28. Zelphati O, Szoka FC, Jr. Intracellular distribution and mechanism of delivery of oligonucleotides mediated by cationic lipids. Pharm Res 1996;13:1367–1372.
29. Zelphati O, Szoka FC, Jr. Mechanism of oligonucleotide release from cationic liposomes. Proc Natl Acad Sci USA 1996;93:11493–11498.
30. Hafez IM, Ansell S, Cullis PR. Tunable pH-sensitive liposomes composed of mixtures of cationic and anionic lipids. Biophys J 2000;79:1438–1446.
31. Sperinde JJ, Choi SJ, Szoka FC, Jr. Phage display selection of a peptide DNase II inhibitor that enhances gene delivery. J Gene Med 2001;3:101–108.
32. Suh J, Wirtz D, Haynes J. Efficient active transport of gene nanocarriers to the cell nucleus. Proc Natl Acad Sci USA 2003;100:3878–3882.
33. Smyth Templeton N. Liposomal delivery of nucleic acids in vivo. DNA Cell Biol 2002;21:857–867.
34. Zuhorn IS, Hoekstra D. On the mechanism of cationic amphiphile-mediated transfection. To fuse or not to fuse: is that the question? J Membr Biol 2002;189:167–179.
35. Takakura Y, Nishikawa M, Yamashita F, Hashida M. Influence of physicochemical properties on pharmacokinetics of non-viral vectors for gene delivery. J Drug Target 2002;10:99–104.
36. de Lima MC, Simoes S, Pires P, et al. Gene delivery mediated by cationic liposomes: from biophysical aspects to enhancement of transfection. Mol Membr Biol 1999;16:103–109.
37. Koping-Hoggard M, Mel'nikova YS, Varum KM, et al. Relationship between the physical shape and the efficiency of oligomeric chitosan as a gene delivery system in vitro and in vivo. J Gene Med 2003;5:130–141.
38. Lesage D, Cao A, Briane D, et al. Evaluation and optimization of DNA delivery into gliosarcoma 9L cells by a cholesterol-based cationic liposome. Biochim Biophys Acta 2002;1564:393–402.
39. Shah DS, Sakthivel T, Toth I, et al. DNA transfection and transfected cell viability using amphipathic asymmetric dendrimers. Int J Pharm 2000;208:41–48.
40. Evett J, Isenberg I. DNA-polylysine interaction as studied by polarization of fluorescence. Ann NY Acad Sci 1969;158:210–222.
41. Wu GY, Wu CH. Receptor-mediated in vitro gene transformation by a soluble DNA carrier system. J Biol Chem 1987;262:4429–4432.
42. Brown MD, Schatzlein A, Brownlie A, et al. Preliminary characterization of novel amino acid based polymeric vesicles as gene and drug delivery agents. Bioconjug Chem 2000;11:880–891.
43. Cotten M, Langle-Rouault F, Kirlappos H, et al. Transferrin-polycation-mediated introduction of DNA into human leukemic cells: stimulation by agents that affect the survival of transfected DNA or modulate transferrin receptor levels. Proc Natl Acad Sci USA 1990;87:4033–4037.
44. Fajac I, Allo JC, Souil E, et al. Histidylated polylysine as a synthetic vector for gene transfer into immortalized cystic fibrosis airway surface and airway gland serous cells. J Gene Med 2000;2:368–378.
45. Midoux P, Monsigny M. Efficient gene transfer by histidylated polylysine/pDNA complexes. Bioconjug Chem 1999;10:406–411.

46. Jeong JH, Park TG. Poly(L-lysine)-g-poly(D,L-lactic-co-glycolic acid) micelles for low cytotoxic biodegradable gene delivery carriers. J Control Release 2002;82:159–166.
47. Merlin JL, N'Doye A, Bouriez T, Dolivet G. Polyethylenimine derivatives as potent nonviral vectors for gene transfer. Drug News Perspect 2002;15:445–451.
48. Kircheis R, Wightman L, Wagner E. Design and gene delivery activity of modified polyethylenimines. Adv Drug Deliv Rev 2001;53:341–358.
49. Fischer D, Bieber T, Li Y, et al. A novel non-viral vector for DNA delivery based on low molecular weight, branched polyethylenimine: effect of molecular weight on transfection efficiency and cytotoxicity. Pharm Res 1999;16:1273–1279.
50. Godbey WT, Wu KK, Mikos AG. Size matters: molecular weight affects the efficiency of poly(ethylenimine) as a gene delivery vehicle. J Biomed Mater Res 1999;45:268–275.
51. Kircheis R, Kichler A, Wallner G, et al. Coupling of cell-binding ligands to polyethylenimine for targeted gene delivery. Gene Ther 1997;4:409–418.
52. Kursa M, Walker GF, Roessler V, et al. Novel shielded transferrin-polyethylene glycol-polyethylenimine/DNA complexes for systemic tumor-targeted gene transfer. Bioconjug Chem 2003;14:222–231.
53. Ferrari S, Moro E, Pettenazzo A, et al. ExGen 500 is an efficient vector for gene delivery to lung epithelial cells in vitro and in vivo. Gene Ther 1997;4:1100–1106.
54. Boletta A, Benigni A, Lutz J, et al. Nonviral gene delivery to the rat kidney with polyethylenimine. Hum Gene Ther 1997;8:1243–1251.
55. Coll JL, Chollet P, Brambilla E, et al. In vivo delivery to tumors of DNA complexed with linear polyethylenimine. Hum Gene Ther 1999;10:1659–1666.
56. Bragonzi A, Boletta A, Biffi A, et al. Comparison between cationic polymers and lipids in mediating systemic gene delivery to the lungs. Gene Ther 1999;6:1995–2004.
57. Nguyen HK, Lemieux P, Vinogradov SV, et al. Evaluation of polyether-polyethyleneimine graft copolymers as gene transfer agents. Gene Ther 2000;7:126–138.
58. Ogris M, Brunner S, Schuller S, et al. PEGylated DNA/transferrin-PEI complexes: reduced interaction with blood components, extended circulation in blood and potential for systemic gene delivery. Gene Ther 1999;6:595–605.
59. Girod A, Ried M, Wobus C, et al. Genetic capsid modifications allow efficient re-targeting of adeno-associated virus type 2. Nat Med 1999;5:1052–1056.

6

TAT and TAT-Like Peptides for Protein Transduction and Intracellular Drug Delivery

Min Jae Lee, Dexi Liu, Guisheng Zhang, and Xiang Gao

1. Introduction

Advances in biomedical research have significantly enhanced our capability to identify drug targets for medical intervention. At same time, major technological breakthroughs such as combinatorial chemistry, computer-assisted drug design, robotic screening, and molecular cloning have made thousands of small compounds and macromolecules such as proteins and nucleic acids available for therapeutic consideration. Despite this progress, two major obstacles need to be overcome before we are able to move these products from basic research into clinical application. The first of these obstacles is the lack of specificity of drug compounds toward target cells. Although exposure of intended cellular targets to a drug results in a therapeutic effect, exposure of unintended nontarget cells often results in undesirable toxic effects. The other obstacle is the inability of most drugs, especially those that are hydrophilic and those of high molecular weight, to pass through the cell membrane and enter cells. The cell membrane imposes a significant physical barrier to the entrance of drug molecules into cells. The current strategy for overcoming the first obstacle is to develop drugs that are cell specific, or pro-drugs that remain inactive until they are inside the intended target cells. A common approach to resolve the second limitation is to employ receptor-mediated endocytosis, pinocytosis, or phagocytosis to enhance drug internalization. Despite being extensively studied and well-documented, these approaches have nevertheless had limited success and have met with serious obstacles, including the lack of proper receptors, insufficient endosomal release or avoidance of lysosomal degradation, and low delivery efficiency. At present, only a handful of delivery systems have the potential to overcome these obstacles and none are capable of high-efficiency delivery of hydrophilic molecules, especially macromolecules. Therefore, development of a highly effective carrier for intracellular drug delivery has been the goal of much research effort in both the pharmaceutical industry and academic institutions.

Recently, application of a short peptide derived from the trans-activating transcriptional activator (TAT) of the human immunodeficient virus (HIV), to the delivery of bioactive molecules into a variety of cell types has drawn significant interest from the scientific community, particularly in the area of drug delivery *(1)*. By covalently link-

From: *Cellular Drug Delivery: Principles and Practice*
Edited by: D. R. Lu and S. Øie © Humana Press Inc., Totowa, NJ

ing this peptide (commonly called the TAT peptide) to a variety of molecular cargoes such as fluorescent probes *(2,3)*, peptides *(4)*, enzymes *(2,5)*, nucleic acids *(6,7)*, nanoparticles *(8)*, or liposomes *(9)*, delivery of these attached cargoes into cells in culture *(2–9)* and in animal tissues *(2)* was accomplished. TAT-mediated intracellular delivery is unprecedented in efficiency and potency that cannot be easily duplicated by any of the known cell-entry mechanisms, including receptor-mediated endocytosis. Although the mechanism of TAT-mediated internalization remains unclear, evidence accumulated so far appears to suggest that it occurs in a nonspecific fashion without involvement of receptors or transporters *(10–12)*.

The apparent efficiency with which TAT is able to enter cells has raised the possibility of its utilization to mediate the delivery of various molecules into cells for basic research and for drug therapy. In this chapter, we will detail major aspects of the TAT-based delivery system and provide readers with our perspectives for future development.

2. TAT and TAT-Like Peptides

TAT is an essential protein involved in transcriptional regulation in HIV replication. The size of the protein varies between 86 and 102 amino acid residues depending on the origin of viral strain. The molecule contains three functional domains: the N-terminal acidic domain involved in *trans*-activation; a cystein-rich DNA binding domain; and a basic domain that is thought to be the domain for nuclear localization *(13)*. Green and Loewenstein *(14)* and Frankel and Pabo *(15)* independently discovered that a full length of TAT protein could cross the cell membrane and transactivate the long terminal repeat promoter of HIV. Subsequent to this discovery, several other proteins with membrane-penetrating capability were also identified (for review, *see* ref. *16*), including the *Drosophila* antennapedia transcription factor ANTP (also called penetratin-1) *(17)*, and the herpes simplex virus type-1 (HSV-1) VP22 transcription factor *(18)*. However, interest in using these transcription factors for intracellular delivery of biologically active compounds did not become widespread until 1994 when Fawell and colleagues *(5)* discovered that heterogeneous proteins chemically cross-linked to several short peptides derived from TAT protein were able to enter cells. Additional studies by Vives and colleagues *(19)* led to the identification of an 11 amino acid long peptide spanning the region from amino acid residue 47 to 57 in the TAT protein as the functional domain for protein translocation *(19)*. Structure–function relationship studies on this TAT peptide revealed that a cluster of six or more basic amino acid residues is critical for its membrane penetrating activity. Deletion or substitution of the basic amino acid residues of the TAT peptide impaired membrane translocation activity. The nuclear localization signal (NLS) of TAT, although also enriched with basic amino acid residues and essential for HIV trans-activation, does not have transduction activity by itself *(19)*. Other NLS sequences such as the NLS from Simian Virus 40 (SV40) large T-antigen exhibited no translocation activity *(5)*. Interestingly, it has been reported that NLS sequence containing TAT peptides can penetrate into the nucleus through the nuclear pores, whereas those without an NLS remain in the cytosol after internalization *(20)*.

The domains responsible for protein translocation in ANTP and VP22 have also been identified. The minimal amino acid residues from 267 to 300 of VP22 *(18)* and those from 43 to 58 in the third alpha helix of ANTP homeodomain are required for protein transduction *(17)*. Encouraged by the success of TAT-mediated protein delivery, the transduction activity of various homopolymers of lysine, arginine, or histidine has also been examined. Various arginine rich peptides including arginine substitutes in TAT *(3,21)*, the RNA binding peptides derived from virus proteins, and the DNA binding segment of the leucine zipper also exhibited translocation activity at levels similar to or even better than that of TAT peptides *(22)*. Among the polyarginine peptides examined, Arg_8 exhibited optimal translocation efficiency. Futaki attributed the activity of polyarginine peptides to their ability to form hydrogen bonds between guanidine group of the arginine residues and phosphate or sulfate groups on the cell surface. Using diphtheria toxin A chain as a model, Branke and coworkers *(23)* demonstrated that fusion proteins of the A chain with 6 lysine residues (Lys_6) at the N-terminal was approx 100-fold more active than similar fusion proteins with Arg_6 or His_6. The translocation efficiency appeared to depend on peptide chain length: $Lys_8 > Lys_6 > Lys_3$. In agreement with this work, Mai and colleagues have recently demonstrated that the transduction efficiency of homeolysine-based peptides consistently exceeded that of previously identified TAT-derived peptides and polyarginines *(24)*. Table 1 provides a list of important TAT-like peptides that have been used for intracellular delivery. Although varying in amino acid composition, peptide sequence, and length, all of these peptides contain a cluster of basic amino acid residues. Although the possibility exists, the secondary structure of these cationic peptides does not seem to be critical for their activity. To date, fusion proteins containing TAT peptides have demonstrated better transduction activity than similar fusion proteins containing the protein-transduction domains of ANTP or VP22, although the recently devised peptide sequence of the retro-inverse form of TAT(57-47), and polyarginines appear to increase cellular uptake several-fold *(3,25)*. D-amino acid-substituted TAT peptides were equally active *(26)*. Using biotinylated peptides and a biotin-streptavidin-based bridge for attachment of the cargo molecules, Mai and colleagues demonstrated that the minimum number of basic amino-acid residues needed for an intracellular delivery is 6 *(24)*.

3. Linking the Cargo to TAT or TAT-Like Peptides

There are many choices available for the attachment of a TAT peptide to its cargo, depending on the size and chemical properties of a particular cargo. For example, when the cargo is a peptide of a moderate size, one can simply design and then synthesize a TAT-cargo fusion peptide by solid-phase peptide synthesis. For larger proteins, a TAT-fusion protein can be made using recombinant DNA techniques and appropriate expression systems.

Physical association between cationic TAT peptides and anionic molecules through electrostatic interaction may also be used to deliver the complexes into cells. Connection through a biotin-avidin-biotin sandwich structure has also been utilized, where biotinylated peptides and cargo molecules are connected by avidin *(24)*.

For those cargoes that do not fall within the aforementioned categories, chemical conjugation approaches can be used (Fig. 1). Depending on reaction type and available functional groups on the cargo, these can be achieved through the formation of amide (a, b), carbamate (c), isourea (d), disulfide (e), thioether (f), or thiazolidine (g) bonds, as well as Schiff base formation (h–j) *(28–31)*.

Table 1
Most Commonly Used TAT and TAT-Like Peptides

Peptide name		Amino acid sequence	Reference
TAT		TGRKKRRQRRR	*(19)*
D-TAT (D-amino acids in italics)		G*RKKRRQRRRPPQ*	*(26)*
Retro-inverso form of TAT			
		RRRQRRKKRGT	*(25)*
VP22		DAATATRGRSAASRPTERPR-APARSASRPRRPVE	*(18)*
ANTP		RQIKIWFQNRRMKWKK	*(17)*
Polyarginine	R6	RRRRRR-GC	*(3)*
	R8	RRRRRRRR-GC	
	R10	RRRRRRRRRR-GC	
	R12	RRRRRRRRRRRR-GC	
	R16	RRRRRRRRRRRRRRRR-GC	
Poly Lys	6K	KKKKKK	*(24)*
	8K	KKKKKKKK	
	10K	KKKKKKKKKK	
	12K	KKKKKKKKKKKK	
LMWP		VSRRRRRGGRRR	*(27)*

4. TAT-Mediated Intracellular Delivery

Since the first report by Fawell and colleagues *(5)* on the use of the TAT peptide for intracellular protein delivery, many reports have appeared in the literature documenting its activity in intracellular delivery of variety of molecules. Table 2 lists some representative studies in this area. For more information, readers are referred to a special issue of *Current Protein and Peptide Science (34a)*, focusing solely on TAT-related topics. The following three subsections were prepared to provide a brief discussion on issues that are most relevant to the theme of this book.

4.1. TAT-Mediated Protein Delivery

TAT-mediated intracellular delivery of proteins is commonly called protein transduction. This process is found to be most efficient when the proteins are covalently linked to TAT peptides. Becker and colleagues have developed a bacterial-expression system that permits in-frame insertion of the gene sequence for a fusion protein containing the 11 amino acid transduction domain at the N-terminal of the fusion protein *(35)*. Sequence coding for 6xHis was also included into a gene-expression cassette that produces His_6-TAT-X fusion proteins in *Escherichia coli* (X representing the cargo protein). Inclusion of 6xHis motif provides a convenient means for purification of the expressed protein using metal-affinity chromatography and commercially available kits. Using this and other protein-expression systems, a large number of fusion proteins with biological activity have been produced and effectively delivered into cells (for a recent review, *see* ref. *36*). For example, Vocero-Akbani and colleagues engineered a TAT-fusion protein with apoptosis-promoting caspase-3 protein (TAT-Casp3) that contains an HIV proteolytic cleavage site *(37)*. In HIV-infected cells, TAT-Casp3 is pro-

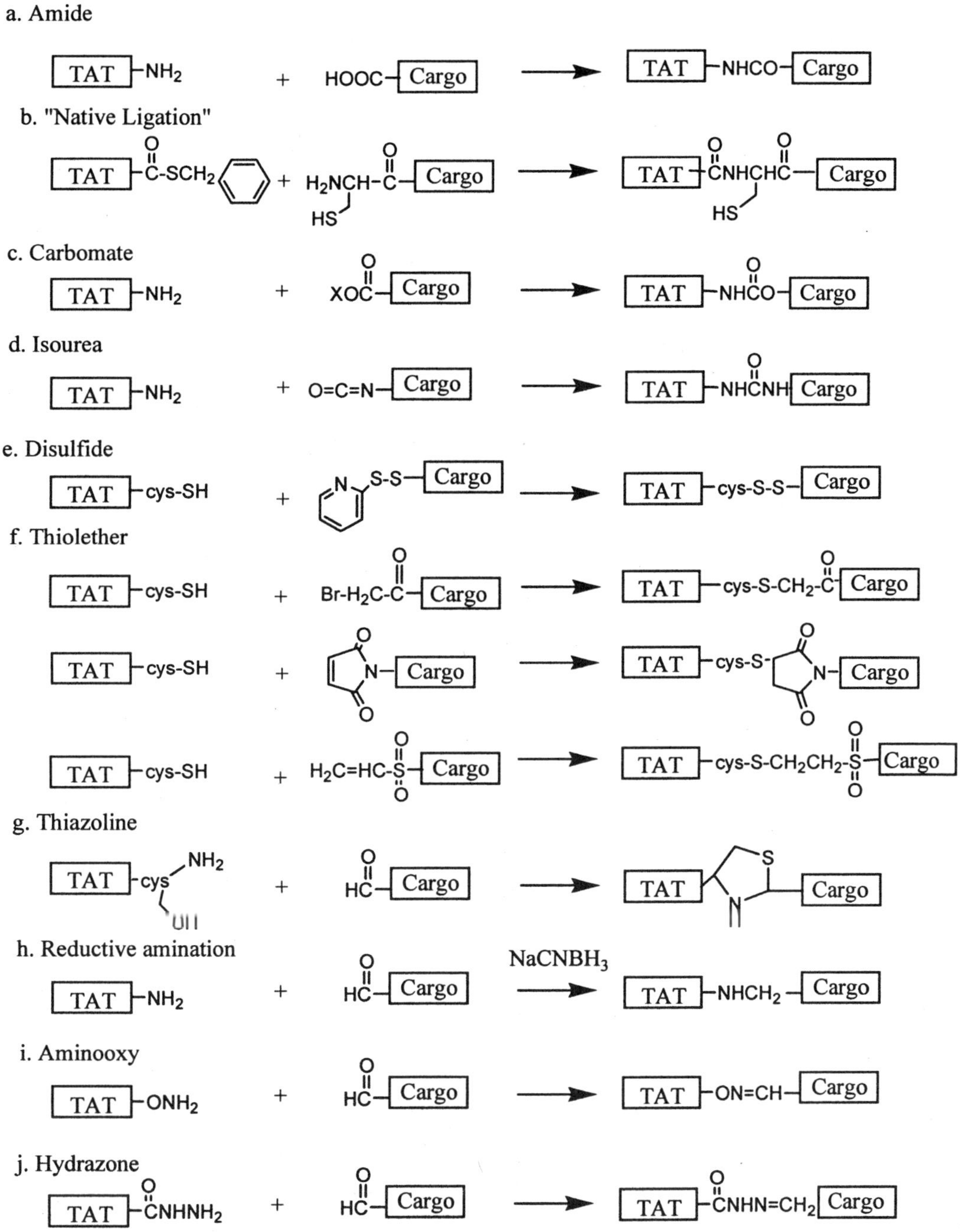

Fig. 1. Conjugation chemistry for linking TAT peptide to its cargo.

cessed into an active form of Casp3 by HIV protease, with resulting apoptosis of the infected cells, whereas in uninfected cells it remained in inactive zymogen form. Nagahara et al. have reported that direct transduction of the denatured TAT fusion protein, TAT-p27[Kip1], could regulate intracellular processes in experimental systems by arresting the cell cycle in the G_1 phase *(38)*. Additionally, ovalbumin (OVA) conju-

Table 2
Intracellular Delivery of Various Molecules by TAT or TAT-Like Peptides

Cargo molecules delivered	Peptides	Linkage	References
Proteins	TAT, PolyR,	Chemical conjugation	*(19)*
	PolyK, ANTP	Genetic, biotin-avidin bridge	*(24)*
Synthetic peptides	TAT, ANTP,	Peptide bond, Disulfide	*(32)*
	PolyR, PolyK		*(4)*
Pharmaceutical drug	TAT, PolyR	Disulfide	*(33)*
	PolyK		
Protein nucleic acid (PNA)	ANTP	Chemical conjugation	*(34)*
Fluorescence molecules	TAT	Chemical conjugation	*(2,3)*
Antisense oligo nucleotide	ANTP, TAT	Disulfide	*(6,7)*
Radioisotopes	TAT	Chelate	
Magnetic beads	TAT	Chemical conjugation	*(8)*
Liposomes	TAT	Chemical conjugation	*(9)*

gated with the TAT peptide was found to be an effective antigen for major histocompatibility complex (MHC) class I-restricted cytotoxic T lymphocytes (CTL) responses and could be used to stimulate CD4 positive T cells, whereas polylysine-conjugated OVA induced no stimulation *(39)*. In this study, antigen-specific CTL immunity was also induced in vivo by treating mice with histocompatible dendritic cells (DCs) that had been exposed to TAT conjugates.

Interestingly, Schwarze and colleagues reported that it is much more efficient for TAT-proteins to enter cells when they are in the denatured form than in their native form *(2)*. Whereas TAT fusion protein failed to induce significant transduction, the same protein misfolded by urea denaturation transduced approx 100% cells in a rapid, concentration-dependent manner. One explanation for the successful delivery of denatured proteins by TAT is that the misfolded proteins may have reduced structural constraints and become more flexible. It remains to be seen, however, whether this is true for other proteins. In general, in-frame fusion proteins generated from genetically engineered TAT expression vectors showed better transduction efficiency, compared with chemically cross-linked TAT peptides *(2,38)*. Apparently, more work is needed to understand the mechanisms by which TAT peptides are able to deliver proteins with various chemical and physical properties.

4.2. Intracellular Delivery of Small Molecules

TAT-mediated intracellular delivery is not limited only to proteins. The utility of this delivery system has been explored with various oligonucleotides and other pharmaceutically interesting molecules. A good example of intracellular delivery of small molecules is the delivery of cyclosporin A, a drug that is ineffective on topical application because of its poor percutaneous adsorption. In a study published by Rothbard and colleagues *(33)*, cyclosporin A was conjugated to a heptamer of arginine through a linker designed to release the active cyclosporin A at physiologic pH. Topical application of the TAT-cyclosporin A conjugates resulted in an efficient transfer into cells in mouse and human skin. More importantly, cyclosporin A reached dermal T lymphocytes and inhibited cutaneous inflammation. Considering that the topical approach is the simplest and most convenient route for drug administration and that skin is a formidable barrier to drug adsorption, the use of TAT or TAT-like peptides to enhance drug bioavailability provides a potential new clinical approach not only to the treatment of skin disorders but also to the treatment of systemic diseases. It will be interesting to see whether the same approach is effective in enhancing the penetration of therapeutic agents across the intestinal epithelium. Enhanced permeability of drug molecules across the intestinal epithelium will have significant impact on the pharmaceutical industry.

Application of TAT peptide to the delivery of nucleic acids into cells has also been explored. In the studies published by Astriab-Fisher and colleagues *(6,7)*, TAT peptide was conjugated to oligonucleotides through disulfide bonds and tested for an enhanced intracellular delivery. Both phosphodiester and phosphorothioate antisense oligonucleotides were evaluated. The results of these studies showed that peptide-antisense oligonucleotides conjugates, but not mismatched control conjugates, were able to inhibit the expression of MDR1, the target gene for the antisense oligonucleotides tested in the study. However, the rate of intracellular accumulation of the peptide-oligonucleotide conjugates was considerably slower than that of the peptide alone or that of oligonucleotides complexed with cationic liposomes, presumably owing to the charge shielding effect of the highly anionic oligonucleotide moiety. These results suggest that conjugation of TAT or TAT-like peptide to antisense oligonucleotides may offer a promising means for gene-function studies and for disease treatment.

4.3. Delivery of Drug Carriers Into Cells

The utility of TAT peptides for intracellular delivery of nanoparticles was initially evaluated by Lewin and coworkers *(8)*. In an effort to monitor the distribution and differentiation of progenitor and stem cells in vivo, TAT peptides were conjugated to the surface of a superparamagnetic iron oxide core that was sterically shielded by cross-linked dextran. The overall diameter of the particles was approx 50 nm. Addition of these particles to hematopoietic and neural progenitor cells resulted in intracellular accumulation of up to 10–30 pg of the superparamagnetic ion per cell. The labeled cells retained their ability to differentiate or proliferate. In a series of studies, Torchilin and colleagues *(9)* explored the possibility of intracellular delivery of liposomes by TAT peptides. It was shown that liposomes as large as 200 nm in diameter could be delivered into cells in a manner identical to those of TAT fusion proteins. They also demonstrated that direct contact of TAT peptides with the cell surface is critical. Steric hindrance to TAT peptide-cell interaction induced by attaching polyethylene glycol (PEG) to the liposome surface abolished liposome internalization, confirming the notion that direct binding of TAT peptides to the cell surface is essential for TAT-mediated internalization. More recent work by the same group showed that TAT-containing liposomes are capable of delivering plasmid DNA into cells *(40)*.

5. Mechanism of TAT-Mediated Intracellular Delivery

By genetic or chemical conjugation with a variety of cargo molecules such as hydrophilic fluorescence probes *(2)*, metal chelators *(41)*, oligonucleotides *(6,7)*, a large number of proteins (for recent review, *see* ref. *36*), magnetic iron oxide nanoparticles *(8)*, and even liposomes as large as 200 nm in diameter *(9)*, TAT peptide was shown to be able to transport these cargoes into cells. The mechanism by which TAT brings the cargo molecules into the cells is not fully understood. It was originally believed that cellular attachment to TAT protein might be mediated through a previously unknown receptor, a 90-kDa cell surface protein that binds to the transduction domain of TAT proteins *(42,43)*. Histological analysis from TAT-β-galactosidase transduction experiments disclosed punctate surface-associated staining at early time points, suggesting the involvement of the endocytosis pathway. However, more recent studies demonstrated that the translocation activity of TAT or TAT-like peptides is not crucially inhibited even at 4°C or in the presence of metabolic inhibitors, as in case of ANTP homeoprotein *(3,9,19)*. A D-amino acid substituted analog of TAT was internalized into cells as efficiently as the corresponding peptide sequences of TAT *(3,26)* and TAT-like peptides, suggesting no structure–activity relationship. Moreover, the fact that certain cationic polypeptides (polyarginines or polylysines) exhibited similar ability in intracellular delivery suggests that typical endocytosis does not play a critical role. Dowdy and colleagues demonstrated enzyme activity in many internal organs in mice injected intraperitoneally with TAT-galactosidase fusion protein *(2)*. This lack of selectivity of protein transduction in vivo serves as additional evidence for a receptor-independent mechanism.

On the other hand, TAT-mediated intracellular delivery appears to require adsorption of the cationic TAT peptides to the negatively charged cell surface, most likely through binding to anionic heparin sulfate on cell surface, because the uptake of TAT was significantly inhibited by heparin sulfate or chrondroitin sulfate A, B, and C *(10)*. In addition, treatment of cells with anti-heparin sulfate antibody or heparinase III clearly reduced the degree of translocation mediated by TAT or TAT-like peptides. These results suggest that although TAT-mediated intracellular delivery may not involve receptor-mediated endocytosis, the electrostatic adsorption of TAT onto cell surface is essential.

A hypothesis was put forward by Prochiantz to explain the mechanism underlying TAT-mediated translocation *(12)*. Figure 2 is a schematic presentation of TAT-based intracellular delivery based on Prochiantz's model. The TAT-based delivery system consists of two parts, the TAT peptide and the cargo, which are connected by various covalent or noncovalent bonds. Upon addition of cargo-carrying TAT peptide to cells, the positively charged basic amino acid residues of the TAT peptides electrostatically bind to the anionic cell surface and induce membrane rearrangement and the formation of an inverted micelle through a regional change of the lipid bilayer into a hexagonal phase. The central micellar structure within the membrane accommodates the peptide-cargo complex in its hydrophilic cavity. Consequently, the TAT-cargo complex enters the cell and reaches the cytosol. Depending on the sequence of the peptide, the conjugates could reach the nucleus through the activity of NLS or simply stay in the cytoplasm. In principle, proteolytic degradation of TAT peptide or breakdown of the linkage bond between the peptide and the cargo will result in cargo release.

Despite mounting evidence supporting the proposed model, it must be pointed out that the conclusive evidence for a ubiquitous internalization mechanism has not yet been found. Cellular translocation of a TAT fusion protein likely is dependent not only on the cationic characteristics of the TAT peptide, but also the cargo molecule carried. For instance, a recent study showed that the cellular accumulation of TAT-oligonucleotides was temperature-dependent and that the oligonucleotides seemed to accumulate

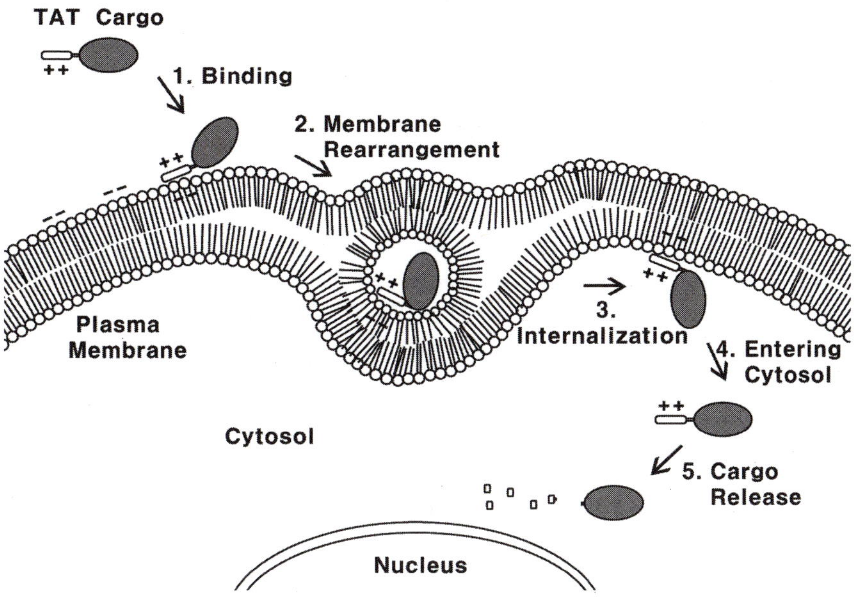

Fig. 2. Proposed mecahnism for TAT-mediated intracellular delivery.

in endosomal compartments before entering the cytoplasm *(6)*, which was in contrast to the delivery of TAT-conjugated proteins.

6. Summary and Perspectives

Intracellular delivery mediated by TAT peptides is so efficient that it can hardly be matched by any of the existing entry mechanisms, including the thoroughly studied receptor-mediated endocytosis or transporters. Early work by Fawell and colleagues demonstrated that a biological response could be detected at a concentration of 1 n*M* and that approx 10^7 TAT molecules were able to enter a single cell *(15)*. Such ability of TAT peptides to deliver molecules across the cell membrane makes the TAT-peptides ideal vehicles for the intracellular delivery of hydrophilic molecules, especially macromolecules such as proteins. The unique characteristics of TAT peptides including their high efficiency, low toxicity, ease of manipulation, minimal susceptibility to the serum , and ability to cross both blood–brain barrier and the cutaneous barrier, provide an invaluable tool for pharmaceutical applications as well as biomedical research.

TAT-mediated intracellular delivery is not limited by cell type or physiological conditions. The same delivery efficiency was seen in cells in primary culture, established cell lines and normal cells in animals. Although it has been reported that TAT protein stimulates both HIV-1 gene expression and the development of Kaposi's sarcoma lesions *(44)*, no significant toxicity has been detected from various in vitro and in vivo studies with TAT peptide carriers. Moreover, several in vivo experiments showed that TAT peptides are able to deliver their conjugates to all tissues in mice including the liver, kidney, lung, heart, spleen, and even the brain, by either intravenous or intraperitoneal administration *(2)*. In most cases, the concentration of TAT-conjugates in transduced cells reached maximal in a few minutes.

The therapeutic potential of the TAT system encompasses both ex vivo and in vivo applications. The ex vivo approach is obvious because much work has already been performed in cell-culture systems. Experimental conditions for loading TAT-conjugates into cells have already been established and can be applied to therapeutic and cell-marking studies. For in vivo applications, although more practical and desirable, possible side effects may hamper clinical development. The primary concern is that although it is an advantage that the TAT-based system is capable of delivering its cargo to any cell type, the lack of selectivity in delivery may lead to unwanted exposure of normal cells to drug. However, the application of TAT to therapies where local administration can be employed would be an advantage. To achieve selective entry of TAT-drug conjugates without diminishing their high cell-entering activity, one may need to hide the cationic groups so that they will become available only at the surface of target cells. Reversibly shielding the positive charges of the TAT for this purpose may prove to be a valuable approach for future development. In its current form, the most appropriate therapeutic application of the TAT-based system is in the area of topical application to enhance the bioavailability of drugs with poor adsorption rates. From the technological standpoint, protein transduction by TAT-mediated protein delivery may prove extremely useful in studying gene or protein function in cells.

Acknowledgment

The authors thank Dr. Joseph E. Knapp for help in critically reading this manuscript.

References

1. Lindgren M, Hallbrink M, Prochiantz A, Langel U. Cell-penetrating peptides. Trends Pharmacol Sci 2000;21:99–103.
2. Schwarze SR, Ho A, Vocero-Akbani A, Dowdy SF. In vivo protein transduction: delivery of a biologically active protein into the mouse. Science 1999;285:1569–1572.
3. Futaki S, Suzuki T, Ohashi W, et al. Arginine-rich peptides: an abundant source of membrane-permeable peptides having potential as carriers for intracellular protein delivery. J Biol Chem 2001;276:5836–5840.
4. Dostmann WR, Tayler MS, Nickl CK, et al. Highly specific, membrane-permeant peptide blockers of cGMP-dependent protein kinase Ialpha inhibit NO-induced cerebral dilation. Proc Natl Acad Sci USA 2000;97:14772–14777.
5. Fawell S, Seery J, Daikh Y, et al. Tat-mediated delivery of heterologous proteins into cells. Proc Natl Acad Sci USA 1994;91:664–668.
6. Astriab-Fisher A, Sergueev D, Fisher M, et al. Conjugates of antisense oligonucleotides with the Tat and Antennapedia cell-penetrating peptides: effects on cellular uptake, binding to target sequences, and biological actions. Pharm Res 2002;19:744–754.
7. Astriab-Fisher A, Sergueev DS. Antisense inhibition of P-glycoprotein expression using peptide-oligonucleotide conjugates. Biochem Pharm 2000;60:83–90.
8. Lewin M, Carlesso N, Tung CH, et al. Tat peptide-derivatized magnetic nanoparticles allow in vivo tracking and recovery of progenitor cells. Nat Biotech 2000;18:410–414.
9. Torchilin V, Rammohan R, Weissig V, Levchenko TS. TAT peptide on the surface of liposomes affords their efficient intracelluar delivery even at low temperature and in the presence of metabolic inhibitors. Proc Natl Acad Sci USA 2001;98:8786–8791.
10. Suzuki T, Futaki S, Niwa M, et al. Possible existence of common internalization mechanisms among arginine-rich peptides. J Biol Chem 2002;277:2437–2443.
11. Mann DA, Frankel AD. Endocytosis and targeting of exogenous HIV-1 Tat protein. EMBO J 1991;10:1733–1739.

12. Prochiantz A. Homeodomain-derived peptides. In and out of the cells. Ann NY Acad Sci 1999;866:172–179.
13. Ruben S, Perkins A, Purcell R, et al. Structural and functional characterization of human immunodeficiency virus Tat protein. J Virol 1989;63:1–18.
14. Green M, Loewenstein PM. Autonomous functional domains of chemically synthesized human immunodeficiency virus tat trans-activator protein. Cell 1988;55:1179–1188.
15. Frankel AD, Paco CO. Cellular uptake of the tat protein from human immunodeficiency virus. Cell 1988;55:1189–1193.
16. Prochiantz A. Messenger proteins: homeoproteins, TAT and others. Curr Opin Cell Biol 2000;12:400–406.
17. Derossi D, Joliot AT, Chassaing G, Prochiantz A. The third helix of the antennapedia homeodomain translocates through biological membranes. J Biol Chem 1994;269:10444–10450.
18. Elliott G, O'Hara P. Intercellular trafficking and protein delivery by a herpes virus structural protein. Cell 1997;88:223–233.
19. Vives E Brodin P, Lebleu B. Truncated HIV-1 TAT protein basic domain rapidly translocates through the plasma membrane and accumulates in the cell nucleus. J Biol Chem 1997;272:16010–16017.
20. Chaloin L, Vidal P, Lory P, et al. Design of carrier peptide-oligonucleotide conjugates with rapid membrane translocation and nuclear localization properties. Biochem Biophys Res Commun 1998;243:601–608.
21. Futaki S, Nakase I, Suzuki T, et al. Translocation of branched-chain arginine peptides through cell membrane: flexibility in the spatial disposition of positive charges in membrane-permeable peptides. Biochemistry 2002;41:7925–7930.
22. Futaki S. Arginine-rich peptides: potential for intracellular delivery of macromolecules and the mystery of the translocation mechanisms. Int J Pharm 2002;245:1–7.
23. Branke SR, Milne JC, Benson EL, Collier RJ. Fused polycationic peptide mediates delivery of diphtheria toxin A chain to the cytosol in the presence of anthrax protective antigen. Proc Natl Acad Sci USA 1996;93:8437–8442.
24. Mai JC, Shen H, Watkins SC, et al. Efficiency of protein transduction is cell type-dependent and is enhanced by dextran sulfate. J Bio Chem 2002;277:30208–30218.
25. Wender PA, Mitchell DJ, Pattabiramman K, et al. The design, synthesis, and evaluation of molecules that enable or enhance cellular uptake: peptoid molecular transporters. Proc Natl Acad Sci USA 2000;97:13003–13008.
26. Huq I, Ping Y-H, Tamilarasu N, Rana TM. Controlling human immunodeficiency virus type 1 gene expression by unnatural peptides. Biochemistry 1999;38:5172–5177.
27. Chang LC, Lee HF, Yang ZQ, Yang VC. Low molecular weight protamine (LMWP) as non-toxic heparin/LMWH antidote: Preparation and characterization. AAPS PharmSci 2001;3(2):Article 17.
28. Hermanson GT (1996) Bioconjugate Techniques. Academic Press, San Diego, CA.
29. Dawson PE, Muir TW, Clark-Lewis I, Kent SBN. Syntheisis of proteins by native chemical ligation. Science 1994;266:776–779.
30. Zhang L, Torgerson TR, Liu XY, et al. Preparation of functionally active cell-permeable peptides by single-step ligation of two peptide modules. Proc Natl Acad Sci USA 1998;95:9184–9189.
31. Vilaseca LA, Rose K, Werlen R, et al. Protein conjugates of defined structure: synthesis and use of a new carrier molecule. Bioconjug Chem 1993;4:515–520.
32. Kato D, Miyazawa K, Ruas M, et al. Features of replicative senescence induced by direct addition of antennapedia-p16INK4A fusion protein to human diploid fibroblasts FEBS Lett 1998;427:203–208.
33. Rothbard JB, Garlington S, Lin Q, et al. Conjugation of arginine oligomers to cyclosporin A facilitates topical delivery and inhibition of inflammation. Nat Med 2000;6:1253–1257.
34. Pooga M, Soomets U, Hallbrink M, et al. Cell penetrating PNA constructs regulate galanin receptor levels and modify pain transmission in vivo. Nat Biotech 1998;16:857–861.
34a. Curr Protein Peptide Sci 2003;4(2).

35. Becker-Hapak M, McAlliater SS, Dowdy SF. TAT-mediated protein transduction into mammalian cells. Methods 2001;24:247–256.
36. Wadia JS, Dowdy SF. Modulation of cellular functions by TAT mediated transduction of full length proteins Curr Protein Peptide Sci 2003;4:97–104.
37. Vocero-Akbani AM, Heyden NV, Lissy NA, et al. Killing HIV-infected cells by transduction with an HIV protease-activated caspase-3 protein. Nat Med 1999;5:29–33.
38. Nagahara H, Vocero-Akbani AM, Snyder EL, et al. Transduction of full-length TAT fusion proteins into mammalian cells: TAT-p27^{Kip1} induces cell migration. Nat Med 1998;4:1449–1452.
39. Lindgren M, Hallbrink MAP, Langel U. Cell-penetrating peptides. Trends Pharmacol Sci 2000;21:99–103.
40. Torchilin VP, Levchenko TS. TAT-liposomes: a novel intracellular drug carrier. Curr Protein Pept Sci 2003;4:133–140.
41. Polyakov V, Sharma V, Dahlheimer JL, et al. Novel Tat-peptide chelates for direct transduction of technetium99m and rhenium into human cells for imaging and radiotherapy. Bioconj Chem 2000;11:762–771.
42. Weeks BS, Desai K, Loewenstein PM, et al. Identification of a novel cell attachment domain in the HIV-1 Tat protein and its 90-kDa cell surface binding protein. J Biol Chem 1993;268:5279–5284.
43. Vogel BE, Lee SJ, Hildebrand A, et al. A novel integrin specificity exemplified by binding of the alpha v beta 5 integrin to the basic domain of the HIV Tat protein and vitronectin. J Cell Biol 1993;121:461-468.
44. Ensoli B, Buonaguro L, Barillari G, et al. Release, uptake, and effects of extracellular human immunodeficiency virus type 1 Tat protein on cell growth and viral transactivation. J Virol 1993;67:277–287.

7

Carrier-Mediated Mechanisms for Cellular Drug Transport

Antara Banerjee, Jeffrey S. Johnston, and Peter W. Swaan

1. Introduction to Transport Proteins

Transporters are a group of polytopic membrane proteins that are involved in almost all biological functions in the cell. In brief, they mediate the entry of nutrients; facilitate uptake and release of hormones, neurotransmitters, and various signaling molecules; participate in transfer of nucleic acids across the cell membrane; and participate in secretion of proteins, lipids, and carbohydrates into and beyond the cytoplasmic membrane. These membrane-transport proteins also represent key targets for therapeutic applications. Implementation of membrane transporters as therapeutic targets for drug delivery requires an understanding of their normal physiological role and the means by which they recognize ligands at the molecular level.

According to the Transporter Commission (TC) nomenclature, membrane transporters can be classified primarily into four main types, i.e., channels, porters, primary active transporters, and group translocators. The latter three types can be grouped together as carriers *(1)*. Channels are integral membrane proteins that form a hydrophilic pore that allows specific inorganic ions and molecules to pass through by the formation of a concentration gradient between the extracellular and intracellular contents. Ion channels are said to be "gated" if they can be opened or closed in response to external stimuli or environmental changes. The rate of transport through a channel is many magnitudes higher (10^8 ions/s) than that of a carrier (10^3 ions/s). Based on the source of energy, carriers can be further subdivided into three groups: (1) uniporters, which are also called facilitated diffusion carriers. They transport a single substrate that would not normally be able to pass through the lipid bilayer membrane (e.g., charged or very hydrophilic molecules), in a passive process driven by the substrate's own electrochemical gradient. (2) Primary active transporters, which transport their substrates across the membrane against their concentration gradient using energy derived from adenosine 5'-triphosphate (ATP) hydrolysis, e.g., ATP-binding cassette (ABC) transporters such as P-glycoprotein (P-gp). (3) Secondary active transporters, which also transport molecules against their concentration gradient, but are coupled to the transport of a second substrate moving down its electrochemical gradient, usually ions such as H^+, Na^+, or Cl^-, which forms the driving force. All active transporters can be

From: *Cellular Drug Delivery: Principles and Practice*
Edited by: D. R. Lu and S. Øie © Humana Press Inc., Totowa, NJ

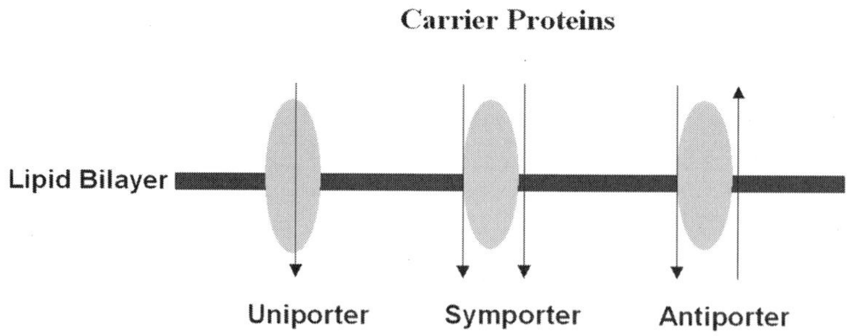

Fig. 1. Active carriers. All active transporters can be symporters, antiporters, or uniporters. Uniporters transport a single molecular species opposite its concentration gradient. A symporter cotransports a solute with an ion moving down its electrochemical gradient, whereas an antiporter countertransports a solute in the opposite direction of an ion that flows down its concentration gradient.

symporters, antiporters, or uniporters (Fig. 1). This chapter will focus briefly on the distinct transporter superfamilies, some of the transporters and the prodrug approaches that have been used to make these transporters pharmaceutically relevant.

1.1. Transporter SuperFamilies

The genome-sequencing data from the human-gene nomenclature committee and a number of biochemical and molecular genetic investigations have led to the classification of the transporters into distinct transporter superfamilies. Currently there exist about 250 entries, each of which may correspond to a single family. Superfamilies form when these individual families are very large. Some examples of superfamilies are the voltage-gated ion channel family (VIC), major facilitator superfamily (MFS), and the ATP-binding cassette superfamily (ABC). The latter two families will be discussed below.

1.1.1. The Major Facilitator Superfamily

This group of transporters primarily comprises single-polypeptide secondary carriers that are capable of transporting small solutes in response to a chemiosmotic ion gradient. The MFS transporters have been classified into 17 families (Table 1). They were originally believed to function as sugar transporters, but later studies revealed that they could transport a variety of small solutes in response to a chemiosmotic gradient. The members of each family share 25–30% sequence similarity *(1)*. The MFS members have been classified by: (1) recognition of the significance of the family to cell physiology; (2) the application of biochemical, biophysical, and molecular genetic information from a few members to all members of the family; (3) use of a rational system for MFS protein classification; (4) evolutionary pathways responsible for the development of structural and functional diversity; and (5) unification of various mechanistic models in order to extend it to the maximum number of transporters possible *(1)*. Almost all MFS transporters transport a wide array of small molecules. These include sugars, drugs, amino acids, metabolites, vitamins, nucleosides, and both organic and inorganic cations and anions. The MFS proteins possess either 12 or 14 putative transmembrane spanning regions and are one of the oldest protein families from an evolutionary standpoint.

1.1.2. The ABC Transporter Superfamily

The ABC superfamily consists of the largest number of transmembrane proteins. To date, 48 known ABC transporters are present and have been grouped into seven subfamilies *(2)*. These proteins bind ATP and use the resultant energy as a driving force to transport various molecules across the cell membrane. The ABC proteins transport a variety of molecules, including large protein molecules and phospholipids. All ABC transporters contain an ATP binding domain, also called the nucleotide binding domain (NBF) and two sets of transmembrane domains. Each set of the transmembrane domain consists of approximately six transmembrane spanning α-helices. The ATP-binding domain has three conserved regions: Walker A and B motifs and a signature sequence LSGGQ, which is a feature distinct to the ABC transporters and distinguishes them from other ATP-binding proteins *(3)*. The signature sequence consists of a loop that forms the point of interaction with the transmembrane domains. The Walker A motif is present just before the first three residues of helix 1 and has an α structure and interacts with the phosphate moieties of ATP, whereas the Walker B motif is located in strand 9 and has a β-sheet conformation (Fig. 2).

The various ABC genes are grouped into the different subfamilies based on gene-structure similarity, sequence homology with regard to the ATP-binding domain, and transmembrane regions. Table 2 lists in all known human ABC gene subfamilies.

Because these transporters are known to perform important physiological functions, mutations in their genes result in a variety of genetic disorders. Most of these disorders are allele-recessive and result in severe reduction or a complete loss of protein function. One such disorder, cystic fibrosis (CF), the most fatal childhood disease among Caucasian children, is the result of a defective CFTR (cystic fibrosis TM conductance regulator) protein. This ABC protein is a cAMP-regulated chloride ion channel. Patients with CF mutations fail to secrete chloride ions in response to a number of stimulations *(4)*. The majority of CF chromosomes (70%) have a single codon deletion that translates to a missing phenylalanine residue at position 508 (δ F508) of the protein *(5)*.

Unfortunately, the lack of sufficient crystal structure information limits our understanding of such transporters from their functional and organizational point of view. Such information may provide crucial insight into the treatment of such diseases. To this end, Locher and colleagues *(6)* recently obtained the structure of the vitamin B_{12} transporter (an *Escherichia coli* ABC transporter) at 3.2Å, wherein they have described the structure and potential transport mechanism of the transporter (Fig. 2). Improved insight of the actual transport process and the residues involved should result from such studies.

1.1.3. The Solute Carrier (SLC) Superfamily

The ABC and MFS superfamilies account for nearly half of the solute transporters *(1)*. Originally, MFS and ABC were two distinct superfamilies each containing SLCs, but the increasing number of solute carriers has led HGNC (HUGO Gene Nomenclature Committee) to classify these proteins into a separate superfamily, SLC. The SLC proteins comprise integral membrane proteins involved in the transport of various solutes like amino acids, sugars, vitamins, and bile acids across the cell membrane. They are mainly secondary active carriers (co-transporters) that catalyze, (1) transport of single molecular species (uniport), (2) cotransport of a solute with a cation (symport, e.g., ASBT), (3) countertransport of a solute against a cation (antiport), or (4) bring

Table 1
The Major Facilitator Superfamily

#	Family name	Abbrev	Members (#)	Source(s)[a]	Size range	TM#	Mechanism(s)	Polarity	Substrates	Representative example(s)
1	Sugar porter	SP	133	Bac, Ar, Pr, Y, An, Pl	404–818	12	Sugar uniport Sugar:H+ symport Sugar:sugar antiport	None In Both	Monosaccharides (hexoses, pentoses), disaccharides, quinate, organocations, inositols	XylE of *E. coli*
2	Drug:H1 antiporter (14 TMS)	DHA14	30	Bac, Y		14	Drug:H1 antiport	Out	Multiple or single drugs	QacA of *S. aureus*
3	Drug:H1 antiporter (12 TMS)	DHA12	46	Bac, Ar, Y, An		12	Drug:H1 antiport	Out	Multiple or single drugs	Bmr of *B. subtilis*
4	Organophosphate: antiport	OPA	12	Bac, An	439–495	12	Organo phosphate:Pi antiport	Both	Sugar-phosphates, glycerol phosphate, phospho-glycerates, phosphoenolpyruvate	UhpT of *E.coli*
5	Oligosaccharide:H lsymporter	OHS	6	Bac	415–425	12	Sugar:H$^+$ symport Sugar:sugar antiport	In Both	Di- and trisaccharides	LacY of *E. coli*
6	Metabolite:H1 symporter	MHS	16	Bac	425–500	12	Solute:H$^+$ t symport	In	Citrate, a-ketoglutarate, proline, betaine, methylphthalate, dicarboxylates	KgtP of *E. coli*
7	Fucose-galactose-glucose:	FGHS	4	Bac	404–438	12	Hexose uniport Hexose:H$^+$ symport	None In	L-Fucose, glucose, galactose	FucP of *E. coli*
8	H1 symporter	NNP	13	Bac, Y, Pl	395–547	12	Nitrite uniport? Nitrate:H$^+$ symport?	Out In	Nitrite, nitrate	NarK of *E. coli*
9	Nitrate/nitrite proter	PHS	11	Y, Pl	518–587	12	Pi:H_1 symport	In	Inorganic phosphate	Pho-5 of *N. crassa*

(continued on next page)

10	Phosphate:H1 symporter	NHS	2	G2 Bac	418	12	Nucleoside:H^+ symport	In	Nucleosides	NupG of *E. coli*
11	Nucleoside:H1 symporter	OFA	5	Bac, Ar, An	373–470	12	Anion:anion antiport	Both	Oxalate, formate	OxlT of *O. formigenes*
12	Oxalate:formate antiporter	SHS	3	G2 Bac	407–496	14	Substrate:H^+ symport	In	Sialate	NanT of *E. coli*
13	Sialate:H1 symporter Monocarboxylate porter	MCP	13	Y, An	450–808	12	Substrate:H^+ symporter	In	Pyruvate, lactate, mevalonate	Mct of *H. sapiens*
14	Anion:cation symporter	ACS	40	Bac, Y, An	411–596	12	Substrate:H^+ or Na1 symporter	In	Glucarate, hexuronate, tartrate, 4-hydroxyphenyl acetate inorganic phosphate, allantoate	ExtU of *E. coli*
15	Aromatic acid:H1 symporter	AAHS	7	Bac	418–460	12	Substrate:H^+ symport	In	Muconate, benzoate; 4-OH-benzoate;2,4-dichlorophenoxy-acetate, protocatachurate,3-hydroxypropionate	PcaK of *P. putida*
16	Unknown major facilitator	UMF	6	Y	500–637	14	Unknown	N/A	Unknown	Yh1040c of *S. cereisiae*
17	Cyanate permease	CP	3	Bac	393–402	12	Substrate:H^+ symport?	In	NCO	CynX of *E. coli*
18	Proton-dependent	POT	24	Bac, Y, An, Pl	463–783	12	Substrate:H^+ symport	In	Peptides, amino acids, nitrate, chlorate, nitrite	DtpT of *L.lactis*

After ref. *1*.

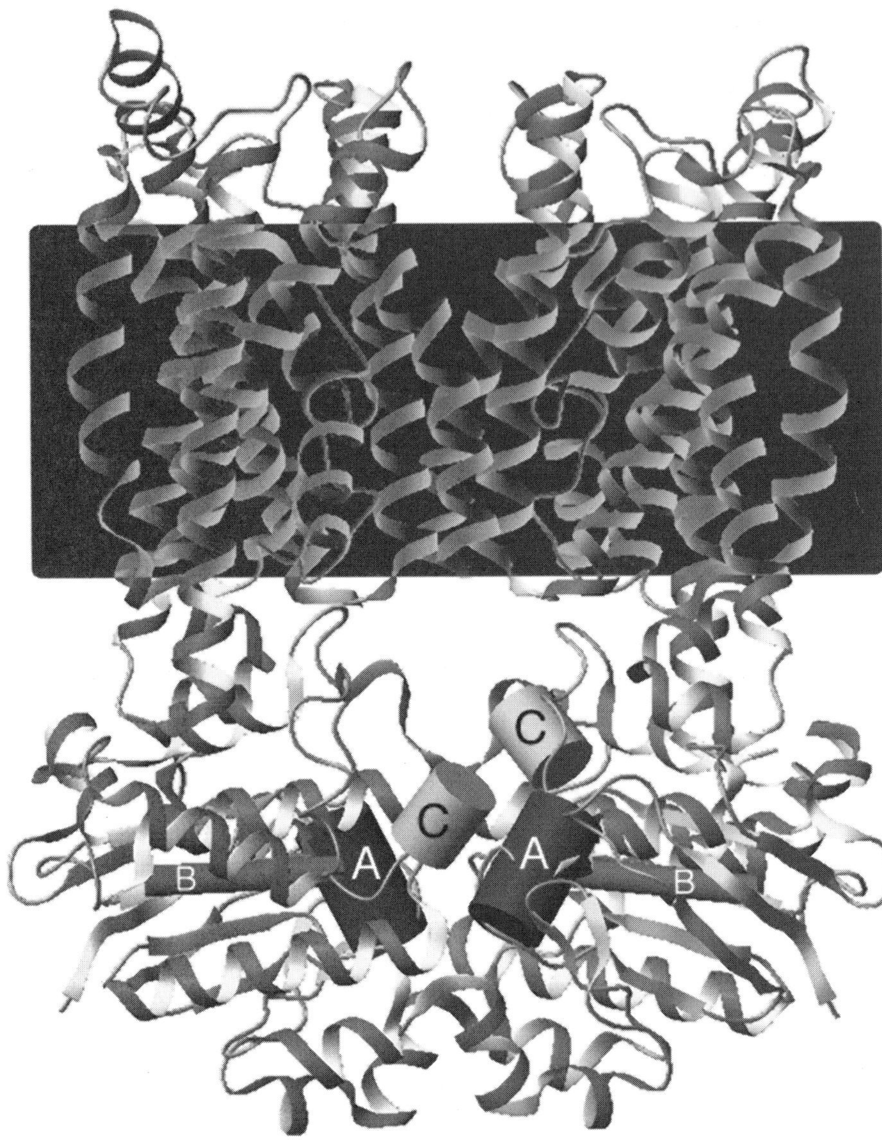

Fig. 2. Crystal structure of the Vitamin B12 transporter. Visualization of protein databank accession number 1L7V (www.rcsb.org) using the molecular-graphics program MolMol. The protein backbone is shown as a ribbon plot; cylinders visualize the following protein motifs: **(A)** Walker A motif; **(B)** Walker B motif; **(C)** ABC signature sequence. The dark background box demarcates the putative boundaries of the cell membrane.

about the exchange of one solute for another (Fig. 1). Currently the SLC superfamily contains 37 families and 237 members. By the end of the Human Genome project, the Human Genome Organization committee anticipates approx 2300 members to make up this class of transporters *(7)*. Because high-resolution structural information is lacking, most SLC proteins have their structures limited to their primary sequence. Nonetheless, the crystal structure of bacteriorhodopsin at 1.55Å resolution was recently obtained and can be used as a starting point for the structural determination of proteins

Table 2
Members of the ATP-Binding Cassette (ABC) Transporter Family

ABC gene subfamily	Description
ABCA (ABC1)	Twelve full transporters. Divided into two subgroups based on intron structure-ABCA1, -2, -3, -4, -7, -12, -13 in one subgroup and ABCA5, -6, -8, -9, -10 in the other. ABCA1 and -4 proteins are involved in cholesterol transport disorder and visual cycle, respectively.
ABCB (MDR/TAP)	Four full and seven half transporters. ABCB1 confers multidrug resistance phenotype to cancer cells. ABCB4 and -11 secrete bile acids and are located in the liver. ABCB2 and -3 are half transporters.
ABCC (CFTR/MRP)	Twelve full transporters involved in ion transport, signal transduction, and toxin secretion. CFTR is a c-AMP regulated chloride ion channel and mutations in this gene cause cystic fibrosis. The MRP proteins transport glutathione and many drug conjugates.
ABCD (ALD)	Four half transporters and are involved in regulation of long-chain fatty-acid transport.
ABCE (OABP)	Not involved in membrane transport functions because they lack the transmembrane domains. Only have ATP-binding domains derived from ABC transporters.
ABCF (GCN20)	Same as ABCE.
ABCG (white)	Six half "reverse" half transporters with ATP-binding domain at the N-terminal and the transmembrane domain at the C-terminal. Involved in cholesterol transport regulation.

Adapted from refs. *2,108*.

with similar membrane topology *(8)*. Recently, Zuniga and colleagues used the topology of bacterial lac permease to prepare a homology model of Glut1 that was used successfully to explain certain amino acid point mutations *(9)*. However, alternate techniques like membrane-insertion scanning, Ala-scanning mutagenesis, and *N*-glycosylation scanning have been used to elucidate the topology of these transporters *(10)*.

SLC proteins have significant therapeutic relevance because they play a significant role in cellular processes. They are able to transport a variety of pharmacologically active compounds, enhancing their oral absorption and bioavailability and influencing their PK/PD. Mutations in these transporters have resulted in disruption of transporter function. For example, a single P290S mutation in the Apical Sodium-dependent bile acid transporter results in a complete loss of the bile acid-uptake function *(11)*.

2. Carrier-Mediated Transport and Prodrugs

To exert their pharmacological activity, drugs must reach their sites of action with certain minimal effective concentration. Most drugs act on specific cellular targets, and in order to reach these targets they must first reach the systemic circulation by penetrating the epithelial barriers covering the absorptive surfaces of the body, such as the skin, intestine, and lung. Transepithelial transport mechanisms for drug molecules include the paracellular route and the transcellular route. In paracellular transport, molecules cross the epithelium through the cells, whereas molecules move across the epithelium through the intercellular junctions between adjacent cells when they use the transcellular pathway. Transcellular transport can be further characterized into passive diffusion and either receptor mediated or active carrier-mediated translocation, the focus of this discussion. In active carrier-mediated transport, a membrane-embedded transport protein transports molecules against a concentration gradient using an energy supply provided by either ATP hydrolysis, or cotransport of ions moving down their concentration gradient, often Na^+ or H^+. In most biological epithelia, drug molecules confront two obstacles in reaching the systemic circulation: (1) a biochemical barrier resulting from enzymatic degradation, and (2) a physical barrier originating from the lipid bilayer. The first obstacle can be overcome by changing the route of administration or drug formulation, for instance, by encapsulating the drugs in vehicles impenetrable to metabolic enzymes. However, for hydrophilic drugs and drugs with high molecular weight, especially macromolecules, epithelial membranes still impose a formidable barrier. Several methods have been employed to increase the transport of a solute across an epithelial membrane: (1) opening the tight junctions or changing the epithelial lipid bilayer membrane (penetration enhancement); (2) oral lipid-based formulations; and (3) design of prodrugs with increased membrane permeability by lipophilization or targeting to a carrier-mediated transport system. Penetration enhancers have had limited success mainly owing to the limited clinical application because of the nonspecific behavior of many tight junctional perturbants. Lipid-based oral drug delivery has been a more promising means of drug delivery with several formulations currently available on the market, including Sandimmum®Neoral (cyclosporine microemulsion), Norvir®) (ritonavir), and Fortovase® (saquinavir). Finally, in recent years significant attention has been placed on the construction of pro-drugs.

Historically, the term prodrug was first introduced by Albert *(12)*, who used the word prodrug or proagent to describe compounds that undergo biotransformation before exhibiting their pharmacological effects. Nowadays, the term prodrug is used to describe a compound that is converted to the pharmacologically active substance after administration *(13)*. Today, rational prodrug design is widely used to overcome problems of absorption, distribution, and biotransformation associated with certain drug molecules. The prodrug concept has been most successfully applied in order to accomplish the following:

- Facilitate absorption and distribution of drugs with poor lipid solubility
- Stabilize against metabolism during oral absorption
- Increase the duration of action of drugs that are rapidly eliminated
- Overcome problems of poor product acceptance by patients

- Eliminate stability and other formulation problems
- Promote site-specific delivery of a drug
- Increase the aqueous solubility
- Lower the toxicity of a drug

In the rational design and synthesis of the ideal prodrug derivative, several factors should be considered: (1) prodrugs should be easily synthesized and purified, (2) prodrugs must be stable in bulk form and dosage form, and (3) neither the prodrug nor its metabolic derivative should be toxic. The vast majority of prodrugs are designed to increase the intestinal absorption of polar drugs—that is, to increase lipophilicity. Prodrug design by increasing lipophilicity has been proven successful. Various ester prodrugs of 6-azauridine (6-AZA), a compound used to treat psoriasis, such as 2',3',5'-triacetyl-6-AZA and 2',3',5'-tribenzoyl-6-AZA have been developed. Orally administered 2',3',5'-triacetyl-6-AZA caused the same clinical effects as an equivalent intravenously administered dose of 6-AZA *(14)*.

Charged hydrophilic compounds and pharmaceutical macromolecules encounter difficulties in permeating the cell membrane. However, the systemic absorption of many water-soluble nutrients (e.g., sugars, vitamins), endogenous proteins (e.g., insulin, growth factors), and toxins (e.g., ricin, cholera toxin) appears to be highly efficient. The effective transcellular movement of these molecules is facilitated by specialized transport processes in the epithelia, that is, carrier-mediated transport and receptor mediated endocytosis/transcytosis. Both processes are operated by specific membrane-associated proteins, which are concentration-, energy-, and temperature-dependent.

Apart from naturally occurring substrates, it is now well-recognized that many drugs can be selectively taken up by active transport processes. For pharmaceutical scientists, these membrane transporters provide alternative routes for the delivery of drugs that would normally be impermeable to the biological barriers. Using a method similar to the conventional prodrug approach, absorption is enhanced by formation of conjugates between drugs and the endogenous ligands of the membrane transporters. Consequently, through specific interactions between the moiety and its transporter, drug candidates can be shuttled across or into the cells and eventually be released from the ligands.

Furthermore, taking advantage of recent advances in molecular biology and computer modeling, scientists are now starting to design prodrugs based on structural requirements of the transporter systems. In general, prodrug strategies involving carrier-mediated pathways have the advantage of high uptake capacity. However, the size of drug conjugates is relatively limited (~1000 Da), probably because larger conjugates fail to be shuttled through the restricted space within the carrier protein. For peptide and protein delivery, carrier-mediated pathways could facilitate peptides only up to four amino acids, although further studies are under way to explore the full potential of this promising pathway. In the following sections, various transporters and prodrug approaches targeting individual transporters will be discussed.

2.1. Sodium-Dependent Multivitamin Transporter (SMVT)

Water-soluble vitamins are not synthesized by the body and therefore must be provided in the diet in order to allow for normal growth and development. Often many of these coenzyme precursors are absorbed into the body through passive diffusion in the intestine. However, uptake of biotin, pantothenate, and the cofactor lipoate occurs by means of an active, carrier-mediated process. Furthermore, it has been reported that this process is dependent on a Na^+-gradient and is not stimulated by an inside-negative membrane potential. Although the name appears to be a misnomer, because the protein transports many other compounds beside vitamins, it has been designated the SMVT *(15)*.

SMVT is a protein of 635 amino acids with 12 transmembrane domains and is responsible for the transport of the water-soluble vitamins biotin, pantothenate, and lipoic acid *(16,17)*. It is expressed in the placenta, intestine, brain, liver, lung, kidneys, and heart in rats, rabbits, and humans *(17)*. Substrate specificity studies indicate that the transporter is specific for the aforementioned compounds as well as biotin analogs with a Na^+:substrate stoichiometry of 2:1. It appears that the carboxylate group of the long side-chain of the substrate molecule serves as the point of interaction for the transporter *(16)*.

The knowledge and characterization of this transporter has led to SMVT targeting as an approach to intestinal drug absorption by chemical modification of drugs to prodrugs and analogs. Recently, Sinko and colleagues characterized biotin conjugates for the treatment of cancer and HIV *(18–20)*. Tat9-C is a 10-amino acid inhibitor of HIV-1 Tat protein that binds to the TAR domain of viral RNA preventing Tat-dependent gene expression in cultured cells. Tat9-C(biotin) and Tat9-K(biotin), biotinylated forms of Tat9-C were taken up 30-fold more efficiently by Jurkat cells than Tat9-C *(21)*. Subsequently, Sinko and colleagues used a retro-inverso derivative of Tat9-K(biotin) denoted R.I.-K(biotin)-Tat9, which was more resistant to proteolysis and had similar TAR RNA binding activity to Tat9-K(biotin) in a mechanistic study of transport to determine its suitability for oral administration *(18)*. They demonstrated both the inhibition of R.I.-K(biotin)-Tat9 transport across Caco-2 cells and the inhibition of R.I.-K(biotin)-Tat9uptake in CHO/hSMVT cells in the presence of biotin, implicating that the compound as a substrate for SMVT. Similar studies confirmed SMVT in the transport of the 29 kDa peptide-loaded bioconjugate (PEG:[R.I-Cys-K{biotin}-Tat9]$_8$).

In addition to the biotinylation of HIV inhibitors, the conjugation of biotin to camptothecin (CPT) has been explored *(20)*. CPT is an anticancer agent whose mechanism of action is the conversion of DNA topoisomerase I into a DNA-damaging agent. Biotinylation of CPT-PEG resulted in a significant increase of cytotoxicity in human ovarian carcinoma cells as compared to CPT-PEG and CPT alone, suggesting the involvement of SMVT in uptake. Thus the targeting of SMVT transporters provides an alternate approach to increasing the bioavailability of compounds especially after oral delivery.

2.2. Sugar Transport

Intestinal sugar transporters are responsible for transporting the monosaccharides glucose, galactose, and fructose from the intestinal lumen to the blood. The sodium ion-dependent glucose transporter (SGLT1) was the first transporter to be cloned and is responsible for the active transport of glucose across the intestinal brush border *(22)*. SGLT1 is a hydrophobic integral membrane protein with approx 12 putative membrane-spanning domains. In addition to SGLT1, there exist other intestinal monosaccharide transporters, including GLUT5 and GLUT2, which are sodium ion-independent facilitative transporters. GLUT5 is responsible for the transport of fructose from the lumen to the cytosol, whereas GLUT2 transports glucose, fructose, and galactose from the cytosol to the blood. Also, a homolog of SGLT1, designated SGLT2 and sharing 59% amino acid homology, has been identified in the human kidney and is responsible for the reabsorption of hexoses *(23,24)*.

The ability to conjugate sugars transported by SGLT1 or GLUT5 to drug molecules provides a means of getting otherwise poorly absorbed drugs into the blood stream. Various research has been performed to test this hypothesis. Using *p*-nitrophenol and 2-naphthol linked to various sugars, it was verified that these compounds moved across everted rat jejunum in a similar manner to that of D-glucose *(25,26)*. Consistent with sodium dependent transport, replacement of sodium ions with potassium ions in the incubation media or the presence of phlorizin significantly inhibited transport. Furthermore, the importance of the type of sugar and the nature of the linkage were investigated, and demonstrated that glucose conjugates were most efficiently transported and β-anomers were more effective than α-anomers.

Conjugation of sugars to peptides was also investigated as a means of increasing absorption. The everted sac technique was used to examine intestinal transport of the tripeptide, tyrosyl-glycyl-glycine (YGG), a component of enkephalin, coupled to glucose and several disaccharides *(27)*. Uncoupled YGG was hydrolyzed by aminopeptidase and no transport was observed. However, sodium ion-dependent transport from the mucosal to the serosal side was observed using the sugar derivatives and was greatest with glucose coupled YGG. Nomoto and colleagues used the tetrapeptide Gly-Gly-Tyr-Arg (GGYR) glycosylated with *p*-(succinylamido)phenyl α or β-D-glucopyranoside (α, β-SAPG) to examine uptake into brush border membrane vesicles (BBMV) and transport through the intestinal membrane using the everted sac method *(28)*. As with YGG, these studies implicated SGLT1 in the uptake and transport of these compounds.

A more clinically relevant approach was used by conjugating the peptide insulin (INS) with succinylamidophenyl (SA)-α-D-glucopyranoside (SAPG), SA-α-D-mannopyranoside, or SA-α-L-arabinopyranoside, and determining the hypoglycemic effect in rats *(29)*. The greatest effect of the modification of insulin occurred with SAPG-INS, resulting in a 56% decrease in blood glucose levels when administered intra-intestinally. Furthermore, the digestion of monosaccharide derivatives by pepsin and chymotrypsin indicated that the resistance of insulin to enzymatic degradation was increased by its modification with monosaccharide. Thus sugar conjugates may be a means to enhance the hypoglycemic effect of insulin by increased stability and targeting the sugar transporter.

In addition to using the prodrug approach for intestinal transport work has been done investigating the transport of peptides across the blood–brain barrier (BBB) with sugar-coupled peptides *(30,31)*. GLUT1 is a facilitated glucose transporter similar in characteristics to GLUT2 and GLUT5 and has been used as target for sugar-coupled enkephalins, endogenous peptides that control brain function. For instance, the δ-selective glycosylated Leu-enkephalin amide 2, H_2N-Tyr-D-Thr-Gly-Phe-Leu-Ser(β-D-Glc)-$CONH_2$, produced analgesic effects similar to morphine. Also, Battaglia and colleagues report the use of D-glucose conjugates of 7-chlorokynurenic acid in rats to protect against seizures *(32)*. The development and use of further such compounds promises enhanced delivery of membrane-impermeable drugs through the BBB for the treatment of migraines as well as neurological disorders.

2.3. Amino Acid Transport

As expected, there are numerous amino acids transporters, especially in the intestine. These include sodium-dependent transporters such as systems B, IMINO, α, A, and ASC, and sodium-independent systems y^+, b, asc, and L. The focus of this section is merely an overview of the targeting of amino acid transporters for drug delivery. For a more detailed characterization of the various amino acid transporters, see refs. *(33,34)*.

System L, including LAT1, which is responsible for the transport of large neutral amino acids in a sodium-independent manner, has been widely studied because of its broad substrate selectivity and its ability to transport amino acids through the intestine as well as the BBB *(35)*. It has been demonstrated that substrates for LAT1 must have a free carboxyl and an amino group. The carbonyl oxygen closer to the amino group needs a computed charge of –0.55 to approx –0.56 and must not participate in hydrogen bonding. Furthermore, the hydrophobic interaction between the substrate side chain and the substrate binding site of LAT1 is crucial for the substrate binding *(36)*.

System L has been used to transport amino acid-analogs such as L-dopa, a therapeutic drug for the treatment of Parkinson's disease *(37)*; the opioid baclofen (38); as well as the anticancer agents melphalan and acivicin *(39,40)*. Yang and colleagues synthesized amino acid ester prodrugs of acyclovir and investigated the nasal uptake of these compounds. They demonstrate improvement of nasal absorption of the L-aspartate β-ester prodrug of acyclovir by the amino acid transporter, demonstrating the potential of designed prodrugs for efficient transmucosal delivery *(41)*. More recently, Balakrishnan provided evidence of the presence of a large, neutral amino acid transport system on the corneal epithelium. The rabbit corneal cell line (SIRC) as well as freshly excised rabbit cornea were utilized as a model, and uptake studies were performed with ^{3}H-L-tyrosine to demonstrate involvement of the amino acid transporter *(42)*.

2.4. Nucleoside Transport

Nucleosides are the precursors of nucleotides and form the building blocks of nucleic acids. Natural nucleosides are hydrophilic molecules *(43)*. and cannot easily cross the cell membrane by nonfacilitated diffusion, and therefore nucleoside transporters are necessary for the permeation of nucleosides across biological membranes. Thus nucleosides permeate most animal cells via multiple transporters present in the plasma membrane of most cell types *(44,45)*. These transporters can be grouped into two broad categories: (1) Na^+-dependent concentrative transporters that mediate only the influx of nucleosides and (2) the Na^+-independent equilibrative transporters that mediate both influx and efflux of nucleosides. The various subtypes of the two major nucleoside transporters are listed below (*see* Table 3).

The equilibrative nucleoside transporters (ENTs) are broadly selective accepting both purines and pyrimidines, have low affinity, and exhibit high rates of transport properties. This group of transporters is further subdivided based on their sensitivities to nitrobenzylmercaptopurineriboside (NBMPR) *(46)* into *es* (equilibrative-sensitive) and *ei* (equilibrative-insensitive) transporters. Equilibrative transporters are widely distributed among various cell types and only very recently have the human *es* and the *ei* subtypes been identified by cloning *(47)*. through the use of placental and HeLa cell cDNA libraries *(48)*. They have been designated as human equilibrative transporters 1 and 2 (hENT1 and hENT2), respectively. Both these transporters contain 456 amino acid residues, share 50% sequence identity, and are thought to contain 11 transmembrane spanning regions with a cytoplasmic N-terminus and an extracellular C-terminus *(49)*. The regions important for inhibitor binding have been identified by creating chimeras from hENT1, rENT1, and rENT2 cDNAs *(50)*. and performing transport-inhibition studies in *Xenopus laevis* oocytes. These studies indicated that transmembrane domains 3–6 comprising amino acids 100–231 in the hENT1 are important for interaction with dilazep and dipyridamole *(51)*.

Table 3
Characteristics of Nucleoside Transport System

Transporter	Property	Substrate selectivity[a]	Tissue Distribution[b]	NBMPR[c]	Na+-dependent?
SPNT	Concentrative	Pu, U	Rat liver, jejunum	No	Yes
rCNT1	Concentrative	Py, A	Intestine, kidney	No	Yes
hCNT1	Concentrative	Pu, U	Kidney, intestine	No	Yes
hCNT2	Concentrative	Py, A	Kidney, brain, heart, liver	No	Yes
hCNT3	Concentrative	Pu, Py	Bone marrow, intestine	No	Yes
N4	Concentrative	Py, G	ND	Yes	Yes
N5	Concentrative	G, FB	ND	Yes	Yes
hENT1	Equilibrative	Pu, Py	ubiquitous	Yes	No
hENT2	Equilibrative	Pu, Py	Skeletal Muscle, heart, pancreas	No	No

Adapted from ref. *44*.
[a]Pu, purine; U, uridine; Py, pyrimidines; A, adenosine; G, guanosine; FB, formycine B.
[b]ND, not determined.
[c]Denotes whether the selected transporter is sensitive to NBMPR.

Na^+-dependent concentrative nucleoside transporters (CNT) are widespread and have been expressed in the intestine, kidney, liver, spleenocytes, and many cultured cell lines *(52)*. The first member of the CNT was cloned from rat jejunum and designated as the rat concentrative nucleoside transporter1 (rCNT1) *(53)*. Since then, the family has expanded rapidly to include rCNT2 from the rat intestine and the human counterparts to both the rat transporters, hCNT1, and hCNT2, respectively *(54)*. This group of transport processes has been further divided into five classes (N1-N5) on the basis of their substrate specificity (*see* Table 3). The SPNT/hSPNT1 (previously named N1 [cif]) system is widely distributed and is purine specific with high selectivity towards uridine, guanosine, and adenosine. hCNT1/rCNT1 (previously N2 [cit]) is more selective towards pyrimidine, especially thymidine, whereas N3 (cib) exhibits broad specificity and transport both purines and pyrimidines. N4 is found in the human kidney and is similar to N2, but additionally transports guanosine. Finally, N5 has been found in leukemic cells, with specificity toward guanosine *(43)*, but is more sensitive to inhibition by NBMPR compared with other subtypes *(55)*. The mammalian CNT family proteins are about 650 amino acids in length and exhibit about 64% identity *(56)*. Recent studies have shown that the mammalian CNTs possess 13 transmembrane segments and a very long glycosylated extracellular C-terminus *(57)*. Residues 319 and 320 in hCNT1 were found to be important for the pyrimidine nucleoside specificity *(58)*.

Structurally modified nucleosides are pharmacologically active molecules that have been used in the treatment of cancer and as antiviral therapies *(59)*. They have the ability to interfere with normal nucleotide metabolism and DNA replication, promoting antiproliferative effects and preventing replication of the viral DNA in infected cells. The transport of nucleoside analogs across the gastrointestinal tract is mediated by the transporters described earlier (i.e., CNT1 and SPNT). Most of these nucleoside analogs inhibit natural nucleoside transport by competitive inhibition *(60,61)*. All nucleoside analogs must be phosphorylated to their triphosphate form in order to be biologically active. This is accomplished by various enzymes including the salvage enzyme deoxyctidine kinase *(62)*. Incorporation of the triphosphate analogs interferes with DNA elongation during replication and repair of DNA damage *(63)*. In addition to their DNA chain terminating effect, it is believed that both expression and activity of nucleoside transporters may be affected by treatment with the analogs *(59)*.

In order to improve the bioavailability of polar drugs such as nucleoside antivirals, membrane transporters located in the intestinal epithelium are usually targeted. However, because the nucleoside transporters themselves are of high-affinity/low-capacity, alternate approaches have to be sought. The intestinal Pept1, which will be discussed in more detail later, is one such transporter. It has been targeted to improve the bioavailability of nucleoside amino acid ester prodrugs of acyclovir and AZT (3'-azido-3'-deoxythymidine), because this transporter is characterized by high capacity and low affinity *(64)*. AZT was the first drug approved for the treatment of AIDS and has an oral bioavailability of up to 70% (65). AZT uptake has been found to be Na^+-dependent in rat jejunum *(54)*, but the primary mode of AZT absorption was found to be by passive diffusion *(66)*. Many chemotherapeutic drugs used in the treatment of cancer are also nucleoside derivatives *(59)*, such as the pharmacologically active fluoropyrimidines, which inhibit thymidylate synthase, causing a reduction of phosphorylated forms of deoxythymidylate, and resulting in apoptosis *(67)*. Gemcitabine (2',2'- difluorodeoxycytidinedFdC), a very recently developed agent in the treatment of solid tumors, has been found to penetrate the cell via *es* nucleoside transporters, and Ara-C is yet another significant anticancer drug that is a substrate for nucleoside transporters *(68)*.

Various mammalian expression systems like the COS-1 cells and oocytes have been used to study the interactions of the nucleoside analogs with the nucleoside transporters *(44,69)*. More recently computer simulations have been utilized to study such actions. Our laboratory recently reported three-dimensional quantitative structure–activity relationship (3D-QSAR) models of each of the nucleoside transporters, based on their available binding profiles. We used comparative molecular field analysis (CoMFA) and GOLPE to generate models that relate the differences among the substrates to their bioactivity *(69a)*. Such information could provide future insight into the design and use of agents targeting the nucleoside transporters. However, further studies need to be carried out to understand completely the interactions between the transporters and the nucleoside derivatives.

2.5. The Intestinal Peptide Transporter

The intestinal oligopeptide transporter (PepT1) is located in the brush-border membrane of the intestine. It provides the main pathway for absorption of dietary nitrogen by mediating the cellular uptake of dipeptides and tripeptides *(70)*. It also transports a number of peptidomimetics such as β-lactam antibiotics, angiotensin-converting enzyme (ACE) inhibitors and, possibly, renin inhibitors *(71)*. As a result of this broad substrate specificity and its ability to transport a variety of peptidomimetic drugs, PepT1 has captured much attention as a target for mucosal drug delivery *(72)*.

PepT1 belongs to a larger family of oligopeptide transporters, the proton/oligopeptide transporter (POT) family, in which only two human members hPepT1 and hPepT2 have been identified *(73)*. It is an integral membrane protein, 707 amino acids long with a molecular weight of 79 kDa and 12 putative membrane spanning regions with a cytoplasmic N- and C-terminal *(74)*. First cloned in rabbit by expression cloning *(75)*, it was later expressed in rat *(76)* and human *(77)*. The protein has several N-glycosylation sites, and the rat and human PepT1 have one and two protein kinase C

(PKC) phosphorylation sites, respectively. Transport studies indicated that uptake was electrogenically coupled to an inward flow of protons with a 1:1 stoichiometry *(78)*. The transporter has a low affinity, with a K_m value in the millimolar range *(79)*, and a very broad substrate specificity *(64,80)*. The high degree of tolerance of the transporter for structurally diverse substrates necessitates the need to investigate its substrate-binding domains. The lack of a crystal structure has led to the development and use of several experimental and computational approaches in order to obtain information about the structural specificity toward its substrates *(72)*. In brief, a systematic investigation of the structural influence on uptake using a number of di- and tripeptides, β-lactam antibiotics in isolated brush-border membrane vesicles *(81)*, molecular modeling to identify the pharmacophore for the substrate binding *(82)*, site-directed mutagenesis *(83)*, and engineered chimeras between PepT1 and PepT2 *(84)* are a few of the methods that have been employed.

The human H^+-coupled oligopeptide transporter has been suggested as an excellent target for the delivery of pharmacologically active compounds because of its ability to accommodate a wide variety of molecules with varied size, hydrophobicity, and charge at its substrate-binding site *(85)*. Most substrates for the peptide transporter, as implied by its nomenclature, are peptide in nature. For example, cephalexin resembles Phe-Cys-Val and enalapril resembles Phe-Ala-Pro, whereas β-lactams are dipeptide analogs. The intestinal peptide transporter has been employed to improve the intestinal absorption of certain drugs by converting them to di- or tripeptide type prodrugs *(86)*. For example the low bioavailability of α-methyldopa, an antihypertensive agent (interestingly, in itself a substrate for the structurally restrictive sodium-coupled amino acid transporter) has been improved by converting it into a dipeptide prodrug, α-methyldopa-L-phenylalanine, which is then efficiently taken up by PepT1 *(87)*. However, in 1998, Balimane and colleagues reported that the drug valacyclovir, which lacks a peptide bond, acts as a substrate for PepT1 *(88)*. Valacyclovir is the water-soluble 5'-L-valyl ester prodrug of acyclovir. Acyclovir, a nucleoside analog, was the first antitherapeutic agent capable of selectively inhibiting the replication of herpes simplex virus (HSV), producing little or no adverse side effects *(43,88)*. However the low oral bioavailability of acyclovir hampered the efficient use of this drug. The prodrug valacyclovir has a mean oral bioavailability that is three to four times greater than that of acyclovir, with a dose-dependent rate of absorption *(89)*. The mechanism of intestinal absorption was studied in three different experimental systems: *in situ* rat perfusion model, CHO/hPEPT1 cells, and Caco-2 cells *(64)*. The results indicated a high affinity of the prodrug to the peptide transporter. Furthermore, uptake was significantly inhibited by the most common substrate (Gly-Sar) of PepT1, implicating it in the transport of the amino acid esters.

Based on this peptidomimetic approach, it may be possible to create other polar antiviral and anticancer therapeutic agents with increased oral bioavaliability. In patients suffering from cytomegalovirus (CMV) infection, clinical trials have shown valganciclovir, the valyl ester of ganciclovir increases the bioavailability of ganciclovir when taken orally. Cloned hPEPT1 was expressed in *X. laevis* oocytes and exposed to varying concentrations of ganciclovir and valganciclovir, to detect currents. Exposure to ganciclovir did not produce any detectable currents, however inward currents were associated with valganciclovir exposure. Their experiments demonstrated that valganciclovir has a high affinity for hPEPT1 and is readily converted to its pharmacologically active parent compound ganciclovir in the intestine, once transported by hPEPT1 through the brush border membrane into the mucosal cells.

2.6. The Apical Sodium-Dependent Bile-Acid Transporter (ASBT)

The metabolism of cholesterol to bile acids is an important route for the elimination of cholesterol from the body. Bile acids are secreted from the gallbladder into the proximal jejunum, where they are passively absorbed by jejunal and ileal enterocytes. At the distal ileum, a sodium-dependent bile-acid transporter (ASBT) exists on the apical surface of the enterocyte. This transporter accepts bile acids, which then enter into the portal system where they complete enterohepatic circulation. ASBT is officially referred to as SLC10A2, an approximately 4 Kb mRNA message that codes for a 348 amino acids protein. ASBT was first cloned from hamster by expression cloning *(90)*. Later human *(11)*, rat, rabbit, and mouse *(91)* were also cloned. The interspecies protein homology is approx 85–90%, however, the K_m values of hamster and human ASBT are 33 and 17 μ*M*, respectively *(90)*. ASBT is an electrogenic active transporter with a 2:1 Na^+: bile-acid coupling stiochiometry *(92)*. The apical sorting signal for ASBT is present at the C-terminal of ASBT *(93)*. The vectorial transport of conjugated bile acids is efficiently regulated by two bile-acid transporters, NTCP (sodium taurocholate transporting protein) and ASBT in the liver and intestine, respectively. These two transporters share a 35% sequence identity and an overall similarity of 63%. Studies have shown that NTCP and ASBT topologies are similar and almost overlap on hydropathy plot analysis *(94)*. Based on hydropathy analysis and an exoplasmic N-terminus and a cytoplasmic C-terminus *(95)*, ASBT was proposed to have a seven-transmembrane model. However, Hallén and colleagues suggested the possibility of a 9TM model based on membrane-insertion scanning analysis *(96)*. This group predicted TM3 and TM 4 region to be made up only 9–10 amino acids each, which suggests that these regions must adopt a β-sheet conformation in order to span the membrane effectively. Also they predicted that TM2 and 3 and TM8-9 are linked by two and four extramembraneous amino acids, respectively, suggesting a very sharp turn in helical structure. To address the topology controversy, our laboratory recently carried out *N*-glycosylation scanning analysis that strongly supports a 7TM topology model for hASBT *(97)*.

In order to elucidate the transport mechanism of ASBT, it is essential to first identify the residues/regions involved in the transport mechanism. It is believed that the charged residues on the extracellular side of the transporter may be involved in the initial substrate recognition and binding. Several studies probed its functional regions by chemical modifications using amino specific reagents *(98)*. Studies using a 3D QSAR pharmacophore model also predicted several important interactions between bile acids and ASBT *(99)*. Their studies on rabbit ASBT led to the development of a 3D pharmacophore model. The model predicts that ring D, the methyl group at position 18 and the α-hydroxyl groups at positions 7 and 12 are essential for ASBT affinity and binding. The hydroxyl group at position 3 allows for certain flexibility and did not interfere with active transport.

ASBT is a transporter with significant pharmaceutical relevance. It has been targeted for drug delivery and has recently been studied as a pharmacological target for cholesterol-lowering therapy *(100,101)*. Kramer and group have shown that bile-acid transporters can efficiently transport bile acid-drug conjugates. Swaan and colleagues conjugated small peptides at the C-24 position, which could also be transported by the bile ASBT *(102)*. Based on these studies, they developed a 3D QSAR model for ASBT using comparative molecular-field analysis *(103)*. The carrier-mediated bile acid transport for targeted drug-delivery purposes can be applied to the liver and gall bladder for directed delivery. It can also be used to enhance oral absorption of various prodrugs and lower serum cholesterol levels *(73)*. Bile acid-transport systems have been used successfully to target various inhibitors such as HMG-Co A reductase inhibitors to the liver. These inhibitors or drug molecules, when conjugated to bile acids, seem to retain their affinity with the bile-acid carrier. ASBT has also been targeted for lowering of serum cholesterol levels. Previously HMG-CoA reductase inhibitors and various bile-acid sequestrants, such as cholestyramine, were used to lower cholesterol levels *(104)*. Bezothiazepine derivative was able to inhibit active ileal bile-acid absorption in mice rats, monkeys, and humans *(105)*. It has also been shown that inhibition of ASBT by SC-435 significantly reduced plasma low-density lipoprotein (LDL) cholesterol levels through enhanced LDL receptor-mediated LDL apoB clearance *(106)*. Inhibition of the intestinal bile-acid transporter is thought to result in an increased fecal bile-acid excretion and lower plasma LDL levels after oral administration of the various drugs.

3. Conclusion

A substantial role has been recognized for transport proteins in oral absorption and drug bioavailability; drug resistance; excretion of drugs and their metabolites, mediated by transporters in the kidney and liver; drug toxicity; and drug pharmacokinetics and pharmacodynamics. Furthermore, the pathophysiology of several hereditary diseases has been attributed to mutations in transport proteins. Overall, solute transporters play an invaluable role in fundamental cellular processes in health and disease, and function as important mediators governing all aspects of drug therapy.

Although the importance of transporters in shuttling drugs across intestinal and other tissues was recognized in the late 1970s, research in the transporter field did not start in earnest until after the initial expression cloning of the sodium-dependent glucose transporter, SGLT1, in 1986 *(22)*. The resulting "transporter revolution" led to the cloning of numerous pharmacologically important transporters, as evidenced by an exponential increase in the number of publications on transporters over the last decade.

Despite the apparent clinical importance of transporter proteins, the knowledge of their structure and mechanism of action has lagged far behind the knowledge of these properties of proteins in general. As a result, we are unable to predict the interaction of drugs with this important class of membrane proteins *a priori*, and detection of drug-transporter interactions remains unacceptably serendipitous. It has been clearly recognized that transporters and their natural substrates may provide an excellent target for increasing the epithelial transfer and systemic availability of therapeutic moieties. The prodrug approach has been applied successfully to target several transporters, but the widespread application of this technique by means of "rational design" is hampered by a lack of structural data. Therefore, it is imperative that scientists direct their attention to investigating structure–function and structure–activity relationships of membrane transporters and, eventually, solve their molecular structure using high-resolution techniques. Clearly, in silico transporter models are a first step in this direction, and the increased availability of data sets from single transporter expression systems should facilitate the construction of more reliable and predictive 3D-QSAR models *(107)*.

Finally, it should be noted that certain transporters can serve as "anti-targets," i.e., systems that negatively impact drug therapy, such as the drug resistance-associated transporters P-gp, MRP, and BCRP. By screening compounds for potential interaction with these systems in silico, one can avoid low response rates and, ultimately, optimize drug therapy.

References

1. Pao SS, Paulsen IT, Saier MH, Jr. Major facilitator superfamily. Microbiol Mol Biol Rev 1998;62:1–34.
2. Dean M, Rzhetsky A, Allikmets R. The human ATP-binding cassette (ABC) transporter superfamily. Genome Res 2001;11:1156–1166.
3. Hyde SC, Emsley P, Hartshorn MJ, et al. Structural model of ATP-binding proteins associated with cystic fibrosis, multidrug resistance and bacterial transport. Nature 1990;346:362–365.
4. Gabriel SE, Clarke LL, Boucher RC, Stutts MJ. CFTR and outward rectifying chloride channels are distinct proteins with a regulatory relationship. Nature 1993;363:263–268.
5. Fuller CM, Benos DJ. Cftr! Am J Physiol 1992;263:C267–C286.
6. Locher KP, Lee AT, Rees DC. The E. coli BtuCD structure: a framework for ABC transporter architecture and mechanism. Science 2002;296:1091–1098.
7. Zhang EY, Knipp GT, Ekins S, Swaan PW. Structural biology and function of solute transporters: implications for identifying and designing substrates. Drug Metab Rev 2002;34:709–750.
8. Luecke H, Schobert B, Richter HT, et al. Structure of bacteriorhodopsin at 1.55 A resolution. J Mol Biol 1999;291:899–911.
9. Zuniga FA, Shi G, Haller JF, et al. A three-dimensional model of the human facilitative glucose transporter Glut1. J Biol Chem 2001;276:44970–44975.
10. Turk E, Kerner CJ, Lostao MP, Wright EM. Membrane topology of the human Na+/glucose cotransporter SGLT1. J Biol Chem 1996;271:1925–1934.
11. Wong MH, Oelkers P, Dawson PA. Identification of a mutation in the ileal sodium-dependent bile acid transporter gene that abolishes transport activity. J Biol Chem 1995;270:27228–27234.
12. Albert A. Chemical aspects of selective toxicity. Nature 1958;182:421–423.
13. Stella V. Prodrugs as novel drug delivery systems, ACS symposium series 14, American Chemical Society, Washington, DC, 1975.
14. Welch AD. Cancer Res 1961;21:1475–1490.
15. Fenstermacher DK, Rose RC. Absorption of pantothenic acid in rat and chick intestine. Am J Physiol 1986;250:G155–G160.
16. Prasad PD, Wang H, Huang W, et al. Molecular and functional characterization of the intestinal Na+-dependent multivitamin transporter. Arch Biochem Biophys 1999;366:95–106.
17. Wang H, Huang W, Fei YJ, et al. Human placental Na+-dependent multivitamin transporter. Cloning, functional expression, gene structure, and chromosomal localization. J Biol Chem 1999;274:14875–14883.
18. Ramanathan S, Pooyan S, Stein S, et al. Targeting the sodium-dependent multivitamin transporter (SMVT) for improving the oral absorption properties of a retro-inverso Tat nonapeptide. Pharm Res 2001;18:950–956.
19. Ramanathan S, Qiu B, Pooyan S, et al. Targeted PEG-based bioconjugates enhance the cellular uptake and transport of a HIV-1 TAT nonapeptide. J Control Release 2001;77:199–212.
20. Minko T, Paranjpe PV, Qiu B, et al. Enhancing the anticancer efficacy of camptothecin using biotinylated poly(ethylene glycol) conjugates in sensitive and multidrug-resistant human ovarian carcinoma cells. Cancer Chemother Pharmacol 2002;50:143–150.
21. Choudhury I, Wang J, Stein S, et al. Translational effects of peptide antagonists of Tat protein of human immunodeficiency virus type 1. J Gen Virol 1999;80(Pt 3):777–782.
22. Hediger MA, Coady MJ, Ikeda TS, Wright EM. Expression cloning and cDNA sequencing of the Na^+/glucose co-transporter. Nature 1987;330:379–381.
23. Wells RG, Mohandas TK, Hediger MA. Localization of the Na+/glucose cotransporter gene SGLT2 to human chromosome 16 close to the centromere. Genomics 1993;17:787–789.
24. Kanai Y, Lee WS, You G, et al. The human kidney low affinity Na+/glucose cotransporter SGLT2. Delineation of the major renal reabsorptive mechanism for D-glucose. J Clin Invest 1994;93:397–404.
25. Mizuma T, Ohta K, Hayashi M, Awazu S. Intestinal active absorption of sugar-conjugated compounds by glucose transport system: implication of improvement of poorly absorbable drugs. Biochem Pharmacol 1992;43:2037–2039.

26. Mizuma T, Ohta K, Awazu S. The beta-anomeric and glucose preferences of glucose transport carrier for intestinal active absorption of monosaccharide conjugates. Biochim Biophys Acta 1994;1200:117–122.
27. Mizuma T, Sakai N, Awazu S. Na(+)-dependent transport of aminopeptidase-resistant sugar-coupled tripeptides in rat intestine. Biochem Biophys Res Commun 1994;203:1412–1426.
28. Nomoto M, Yamada K, Haga M, Hayashi M. Improvement of intestinal absorption of peptide drugs by glycosylation: transport of tetrapeptide by the sodium ion-dependent D-glucose transporter. J Pharm Sci 1998;87:326–332.
29. Haga M, Saito K, Shimaya T, et al. Hypoglycemic effect of intestinally administered monosaccharide-modified insulin derivatives in rats. Chem Pharm Bull (Tokyo) 1990;38:1983–1986.
30. Bilsky EJ, Egleton RD, Mitchell SA, et al. Enkephalin glycopeptide analogues produce analgesia with reduced dependence liability. J Med Chem 2000;43:2586–2590.
31. Polt R, Porreca F, Szabo LZ, et al. Glycopeptide enkephalin analogues produce analgesia in mice: evidence for penetration of the blood-brain barrier. Proc Natl Acad Sci USA 1994;91:7114–7148.
32. Battaglia G, La Russa M, Bruno V, et al. Systemically administered D-glucose conjugates of 7-chlorokynurenic acid are centrally available and exert anticonvulsant activity in rodents. Brain Res 2000;860:149–156.
33. Christensen HN. Distinguishing amino acid transport systems of a given cell or tissue. Methods Enzymol 1989;173:576–616.
34. Ganapathy V. Intestinal transport of amino acids and peptides. In: Johnson LR, ed. Physiology of the Gastrointestinal Tract. New York, Raven Press, 1994, pp. 1773–1794.
35. Christensen HN. Role of amino acid transport and countertransport in nutrition and metabolism. Physiol Rev 1990;70:43–77.
36. Uchino H, Kanai Y, Kim do K, et al. Transport of amino acid-related compounds mediated by L-type amino acid transporter 1 (LAT1): insights into the mechanisms of substrate recognition. Mol Pharmacol 2002;61:729–737.
37. Bodor N, Sloan KB, Higuchi T, Sasahara K. Improved delivery through biological membranes. 4. Prodrugs of L-dopa. J Med Chem 1977;20:1435–1445.
38. Cercos-Fortea T, Polache A, Nacher A, et al. Influence of leucine on intestinal baclofen absorption as a model compound of neutral alpha-aminoacids. Biopharm Drug Dispos 1995;16:563–577
39. Goldenberg GJ, Lam HY, Begleiter A. Active carrier-mediated transport of melphalan by two separate amino acid transport systems in LPC-1 plasmacytoma cells in vitro. J Biol Chem 1979;254:1057–1064.
40. Chikhale EG, Chikhale PJ, Borchardt RT. Carrier-mediated transport of the antitumor agent acivicin across the blood-brain barrier. Biochem Pharmacol 1995;49:941–945.
41. Yang C, Gao H, Mitra AK. Chemical stability, enzymatic hydrolysis, and nasal uptake of amino acid ester prodrugs of acyclovir. J Pharm Sci 2001;90:617–624.
42. Balakrishnan A, Jain-Vakkalagadda B, Yang C, et al. Carrier mediated uptake of L-tyrosine and its competitive inhibition by model tyrosine linked compounds in a rabbit corneal cell line (SIRC): strategy for the design of transporter/receptor targeted prodrugs. Int J Pharm 2002;247:115–125.
43. Balimane PV, Sinko PJ. Involvement of multiple transporters in the oral absorption of nucleoside analogues. Adv Drug Deliv Rev 1999;39:183–209.
44. Wang J, Schaner ME, Thomassen S, et al. Functional and molecular characteristics of Na(+)-dependent nucleoside transporters. Pharm Res 1997;14:1524–1532.
45. Plagemann PG, Wohlhueter R. Permeation of Nucleosides nucleic acids bases and nucleotides in animal cells. Curr Top Membr Transport 1980;14:225.
46. CE C. Nucleoside transport in drug transport. Antimicrobial and anticancer chemotherapy 1995:403.

47. Griffiths M, Beaumont N, Yao SY, et al. Cloning of a human nucleoside transporter implicated in the cellular uptake of adenosine and chemotherapeutic drugs. Nat Med 1997;3:89–93.
48. Crawford CR, Patel DH, Naeve C, Belt JA. Cloning of the human equilibrative, nitrobenzylmercaptopurine riboside (NBMPR)-insensitive nucleoside transporter ei by functional expression in a transport-deficient cell line. J Biol Chem 1998;273:5288–5293.
49. Sundaram M, Yao SY, Ingram JC, et al. Topology of a human equilibrative, nitrobenzylthioinosine (NBMPR)-sensitive nucleoside transporter (hENT1) implicated in the cellular uptake of adenosine and anti-cancer drugs. J Biol Chem 2001;276:45270–45275.
50. Yao SY, Ng AM, Muzyka WR, et al. Molecular cloning and functional characterization of nitrobenzylthioinosine (NBMPR)-sensitive (es) and NBMPR-insensitive (ei) equilibrative nucleoside transporter proteins (rENT1 and rENT2) from rat tissues. J Biol Chem 1997;272:28423–28430.
51. Sundaram M, Yao SY, Ng AM, et al. Chimeric constructs between human and rat equilibrative nucleoside transporters (hENT1 and rENT1) reveal hENT1 structural domains interacting with coronary vasoactive drugs. J Biol Chem 1998;273:21519–215125.
52. Huang QQ, Harvey CM, Paterson AR, et al. Functional expression of Na(+)-dependent nucleoside transport systems of rat intestine in isolated oocytes of Xenopus laevis. Demonstration that rat jejunum expresses the purine-selective system N1 (cif) and a second, novel system N3 having broad specificity for purine and pyrimidine nucleosides. J Biol Chem 1993;268:20613–20619.
53. Huang QQ, Yao SY, Ritzel MW, et al. Cloning and functional expression of a complementary DNA encoding a mammalian nucleoside transport protein. J Biol Chem 1994;269:17757–17760.
54. Ritzel MW, Yao SY, Ng AM, et al. Molecular cloning, functional expression and chromosomal localization of a cDNA encoding a human Na+/nucleoside cotransporter (hCNT2) selective for purine nucleosides and uridine. Mol Membr Biol 1998;15:203–211.
55. Chandrasena G, Giltay R, Patil SD, et al. Functional expression of human intestinal Na+-dependent and Na+-independent nucleoside transporters in Xenopus laevis oocytes. Biochem Pharmacol 1997;53:1909–1918.
56. Cass CE, Young JD, Baldwin SA, et al. Nucleoside transporters of mammalian cells. Pharm Biotechnol 1999;12:313–352.
57. Hamilton SR, Yao SY, Ingram JC, et al. Subcellular distribution and membrane topology of the mammalian concentrative Na+-nucleoside cotransporter rCNT1. J Biol Chem 2001;276:27981–2988.
58. Wang J, Giacomini KM. Molecular determinants of substrate selectivity in Na+-dependent nucleoside transporters. J Biol Chem 1997;272:28845–28848.
59. Pastor-Anglada M, Felipe A, Casado FJ. Transport and mode of action of nucleoside derivatives used in chemical and antiviral therapies. Trends Pharmacol Sci 1998;19:424–430.
60. Baldwin SA, Mackey JR, Cass CE, Young JD. Nucleoside transporters: molecular biology and implications for therapeutic development. Mol Med Today 1999;5:216–224.
61. Pastor-Anglada M, Felipe A, Casado FJ, et al. Nucleoside transporters and liver cell growth. Biochem Cell Biol 1998;76:771–777.
62. Arner ES, Eriksson S. Mammalian deoxyribonucleoside kinases. Pharmacol Ther 1995;67:155–186.
63. Cheson BD, Keating MJ, Plunkett W. Nucleoside Analogs in Cancer Therapy. Marcel Dekker, New York, 1997, p. 470.
64. Han H, de Vrueh RL, Rhie JK, et al. 5'-Amino acid esters of antiviral nucleosides, acyclovir, and AZT are absorbed by the intestinal PEPT1 peptide transporter. Pharm Res 1998;15:1154–1159.
65. Klecker RW Jr, Collins JM, Yarchoan R, et al. Plasma and cerebrospinal fluid pharmacokinetics of 3'-azido-3'-deoxythymidine: a novel pyrimidine analog with potential application for the treatment of patients with AIDS and related diseases. Clin Pharmacol Ther 1987;41:407–412.
66. Fang X, Parkinson FE, Mowles DA, et al. Functional characterization of a recombinant sodium-dependent nucleoside transporter with selectivity for pyrimidine nucleosides (cNT1rat) by transient expression in cultured mammalian cells. Biochem J 1996;317:457–465.

67. Bouffard DY, Momparler RL. Comparison of the induction of apoptosis in human leukemic cell lines by 2',2'-difluorodeoxycytidine (gemcitabine) and cytosine arabinoside. Leuk Res 1995;19:849–856.
68. Mackey JR, Mani RS, Selner M, et al. Functional nucleoside transporters are required for gemcitabine influx and manifestation of toxicity in cancer cell lines. Cancer Res 1998;58:4349–4357.
69. Yao SY, Cass CE, Young JD. Transport of the antiviral nucleoside analogs 3'-azido-3'-deoxythymidine and 2',3'-dideoxycytidine by a recombinant nucleoside transporter (rCNT) expressed in Xenopus laevis oocytes. Mol Pharmacol 1996;50:388–393.
69a. Chang, CA, Swaan PW, Ngo LY, Lum PY, Patil SD, Unadkat JD. Molecular Requirements of the Human Nucleoside Transporters, hCNT1, hCNT2 and hENT1. Mol Pharmacol. In press.
70. Meredith D, Boyd CA. Oligopeptide transport by epithelial cells. J Membr Biol 1995;145:1–12.
71. Dantzig AH, Duckworth DC, Tabas LB. Transport mechanisms responsible for the absorption of loracarbef, cefixime, and cefuroxime axetil into human intestinal Caco-2 cells. Biochim Biophys Acta 1994;1191:7–13.
72. Lee VH. Membrane transporters. Eur J Pharm Sci 2000;11 Suppl 2:S41–S50.
73. Swaan PW. Membrane transport proteins and drug transport. In: Abraham DJ, ed. Burger's Medicinal Chemistry and Drug Discovery, vol. 2, Drug Development. Wiley, New York, 2003, pp. 342–379.
74. Fei YJ, Ganapathy V, Leibach FH. Molecular and structural features of the proton-coupled oligopeptide transporter superfamily. Prog Nucleic Acid Res Mol Biol 1998;58:239–261.
75. Miyamoto Y, Thompson YG, Howard EF, et al. Functional expression of the intestinal peptide-proton co-transporter in Xenopus laevis oocytes. J Biol Chem 1991;266:4742–4725.
76. Tamai I, Tomizawa N, Kadowaki A, et al. Functional expression of intestinal dipeptide/beta-lactam antibiotic transporter in Xenopus laevis oocytes. Biochem Pharmacol 1994;48:881–888.
77. Saito H, Ishii T, Inui K. Expression of human intestinal dipeptide transporter in Xenopus laevis oocytes. Biochem Pharmacol 1993;45:776–779.
78. Mackenzie B, Loo DD, Fei Y, et al. Mechanisms of the human intestinal H+-coupled oligopeptide transporter hPEPT1. J Biol Chem 1996;271:5430–5437.
79. Yang CY, Dantzig AH, Pidgeon C. Intestinal peptide transport systems and oral drug availability. Pharm Res 1999;16:1331–1343.
80. Ganapathy ME, Huang W, Wang H, et al. Valacyclovir: a substrate for the intestinal and renal peptide transporters PEPT1 and PEPT2. Biochem Biophys Res Commun 1998;246:470–475.
81. Daniel H, Adibi SA. Transport of beta-lactam antibiotics in kidney brush border membrane. Determinants of their affinity for the oligopeptide/H+ symporter. J Clin Inves 1993;92:2215–2223.
82. Swaan PW, Stehouwer MC, Tukker JJ. Molecular mechanism for the relative binding affinity to the intestinal peptide carrier. Comparison of three ACE-inhibitors: enalapril, enalaprilat, and lisinopril. Biochim Biophys Acta 1995;1236:31–38.
83. Fei YJ, Liu W, Prasad PD, et al. Identification of the histidyl residue obligatory for the catalytic activity of the human H+/peptide cotransporters PEPT1 and PEPT2. Biochemistry 1997;36:452–460.
84. Doring F, Dorn D, Bachfischer U, et al. Functional analysis of a chimeric mammalian peptide transporter derived from the intestinal and renal isoforms. J Physiol 1996;497(Pt 3):773–779.
85. Rubio-Aliaga I, Daniel H. Mammalian peptide transporters as targets for drug delivery. Trends Pharmacol Sci 2002;23:434–440.
86. Tsuji A, Tamai I. Carrier-mediated intestinal transport of drugs. Pharm Res 1996;13:963–977.
87. Amidon GL, Lee HJ. Absorption of peptide and peptidomimetic drugs. Annu Rev Pharmacol Toxicol 1994;34:321–341.
88. Balimane PV, Tamai I, Guo A, et al. Direct evidence for peptide transporter (PepT1)-mediated uptake of a nonpeptide prodrug, valacyclovir. Biochem Biophys Res Commun 1998;250:246–251.
89. Jacobson MA. Valaciclovir (BW256U87): the L-valyl ester of acyclovir. J Med Virol 1993;Suppl 1:150–153.

90. Wong MH, Oelkers P, Craddock AL, Dawson PA. Expression cloning and characterization of the hamster ileal sodium-dependent bile acid transporter. J Biol Chem 1994;269:1340–1347.
91. Shneider BL, Dawson PA, Christie DM, et al. Cloning and molecular characterization of the ontogeny of a rat ileal sodium-dependent bile acid transporter. J Clin Invest 1995;95:745–754.
92. Weinman SA, Carruth MW, Dawson PA. Bile acid uptake via the human apical sodium-bile acid cotransporter is electrogenic. J Biol Chem 1998;273:34691–34695.
93. Sun AQ, Ananthanarayanan M, Soroka CJ, et al. Sorting of rat liver and ileal sodium-dependent bile acid transporters in polarized epithelial cells. Am J Physiol 1998; 275:G1045–G1055.
94. Dawson PA, Oelkers P. Bile acid transporters. Curr Opin Lipidol 1995;6:109–114.
95. Hagenbuch B, Meier PJ. Molecular cloning, chromosomal localization, and functional characterization of a human liver Na+/bile acid cotransporter. J Clin Invest 1994;93:1326–1331.
96. Hallen S, Branden M, Dawson PA, Sachs G. Membrane insertion scanning of the human ileal sodium/bile acid co-transporter. Biochemistry 1999;38:11379–11388.
97. Zhang EY, Phelps MA, Chang C, et al. Membrane topology and three-dimensional structure of the apical sodium-dependent bile acid transporter (ASBT), SLC10A2. Evidence for a seven transmembrane domain conformation. Submitted.
98. Kramer W, Nicol SB, Girbig F, et al. Characterization and chemical modification of the Na(+)-dependent bile-acid transport system in brush-border membrane vesicles from rabbit ileum. Biochim Biophys Acta 1992;1111:93–102.
99. Baringhaus KH, Matter H, Stengelin S, Kramer W. Substrate specificity of the ileal and the hepatic Na(+)/bile acid cotransporters of the rabbit. II. A reliable 3D QSAR pharmacophore model for the ileal Na(+)/bile acid cotransporter. J Lipid Res 1999;40:2158–2168.
100. Izzat NN, Deshazer ME, Loose-Mitchell DS. New molecular targets for cholesterol-lowering therapy. J Pharmacol Exp Ther 2000;293:315–320.
101. Thompson GR, Naoumova RP. Novel lipid-regulating drugs. Expert Opin Invest Drugs 2000;9:2619–2628.
102. Swaan PW, Hillgren KM, Szoka FC Jr, Oie S. Enhanced transepithelial transport of peptides by conjugation to cholic acid. Bioconjug Chem 1997;8:520–525.
103. Swaan PW, Szoka FC Jr, Oie S. Molecular modeling of the intestinal bile acid carrier: a comparative molecular field analysis study. J Comput Aided Mol Des 1997;11:581–588.
104. Russell-Jones GJ, Westwood SW, Farnworth PG, et al. Synthesis of LHRH antagonists suitable for oral administration via the vitamin B12 uptake system. Bioconjug Chem 1995;6:34–42.
105. Lewis MC, Brieaddy LE, Root C. Effects of 2164U90 on ileal bile acid absorption and serum cholesterol in rats and mice. J Lipid Res 1995;36:1098–1105.
106. Huff MW, Telford DE, Edwards JY, et al. Inhibition of the apical sodium-dependent bile acid transporter reduces LDL cholesterol and apoB by enhanced plasma clearance of LDL apoB. Arterioscler Thromb Vasc Biol 2002;22:1884–1891.
107. Ekins S, Boulanger B, Swaan PW, Hupcey MA. Towards a new age of virtual ADME/TOX and multidimensional drug discovery. J Comput Aided Mol Des 2002;16:381–401.
108. Dean M, Hamon Y, Chimini G. The human ATP-binding cassette (ABC) transporter superfamily. J Lipid Res 2001;42:1007–1017.

8

ABC Transporter Proteins and Cellular Drug Resistance

P-Glycoprotein and Analogous Transporters

William W. Johnson

yea, at that very moment
Consideration, like an angel, came
And whipp'd the offending Adam out of him,
Leaving his body as a paradise,
To envelop and contain celestial spirits...
King Henry V

1. Introduction

Mammalian cells possess a natural battery of defense mechanisms against xenobiotic assault. A class of proteins actively transports an extensive array of structurally unrelated large lipophilic compounds from the cell, providing what is often known as multiple drug resistance (MDR) *(1)*. MDR is characterized by active efflux or pumping of xenobiotics and pharmaceuticals via transmembrane proteins acting as hydrophobic "vacuum cleaners" *(2,3)*. The MDR1 gene encodes a 170 kD integral plasma membrane phosphorylated glycoprotein, P-glycoprotein (P-gp), which is the best known and most extensively studied among these transporters, and which thus far has the largest substrate list. The gross structural features of P-gp are shared by a large family of membrane transporters known as adenosine 5'-triphosphate (ATP)-binding cassette (ABC) transporters, which evidently act as ATP-driven pumps that remove xenobiotics from the interior of cells. Expression of P-gp in normal human tissues—particularly within the cellular membranes of the gastrointestinal tract, liver, blood–brain barrier (BBB), adrenal glands, and kidneys—suggests that the protein plays a role in cellular protection as well as in secretion *(1–4)*. Although the primary function of this protein is unknown, its ability to confer resistance to a wide variety of structurally and chemically unrelated compounds remains impressive. Indeed, the substrate list for this transporter reveals that P-gp shares a similar tolerance or acceptance for chemicals as cytochrome P450 3A4 (CYP3A4), the predominant intestinal and hepatic cytochrome P450 oxygenase enzyme, and may even prove to be more extensive in its substrate recognition and as an avenue of drug elimination *(5)*.

It is becoming evident that drug interactions ostensibly mediated by the cytochrome P450 3A4 oxidative pathway are also the result of P-gp inhibition *(6,7)*. Given the

From: *Cellular Drug Delivery: Principles and Practice*
Edited by: D. R. Lu and S. Øie © Humana Press Inc., Totowa, NJ

enormous number of substrates now known to be recognized by P-gp, and a binding site that is evidently hydrophobic, the substrates for CYP3A4 and P-gp clearly overlap. Among the more grave examples of clinical drug interactions are those of the H_1-receptor antagonist terfenadine (Seldane) with ketoconazole and erythromycin *(8)*, as well as those of simvastatin with itraconazole and mibefradil *(9)*; all are substrates/inhibitors of P-gp *(4–6)*. Additionally, P-gp polymorphisms may cause compound-specific drug reactions to treatment with P-gp substrates. If a polymorphic gene product of MDR1 has inferior selectivity toward a therapy, increased systemic exposure to that erstwhile P-gp substrate could be expected *(10–12)*.

1.1. Function and Mechanism

One of the many intractable aspects of investigating P-gp is an absence of chemical change to its substrates and hence no product formation or gross substrate depletion to monitor or analyze. Because extensive in vitro and in vivo studies indicate that MDR enzymes transport/move lipophiles across a lipid bilayer against a concentration gradient, P-gp has been described as a translocase, a flippase, or even a hydrophobic vacuum cleaner. The purpose of this enzyme ostensibly is: (1) protection against exogenous toxins ingested with food; (2) excretion of (endogenous) metabolites or toxins; (3) prevention of toxic materials from entering the brain, gonads, and fetus; (4) transportation of steroid hormones; and (5) extrusion of peptides (cytokines) not exported via signal/cleavage.

As a member of the ABC superfamily of transporters, P-gp possesses two ATP binding sites and uses ATP (via hydrolysis) as the source of energy for "translocating" substrates *(4)*. The large transmembrane protein P-gp has two homologous domains, each containing a nucleotide binding site (ATP) and a substrate binding site near or within the inner leaflet of the membrane. With ATP hydrolysis providing the energy for function, experimental quantitation of the coupling of ATP hydrolysis to efflux events indicates a ratio of approx 1 or 2 ATP per substrate molecule transported *(13–19)*. Hence, measurement of the rate of ATP hydrolysis serves as an indirect assay of enzymatic activity. The inactivation at only one of the two ATP binding sites is sufficient to abolish entirely the activity of the transporter *(20)*. An allosteric linkage has been shown between one ATP binding site and catalysis at the other homologous site; these sites appear to drive substrate transport in tandem.

The substrates enter the active transport sites from the inner-leaflet, cell-membrane lipid bilayer *(21)* and can bind to two (or more) nonidentical sites *(17)*. Kinetic data indicating noncompetitive inhibition of P-gp-mediated transport by substrates and inhibitors suggest differing substrate specificity between the binding sites *(17)*. Many of these studies show V_{max} changes that are consistent with nonexclusive binding of an inhibitor or alternative substrate, whereas photoaffinity labeling has indicated nonidentical substrate binding sites *(22)*. Indeed, binding studies also show biphasic binding curves and hence two distinct K_d's for many P-gp substrates, further indications of affinities that are unique to the two binding sites *(23)*. Moreover, allosteric and perhaps synergistic effects have been observed for certain substrate combinations and conditions *(24,25)*, with recent evidence even suggesting interaction between the two binding sites *(26,27)* and/or the nucleotide binding sites *(20,28,29)*. Such an allosteric linkage between sites could affect the inhibition characteristics of marker-substrate transport out of the cell by particular inhibitors. Under these circumstances, the inhibition of transport might differ from that observed by simple Henri-Michaelis-Menten hyperbolic saturation plot of inhibition. Moreover, in light of the substantive evidence for two binding sites, it is probable that some substrates are uniquely recognized and/or transported by P-gp. This responsiveness or sensitivity would, in turn, determine the extent of influence (including concentration dependency) a particular inhibitor will have on such substrates *(30)*.

2. Clinical Impact

Numerous examples illuminating the critical and potent physiological impact of P-gp have been described. Striking effects on bioavailability are shown by genetic knock-out (KO) animals as well as the therapeutic intervention of P-gp function. KO animals have no functional mdr1a (P-gp null) and are highly sensitive to the neurotoxin ivermectin and many other P-gp substrates. Dramatic effects on exposure have been observed in particular sanctuaries protected by the activity of P-gp. For example, brain-tissue levels of ivermectin were 87-fold higher, and those of the carcinostatic drug vinblastine were 22-fold higher, in the KO animal vs wild-type *(31)*. Furthermore, paclitaxel and digoxin (cardio-toxin) oral uptake is markedly increased. Indeed, even systemic plasma concentrations are increased 2-, 3.5-, 5.7-, and 4.1-fold, respectively, for cyclosporin A, digoxin, erythromycin, and saquinavir in the P-gp null mouse *(32)*. Co-administration of the potent P-gp inhibitor PSC833 with digoxin in wild-type mouse increased brain exposure to levels almost that of KO mice by abolishing intestinal P-gp function *(33)*. An example of the dramatic impact of P-gp on physiological exposure to xenobiotics in humans is the remarkable decrease of the cyclosporin A plasma area under the curve (AUC)-dependent on the location of absorption in the rank order stomach>jejunum/ileum>colon (intubation) *(34)*. The decrease in absorption exhibited a marked correlation ($r = 0.994$) to expression of mRNA for P-gp over the gastrointestinal (GI) tract (stomach<jejunum<colon).

2.1. Tissue Distribution and Toxicity

The evidence thus far shows that P-gp has been found virtually wherever investigators have searched for it. P-gp has been detected in adrenal glands (endothelial cells), kidney (brush border of proximal renal tubule); liver (biliary canaliculi); intestine (columnar epithelium); jejunum, ileum, pancreas (epithelia); colon, central nervous system (CNS; endothelial cells); pregnant uterus (glandular epithelial cells of the endometrium); $CD34^+$ bone marrow cells, circulating lymphocytes, and haematopoietic stem cells *(1–4)*.

2.2. Blood–Brain Barrier and Neurotoxicity

P-gp is expressed at the apical surface of the capillary endothelial cells that form the BBB, where it seems to play a dramatic role in the exclusion of many drugs from the brain. Studies have shown that the sensitivity of homozygous P-gp KO mice to the neurotoxicity (and CNS concentration) of ivermectin and vinblastine is increased 100-fold compared with that of wild-type animals *(31)*. In addition to MDR1, another isoenzyme of the drug/xenobiotic efflux family, multidrug resistance-associated protein (MRP1), has been detected in about 30% of head and neck squamous cell carcinomas by immunohistochemistry.

A comparative analysis of 18 physicochemical properties revealed that drugs for CNS indications had fewer hydrogen bond donors, fewer positive charges, greater lipophilicity, lower polar surface area, and reduced flexibility compared with the non-

CNS group (properties that enhance membrane permeability) *(35)*. Because a CNS drug should ideally have high passive permeability and not be a substrate of P-gp, it is not surprising that there was a threefold lower incidence of P-gp-mediated efflux of CNS drugs (n = 7 of 48, 14.6%) than of non-CNS drugs (n = 19 of 45, 42%) *(35)*.

The striking differences in brain concentrations of P-gp substrate drugs between wild-type and KO mice is impressive even compared with plasma differences, indicating P-gp is more critical to brain exposure than intestinal absorption (*see* below). Administration of ivermectin (oral), digoxin intravenous, or cyclosporin intravenous in wild-type and KO mice resulted in 87-, 27-, or 55-fold higher brain levels in the KO over wild-type *(32)*, yet the increases were less than fourfold in liver, kidney, and plasma. These results indicate that P-gp inhibitors should be used with caution to avoid potential neurotoxicity.

2.3. Intestinal Absorption and Bioavailability

It is noteworthy that poor pharmacokinetic (PK) properties, such as poor oral bioavailability or duration of action (clearance), account for nearly half of the failures in clinical development *(36)*. MDR1 (P-gp) and MRP1 are constitutively expressed in epithelia throughout the GI tract and are often overexpressed in carcinomas originating from these tissues. Double MDR1 genetic KO mice have shown decreased elimination of drugs resulting from impaired excretion by liver, kidney, or gut (depending on which route is important in the P-gp-mediated excretion of the drug), and/or enhanced reabsorption of drug from bile, gut lumen, or urine prior to elimination from the body *(32)*.

For drugs that are excreted unaltered or as a conjugate that can be hydrolyzed after secretion, P-gp in the epithelial surfaces of bile ducts, gut, and kidney proximal tubule may prevent reabsorption of the excreted drug. P-gp generally has greater impact on drug uptake than on drug excretion. Clinical results with the (relatively low potency) inhibitors tested so far demonstrate, for instance, that these agents interfere with the elimination of anthracyclines, a process that results in considerable increases in AUC and concomitant toxicity *(37)*. Moreover, oral bioavailability of paclitaxel (MDR substrate) increased from 9.3% to 67% with either intravenous or oral co-administration of the MDR inhibitor cyclosporin A *(38)*. Similarly, high levels of bioavailability were observed with the P-gp KO mice *(32)*.

There are many instances where absorption from the small intestine may be complete but bioavailability is poor owing to enterocyte recycling via P-gp *(34,39–41)*. Reabsorbtion after ejection increases the compound exposure to enterocyte drug-metabolising enzymes, and P-gp may enhance intestinal metabolism of drugs. Midazolam *(42)* and cyclosporin *(43)* endure extensive first-pass gut elimination owing to active transport and recycling. The efficient P-gp substrate verapamil has a low oral bioavailability of about 20% at doses of at least 120 mg, and propranolol has a low bioavailability of about 26%.

The poor bioavailability of HIV protease inhibitors *(44)*, digoxin (to the brain *[45]*), and taxol *(46)* is apparently owing to efficient transport by P-gp, with paclitaxel oral bioavailability of less than 5% *(47)*. The tacrolimus (bioavailability 18%) concentration/dose ratios in a recipient of a small-bowel transplant correlated well with the levels of MDR1 mRNA, but not with CYP3A4, indicating that P-gp determines intraindividual variability in tacrolimus pharmacokinetics *(48)*. Moreover, the pharmacokinetics of the β-blocking agents celiprolol, pafenolol, and talinolol are also significantly affected by P-gp in the intestine *(49–53)*; the H2 receptor antagonists ranitidine and cimetidine are similarly affected *(54)*, though 50% absorbed at the high therapeutic doses.

However, intestinal P-gp may be saturated when drug concentrations in the intestinal lumen exceed the K_m values after high oral doses. Whereas P-gp substrate drugs, such as digoxin, given at low doses result in low and variable absorption, many of these drugs (talinolol, indinavir, etc.) exhibit dose-dependent absorption owing to P-gp saturation. Chiou et al. *(55)* recently concluded that 13 P-gp substrate drugs are not significantly impeded by efflux transport in vivo. Yet the absorption of some drugs that are administered at high doses is still significantly affected by intestinal P-gp. Despite the high clinical oral dose of cyclosporin and paclitaxel (200–700 mg and 100–200 mg, respectively), P-gp significantly limits their oral absorption *(34,46)*, perhaps owing to poor solubility. Therefore, drug absorption is unlikely to be quantitatively limited by active efflux transport unless a very small oral dose is given, or the dissolution and/or membrane diffusion rates of the drug are very slow. To highlight further the importance of oral absorption, poor absorption is asserted to be causing a new public health issue because it increases the chemical burden on municipal waste treatment facilities *(56)*.

2.4. Excipients Affect Permeability by Inhibiting Transporters

Nonabsorbable pharmaceutical excipients such as Tween-20, Tween-80, pluronic P85, or TPGS have been shown to be potent modulating agents of membrane transporters *(57,58)*. Other nonionic surfactants affecting transport pumps include Cremophors, pluronic block copolymers, Nonidet P-40, and Spans. Surfactants that are commonly used as vehicles for solubilizing certain drugs can inhibit MDR in resistant cells at clinically achievable concentrations *(59)*. Addition of the surfactants Solutol HS-15, Tween-40, and Cremophor EL (10 mg/mL each) decreased lipid fluidity of isolated crude plasma membranes of resistant cells *(60)*, whereas noninhibiting surfactants (octylgucoside, hecameg) did not affect membrane fluidity. Furthermore, Tween-80 and Cremophor EL fluidized cell-lipid bylayers, whereas vitamin E TPGS rigidized lipid bilayers reducing the BL–AP permeability of rhodamine 123, and the noninhibitor *N*-octyl gulcoside did not modulate membrane fluidity *(61)*. PEG-300 (polyethylene glycol) inhibited efflux-transporter activity in Caco-2 cell monolayers, probably caused by changes in the microenvironment of the cell membranes, perturbing the ability of these transporters to efflux substrates such as taxol and doxorubicin *(62)*. Inhibition is observed at concentrations below the critical micelle concentration (cmc) for the different surfactants, suggesting that the monomer is responsible and may be partitioning into the membrane and inhibiting P-gp through a membrane-fluidizing mechanism. Indeed, low concentrations of the nonionic surfactant Triton X-100 inhibited azidopine binding to P-gp in vinblastine-resistant human lymphoma *(63)*.

2.5. Liver, Kidney, and Excretion

P-gp plays a significant role in the biliary excretion of digoxin, doxorubicin, vincristine, and vinblastine in mice *(64)*. The biliary clearance of digoxin is substantially greater in wild-type mice (2.3 mL/min/kg) than in KO mice (0.84 mL/min/kg) *(65)*. Approximately 45% of digoxin is excreted in the bile of wild-type mice.

Digoxin is also actively secreted in the isolated perfused rat kidney with the P-gp inhibitors quinidine and verapamil inhibiting tubular secretion *(66)*. The renal clearance of digoxin in wild-type mice was three times greater than that in KO mice *(65)*. Therefore, digoxin appears to be actively secreted into the renal tubular lumen by P-gp.

2.6. Placenta and Teratogenicity

MDR1 functions as a critical component of the maternal blood–placental barrier, protecting the fetus from exposure to various maternal blood-borne chemicals. Pregnant dams of a mouse CF-1 subpopulation, known to lack P-gp, exposed to an avermectin were highly sensitive to fetal cleft palate *(67)*. The degree of chemical exposure of fetuses within each litter was inversely related to expression of placental P-gp, which was determined by the fetal genotype.

3. Substrates, Their Diversity, and Drug–Drug Interactions

It is, as yet, unclear how P-gp can recognize and transport such a structurally diverse spectrum of compounds ranging in size from less than 250 Da (cimetidine) to more than 1800 Da (Gramicidin D). The only structural common denominator identified so far is that all transported substrates are at least somewhat hydrophobic and/or amphipathic in nature, containing a hydrophobic and often a polar or even a (generally positively) charged domain *(1–4)*.

The ability of transported substrates to insert into biological membranes may be an essential requirement for recognition of the compound by P-gp. Favored partitioning into the lipid membrane would increase the effective substrate concentration at the transport binding site. Distinct but overlapping specificities of the drug binding/transport sites may help explain the broad substrate tolerance or lack of specificity *(30)*.

3.1. Substrate Recognition

The substrate recognition abilities of P-gp are broad and tolerant, even among most xenobiotic-defense enzymes. The purported xenobiotic-protection role of P-gp *(1,2)* mimics that of CYP3A4. Both enzymes provide a protective role to many of the same cells and defend against a generally shared list of xenobiotic substrates. The extensive overlap between these two enzymes is probably fortuitous, as opposed to concerted, because of their great tolerance for and acceptance of large lipophilic substrates. Both enzymes appear to have large accommodating hydrophobic binding sites that do not discriminate among many lipophilic compounds. However, substrate recognition and preference are not this simple, because both enzymes have shown cooperativity and a role for decisively oriented hydrogen bonding in the substrate binding sites *(7,68)*.

Most P-gp pharmacophore models can only address very general properties such as lipophilicity and size owing to the multiple binding sites, different assays, binding tolerance, and other complications. General properties appear to converge around hydrophobicity, presence of rings, size, and in particular, tertiary amines (extensively reviewed in refs. *4,69,70*). Indeed, it appears that P-gp recognizes its substrates directly from the lipid phase *(21,71–74)*, where they are expected to be much more concentrated owing to partitioning of the lipophilic compounds *(75)*. However, the lipophili-city factor logP (a partition coefficient phase preference) often is not correlated with P-gp binding affinity, and certainly not across compound classes or series *(70,76)*. Structure activity relationships have shown direct correlation of MDR inhibition to logP only for compounds within a closely related series *(4)*. Litman et al. *(77)* showed that 34 inhibitors from different pharmacological classes have no significant correlation with calculated partition coefficients and that the size of the molecule (van der Waals surface area) was a better corollary. In fact, a P-gp inhibitor has been has been defined as a compound containing at least two aromatic rings separated by a basic chain with a secondary or tertiary amine *(78–80)*, and even stereospecific interactions have often been observed for pairs of chiral compounds (81–85). Furthermore, the contribution of hydrogen bonding has been shown for P-gp substrates (86–88). Important features of molecular recognition of substrates include multiple hydrophobic and hydrogen-bond acceptor features (89). Indeed, Seelig has described a pharmacophore with two general patterns for substrate recognition: the "type I unit" of two electron-donor groups (hydrogen-bond acceptors) with a spatial separation of 2.5 ± 0.3 Å and the "type II unit" of three electron-donor groups with a spatial separation of 4.6 ± 0.6 Å (87). A recent computational ensemble pharmacophore model supports these recognition patterns (90), although the less restrictive van der Waal's interactions, stacking interactions, and the hydrophobic effect may generally combine to provide affinity with multiple diverse compounds (91). Although P-gp does not possess acidic residues in their membrane domains, it transports cationic amphipathic compounds. Therefore, another physical quality must provide this selectivity and the face of the aromatic ring structures of tyrosine, phenylalanine, and tryptophan residues can bind to cations (92,93). Furthermore, binding interactions are modulated by the membrane-lipid environment (75,94,95).

3.2. P-gp Inhibition and Pharmacokinetic Drug Interactions

Because of the likelihood of co-administered drugs sharing recognition by the transport site, the inhibition of P-gp causes many PK interactions (vide supra), such as the increase of the oral bioavailability of paclitaxel from 9.3% to 67% with co-administration of cyclosporin A (38). Cyclosporin A also inhibits the renal secretion of vincristine and vinblastine, and other P-gp inhibitors reduce the active biliary excretion of colchicine, doxorubicin, and etoposide by the liver (96,97). Intravenous administration of potent P-gp inhibitors resulted in up to 37-fold increase in HIV-1 protease inhibitor concentrations in the brain of mice (98); and the P-gp inhibitor GF120918 raised the HIV drugs' brain–plasma ratio about 100-fold (99). These and many further examples of significant clinical drug–herb interactions mediated by P-gp indicate that this transporter should be routinely examined in drug development for binding and inhibition caused by proposed therapies.

4. P-gp Elevation

Increased P-gp expression and/or activity will naturally have the opposite effect of thwarting the P-gp activity described earlier. Dramatic examples of PK interactions mediated by P-gp induction have been reported. Significant elevation of intestinal P-gp quantity and the suppression of talinolol (100) or digoxin (101) exposure with co-administration of rifampin, a P-gp inducer, were remarkably well-correlated. P-gp and CYP3A regulation appear to respond similarly to PXR binding and share some molecular-regulation signals (102). Following chemotherapy, tumor cells may mutate to present supernormal quantities of P-gp, although many malignancies are already MDR-positive at diagnosis (chemotherapy naïve). For example, in a clinical study, P-gp levels increased 3- to 15-fold, showing that tumors adjust rapidly to anticancer drugs (103). A major obstacle for successful chemotherapy of cancer is the resistance of tumors to multiple anticancer drugs (MDR). Because P-gp can account for up to a 100-fold increase in drug resistance, overexpression or upregulation of this transporter can be applied as a prognostic marker in certain diseases, such as leukemia, breast cancer, neuroblastoma, pancreatic cancer, or ovarian cancer. Indeed, many compounds are in clinical trials to inhibit P-gp, with the goal of overcoming MDR (*vida infra*).

Many other cytotoxic (xenobiotic) compounds are also inducers of P-gp quantity; for example, verapamil, nifedipine, nicardipine, diltiazem, rifampicin, cyclosporin A, progesterone, estradiol, phenobarbital, insulin, clotrimazole, reserpine, isosafrole, St. John's Wort, hyperforin, dexamethasone, androstanol, troglitazone, ecteinascidin, digoxin, some PAHs, 2-acetylaminofluorene, and anthracyclins *(104,105)*. Many results indicate a tissue-dependent inductive response of P-gp to inducer exposure *(106)*.

5. Genetics and Variability

5.1. Polymorphisms

As with the cytochrome P450 superfamily, MDR1 genetic polymorphism might result in observed outcomes in unique subpopulations, with naturally occurring MDR1 single-nucleotide polymorphisms (SNPs) having clinical and pharmacological relevance. In the 28 exons of MDR1 genomic DNA of healthy Caucasians, 15 SNPs were detected, including six in the coding region *(11)*. Three of these altered the primary amino acid sequence of the protein. Phenotypical consequences for C3435T in exon 26 correlated with intestinal P-gp expression and uptake of orally administered P-gp substrates *(12,107)*. Individuals homozygous for this polymorphism (TT, ~25%, $n = 188$) showed significantly lower duodenal P-gp expression, lower in vivo activity of P-gp (approximately twofold), and increased digoxin plasma levels. However, C3435T is located at a noncoding, nonpromoter position in the MDR1 gene and is unlikely to influence P-gp expression. It is more likely linked to other as-yet-unidentified changes in regions of the MDR1 gene that control expression, e.g., in the promoter or enhancer region, or in sequences that are important for mRNA processing. Serving as a surrogate for the estimation of other tissue levels, the concentration of P-gp in a subset of lymphoid cells ($CD56^+$ natural killer [NK] cells) is also substantially lower in the T/T genotypes.

The frequency of C/C genotype (higher activity or function) in West Africans and African Americans is 83% ($n = 172$) and 61 % ($n = 41$), respectively, whereas in Caucasians it is 26% ($n = 537$) *(108)* (*see* Table 1).

Higher doses of tacrolimus or cyclosporine were required in African Americans than Caucasians to attain similar plasma levels. Conversely, the T/T genotype patients attain lower plasma concentrations of the P-gp substrates nelfinavir and efavirenz—anti-HIV drugs (despite low expression of the MDR1 transcript and P-gp) *(110)*—although the T/T genotype responded better and faster (greater rise in CD4-cell count) to therapy. The 3435C/T polymorphism is noncoding and could be in linkage disequilibrium with a polymorphism elsewhere in the genome that modifies MDR1 expression or function. In another study, the T/T genotype (in the context of a C1236T, G2677T haplotype) was associated with high P-gp expression in vitro and low plasma concentrations of fexofenadine *(111)*. The reason for this discrepancy with the C3435T allele subject observations *(12)* described earlier is currently unclear. There could be an indirect effect of the 3435 genotype, i.e., low P-gp could be compensated for by induction of other transporters (or CYP3A4) and dietary/environmental differences could contribute. The allelic variant MDR1*2 (haplotype) exhibits enhanced efflux of digoxin, is statistically associated with lower fexofenadine exposure, and includes C1236T, C3435T, and G2677T[Ala893Ser] (62% of European Americans, 13% of African Americans). A recent haplotype analysis has been able to reconcile conflicting results of studies whose analysis is based solely in individual SNPs *(112)*. Haplotype 12 (2677G/3435T) codes

Table 1
Prevalence of C/C Genotype in Different Ethnic Groups

Ethic Group	C/C Genotype (%)
Ghanaian	83
Kenyan	83
African-American	84
Sudanese	73
British Caucasian	48
Portuguese	43
Southwest Asian	34
Chinese	53
Filipino	59
Saudi	55

Adapted from ref. *109*.

for elevated concentrations of digoxin after oral dose and the superior haplotype analysis results match data of other MDR1 studies *(112)*.

However, 20+ million humans in Africa and South America have been treated with ivermectin (with no evidence of neurotoxicity), an antiparasitic P-gp substrate and potent neurotoxin in P-gp-null genotype KO mouse. This suggests that: (1) MDR1 P-gp expression is highly conserved in humans overall and (2) defining a subgroup of humans with complete absence of P-gp expression is unlikely. Indeed, 10 SNPs do not result in amino acid changes or are in noncoding regions, and three have an unknown effect on function.

5.2. Expression Variability

Humans exhibit wide variation in liver expression of MDR1 mRNA and P-gp protein. The variability of enterocyte P-gp concentration is about 10-fold in transplant patients and a bit less in normal nonmedicated adults (about fourfold, interpatient); males expressed twofold higher amounts of P-gp than females *(113,114)*. There is even an indication of up to eightfold interindividual variability in P-gp content *(115)*, with more than eightfold differences in the P-gp expression observed in a small population (25 patients *[116]*). This variability is roughly similar to, or perhaps more than, what is observed for CYP3A4, though these studies are from a limited sample population. Temporal variation of P-gp levels is expected (*vide supra*), such as a threefold intrapatient variability that was observed in a transplant case study *(48)*!

6. Role of P-gp Polymorphisms and Mutation in Cancer

Many cancer types have provided examples of gene modifications associated with drug resistance and P-gp primary structure. Naturally occurring mutations in the MDR1 gene associated with colorectal cancers with high microsatellite instability (MSI-H) were found in both the coding and promoter regions *(117)*. A mutation in the promoter of the MDR1 gene in human hematological malignancies may contribute to the pathogenesis of resistant disease *(118)*. Similarly, point mutations in the MDR1 promoter have been found in osteogenic sarcoma and various types of leukemia *(119)* and are associated with diminished in vitro responsiveness to MDR relevant drugs. Moreover, DNA methylation and hypermethylation can affect transcription and gene-product levels *(120)*.

7. In Vitro to In Vivo Correlation

Although observations must be carefully judged in the context of contrived elemental systems, the correlation of in vitro experiments to clinical observations generally is quite good owing to the dramatic role or influence of P-gp. The significant impact of P-gp inhibition has been illustrated ealier, including some corollary to in vitro experiments. There are many further examples, including the increased digoxin concentrations in the brain of PSC833 (very potent P-gp inhibitor in vitro) orally treated, wild-type mice near the levels found in mdr1a/1b(–/–) double KO mice *(33)*. The relative P-gp inhibition potencies of many herbs and drugs such as the azole antifungals (itraconazole, etc.) and statins (HMG-CoA reductase inhibitors) are consistent with their observed PK effects *(121–124)*. Additionally, correlation between in vitro transcellular molecule transport ratios from transfected (L-mdr1a) cells (efficiency as a P-gp substrate) and brain concentration ratios of mdr1a (–/–) to mdr1a (+/+) CF-1 mice is remarkable and predictive (r^2 = 0.93; *125*). The cellular accumulation ratio and transcellular transport ratio methods for substrate characterization/quantification also correlated well and consistently *(125)*. Sandwich-Cultured (SC) rat hepatocytes have been shown to be an in vitro model to assess and predict the biliary excretion of xenobiotics with notable correlation *(126)*.

To evaluate substrate-transport kinetics, ATP hydrolysis is a useful assay for P-gp substrates *(17,127)*, although some substrates do not significantly alter P-gp-mediated ATP hydrolysis (from baseline activity ostensibly caused by co-purified endogenous substrate) in the presence of standard lipid constituents *(128)*. However, Caco-2 cells apparently do not afford good general correlation with gastrointestinal (GI) absorption owing to variable expression of P-gp, other ABC transporters, plasma protein binding, rates of passive diffusion, luminal saturation of P-gp, and so forth *(55)*. Moreover, physiological factors such as gastric emptying, GI motility, mucus dissolution, intestinal pH, and blood flow, and lymph flow can uniquely impact each transport mechanism. Yet Caco-2 cells can be a useful indicator of jejunal drug efflux, if the low expression of BCRP (ABC-G2) and CYP3A is accounted for *(129)*. Caco-2, HT-29, MDCK, TC7 can reproducibly display some properties of differentiated intestinal cells *(130)* and, therefore, suitable for qualitative predictions and molecular-permeability screening studies. Indeed, the in vivo to in vitro Caco-2 drug transport permeability measurements correlate well for passively or highly absorbed drugs (r^2 = 85%) *(131,132*; also see below) or small molecules *(133)*. However, even some studies of peptidomimetics have resulted in reasonable correlations for slowly and incompletely absorbed drugs *(134,135)*.

7.1. Overall Efficiency of Transport

The implications of P-gp activity and function must account for the ratio of permeability/active-transport for the distinction of a substrate from an inhibitor *(136)*. Although a substrate typically competes with alternate substrates for the active site of a xenobiotic-disposition enzyme, many P-gp substrates are not competitive inhibitors for overall efflux. We have shown that many P-gp substrates have no effect on the ultimate ability of P-gp to cause removal of the marker substrates from a viable cell *(136)*, and others have indicated a potentially similar distinction for other compounds *(137–139)*. It is very important to properly define a compound as a substrate, inhibitor, or both in the context of evaluating the potential for drug interactions and drug–herb interactions as well as exposure to toxins and drugs. This disconnect between a substrate and its ability to inhibit P-gp can be explained by an element of the natural system: P-gp exports its substrates across a lipid bilayer that is intended to preclude permeation of adverse xenobiotics. A requirement for a P-gp substrate also to be an inhibitor is the rapid passive transbilayer movement across the membrane bilayer, a process that allows the substrate to re-enter the cell quickly and hence effectively occupy the P-gp active site.

Many studies indicate the critical role of membrane permeation or passive transmembrane movement rate of a substrate toward the inhibition of P-gp *(140,141)*. Although P-gp-mediated active rate of compound transfer or efflux is slow (~ 900 min^{-1}; *142*), the "flip-flop" rate of its many substrates is even slower ($t_{1/2}$ > 2 min) against those for which there is "resistance." In other words, to create a concentration gradient, the P-gp substrate must have a relatively slow transmembrane passive transport rate. Conversely, P-gp substrates with relatively rapid permeation will overcome the pace of P-gp to re-enter the binding site and competitively inhibit function. The rate of active efflux transport relative to the rate of passive permeation or influx determines the net movement of drug from inside to outside of the cell membrane; a substrate cycling rapidly back into cytosol will compete for P-gp binding site access. MDR-type drugs are amphipathic (hydrophobic and positively charged) and as such bind readily to negatively charged phospholipid head groups of the membrane. Therefore, transmembrane movement often requires a "flip-flop" through the membrane bilayer and is slow. Indeed, the passive transbilayer diffusion of phosphatidylcholine (PC), the most abundant membrane lipid, is very slow ($t_{1/2}$ ~ d) in both artificial and natural membranes *(143)*. By measuring the transport rate and passive transbilayer permeation rate of five inhibitors and five substrates, it was shown that P-gp inhibitors cross the bilayer membrane faster than the egress rate of P-gp, thereby resulting in rapid equilibration rates *(139)*. At an approximated turnover rate of 900 min^{-1}, P-gp can keep pace with a compound like Rho with a transbilayer movement lifetime of minutes. Conversely, P-gp is inefficient in protecting MDR cells against molecules rapidly permeating through lipid bilayer membranes; for example, potent P-gp inhibitors such as the carrier-type ionophore valinomycin, which traverse membranes within microseconds ($k > 25 \times 10^4\ s^{-1}$). Conversely, gramicidin, effectively excluded by P-gp, has a transmembrane "flip-flop" rate with a lifetime of minutes. Additionally, the transmembrane movement rate is critical to the overall efficiency of P-gp removal of rhodamine dyes from MDR cells. Indeed, rhodamine B and tetramethylrosamine exhibit high affinity for P-gp, but rhodamine B was the fastest membrane-traversing dye and the least efficiently excluded from the cell. There was a similar corollary for all of the related dyes tested *(144)*. Another efficient marker of efflux, doxorubicin exhibits a "flip-flop" rate with a $t_{1/2}$ of approx 1.7 min, and 30% cholesterol addition to the vesicles to reduce membrane fluidity decreased the rate sevenfold *(145)*. The fluidizer benzyl alcohol accelerated the rate, consistent with the role of membrane bilayer permeability dictating the rate of equilibration. Consequently, of compounds ejected by P-gp, those that are relatively slow to cross the lipid bilayer are efficient "substrates," and those that rapidly permeate the membrane are (also) good "inhibitors." This effect has been modeled using the highly permeable compound nicardipine and further supports this conclusion *(39,146)*, as does ranitidine, which exhibits very low passive permeability *(147)* and is not a P-gp inhibitor.

Lipophilicity could sometimes be used to discriminate between P-gp substrates and inhibitors. Among a series of anthracyclines, the less lipophilic derivatives were correlated with slower passive diffusion and resulted in lower intracellular accumulation *(140,141,148,149)*. Indeed, lipophilicity is inversely correlated with P-gp efflux efficiency of some vinca alkaloids *(150)*, steroids *(151)*, and peptides *(152)*; and directly relates to P-gp inhibition for cyclosporins *(153)*, steroids *(151)*, and linear hydrophobic peptides *(154)*. Although more lipophilic compounds appear to be better inhibitors generally (as opposed to ejected efficiently as substrates), this correlation does not apply outside of a particular set of analogs *(155,156)*. This result is probably owing to the fact that lipophilicity parameters are inadequate for judging the interaction of structurally diverse compounds with the complexities of membranes *(76)*.

We can now appreciate that the number of P-gp substrates that are not also inhibitors is striking, and many substrates could be expected to have no substantial effect on P-gp function in the viable cell. Despite this absence of effect on P-gp overall function by many substrates, these P-gp substrates can still be affected pharmacokinetically by the modulators of P-gp activity. It is therefore important to characterize both the efflux-rate parameters and those of inhibition.

8. Pharmacological Modulation

Of the types of cellular multidrug resistance (resistance to unrelated drugs), the ATP-dependent efflux pumps are the dominant factor *(157)*. Early studies showed that P-gp was highly expressed in colon, kidney, adrenocortical, and hepatocellular cancers *(158)*, and P-gp expression is correlated with a reduced complete remission rate of acute myelogenous leukemia (AML) and a higher incidence of refractory disease *(159)*. High P-gp expression is also well-correlated with the poor clinical outcome in childhood acute lymphoblastic leukemia (ALL) *(119)*. Additionally, breast tumors, and possibly ovarian cancers and lung cancers, that expressed increased levels of P-gp after therapy were associated with over a threefold greater likelihood of treatment failure *(160)*. After chemotherapy, a significant proportion of breast cancer patients express increased tumor levels of P-gp *(161)*. In ovarian cancer samples, 16–47% were found to express P-gp, as measured by immunohistochemistry *(162)*. Studies using mRNA detection or immunohistochemistry methods, ex vivo functional assays, or in vivo tumor imaging all show a strong association between therapy with MDR drugs, intrinsic or acquired expression of P-gp, reduced tumor-cell drug retention, and a poor treatment response in breast cancer patients *(163,164)*. Owing to significant resistance to anticancer drugs, the mitigation of P-gp has been ardently sought for more than 20 yr *(165)*. As mentioned ealier, PSC833 (cyclosporin analog) is a potent P-gp inhibitor and has been investigated in the clinic as an adjuvant therapy toward overcoming chemotherapy resistance *(166)*, although it also causes profound PK effects *(33)*. Most of these compounds, however, emerged as weak inhibitors that were toxic at high doses. The poor activity of current P-gp-inhibition compounds in patients has also been attributed to the presence of resistance factors in addition to P-gp (e.g., other ABC transporters), inappropriate design of clinical trials, toxicity, bioavailability, and/or lack of specificity of anti-P-gp reagents. Yet a Phase I/II trial of cyclosporin A added to daunorubicin and cytarabine in "poor-risk" patients with AML resulted in 62% complete remissions and a 69% overall response rate *(167)*. Combinations of suboptimal doses of P-gp inhibitors were shown to be effective at 15–100 times less than the optimal doses, suggesting the possibility of avoiding associated toxicities of these agents *(168)*.

An intriguing possibility of a dual-acting chemotherapy is found in SCH66336 (lonafarnib), an orally active, potent, and selective inhibitor of the farnesyl protein transferase (FPT) enzyme *(169)*. This novel therapeutic agent has activity against a wide variety of human tumor xenografts and also causes regression of tumors in wap-H-ras transgenic mice. Enhanced antitumor activity has been reported in preclinical cancer models when SCH66336 is combined with cyclophosphamide, 5-fluorouracil (5-FU), or vincristine *(169,170)*, all substrates of P-gp. Furthermore, a synergistic effect on antitumor activity of SCH66336 and taxanes (also P-gp substrates) has recently been described *(171)*. Moreover, a recent report shows a synergy with co-administration of SCH 66336 with paclitaxel or docetaxel (two known substrates of P-gp) in vitro and in vivo *(171)*. SCH66336 significantly enhanced the effect of paclitaxel in the NCIH460 lung-cancer xenograft model and was able to sensitize wap-ras/F mammary tumors as well as tumor cell lines to paclitaxel *(171)*. We have directly characterized and quantified a specific synergy on P-gp function between SCH66336 and either tamoxifen, paclitaxel, or vinblastine *(124)*. Relatively small concentrations of SCH66336 can increase the affinity (potency) of these additional compounds as inhibitors of P-gp function. Treatment with SCH66336 would be predicted to be synergistic with co-administered cancer therapeutics that are substrates of P-gp. A further benefit of co-administration of SCH66336 could be reduced chemotherapy dosage, hence, lower exposure to normal cells, and therefore less undesired toxicity.

8.1. Classes of Modulators

Many of the characterized MDR modulators (inhibitors) can be categorized in various compound classes and this has been very well-reviewed *(4,172,173)*. The classes include: (1) calcium-channel blockers (dihydropyridine analogs, i.e., verapamil); (2) calmodulin antagonists (phenothiazines and thioxanthenes, i.e., trifluoperazine); (3) cyclic peptides (cyclosporin A, PSC833); (4) steroids and hormonal analogs (progesterone, tamoxifen); (5) dipyridamole; (6) anthracycline/vinca alkaloid analogs; and (7) miscellaneous other compounds.

8.2. Toxicity and Pharmacokinetic Interactions

Dose-limiting toxicities of the MDR modulators have often precluded their further clinical development *(174)*. Ventricular arrhythmia (verapamil), myelosuppression, cerebellar ataxia (PSC833, valspodar), and hypertension (cyclosporin A), have been observed, but may be owing to PK interactions caused by P-gp inhibition *(6)*. The antidiarrheal agent loperamide is a P-gp substrate, hence brain exposure is very limited normally; yet in the presence of quinidine (potent P-gp inhibitor), the brain loperamide concentrations increase resulting in serious neurotoxicity *(175)*. Moreover, PK interaction could be caused by inhibition of other xenobiotic-defense enzymes such as the cytochrome P450s, and could also enhance exposure of the chemotherapy to "sanctuary" sites such as CNS and testis, thus unveiling new toxicities not previously seen with the cytotoxin alone. This suggests that a significant fraction of patients have been under-dosed, thus making efficacy interpretations difficult.

8.3. Clinical Success or "Proof of Principle"

Clinical trials with some of the second-generation modulators are in progress, and some studies show clinical benefit from the use of modulators such as Valspodar

(PSC833) *(176)*, although Valspodar is saddled with enhancement of toxicity owing to PK interactions. Furthermore, the more potent and specific modulators (GF120918 [elacridar], LY335979 [zosuquidar]) do not show significant PK interaction with doxorubicin, etoposide, and paclitaxel in animal studies *(177–179)*. In a human Phase II study, a subgroup of paclitaxel-refractory advanced breast-cancer patients was resensitized by biricodar (VX-710) with adequate safety margins and PK parameters *(180)*. Trials in multiple myeloma, non-Hodgkin's lymphoma (NHL), and acute leukemia have shown that positive responses to P-gp modulators may occur in patients who are refractory to standard chemotherapy regimens *(180a–182)*.

These studies clearly indicate that the development of potent and selective P-gp inhibitors is an important approach to reversing MDR in the clinic. MDR of cancer cells is a potentially surmountable obstacle to effective chemotherapy of cancer.

9. MDR Pharmacological Inhibitors and Development Phase

The extensive list of MDR modulators cataloged in Table 2 serves as a strong illustration of the pursuit of an adjuvant cancer therapy to mitigate the action of P-gp. This table includes the pharmacological product classifications, compound name, pharmacological activity with parameter value, the evaluation stage as development phase, and the reference. This catalog was generated from the MF-line database and data management software.

10. Other ABC Transporters and the MDR Protein Multi-Gene Family

Much like the genetic superfamilies for cytochromes P450 and glutathione S-transferases, the xenobiotic transport enzymes are significantly contributing members of an extensive family that affects the overall disposition of many compounds. MDR1 and MRP1 exhibit much overlapping substrate specificity, although MDR1 currently seems to be broader in scope. However, MRP1 can act as a GS-X pump, i.e., it can transport drugs conjugated with GSH and glucuronide. MRP2 (also known as cMOAT; canalicular multispecific organic anion transporter) almost exclusively pumps out conjugates. Rats deficient in MRP2 show a chronic conjugated hyperbilirubinemia (model for the human disorder Dubin-Johnson syndrome). MRP homologs recently identified are MRP3, MRP4, MRP5, MRP6, MRP7, MDR3, BSEP, and MXR *(205–208)*. MDR2 apparently translocates only phospholipids, e.g., phosphatidylcholine. It is believed that humans possess 48 genes encoding ABC transporters.

The presence of other cellular drug-resistance mechanisms in addition to P-gp is most likely responsible for the apparent ineffectiveness of some P-gp modulators to date. The expression of MRP, P-gp, or both can account for diminished accumulation and retention of daunorubicin in blast cells from AML patients *(209)*. MRP2 is capable of mediating drug efflux, and a recent study showed increased bioavailability of several drugs and carcinogens in Mrp2-null rats *(210)*.

10.1. MRP1, MRP2, and MRP3

The multi-drug resistance protein (MRP) transporters are a subfamily of ABC transporters (ABC-C) related further to P-gp as expellers of various xenobiotics such as chemotherapy *(194,205,208,211,212)*. Extruding many anionic amphipathic compounds and conjugates from cells, the MRPs have become well-appreciated for their significant role in chemotherapy resistance *(213–215)*. MRP1 (ABC-C1) and MRP2

Table 2
P-Glycoprotein (P-gp) Interaction of Selected Multidrug-Resistance Modulators (MDRMs) and Other Selected Compounds With Significant Activity on P-gp

Compound	P-gp inhibition[a] $IC_{50}[K_i]^*(\mu M)$/References		P-gp affinity[b] $IC_{50}[K_i]^*(\mu M)$/References	
MDRMs				
Biricodar dicitrate (VX-710)	2.5	*(183)*	0.75	*(183)*
B9109-012			0.0107*	*(184)*
CP-100356			3.50	*(186)*
Dexniguldipine(HCl	0.45	*(184)*	2.5	*(185)*
			0.0112*	*(184)*
Elacridar, GF120918	0.044–0.51	*(187,191)*	0.0014–0.0025*	*(187)*
Laniquidar	0.2	*(192)*		
MS-209	0.4*	*(193)*		
ONT-093	0.03	*(194,195)*		
OC144-093				
Progesterone derivative	0.6–0.8	*(196)*		
S-9788	0.3	*(197)*	60	*(92)*
Tariquidar (XR9576)	0.025–0.49	*(187,191,198)*	0.55	*(191)*
			0.0026-0.0042*	*(187)*
Timcodar·$2CH_3SO_3H$	0.2–0.4	*(199)*		
Valspodar	0.29–1.06	*(200)*		
XR-9051	0.70	*(188,189,190)*	0.0018–0.0106*	*(190)*
XR-9576			0.0018–0.0106*	*(190)*
Zosuquidar(3HCl (LY-335979)	0.059*	*(178)*	0.03–0.06	*(179)*
Oncolytic drugs				
Paclitaxel			0.335*	*(187)*
Vinblastine·H_2SO_4			0.02*	*(187)*
Mitotane	10–30	*(180a)*		
Immunosuppressants				
Cyclosporine	0.44–5.10	*(200,201,203)*	0.6–1.17	*(191)*
FR-901459	6.0	*(201)*		
Antifungal agents				
Itraconazole	1.7	*(202)*		
Ketoconazole	5.6	*(202)*		
Calcium antagonists				
Fantofarone (SR33557)	2.5–20	*(203)*	0,0083	*(203)*
Nicardipine(HCl			0.011–0.182*	*(187)*
Verapamil	0.58–15.3	*(124,196,198,203)*	1.27–110	*(124,179,203)*
Farnesyltransferase inhibitors				
Lonafarnib (SCH-66336)	2.7	*(124)*		
Calmodulin antagonists, Dopamine antagonists				
Trifluoperazine(HCl	7.2	*(124)*		
Others				
Silybin			6.8*	*(204)*
Taxifolin			37.4*	*(204)*
CBT-1 (NSC-77037)				*(204a)*

[a]Inhibitory activity measured in different cell lines expressing P-gp and/or exhibiting multidrug resistance by means of different assays.
[b]Affinity to P-gp evaluated by displacement of different radioligands in different cell lines expressing P-gp and/or exhibiting MDR.
Data used with permission from Prous Science Integrity®.

(ABC-C2) have been shown to be expressed in cancers such as lung *(216,217)*, leukemia, bladder, neuroblastoma, breast, ovarian, hepatic, gastric, and prostate *(211,212,218–220)* and have conferred resistance against various chemotherapeutics. Accordingly, MRP1 is ubiquitously distributed among many organs predominantly in the basal and lateral plasma membranes, whereas MRP2 is in intestine, liver, kidney, placenta, BBB, pancreas, spleen, and choroid plexus and localized in the apical membrane of these cells *(211,212)*. The role in these locations is exemplified by the significant contribution of MRP2 to BBB function that was recently reported *(221)*. MRP1 and P-gp share only modest amino acid sequence identity (15%), whereas MRP1 and MRP2 share 50% amino acid identity, with MRP1 and MRP3 sharing even more (58%)

Because MRP1 secretes compounds into the system or body on the basolateral side, its role is more of a cellular defense than one of total organism defense such as P-gp and MRP2 (eliminated drugs from the body). Absence of MRP1 causes etoposide levels to increase 10-fold in the cerebrospinal fluid (CSF) after intravenous administration of the drug *(222)*. The body must have a basolateral transporter to protect sperm or CSF, because P-gp or other apical side efflux transporters would extrude drug into the sanctuary.

Substrate recognition by MRP1 and MRP2 is generally similar, with known examples being glucuronosyl-etoposide, estrone-3-sulfate, estradiol-17β-glucuronide, etoposide, vincristine (vinca alkaloids), sulfinpyrazone, methotrexate, leukotriene C4 (perhaps as a GSH conjugate), and anthracyclines. MRP1 can, however, exhibit remarkable selectivity, which probably contributes to cancer-therapy resistance. Whereas estradiol 17β-glucuronide is a good substrate, the 3-isomer is not *(223)*. Furthermore, MRP2 transports HIV protease inhibitors, whereas MRP1 does not *(224)*.

Cisplatin is an actively transported substrate of MRP2 *(214,225–232)*. The introduction of MRP2 antisense cDNA into human hepatic cancer HepG2 cells results in increased sensitivity to cisplatin, vincristine, doxorubicin, and the camptothecin derivatives *(231)*, with re-inoculated rats confirming an in vivo MRP2-mediated resistance to cisplatin *(229)*. MRP2 mRNA expression is significantly associated with the resistance of colorectal cancer to cisplatin *(227)*. In transfected cells, the overexpression of MRP2 resulted in resistance to cisplatin (10-fold), etoposide, doxorubicin, epirubicin, and MTX *(225)*. Moreover, MRP2 mRNA levels correlate with cisplatin resistance in a subset of resistant cell lines *(214)*, whereas cellular accumulation and drug sensitivity to cisplatin in human MRP2 transfectants decreased *(214)*.

5-FU is a possible substrate for the MRP1 transporter. HL-60 cells selected for overproduction of MRP1 (10× the level of parent HL-60 cells) showed resistance to 5-FU *(233)*, and MRP1 expression is correlated with 5-FU resistance in seven GI cancer-cell lines *(234)*. Treatment of cisplatin-resistant cell line with 5-FU increased the cytotoxicity of cisplatin fourfold, indicating that 5-FU may interact with MRP2 as well *(235)*.

Tyrosine kinase inhibitors, particularly STI571 (imatinib mesylate; Gleevec), have been shown to interact directly with MRP1 and P-gp *(236)*. In transfected cell lines expressing high levels of either MRP1 or P-gp, several tyrosine kinase inhibitors can inhibit transport function as well as substrate-stimulated ATP hydrolysis *(236)*. Moreover, P-gp has been detected in cell lines resistant to STI571 *(237)*.

A substrate of a trans-membrane transporter may not also be an inhibitor; conversely, inhibitors may not also be transported. Therefore, the list of substrates may not equal that of substrates. The known MRP1 inhibitors are nonspecific and include sulfinpyrazine, probenecid, benzbromarone, indomethacin (nonsteroidal anti-inflammatory drugs [NSAID]), some flavonoids, and even ritonavir *(238)*.

MRP3 also transports amphipathic anions on the basolateral side, like MRP1 *(219)*; is found largely in the liver, gut, and kidney, like MRP2; and is also present in the adrenal gland, pancreas, gallbladder, lung, and ovary *(239)*. MRP3 may contribute to enterohepatic recycling of bile salts (e.g., glycocholate) and removal of toxic anions during cholestasis *(240)*. Substrates include estradiol-17β-glucuronide, glucuronosyl-drug conjugates, vinca alkaloids, methotrexate, etoposide, and teniposide and may contribute to chemotherapy resistance. During cholestatic conditions the high MRP2 levels drop, whereas MRP3 is significantly elevated. Therefore, many amphipathic anions (including bile salts) secrete to the basolateral instead of the apical direction, preventing intrahepatic accumulation to toxic concentrations *(240)*.

10.2. MRP4 and MRP5

MRP4 and MRP5 are on the basolateral side of cells in many tissue types and are able to transport therapeutic nucleoside-based compounds *(241,242)*. Indeed, HIV was less effectively inhibited by the modified nucleoside analogs in cells overexpressing MRP4 *(241)*. MRP4 expression may also affect cancer chemotherapy. MRP4-overexpressing cells were resistant to cytotoxic effects of 6-mercaptopurine and 6-thioguanine (and AZT , MTX), important drugs in the treatment of childhood leukemias *(242)*. A role for MRP4 in the transport of DHEA (dehydroepiandrosterone 3-sulphate) and other conjugated steroids has recently been suggested *(243)*.

With MRP5 transporting not only PMEA (adenine nucleotide analog, acyclic nucleoside phosphonate) but also monophosphate diphospho-thiopurines *(244)*, this transporter may have a role in cancer chemotherapy as well. It was recently demonstrated that MRP5 is a cGMP transporter, yet a comparitively poor cAMP transporter *(245)*. Because sildenafil (i.e., Viagra, as well as trequinsin and zaprinast), a potent phosphodiesterase inhibitor, is also a very effective MRP5 inhibitor, speculation suggests that the vasodilatory effects could be owing to inhibition of MRP5-mediated cyclic nucleotide transport. Although MRP4 seems to prefer methylated thio-IMP, MRP5 prefers the unmethylated thioinucleotides. Moreover, MRP4 mediates transport of glucuronate conjugates and methotrexate, whereas MRP5 apparently does not *(246)*.

Many of the ABC transporters have been shown to be temporally regulating, or induced by exposure to xenobiotics. MRP4 transcription regulation may be controlled by its natural substrates, cyclic nucleotides *(247)*, and MRP4 is upregulated in response to elevated levels of hepatic bile acids *(248)*.

10.3. MRP6

MRP6, located in the basolateral membranes of cells in the liver and kidney, transports an anionic cyclopentapeptide BQ-123 and certain glutathione conjugates *(249)*. Possibly a highly selective pump for amphipathic anions *(250)*, MRP6 can cause low levels of resistance to some anticancer agents (etoposide, teniposide, doxorubicin) *(249)*. The absence of MRP6 causes pseudoxanthoma elasticum, a heritable disorder characterized by calcification of elastic fibers in skin, arteries, and retina *(251)*.

10.4. BCRP/MXR/ABCG2

The ABC transporter breast cancer resistance protein (BCRP), also known as mitoxantrone-resistance protein (MXR) *(252)*, is overproduced in MCF7 breast cancer cells *(253)*. As part of the ABC-G subfamily, this transporter has been renamed ABCG2 and apparently functions as a homodimer. It is not known if ABC-G2 can also heterodimerize with other proteins, such as those of the AGC-G class. Though possibly more selective in substrate recognition than P-gp, ABCG2 ejects mitoxantrone, topotecan derivatives, anthracyclines, bisantrene, etoposide, prazosin, and flavopiridol *(254)*, as well as HIV-1 nucleoside reverse-transcriptase inhibitor zidovidine (AZT) *(255)*. The active transport of indolocarbazole compound A was inhibited by indolocarbazole analogs but not by mitoxantrone, suggesting unique binding sites *(256)* reminiscent of P-gp. Other likely substrates include the experimental indolocarbazole topoisomerase inhibitors NB-506 and J-107088 and the active metabolite of the camptothecin analogue irinotecan/CPT-11 *(257)*. However, vincristine, vinblastine, paclitaxel, cisplatin, colchicine, verapamil, calcein-AM, rhodamine 123, and doxorubicin are not significantly transported by AGC-G2 *(258)*. It appears that P-gp, MRP1, and ABCG2 can account for most of the known active MDR *(259)*, with relatively high expression of BCRP mRNA observed in approx 30% of AML cases *(260)*. ABC-G2 mRNA also increased significantly from diagnosis to relapse or refractory disease, indicating that ABC-G2 levels may correlate with clinical resistance in AML *(261)*. ABCG2 has been detected in the apical membranes of placental syncytiotrophoblasts, hepatocytes, the epithelial lining of the small intestine and colon, brain microvessel endothelium *(262)*, the ducts and lobules of the mammary gland *(263)*, and hematopoietic progenitor cells *(264)*. Drug-resistant cell lines overexpressing ABC-G2 are derived from parent cells of fibroblasts, breast, colon, gastric, lung, or ovarian carcinomas, fibrosarcomas, and myelomas, which suggests that ABC-G2 may contribute to drug resistance in tumors of various tissue types.

ABC-G2 inhibitors may be useful to improving chemotherapy response, analogous to P-gp inhibitors. The P-gp inhibitors reserpine, GF120918, are potent inhibitors of the ABC-G2. Other inhibitors include fumitremorgin C and the tyrosine kinase inhibitor CI1033, whereas verapamil, cyclosporin A, PSC833 and some other P-gp inhibtors have little effect on ABC-G2.

Similar to P-gp, ABC-G2 could limit oral bioavailability. Indeed, the mRNA level of BCRP is significantly higher than that of MDR1 in jejunum *(125)*. Morever, inhibtion of ABC-G2 by oral dose of GF120918 has been shown to cause a drug interaction with oral topotecan, rasing plasma concentrations approximately sixfold. Biliary ABC-G2 also seems to contribute bioavailability and drug interactions because the GF120918 dose decreased hepatobiliary excretion of intravenously administered topotecan by approximately twofold *(265)*. Furthermore, brain exclusion of xenobiotics may be significantly dependent on AGC-G2, with its mRNA more highly expressed than P-gp and MRP1 in porcine brain *(266)*.

Differences in ABC-G2 function or expression among the population may result in diverse (and possibly dangerous) clinical exposure to substrate drugs, analogous to other polymorphic xenobiotic-metabolism enzymes. Analysis of DNA from 11 different ethnic populations revealed that there are several common natural allelic variants of ABC-G2, but their effect on function is yet to be examined *(267)*. A 78-fold variation in expression of BCRP mRNA and significant variation in protein expression in human intestine could not be accounted for by one of the common allelic variants *(267)* (*see* Table 3).

10.5. Sister-P-gp (SPGP)/Bile Salt Export Protein (BSEP)/ABC-B11

Sharing 50% amino acid identity with P-gp, sister-P-gp (SPGP), also known as the Bile Salt Export Pump (BSEP, ABC-B11), is significantly more selective with respect to substrate recognition. Expressed exclusively in the liver *(268)*, BSEP appears to have a role in efflux of endogenous compounds (bile acids) and exogenous compounds (xenobiotics) into the bile *(269,270)*. Some examples of endogenous substrates thought to be exported by BSEP include taurocholate, estradiol-17β-glucuronide, cholic acid, muricholates, and other monoanionic bile salts. Canalicular secretion of bile acids from the liver in the form of bile facilitates the emulsification of dietary lipids and fat-soluble vitamins. Defective bile secretion results in cholestasis with accumulation of bile salts and other toxic bile constituents within hepatocytes and blood plasma. Mutations in the *BSEP* gene can result in the absence of BSEP expression and are the cause of certain forms of progressive familial intrahepatic cholestasis (PFIC-2) *(271,272)*. PFIC manifestations are jaundice, fibrosis, cirrhosis (caused by <1% of normal biliary bile salts), hyperbilirubinemia, suppressed lipid and cholesterol metabolism, and intestinal malabsorption of fat and fat-soluble vitamins.

BSEP has recently been characterized as a transporter that interacts with drugs and xenobiotics, including vinblastine, ditekiren, troglitazone, troglitazone sulfate, cyclosporin, rifamycin, glibenclamide *(273)*, sulindac, and taxol *(274)*. Moreover, the administration of troglitazone *(275,276)*, cyclosporin, rifampicin, and bosentan *(277)*—all inhibitors of BSEP—has been linked with cholestasis. Xenobiotic-induced cholestasis is a significant clinical problem, though drug interactions mediated by BSEP may also have dangerous consequences. Because P-gp and BSEP are both expressed in the liver, it is clear that the extent of overlap between P-gp and BSEP drug substrates and inhibitors needs to be established to estimate the BSEP role in drug disposition. The importance of BSEP interactions at the level of hepatobiliary export processes should thus be considered in the evaluation of drug interactions.

Several cholestatic drugs have already been shown to potently inhibit BSEP: cyclosporine A, rifampicin, glibenclamide, estradiol 17β glucuronide, bosentan, troglitazone, and sulindac all can cause increased bile-salt concentrations in serum and eventually cholestatic liver injury and are BSEP inhibitors *(273,275,278)* and probably substrates *(279)*. Thus far, the most potent inhibitors seem to be cyclosporin, tamoxifen, and valinomycin *(280)*. There are likely to be much more potent inhibitors, and other substrates also may be more sensitive to inhibition of transport.

MDR3 is also involved in bile transport and another form of progressive familial intrahepatic cholestasis (PFIC-3). Sharing 75% amino acid identity with P-gp, it is inhibited by some known P-gp inhibitors and can also transport some amphipathic drugs *(281)*. However, it is generally considered selective for efflux of biliary phospholipids (phosphorylcholine).

11. Conclusion

Because of the wide tolerance of substrate recognition, P-gp can often be the mechanism for significant pharmacokinetic drug interactions when two or more drugs are competing for the P-gp transport site. P-gp levels are also inducible and can be even further elevated in cancer cells, thus contributing to the confounding pleiotropic resistance to chemotherapy and poor treatment prognosis. Consequently, a broad scope of research over 20 years has led to the evaluation of co-therapies intended to augment

Table 3
ABC Transporters Relevant to Xenobiotic Disposition

Name	AKA	Tissue	Side	Substrates	Notes
MDR1	P-gp, ABC-B1	Ubiquitous	Apical	Hydrophobic	Induced cholestasis
MDR2	MDR3, ABC-B4	?	Apical	PL	xenobiotics
BSEP	SPGP, ABC-B11	Liver	Apical	Bile salts	PFIC
MRP1	ABC-C1	Ubiquitous (low liver)	Lateral	Lipophilic anions, leukotriene, conjugates	Induced cholestasis
MRP2	CMOAT ABC-C2	Liver, kidney, gut, placenta, brain	Apical	Conjugates, lipophilic anions, Bilirubin	Suppressed cholestasis
MRP3	ABC-C3	Liver, brain, adrenals, pancreas kidney, gut, ovary, lung	Basolateral	Etoposide, vincristine, methotrezate, glycocholate, conjugates (anionic)	Induced cholestasis Conjugates
MRP4	ABC-C4	Ubiquitous	Basolateral	Anti HIV	Nucleoside antiviral
MRP5	ABC-C5	Ubiquitous	Basolateral	Anti HIV conjugates	Nucleoside antiviral
MRP6	ABC-C6	Liver, kidney, brain	Lateral	Amphipathic anions, conjugates	
CFTR	ABC-C7	Exocrine tissue		Anions bicarbonate	Cystic fibrosis
BCRP/MXR	ABC-G2	Ubiquitous	Apical	Amphipathic, Chemotherapy	Chemotherapy resistance

chemotherapy by inhibiting P-gp. This review includes a list of the currently known P-gp inhibiting adjuvant candidates described in the literature, with associated references and summary data. The summary catalog of P-gp modulators illustrates the ardent pursuit to overcome this form of therapy resistance and gives examples of clinical success and failure. Significant in vivo and in vitro experimental observations as well as the extensive catalog of P-gp inhibitors shown earlier illuminate the critical pursuit of impeding MDR by inhibiting P-gp. However, there remain many difficulties and hurdles to effective and safer therapies intended to block the active efflux provided by P-gp owing to its broad selectivity and tissue distribution as well as clinical liabilities of the compounds.

Acknowledgments

We are very grateful to Prof. Adriane L. Stewart for editorial assistance.

References

1. Pastan I, Gottesman M. Multiple-drug resistance in human cancer. N Engl J Med 1987;316(22):1388–1393.
2. Gottesman MM, Pastan I. Biochemistry of multidrug resistance mediated by the multidrug transporter. Annu Rev Biochem 1993;62:385–427.
3. Gottesman MM, Pastan I, Ambudkar SV. P-glycoprotein and multidrug resistance. Curr Opin Genet Dev 1996;6:610–617.
4. Ambudkar SV, Dey S, Hrycyna CA, et al. Biochemical, cellular, and pharmacological aspects of the multidrug transporter. Annu Rev Pharmacol Toxicol 1999;39:361–398.
5. Fisher GA, Lum BL, Hausdorff J, Sikic BI. Pharmacological considerations in the modulation of multidrug resistance. Eur J Cancer 1996;32A(6):1082–1088.
6. Siegsmund MJ, Cardarelli C, Aksentijevich I, et al. Ketoconazole effectively reverses multidrug resistance in highly resistant KB cells. J Urol 1994;151:485–491.
7. Wang E-J, Lew K, Barecki, ME, et al. Quantitative distinctions of active site molecular recognition by P-glycoprotein and cytochrome P450 3A4. Chemical Res Tox 2001;14(12):1596–1603.
8. Monahan BP, Ferguson CL, Killeavy ES, et al. Torsades de pointes occurring in association with terfenadine use. JAMA 1990;264(21):2788–2790.
9. Neuvonen PJ, Kantola T, Kivisto KT. Simvastatin but not pravastatin is very susceptible to interaction with the CYP3A4 inhibitor itraconazole. Clin Pharmacol Ther 1998;63(3):332–341.
10. Brinkmann U, Eichelbaum M. Polymorphisms in the ABC drug transporter gene MDR1. Pharmacogenomics J 2001;1(1):59–64.
11. Kerb R, Hoffmeyer S, Brinkmann U. ABC drug transporters: hereditary polymorphisms and pharmacological impact in MDR1, MRP1 and MRP2. Pharmacogenomics 2001;2(1):51–64.
12. Hoffmeyer S, Burk O, von Richter O, et al. Functional polymorphisms of the human multidrug-resistance gene: multiple sequence variations and correlation of one allele with P-glycoprotein expression and activity in vivo. Proc Natl Acad Sci USA 2000;28:97.
13. Eytan GD, Regev R, Assaraf YG. Functional reconstitution of P-glycoprotein reveals an apparent near stoichiometric drug transport to ATP hydrolysis. J Biol Chem 1996;271:3172–3178.
14. Ambudkar SV, Cardarelli CO, Pashinsky E, Stein WD. Relation between the turnover number for vinblastine transport and for vinblastine-stimulated ATP hydrolysis by human P-glycoprotein. J Biol Chem 1997;272:21160–21166.
15. Stein WD. Kinetics of the multidrug transporter (P-glycoprotein) and its reversal. Physiol Rev 1997;77:545–590.
16. Shapiro AB, Ling V. Stoichiometry of coupling of rhodamine 123 transport to ATP hydrolysis by P-glycoprotein. Eur J Biochem 1998;254:189–193.
17. Wang E-J, Casciano CN, Clement RP, Johnson WW. Two transport binding sites of P-glycoprotein are unequal yet contingent: initial rate kinetic analysis by ATP hydrolysis demonstrates intersite dependency. Biochim Biophys Acta 2000;1481:63–74.

18. Sauna ZE, Ambudkar SV. Evidence for a requirement for ATP hydrolysis at two distinct steps during a single turnover of the catalytic cycle of human P-glycoprotein. Proc Natl Acad Sci USA 2000;97:2515–2520.
19. Sauna ZE, Ambudkar SV. Characterization of the catalytic cycle of ATP hydrolysis by human P-glycoprotein: The two ATP hydrolysis events in a single catalytic cycle are kinetically similar but affect different functional outcomes. J Biol Chem 2001;276:11653–11661.
20. Senior AE, Bhagat S. P-glycoprotein shows strong catalytic cooperativity between the two nucleotide sites. Biochemistry 1998;37:831-856.
21. Shapiro AB, Ling V. Extraction of Hoechst 33342 from the cytoplasmic leaflet of the plasma membrane by P-glycoprotein. Eur J Biochem 1997;250:122–129.
22. Dey S, Ramachandra M, Pastan I, Gottesman MM. Evidence for two nonidentical drug-interaction sites in the human P-glycoprotein. Proc Natl Acad Sci USA 1997;94:10594–10599.
23. Sharom FJ, Liu R, Qu Q, Romsicki Y. Exploring the structure and function of the P-glycoprotein multidrug transporter using fluorescence spectroscopic tools. Semin Cell Dev Biol 2001;12(3):257–265.
24. Shapiro AB, Ling V. Positively cooperative sites for drug transport by P-glycoprotein with distinct drug specificities. Eur J Biochem 1997;250:130–137.
25. Wang E-J, Casciano CN, Clement RP, Johnson WW. Cooperativity in the inhibition of P-glycoprotein-mediated daunorubicin transport: evidence for half-of-the-sites reactivity." Arch Biochem Biophys 2000;383(1):91–98.
26. Wang G, Pincheira R, Zhang J-T. Dissection of drug-binding-induced conformational changes in P-glycoprotein. Eur J Biochem 1998;255:383–390.
27. Pascaud C, Garrigos M, Orlowski S. Multidrug resistance transporter P-glycoprotein has distinct but interacting binding sites for cytotoxic drugs and reversing agents. Biochem J 1998;333:351–358.
28. Hrycyna CA, Ramachandra M, Ambudkar SV, et al. Mechanism of action of human P-glycoprotein ATPase activity. Photochemical cleavage during a catalytic transition state using orthovanadate reveals cross-talk between the two ATP sites. J Biol Chem 1998;273:16631–16634.
29. Loo TW, Clarke DM. Covalent modification of human P-glycoprotein mutants containing a single cysteine in either nucleotide-binding fold abolishes drug-stimulated ATPase activity. J Biol Chem 1995;270:22957–22961.
30. Wang EJ, Casciano CN, Clement RP, Johnson WW. Active transport of fluorescent P-glycoprotein substrates: evaluation as markers and interaction with inhibitors. Biochem Biophys Res Commun 2001;289(2):580–585.
31. Schinkel AH, Smit JJ, van Tellingen O, et al. Disruption of the mouse mdr1a P-glycoprotein gene leads to a deficiency in the blood-brain barrier and to increased sensitivity to drugs. Cell 1994;77(4):491–502.
32. Schinkel AH, Mayer U, Wagenaar E, et al. Normal viability and altered pharmacokinetics in mice lacking mdr1-type (drug-transporting) P-glycoproteins. Proc Natl Acad Sci USA 1997;94(8):4028–4033.
33. Mayer U, Wagenaar E, Dorobek B, et al. Full blockade of intestinal P-glycoprotein and extensive inhibition of blood-brain barrier P-glycoprotein by oral treatment of mice with PSC833. J Clin Invest 1997;100(10):2430–2436.
34. Fricker G, Drewe J, Huwyler J, et al. Relevance of p-glycoprotein for the enteral absorption of cyclosporin A: in vitro-in vivo correlation. Br J Pharmacol 1996;118(7):1841–1847.
35. Doan KM, Humphreys JE, Webster LO, et al. Passive permeability and P-glycoprotein-mediated efflux differentiate central nervous system (CNS) and non-CNS marketed drugs. J Pharmacol Exp Ther 2002;303(3):1029–1037.
36. Kennedy T. Managing the drug discovery/development interface. Drug Disc Today 1997;2(10):436–444.
37. Gonzalez O, Colombo T, De Fusco M, et al. Changes in doxorubicin distribution and toxicity in mice pretreated with the cyclosporin analogue SDZ PSC 833. Cancer Chemother Pharmacol 1995;36(4):335–340.

38. van Asperen J, van Tellingen O, van der Valk MA, et al. Enhanced oral absorption and decreased elimination of paclitaxel in mice cotreated with cyclosporin A. Clin Cancer Res 1998;4(10):2293–2297.
39. Ito K, Kusuhara H, Sugiyama Y. Effects of intestinal CYP3A4 and P-glycoprotein on oral drug absorption—theoretical approach. Pharm Res 1999;16:225–231.
40. Hochman JH, Chiba M, Nishime J, et al. Influence of P-glycoprotein on the transport and metabolism of indinavir in Caco-2 cells expressing cytochrome P-450 3A4. J Pharmacol Exp Ther 2000;292(1):310–318.
41. Hochman JH, Chiba M, Yamazaki M, et al. P-glycoprotein-mediated efflux of indinavir metabolites in Caco-2 cells expressing cytochrome P450 3A4. J Pharmacol Exp Ther 2001;298(1):323–330.
42. Thummel KE, O'Shea D, Paine MF, et al. Oral first-pass elimination of midazolam involves both gastrointestinal and hepatic CYP3A-mediated metabolism. Clin Pharmacol Ther 1996;59(5):491–502.
43. Wu CY, Benet LZ, Hebert MF, et al. Differentiation of absorption and first-pass gut and hepatic metabolism in humans: studies with cyclosporine. Clin Pharmacol Ther 1995;58(5):492–497.
44. Lee CG, Gottesman MM, Cardarelli CO, et al. HIV-1 protease inhibitors are substrates for the MDR1 multidrug transporter. Biochemistry 1998;37(11):3594–3601.
45. Mayer U, Wagenaar E, Beijnen JH, et al. Substantial excretion of digoxin via the intestinal mucosa and prevention of long-term digoxin accumulation in the brain by the mdr 1a P-glycoprotein. Br J Pharmacol 1996;119(5):1038–1044.
46. Sparreboom A, van Asperen J, Mayer U, et al. Limited oral bioavailability and active epithelial excretion of paclitaxel (Taxol) caused by P-glycoprotein in the intestine. Proc Natl Acad Sci USA 1997;94(5):2031–2035.
47. Meerum Terwogt JM, Malingre MM, Beijnen JH, et al. Coadministration of oral cyclosporin A enables oral therapy with paclitaxel. Clin Cancer Res 1999;5(11):3379–3384.
48. Masuda S, Uemoto S, Hashida T, et al. Effect of intestinal P-glycoprotein on daily tacrolimus trough level in a living-donor small bowel recipient. Clin Pharmacol Ther 2000;68(1):98–103.
49. Karlsson J, Kuo SM, Ziemniak J, Artursson P. Transport of celiprolol across human intestinal epithelial (Caco-2) cells: mediation of secretion by multiple transporters including P-glycoprotein. Br J Pharmacol 1993;110(3):1009–1016.
50. Lennernas H, Regardh CG. Evidence for an interaction between the beta-blocker pafenolol and bile salts in the intestinal lumen of the rat leading to dose-dependent oral absorption and double peaks in the plasma concentration-time profile. Pharm Res 1993;10(6):879–883.
51. Gramatte T, Oertel R, Terhaag B, Kirch W. Direct demonstration of small intestinal secretion and site-dependent absorption of the beta-blocker talinolol in humans. Clin Pharmacol Ther 1996;59(5):541–549.
52. Wetterich U, Spahn-Langguth H, Mutschler E, et al. Evidence for intestinal secretion as an additional clearance pathway of talinolol enantiomers: concentration- and dose-dependent absorption in vitro and in vivo. Pharm Res 1996;13(4):514–522.
53. Gramatte T, Oertel R,. Intestinal secretion of intravenous talinolol is inhibited by luminal R-verapamil. Clin Pharmacol Ther 1999;66:239–245.
54. Collett A, Higgs NB, Sims E, et al. Modulation of the permeability of H2 receptor antagonists cimetidine and ranitidine by P-glycoprotein in rat intestine and the human colonic cell line Caco-2. J Pharmacol Exp Ther 1999;288(1):171–178.
55. Chiou WL, Chung SM, Wu TC, Ma C. A comprehensive account on the role of efflux transporters in the gastrointestinal absorption of 13 commonly used substrate drugs in humans. Int J Clin Pharmacol Ther 2001;39(3):93–101.
56. Zuccato E, Calamari D, Natangelo M, Fanelli R. Presence of therapeutic drugs in the environment. Lancet 2000;355(9217):1789–1790.
57. Batrakova EV, Li S, Miller DW, Kabanov AV. Pluronic P85 increases permeability of a broad spectrum of drugs in polarized BBMEC and Caco-2 cell monolayers. Pharm Res 1999;16(9):1366–1372.

58. Dintaman JM, Silverman JA. Inhibition of P-glycoprotein by D-alpha-tocopheryl polyethylene glycol 1000 succinate (TPGS). Pharm Res 1999;16(10):1550–1556.
59. Coon JS, Knudson W, Clodfelter K, et al. Solutol HS 15, nontoxic polyoxyethylene esters of 12-hydroxystearic acid, reverses multidrug resistance. Cancer Res 1991;51(3):897–902.
60. Dudeja PK, Anderson KM, Harris JS, et al. Reversal of multidrug resistance phenotype by surfactants: relationship to membrane lipid fluidity. Arch Biochem Biophys 1995;319(1):309–315.
61. Rege BD, Kao JP, Polli JE. Effects of nonionic surfactants on membrane transporters in Caco-2 cell monolayers. Eur J Pharm Sci 2002;16(4–5):237–246.
62. Hugger ED, Audus KL, Borchardt RT. Effects of poly(ethylene glycol) on efflux transporter activity in Caco-2 cell monolayers. J Pharm Sci 2002;91(9):1980–1990.
63. Zordan-Nudo T, Ling V, Liu Z, Georges E. Effects of nonionic detergents on P-glycoprotein drug binding and reversal of multidrug resistance. Cancer Res 1993;53(24):5994–6000.
64. Smit JW, Schinkel AH, Weert B, Meijer DK. Hepatobiliary and intestinal clearance of amphiphilic cationic drugs in mice in which both mdr1a and mdr1b genes have been disrupted. Br J Pharmacol 1998;124(2):416–424.
65. Kawahara M, Sakata A, Miyashita T, et al. Physiologically based pharmacokinetics of digoxin in mdr1a knockout mice. J Pharm Sci 1999;88(12):1281–1287.
66. Hori R, Okamura N, Aiba T, Tanigawara Y. Role of P-glycoprotein in renal tubular secretion of digoxin in the isolated perfused rat kidney. J Pharmacol Exp Ther 1993;266(3):1620–1625.
67. Lankas GR, Cartwright ME, Umbenhauer D. P-glycoprotein deficiency in a subpopulation of CF-1 mice enhances avermectin-induced neurotoxicity. Toxicol Appl Pharmacol 1997;143(2):357–365.
68. Ueng YF, Kuwabara T, Chun, YJ, Guengerich FP. Cooperativity in oxidations catalyzed by cytochrome P450 3A4. Biochemistry 1997;36(2):370–381.
69. Stouch TR, Gudmundsson O. Progress in understanding the structure-activity relationships of P-glycoprotein. Adv Drug Deliv Rev 2002;54(3):315–328.
70. Wiese M, Pajeva IK. Structure-activity relationships of multidrug resistance reversers. Curr Med Chem 2001;8(6):685–713.
71. Sharom FJ, Liu R, Romsicki Y, Lu P. Insights into the structure and substrate interactions of the P-glycoprotein multidrug transporter from spectroscopic studies. Biochim Biophys Acta 1999;1461(2):327–345.
72. Raviv Y, Pollard HB, Bruggemann EP, et al. Photosensitized labeling of a functional multidrug transporter in living drug-resistant tumor cells. J Biol Chem 1990;265(7):3975–3980.
73. Higgins CF, Gottesman MM. Is the multidrug transporter a flippase? Trends Biochem Sci 1992;17(1):18–21.
74. Homolya L, Hollo Z, Germann UA, et al. Fluorescent cellular indicators are extruded by the multidrug resistance protein. J Biol Chem 1993;268(29):21493–21496.
75. Lu P, Liu R, Sharom FJ. Drug transport by reconstituted P-glycoprotein in proteoliposomes. Effect of substrates and modulators, and dependence on bilayer phase state. Eur J Biochem 2001;268(6):1687–1697.
76. Ferte J. Analysis of the tangled relationships between P-glycoprotein-mediated multidrug resistance and the lipid phase of the cell membrane. Eur J Biochem 2000;267:277–294.
77. Litman T, Zeuthen T, Skovsgaard T, Stein W. Competitive, non-competitive and cooperative interactions between substrates of P-glycoprotein as measured by its ATPase activity. Biochim Biophys Acta 1997;1361:169–176.
78. Ramu A, Ramu N. Reversal of multidrug resistance by bis (phenylalkyl) amines and structurally related compounds. Cancer Chemother Pharmacol 1994;34:423–430.
79. Klopman G, Shi LM, Ramum A. Quantitative structure-activity relationship of multidrug resistance reversal agents. Mol Pharmacol 1997;52:323–334.
80. Pearce HL, Winter MA, Beck WT. Structural characteristics of compounds that modulate P-glycoprotein-associated multidrug resistance. Adv Enzyme Reg 1990;30:357–373.
81. Ford JM, Hait WN. Pharmacology of drugs that alter multidrug resistance in cancer. Pharmacol Rev 1990;42:155–199.
82. Ichikawa-Haraguchi M, Sumizawa T, Yoshimura A, et al. Progesterone and its metabolites: the potent inhibitors of the transporting activity of P-glycoprotein in the adrenal gland. Biochim Biophys Acta 1993;1158(3):201–208.

83. Neuhoff S, Langguth P, Dressler C, et al. Affinities at the verapamil binding site of MDR1-encoded P-glycoprotein: drugs and analogs, stereoisomers and metabolites. Int J Clin Pharmacol Ther 2000;38(4):168–179.
84. Lu L, Leonessa F, Baynham MT, et al. The enantioselective binding of mefloquine enantiomers to P-glycoprotein determined using an immobilized P-glycoprotein liquid chromatographic stationary phase. Pharm Res 2001;18(9):1327–1330.
85. Prueksaritanont T, Meng Y, Ma B, et al. Differences in the absorption, metabolism and biliary excretion of a diastereomeric pair of alphavbeta3-antagonists in rat: limited role of P-glycoprotein. Xenobiotica 2002;32(3):207–220.
86. Tmej C, Chiba P, Huber M, et al. A combined Hansch/Free-Wilson approach as predictive tool in QSAR studies on propafenone-type modulators of multidrug resistance. Arch Pharm (Weinheim) 1998;331(7–8):233–240.
87. Seelig A. A general pattern for substrate recognition by P-glycoprotein. Eur J Biochem 1998;251:252–261.
88. Ecker G, Huber M, Schmid D, Chiba P. The importance of a nitrogen atom in modulators of multidrug resistance. Mol Pharmacol 1999;56:791–796.
89. Ekins S, Kim RB, Leake BF, et al. Three-dimensional quantitative structure-activity relationships of inhibitors of P-glycoprotein. Mol Pharmacol 2002;61(5):964–973.
90. Penzotti JE, Lamb ML, Evensen E, Grootenhuis PD. A computational ensemble pharmacophore model for identifying substrates of P-glycoprotein. J Med Chem 2002;45(9):1737–1740.
91. Vazquez-Laslop N, Zheleznova EE, Markham PN, et al. Recognition of multiple drugs by a single protein: a trivial solution of an old paradox. Biochem Soc Trans 2000;28(4):517–520.
92. Xiao G, Liu S, Ji X, et al. First-sphere and second-sphere electrostatic effects in the active site of a class mu gluthathione transferase. Biochemistry 1996;35(15):4753–4765.
93. Dougherty DA. Cation-pi interactions in chemistry and biology: a new view of benzene, Phe, Tyr, and Trp. Science 1996;271(5246):163–168.
94. Romsicki Y, Sharom FJ. The ATPase and ATP-binding functions of P-glycoprotein: modulation by interaction with defined phospholipids. Eur J Biochem 1998;256(1):170–178.
95. Romsicki Y, Sharom FJ. The membrane lipid environment modulates drug interactions with the P-glycoprotein multidrug transporter. Biochemistry 1999;38(21):6887–6896.
96. de Lannoy IA, Mandin RS, Silverman M. Renal secretion of vinblastine, vincristine and colchicine in vivo. J Pharmacol Exp Ther 1994;268(1):388–395.
97. Lum BL, Fisher GA, Brophy NA, et al. Clinical trials of modulation of multidrug resistance. Pharmacokinetic and pharmacodynamic considerations. Cancer 1993;72(11 Suppl):3502–3514.
98. Choo EF, Leake B, Wandel C, et al. Pharmacological inhibition of P-glycoprotein transport enhances the distribution of HIV-1 protease inhibitors into brain and testes. Drug Metab Dispos 2000;28(6):655–660.
99. Savolainen J, Edwards JE, Morgan ME, et al. Effects of a p-glycoprotein inhibitor on brain and plasma concentrations of anti-human immunodeficiency virus drugs administered in combination in rats. Drug Metab Dispos 2002;30(5):479–482.
100. Westphal K, Weinbrenner A, Zschiesche M, et al. Induction of P-glycoprotein by rifampin increases intestinal secretion of talinolol in human beings: a new type of drug/drug interaction. Clin Pharmacol Ther 2000;68(4):345–355.
101. Greiner B, Eichelbaum M, Fritz P, et al. The role of intestinal P-glycoprotein in the interaction of digoxin and rifampin. J Clin Invest 1999;104(2):147–153.
102. Geick A, Eichelbaum M, Burk O. Nuclear receptor response elements mediate induction of intestinal MDR1 by rifampin. J Biol Chem 2001;276(18):14581–14587.
103. Abolhoda A, Wilson AE, Ross H, et al. Rapid activation of MDR1 gene expression in human metastatic sarcoma after in vivo exposure to doxorubicin. Clin Cancer Res 1999;5(11):3352–3356.
104. Ayrton A, Morgan P. Role of transport proteins in drug absorption, distribution and excretion. Xenobiotica 2001;31(8–9):469–497.
105. Kim RB. Drugs as P-glycoprotein substrates, inhibitors, and inducers. Drug Metab Rev 2002;34(1–2):47–54.
106. Jette L, Beaulieu E, Leclerc JM, Beliveau R Cyclosporin A treatment induces overexpression of P-glycoprotein in the kidney and other tissues. Am J Physiol 1996;270(5 Pt 2):F756–F765.

107. Drescher S, Schaeffeler E, Hitzl M, et al. MDR1 gene polymorphisms and disposition of the P-glycoprotein substrate fexofenadine. Br J Clin Pharmacol 2002;53(5):526–534.
108. Schaeffeler E, Eichelbaum M, Brinkmann U, et al. Frequency of C3435T polymorphism of MDR1 gene in African people. Lancet 2001;358(9279):383–384.
109. Ameyaw MM, Regateiro F, Li T, et al. MDR1 pharmacogenetics: frequency of the C3435T mutation in exon 26 is significantly influenced by ethnicity. Pharmacogenetics 2001;11(3):217–221.
110. Fellay J, Marzolini C, Meaden ER, et al. The Swiss HIV Cohort Study. Response to antiretroviral treatment in HIV-1-infected individuals with allelic variants of the multidrug resistance transporter 1: a pharmacogenetics study. Lancet 2002;359(9300):30–36.
111. Kim RB, Leake BF, Choo EF, et al. Identification of functionally variant MDR1 alleles among European Americans and African Americans. Clin Pharmacol Ther 2001;70(2):189–199.
112. Johne A, Kopke K, Gerloff T, et al. Modulation of steady-state kinetics of digoxin by haplotypes of the P-glycoprotein MDR1 gene. Clin Pharmacol Ther 2002;72(5):584–594.
113. Schuetz EG, Furuya KN, Schuetz JD. Interindividual variation in expression of P-glycoprotein in normal human liver and secondary hepatic neoplasms. J Pharmacol Exp Ther 1995;275(2):1011–1018.
114. Fontana RJ, Lown KS, Paine MF, et al. Effects of a chargrilled meat diet on expression of CYP3A, CYP1A, and P-glycoprotein levels in healthy volunteers. Gastroenterology 1999;117(1):89–98.
115. Wacher VJ, Wu CY, Benet LZ. Overlapping substrate specificities and tissue distribution of cytochrome P450 3A and P-glycoprotein: implications for drug delivery and activity in cancer chemotherapy. Mol Carcinog 1995;13(3):129–134.
116. Lown KS, Mayo RR, Leichtman AB, et al. Role of intestinal P-glycoprotein (mdr1) in interpatient variation in the oral bioavailability of cyclosporine. Clin Pharmacol Ther 1997;62(3):248–260.
117. Potocnik U, Glavac MR, Golouh R, Glavac D. The role of P-glycoprotein (MDR1) polymorphisms and mutations in colorectal cancer. Pflugers Arch 2001;442(6 Suppl 1):R182–R183.
118. Rund D, Azar I, Shperling O. A mutation in the promoter of the multidrug resistance gene (MDR1) in human hematological malignancies may contribute to the pathogenesis of resistant disease. Adv Exp Med Biol 1999;457:71–75.
119. Dhooge C, De Moerloose B, Laureys G, et al. Expression of the multidrug transporter P-glycoprotein is highly correlated with clinical outcome in childhood acute lymphoblastic leukemia: results of a long-term prospective study. Leuk Lymphoma 2002;43(2):309–314.
120. Tada Y, Wada M, Kuroiwa K, et al. MDR1 gene overexpression and altered degree of methylation at the promoter region in bladder cancer during chemotherapeutic treatment. Clin Cancer Res 2000;6(12):4618–4627.
121. Wang E-J, Casciano CN, Clement RP, Johnson WW. Inhibition of P-glycoprotein transport function by grapefruit juice psoralen. Pharm Res 2001;18(4):432–438.
122. Wang E-J, Casciano CN, Clement RP, Johnson WW. Evaluation of the interaction of loratadine and desloratadine with P-glycoprotein. Drug Metab Dispos 2001;29(8):1080–1083.
123. Wang E-J, Casciano CN, Clement RP, Johnson WW. HMG-CoA reductase inhibitors (statins) characterized as direct inhibitors of P-glycoprotein. Pharm Res 2001;18(6):800–806.
124. Wang E-J, Casciano CN, Clement RP, Johnson WW. The farnesyl protein transferase inhibitor SCH66336 is a potent inhibitor of MDR1 product P-glycoprotein. Cancer Res 2001;61(20):7525–7529.
125. Yamazaki M, Neway WE, Ohe T, et al. In vitro substrate identification studies for p-glycoprotein-mediated transport: species difference and predictability of in vivo results. J Pharmacol Exp Ther 2001;296(3):723–735.
126. Annaert PP, Turncliff RZ, Booth CL, et al. P-glycoprotein-mediated in vitro biliary excretion in sandwich-cultured rat hepatocytes. Drug Metab Dispos 2001;29(10):1277–1283.
127. Megard I, Garrigues A, Orlowski S, et al. A co-culture-based model of human blood-brain barrier: application to active transport of indinavir and in vivo-in vitro correlation. Brain Res 2002;927(2):153–167.

128. Adachi Y, Suzuki H, Sugiyama Y. Comparative studies on in vitro methods for evaluating in vivo function of MDR1 P-glycoprotein. Pharm Res 2001;18(12):1660–1668.
129. Taipalensuu J, Tornblom H, Lindberg G, et al. Correlation of gene expression of ten drug efflux proteins of the ATP-binding cassette transporter family in normal human jejunum and in human intestinal epithelial Caco-2 cell monolayers. J Pharmacol Exp Ther 2001;299(1):164–170.
130. Le Ferrec E, Chesne C, Artusson P, et al. In vitro models of the intestinal barrier. The report and recommendations of ECVAM Workshop 46. European Centre for the Validation of Alternative methods. Altern Lab Anim 2001;29(6):649–668.
131. Artursson P, Palm K, Luthman K. Caco-2 monolayers in experimental and theoretical predictions of drug transport. Adv Drug Deliv Rev 2001;46(1-3):27–43.
132. Sun D, Lennernas H, Welage LS, et al. Comparison of human duodenum and Caco-2 gene expression profiles for 12,000 gene sequences tags and correlation with permeability of 26 drugs. Pharm Res 2002;19(10):1400–1416.
133. Stewart BH, Chan OH, Lu RH, et al. Comparison of intestinal permeabilities determined in multiple in vitro and in situ models: relationship to absorption in humans. Pharm Res 1995;12(5):693–699.
134. Conradi RA, Wilkinson KF, Rush BD, et al. In vitro/in vivo models for peptide oral absorption: comparison of Caco-2 cell permeability with rat intestinal absorption of renin inhibitory peptides. Pharm Res 1993;10(12):1790–1792.
135. Walter E, Kissel T, Raddatz P. Transport of peptidomimetic renin inhibitors across monolayers of a human intestinal cell line (Caco-2): evidence for self-enhancement of paracellular transport route. Pharm Res 1995;12(11):1801–1805.
136. Barecki-Roach M, Wang EJ, Johnson WW. Many P-glycoprotein substrates do not inhibit the transport process across cell membranes. Xenobiotica 2003;33(2):131–140.
137. Scala S, Akhmed N, Rao US, et al. P-glycoprotein substrates and antagonists cluster into two distinct groups. Mol Pharmacol 1997;51:1024–1033.
138. Polli JW, Wring SA, Humphreys JE, et al. Rational use of in vitro P-glycoprotein assays in drug discovery. J Pharmacol Exp Ther 2001;299(2):620–628.
139. Eytan GD, Regev R, Oren G, Assaraf YG. The role of passive transbilayer drug movement in multidrug resistance and its modulation. J Biol Chem 1996;271:12897–12902.
140. Sonveaux N, Vigano C, Shapiro AB, et al. Ligand-mediated tertiary structure changes of reconstituted P-glycoprotein. A tryptophan fluorescence quenching analysis. J Biol Chem 1999;274:17649–17654.
141. Wielinga PR, Westerhoff HV, Lankelma J. The relative importance of passive and P-glycoprotein mediated anthracycline efflux from multidrug-resistant cells. Eur J Biochem 2000;267:649–657.
142. Urbatsch IL, al-Shawi MK, Senior AE. Characterization of the ATPase activity of purified Chinese hamster P-glycoprotein. Biochemistry 1994;33:7069–7076.
143. Zachowski A. Phospholipids in animal eukaryotic membranes: transverse asymmetry and movement. Biochem J 1993;294(Pt 1):1–14.
144. Eytan GD, Regev R, Oren G, et al. Efficiency of P-glycoprotein-mediated exclusion of rhodamine dyes from multidrug-resistant cells is determined by their passive transmembrane movement rate. Eur J Biochem 1997;248:104–112.
145. Regev R, Eytan GD. Flip-flop of doxorubicin across erythrocyte and lipid membranes. Biochem Pharmacol 1997;54:1151–1158.
146. Doppenschmitt S, Spahn-Langguth H, Regardh CG, Langguth P. Role of P-glycoprotein-mediated secretion in absorptive drug permeability: an approach using passive membrane permeability and affinity to P-glycoprotein. J Pharm Sci 1999;88,1067–1072.
147. Lentz KA, Polli JW, Wring SA, et al. Influence of passive permeability on apparent P-glycoprotein kinetics. Pharm Res 2000;17:1456–1460.
148. Mankhetkorn S, Dubru F, Hesschenbrouck J, et al. Relation among the resistance factor, kinetics of uptake, and kinetics of the P-glycoprotein-mediated efflux of doxorubicin,

daunorubicin, 8-(S)-fluoroidarubicin, and idarubicin in multidrug-resistant K562 cells. Mol Pharmacol 1996;49:532–539.
149. Mankhetkorn S, Garnier-Suillerot A. The ability of verapamil to restore intracellular accumulation of anthracyclines in multidrug resistant cells depends on the kinetics of their uptake. Eur J Pharmacol 1998;343:313–321.
150. Borrel MN, Fiallo M, Veress I, Garnier-Suillerot A. The effect of crown ethers, tetraalkylammonium salts, and polyoxyethylene amphiphiles on pirarubicin incorporation in K562 resistant cells. Biochem Pharmacol 1995;50:2069–2076.
151. Barnes KM, Dickstein B, Cutler GB Jr, et al. Steroid treatment, accumulation, and antagonism of P-glycoprotein in multidrug-resistant cells. Biochemistry 1996;35:4820–4827.
152. Borgnia MJ, Eytan GD, Assaraf YG. Competition of hydrophobic peptides, cytotoxic drugs, and chemosensitizers on a common P-glycoprotein pharmacophore as revealed by its ATPase activity. J Biol Chem 1996;271:3163–3171.
153. Tanaka K, Hirai M, Tanigawara Y, et al. Effect of cyclosporin analogues and FK506 on transcellular transport of daunorubicin and vinblastine via P-glycoprotein. Pharm Res 1996;13:1073–1077.
154. Sarkadi B, Muller, M, Homolya L, et al. Interaction of bioactive hydrophobic peptides with the human multidrug transporter. FASEB J 1994;8:766–770.
155. Chiba P, Ecker G, Schmid D, et al. Structural requirements for activity of propafenone-type modulators in P-glycoprotein-mediated multidrug resistance. Mol Pharmacol 1996:49:1122–1130.
156. Ferte J, Kuhnel JM, Chapuis G, et al. Flavonoid-related modulators of multidrug resistance: synthesis, pharmacological activity, and structure-activity relationships. J Med Chem 1999;42:478–489.
157. Gottesman MM, Fojo T, Bates SE. Multidrug resistance in cancer: role of ATP-dependent transporters. Nature Rev Cancer 2002;2(1):48–58.
158. Fojo AT, Ueda K, Slamon DJ, et al. Expression of a multidrug-resistance gene in human tumors and tissues. Proc Natl Acad Sci USA 1987;84(1):265–269
159. Leith CP, Kopecky KJ, Chen IM, et al. Frequency and clinical significance of the expression of the multidrug resistance proteins MDR1/P-glycoprotein, MRP1, and LRP in acute myeloid leukemia: a Southwest Oncology Group Study. Blood 1999;94(3):1086–1099.
160. Trock BJ, Leonessa F, Clarke R. Multidrug resistance in breast cancer: a meta-analysis of MDR1/gp170 expression and its possible functional significance. J Natl Cancer Inst 1997;89(13):917–931.
161. Mechetner E, Kyshtoobayeva A, Zonis S, et al. Levels of multidrug resistance (MDR1) P-glycoprotein expression by human breast cancer correlate with in vitro resistance to taxol and doxorubicin. Clin Cancer Res 1998;4(2):389–398
162. Baekelandt MM, Holm R, Nesland JM, et al. P-glycoprotein expression is a marker for chemotherapy resistance and prognosis in advanced ovarian cancer. Anticancer Res 2000;20(2B):1061–1067.
163. Ciarmiello A, Del Vecchio S, Silvestro P, et al. Tumor clearance of technetium 99m-sestamibi as a predictor of response to neoadjuvant chemotherapy for locally advanced breast cancer. J Clin Oncol 1998;16(5):1677–1683.
164. Yu D, Liu B, Jing T, S et al. Overexpression of both p185c-erbB2 and p170mdr-1 renders breast cancer cells highly resistant to taxol. Oncogene 1998;16(16):2087–2094.
165. Tsuruo T, Iida H, Tsukagoshi S, Sakurai Y. Overcoming of vincristine resistance in P388 leukemia in vivo and in vitro through enhanced cytotoxicity of vincristine and vinblastine by verapamil. Cancer Res 1981;41(5):1967–1972.
166. Twentyman PR, Bleehen NM. Resistance modification by PSC-833, a novel non-immunosuppressive cyclosporin. Eur J Cancer 1991;27(12):1639–1642.
167. List AF, Spier C, Greer J, et al. Phase I/II trial of cyclosporine as a chemotherapy-resistance modifier in acute leukemia. J Clin Oncol 1993;11(9):1652–1660.

168. Aszalos A, Ross DD. Biochemical and clinical aspects of efflux pump related resistance to anti-cancer drugs. Anticancer Res 1998;18(4C):2937–2944.
169. Liu M, Bryant MS, Chen J, et al. Antitumor activity of SCH 66336, an orally bioavailable tricyclic inhibitor of farnesyl protein transferase, in human tumor xenograft models and wap-ras transgenic mice. Cancer Res 1998;58(21):4947–4956.
170. Nielsen LL, Shi B, Hajian G, et al. Combination therapy with the farnesyl protein transferase inhibitor SCH66336 and SCH58500 (p53 adenovirus) in preclinical cancer models. Cancer Res 1999;59(23):5896–5901.
171. Shi B, Yaremko B, Hajian G, et al. The farnesyl protein transferase inhibitor SCH66336 synergizes with taxanes in vitro and enhances their antitumor activity in vivo. Cancer Chemother Pharmacol 2000;46(5):387–393.
172. Ford JM. Experimental reversal of P-glycoprotein-mediated multidrug resistance by pharmacological chemosensitisers. Eur J Cancer 1996;32A(6):991–1001.
173. Sonneveld P, Wiemer E. Inhibitors of multidrug resistance. Curr Opin Oncol 1997;9(6):543–548.
174. Stupp R, Bauer J, Pagani O, et al. Ventricular arrhythmia and torsade de pointe: dose limiting toxicities of the MDR-modulator S9788 in a phase I trial. Ann Oncol 1998;9(11):1233–1242.
175. Sadeque AJ, Wandel C, He H, et al. Increased drug delivery to the brain by P-glycoprotein inhibition. Clin Pharmacol Ther 2000;68(3):231–237.
176. Fields A, Hochster H, Runowicz C, et al. PSC833: initial clinical results in refractory ovarian cancer patients. Curr Opin Oncol 1998;10(Suppl 1):S21.
177. Hyafil F, Vergely C, Du Vignaud P, Grand-Perret T. In vitro and in vivo reversal of multidrug resistance by GF120918, an acridonecarboxamide derivative. Cancer Res 1993;53(19):4595–4602.
178. Dantzig AH, Shepard RL, Cao J, et al. Reversal of P-glycoprotein-mediated multidrug resistance by a potent cyclopropyldibenzosuberane modulator, LY335979. Cancer Res 1996;56(18):4171–4179.
179. Dantzig AH, Law KL, Cao J, Starling JJ. Reversal of multidrug resistance by the P-glycoprotein modulator, LY335979, from the bench to the clinic. Curr Med Chem 2001;8(1):39–50.
180. Toppmeyer D, Seidman AD, Pollak M, et al. Safety and efficacy of the multidrug resistance inhibitor Incel (biricodar; VX-710) in combination with paclitaxel for advanced breast cancer refractory to paclitaxel. Clin Cancer Res 2002;8(3):670–678.
180a. Abraham J, Bakke S, Rutt A, et al. A phase II trial of combination chemotherapy and surgical resection for the treatment of metastatic adrenocortical carcinoma: continuous infusion doxorubicin, vincristine, and etoposide with daily mitotane as a P-glycoprotein antagonist. Cancer 2002;94(9):2333–2343.
181. Sonneveld P. Reversal of multidrug resistance in acute myeloid leukaemia and other haematological malignancies. Eur J Cancer 1996;32A(6):1062–1069.
182. Sonneveld P, List AF. Chemotherapy resistance in acute myeloid leukaemia. Best Pract Res Clin Haematol 2001;14(1):211–233.
183. Germann UA, Shlyakhter K, Mason VS, et al. Cellular and biochemical characterization of VX-710 as a chemosensitizer: Reversal of P-glycoprotein-mediated multidrug resistance in vitro. Anti-Cancer Drugs 1997;8(2):125.
184. Malkhandi PJ, Ferry DR, Boer R, Kerr DJ. P-Glycoprotein has a drug acceptor site for 1, 4-dihydropyridines which is localised on an intracellular domain. Proc Amer Assoc Cancer Res 1995;36:Abst 1977.
185. Boer R, Ulrich WR, Haas S, et al. Interaction of cytostatics and chemosensitizers with the dexniguldipine binding site on P-glycoprotein. Eur J Pharmacol 1996;295(2–3):253–260.
186. Safa AR, et al. The relationship between modulation of multidrug resistance by novel, potent agents and their affinities for P-glycoprotein. 9th NCI-EORTC Symp New Drugs Cancer Ther (March 12–15) 1996;Abst 419.
187. Malingre MM, Beijnen JH, Rosing H, et al. Co-administration of GF120918 significantly increases the systemic exposure to oral paclitaxel in cancer patients. Br J Cancer 2001;84(1):42–47.

188. Mistry P, Plumb J, Eccles S, et al. In vivo efficacy of XR9051, a potent modulator of P-glycoprotein mediated multidrug resistance. Br J Cancer 1999;79(11–12):1672–1678.
189. Martin C, Berridge G, Mistry P, et al. The molecular interaction of the high affinity reversal agent XR9576 with P-glycoprotein. Br J Pharmacol 1999;128(2):403–411.
190. Mistry P, Stewart AJ, Dangerfield W, et al. In vitro and in vivo reversal of P-glycoprotein-mediated multidrug resistance by a novel potent modulator, XR9576. Cancer Res 2001;61(2):749–758.
191. Dodic N, Dumaitre B, Daugan A, Pianetti P. Synthesis and activity against multidrug resistance in Chinese hamster ovary cells of new acridone-4-carboxamides. J Med Chem 1995;38(13):2418.
192. van Zuylen L, Sparreboom A, van der Gaast A, et al. Disposition of docetaxel in the presence of P-glycoprotein inhibition by intravenous administration of R101933. Eur J Cancer 2002;38(8):1090–1099.
193. Kimura Y, Aoki J, Kohno M, et al. P-glycoprotein inhibition by the multidrug resistance-reversing agent MS-209 enhances bioavailability and antitumor efficacy of orally administered paclitaxel. Cancer Chemother Pharmacol 2002;49(4):322–328.
194. Dixon R, Toyonaga B. P-glycoprotein inhibitor OC144-093: Phase 1 intravenous and oral pharmacokinetics of a novel multidrug resistance modulator. Proc Amer Assoc Cancer Res 2000;41:Abst 3863.
195. Guns ES, Denyssevych T, Dixon R, et al. Drug interaction studies between paclitaxel (Taxol) and OC144-093—a new modulator of MDR in cancer chemotherapy. Eur J Drug Metab Pharmacokinet 2002;27(2):119–126.
196. Leonessa F, Kim JH, Ghiorghis A, et al. C-7 analogues of progesterone as potent inhibitors of the P-glycoprotein efflux pump. J Med Chem 2002;45(2):390–398.
197. Tranchand B, Catimel G, Lucas C, et al. Phase I clinical and pharmacokinetic study of S9788, a new multidrug-resistance reversal agent given alone and in combination with doxorubicin to patients with advanced solid tumors. Cancer Chemother Pharmacol 1998;41(4):281–291.
198. Roe M, Folkes A, Ashworth P, et al. Reversal of P-glycoprotein mediated multidrug resistance by novel anthranilamide derivatives. Bioorg Med Chem Lett 1999;9(4):595–600.
199. Harding MW, Ford, PJ, Mason VS, et al. VX-853: A novel bispecific chemosensitizer which reverses P-glycoprotein- and MRP-mediated multidrug resistance. Proc Amer Assoc Cancer Res 1996;37: Abst. 2283.
200. Kusunoki N, Takara K, Tanigawara Y, et al. Inhibitory effects of a cyclosporin derivative, SDZ PSC 833, on transport of doxorubicin and vinblastine via human P-glycoprotein. Jpn J Cancer Res 1998;89(11):1220–1228.
201. Tiberghien F, Wenandy T, Loor F. The potent immunosuppressive cyclosporin FR901459 inhibits the human P-glycoprotein and formyl peptide receptor functions. J Antibiot (Tokyo) 2000;53(5):509–515.
202. Wang E-J, Lew K, Casciano CN, et al. Characterization of common azole antifungals contrasted for interaction with P-glycoprotein. Antimicrob Agents Chemother 2002;46(1):160–165.
203. Martin C, Berridge G, Higgins CF, Callaghan R. The multi-drug resistance reversal agent SR33557 and modulation of vinca alkaloid binding to P-glycoprotein by an allosteric interaction. Br J Pharmacol 1997;122(4):765–771.
204. Maitrejean M, Comte G, Barron D, et al. The flavanolignan silybin and its hemisynthetic derivatives, a novel series of potential modulators of P-glycoprotein. Bioorg Med Chem Lett 2000;10(2):157–160.
204a. Oldham RK, Reid WK, Barnett D. Phase I study of CBT-1 and Taxol in patients with Taxol resistant cancers. Cancer Biother Radiopharm 2000;15(2):153–159.
205. Borst P, Evers R, Kool M, Wijnholds J. A family of drug transporters: the multidrug resistance-associated proteins. J Natl Cancer Inst 2000;92(16):1295–1302.
206. Kruh GD, Zeng H, Rea PA, et al. MRP subfamily transporters and resistance to anticancer agents. J Bioenerg Biomembr 2001;33(6):493–501.

207. Suzuki T, Nishio K, Tanabe S. The MRP family and anticancer drug metabolism. Curr Drug Metab 2001;2(4):367–377.
208. Borst P, Elferink RO. Mammalian ABC transporters in health and disease. Annu Rev Biochem 2002;71:537–592.
209. Ross DD, Doyle LA, Schiffer CA, et al. Expression of multidrug resistance-associated protein (MRP) mRNA in blast cells from acute myeloid leukemia (AML) patients. Leukemia 1996;10(1):48–55.
210. Dietrich CG, de Waart DR, Ottenhoff R, et al. Increased bioavailability of the food-derived carcinogen 2-amino-1-methyl-6-phenylimidazo[4,5-b]pyridine in MRP2-deficient rats. Mol Pharmacol 2001;59(5):974–980.
211. Borst P, Kool M, Evers R. Do cMOAT (MRP2), other MRP homologues, and LRP play a role in MDR? Semin Cancer Biol 1997;8(3):205–213.
212. Konig J, Nies AT, Cui Y, et al. Conjugate export pumps of the multidrug resistance protein (MRP) family: localization, substrate specificity, and MRP2-mediated drug resistance. Biochim Biophys Acta 1999;1461(2):377–394.
213. Kawabe T, Chen ZS, Wada M, et al. Enhanced transport of anticancer agents and leukotriene C4 by the human canalicular multispecific organic anion transporter (cMOAT/MRP2). FEBS Lett 1999;456(2):327–331
214. Allen JD, Brinkhuis RF, van Deemter L, et al. Extensive contribution of the multidrug transporters P-glycoprotein and Mrp1 to basal drug resistance. Cancer Res 2000;60(20):5761–5766.
215. Kool M, de Haas M, Scheffer GL, et al. Analysis of expression of cMOAT (MRP2), MRP3, MRP4, and MRP5, homologues of the multidrug resistance-associated protein gene (MRP1), in human cancer cell lines. Cancer Res 1997;57(16):3537–3547.
216. Zaman GJ, Versantvoort CH, Smit JJ, et al. Analysis of the expression of MRP, the gene for a new putative transmembrane drug transporter, in human multidrug resistant lung cancer cell lines. Cancer Res 1993;53(8):1747–1750.
217. Cole SP, Bhardwaj G, Gerlach JH, et al. Overexpression of a transporter gene in a multidrug-resistant human lung cancer cell line. Science 1992;258(5088):1650–1654
218. Kuwano M, Toh S, Uchiumi T, et al. Multidrug resistance-associated protein subfamily transporters and drug resistance. Anticancer Drug Des 1999;14(2):123–131.
219. Nies AT, Konig J, Pfannschmidt M, et al. Expression of the multidrug resistance proteins MRP2 and MRP3 in human hepatocellular carcinoma. Int J Cancer 2001;94(4):492–499.
220. Zalcberg J, Hu XF, Slater A, et al. MRP1 not MDR1 gene expression is the predominant mechanism of acquired multidrug resistance in two prostate carcinoma cell lines. Prostate Cancer Prostatic Dis 2000;3(2):66–75.
221. Potschka H, Fedrowitz M, Loscher W. Multidrug resistance protein MRP2 contributes to blood-brain barrier function and restricts antiepileptic drug activity. J Pharmacol Exp Ther 2003;306(1):124–131.
222. Wijnholds J, deLange EC, Scheffer GL, et al. Multidrug resistance protein 1 protects the choroid plexus epithelium and contributes to the blood-cerebrospinal fluid barrier. J Clin Invest 2000;105(3):279–285.
223. Loe DW, Almquist KC, Cole SP, Deeley RG. ATP-dependent 17 beta-estradiol 17-(beta-D-glucuronide) transport by multidrug resistance protein (MRP). Inhibition by cholestatic steroids. J Biol Chem 1996;271(16):9683–9689.
224. Huisman MT, Smit JW, Crommentuyn KM, et al. Multidrug resistance protein 2 (MRP2) transports HIV protease inhibitors, and transport can be enhanced by other drugs. AIDS 2002;16(17):2295–22301
225. Cui Y, Konig J, Buchholz JK, et al. Drug resistance and ATP-dependent conjugate transport mediated by the apical multidrug resistance protein, MRP2, permanently expressed in human and canine cells. Mol Pharmacol 1999;55(5):929–937.
226. Hatanaka H, Abe Y, Naruke M, et al. Modulation of multidrug resistance in a cancer cell line by anti-multidrug resistance-associated protein (MRP) ribozyme. Anticancer Res 2001;21(2A):879–885.
227. Hinoshita E, Uchiumi T, Taguchi K, et al. Increased expression of an ATP-binding cassette superfamily transporter, multidrug resistance protein 2, in human colorectal carcinomas. Clin Cancer Res 2000;6(6):2401–2407.

228. Hour TC, Chen J, Huang CY, et al. Characterization of chemoresistance mechanisms in a series of cisplatin-resistant transitional carcinoma cell lines. Anticancer Res 2000;20(5A):3221–3225.
229. Itoh Y, Tamai M, Yokogawa K, et al. Involvement of multidrug resistance-associated protein 2 in in vivo cisplatin resistance of rat hepatoma AH66 cells. Anticancer Res 2002;22(3):1649–1653.
230. Khokhar NZ, She Y, Rusch VW, Sirotnak FM. Experimental therapeutics with a new 10-deazaaminopterin in human mesothelioma: further improving efficacy through structural design, pharmacologic modulation at the level of MRP ATPases, and combined therapy with platinums. Clin Cancer Res 2001;7(10):3199–3205.
231. Koike K, Kawabe T, Tanaka T, et al. A canalicular multispecific organic anion transporter (cMOAT) antisense cDNA enhances drug sensitivity in human hepatic cancer cells. Cancer Res 1997;57(24):5475–5479.
232. Komuro Y, Udagawa Y, Susumu N, et al. Paclitaxel and SN-38 overcome cisplatin resistance of ovarian cancer cell lines by down-regulating the influx and efflux system of cisplatin. Jpn J Cancer Res 2001;92(11):1242–12450.
233. O'Brien ML, Vulevic B, Freer S, et al. Glutathione peptidomimetic drug modulator of multidrug resistance-associated protein. J Pharmacol Exp Ther 1999;291(3):1348–1355.
234. Kirihara Y, Yamamoto W, Toge T, Nishiyama M. Dihydropyrimidine dehydrogenase, multidrug resistance-associated protein, and thymidylate synthase gene expression levels can predict 5-fluorouracil resistance in human gastrointestinal cancer cells. Int J Oncol 1999;14(3):551–556.
235. Zhan M, Liu X. Schedule-dependent reversion of cisplatin resistance by 5-fluorouracil in a cisplatin-resistant human lung adenocarcinoma cell line A549DDP. Chin Med J (Engl) 1999;112(4):336–339.
236. Hegedus T, Orfi L, Seprodi A, et al. Interaction of tyrosine kinase inhibitors with the human multidrug transporter proteins, MDR1 and MRP1. Biochim Biophys Acta 2002;1587(2–3):318–325.
237. Mahon FX, Belloc F, Lagarde V, et al. MDR1 gene overexpression confers resistance to imatinib mesylate in leukemia cell line models. Blood 2003;101(6):2368–2373.
238. Olson DP, Scadden DT, D'Aquila RT, De Pasquale MP. The protease inhibitor ritonavir inhibits the functional activity of the multidrug resistance related-protein 1 (MRP-1). AIDS 2002;16(13):1743–1747.
239. Scheffer GL, Kool M, de Haas M, et al. Tissue distribution and induction of human multidrug resistant protein 3. Lab Invest 2002;82(2):193–201.
240. Ogawa K, Suzuki H, Hirohashi T, et al. Characterization of inducible nature of MRP3 in rat liver. Am J Physiol Gastrointest Liver Physiol 2000;278(3):G438–G446.
241. Schuetz JD, Connelly MC, Sun D, et al. MRP4: a previously unidentified factor in resistance to nucleoside-based antiviral drugs. Nat Med 1999;5(9):1048–1051.
242. Chen ZS, Lee K, Kruh GD. Transport of cyclic nucleotides and estradiol 17-beta-D-glucuronide by multidrug resistance protein 4. Resistance to 6-mercaptopurine and 6-thioguanine. J Biol Chem 2001;276(36):33747–33754.
243. Zelcer N, Reid G, Wielinga P, et al. Steroid and bile acid conjugates are substrates of human multidrug-resistance protein (MRP) 4 (ATP-binding cassette C4). Biochem J 2003;371(Pt 2):361–367.
244. Wielinga PR, Reid G, Challa EE, et al. Thiopurine metabolism and identification of the thiopurine metabolites transported by MRP4 and MRP5 overexpressed in human embryonic kidney cells. Mol Pharmacol 2002;62(6):1321–1331.
245. Jedlitschky G, Burchell B, Keppler D. The multidrug resistance protein 5 functions as an ATP-dependent export pump for cyclic nucleotides. J Biol Chem 2000;275(39):30069–30074.
246. Chen ZS, Lee K, Walther S, et al. Analysis of methotrexate and folate transport by multidrug resistance protein 4 (ABCC4): MRP4 is a component of the methotrexate efflux system. Cancer Res 2002;62(11):3144–3150.
247. Shoshani I, Laux WH, Perigaud C, et al. Inhibition of adenylyl cyclase by acyclic nucleoside phosphonate antiviral agents. J Biol Chem 1999;274(49):34742–3474.

248. Schuetz EG, Strom S, Yasuda K, et al. Disrupted bile acid homeostasis reveals an unexpected interaction among nuclear hormone receptors, transporters, and cytochrome P450. J Biol Chem 2001;276(42):39411–39418.
249. Belinsky MG, Chen ZS, Shchaveleva I, et al. Characterization of the drug resistance and transport properties of multidrug resistance protein 6 (MRP6, ABCC6). Cancer Res 2002;62(21):6172–6177.
250. Madon J, Hagenbuch B, Landmann L, et al. Transport function and hepatocellular localization of mrp6 in rat liver. Mol Pharmacol 2000;57(3):634–641.
251. Bergen AA, Plomp AS, Schuurman EJ, et al. Mutations in ABCC6 cause pseudoxanthoma elasticum. Nat Genet 2000;25(2):228–231.
252. Miyake K, Mickley L, Litman T, et al. Molecular cloning of cDNAs which are highly overexpressed in mitoxantrone-resistant cells: demonstration of homology to ABC transport genes. Cancer Res 1999;59(1):8–13.
253. Doyle LA, Yang W, Abruzzo LV, et al. A multidrug resistance transporter from human MCF-7 breast cancer cells. Proc Natl Acad Sci USA 1998;95(26):15665–15670.
254. Robey RW, Medina-Perez WY, Nishiyama K, et al. Overexpression of the ATP-binding cassette half-transporter, ABCG2 (Mxr/BCrp/ABCP1), in flavopiridol-resistant human breast cancer cells. Clin Cancer Res 2001;7(1):145–152.
255. Wang X, Furukawa T, Nitanda T, et al. Breast cancer resistance protein (BCRP/ABCG2) induces cellular resistance to HIV-1 nucleoside reverse transcriptase inhibitors. Mol Pharmacol 2003;63(1):65–72.
256. Nakagawa R, Hara Y, Arakawa H, et al. ABCG2 confers resistance to indolocarbazole compounds by ATP-dependent transport. Biochem Biophys Res Commun 2002;299(4):669–675.
257. Allen JD, Schinkel AH. Multidrug resistance and pharmacological protection mediated by the breast cancer resistance protein (BCRP/ABCG2). Mol Cancer Ther 2002;1(6):427–434.
258. Litman T, Brangi M, Hudson E, et al. The multidrug-resistant phenotype associated with overexpression of the new ABC half-transporter, MXR (ABCG2). J Cell Sci 2000;113(Pt 11):2011–2021.
259. Allen JD, Brinkhuis RF, Wijnholds J, Schinkel AH. The mouse Bcrp1/Mxr/Abcp gene: amplification and overexpression in cell lines selected for resistance to topotecan, mitoxantrone, or doxorubicin. Cancer Res 1999;59(17):4237–4241.
260. Ross DD. Novel mechanisms of drug resistance in leukemia. Leukemia 2000;14(3):467–473.
261. van den Heuvel-Eibrink MM, Wiemer EA, Prins A, et al. Increased expression of the breast cancer resistance protein (BCRP) in relapsed or refractory acute myeloid leukemia (AML). Leukemia 2002;16(5):833–839.
262. Cooray HC, Blackmore CG, Maskell L, Barrand MA. Localisation of breast cancer resistance protein in microvessel endothelium of human brain. Neuroreport 2002;13(16):2059–2063.
263. Maliepaard M, Scheffer GL, Faneyte IF, et al. Subcellular localization and distribution of the breast cancer resistance protein transporter in normal human tissues. Cancer Res 2001;61(8):3458–3464.
264. Scharenberg CW, Harkey MA, Torok-Storb B. The ABCG2 transporter is an efficient Hoechst 33342 efflux pump and is preferentially expressed by immature human hematopoietic progenitors. Blood 2002;99(2):507–512.
265. Jonker JW, Smit JW, Brinkhuis RF, et al. Role of breast cancer resistance protein in the bioavailability and fetal penetration of topotecan. J Natl Cancer Inst 2000;92(20):1651–1656.
266. Eisenblatter T, Huwel S, Galla HJ. Characterisation of the brain multidrug resistance protein (BMDP/ABCG2/BCRP) expressed at the blood-brain barrier. Brain Res 2003;971(2):221–231.
267. Zamber CP, Lamba JK, Yasuda K, et al. Natural allelic variants of breast cancer resistance protein (BCRP) and their relationship to BCRP expression in human intestine. Pharmacogenetics 2003;13(1):19–28.
268. Childs S, Yeh RL, Georges E. Ling V. Identification of a sister gene to P-glycoprotein. Cancer Res 1995;55(10):2029–2034.
269. Thompson R, Strautnieks S. BSEP: function and role in progressive familial intrahepatic cholestasis. Semin Liver Dis 2001;21(4):545–550.

270. Meier PJ, Stieger B. Bile salt transporters. Annu Rev Physiol 2002;64:635–661.
271. Strautnieks SS, Bull LN, Knisely AS, et al. A gene encoding a liver-specific ABC transporter is mutated in progressive familial intrahepatic cholestasis. Nat Genet 1998;20(3):233–238.
272. Wang R, Salem M, Yousef IM, et al. Targeted inactivation of sister of P-glycoprotein gene (BSEP) in mice results in non-progressive but persistent intrahepatic cholestasis. Proc Natl Acad Sci USA 2001;98(4):2011–2016.
273. Stieger B, Fattinger K, Madon J, et al. Drug- and estrogen-induced cholestasis through inhibition of the hepatocellular bile salt export pump (Bsep) of rat liver. Gastroenterology 2000;118(2):422–430.
274. Childs S, Yeh RL, Hui D, Ling V. Taxol resistance mediated by transfection of the liver-specific sister gene of P-glycoprotein. Cancer Res 1998;58(18):4160–4167.
275. Funk C, Ponelle C, Scheuermann G, Pantze M. Cholestatic potential of troglitazone as a possible factor contributing to troglitazone-induced hepatotoxicity: in vivo and in vitro interaction at the canalicular bile salt export pump (Bsep) in the rat. Mol Pharmacol 2001;59(3):627–635.
276. Kostrubsky VE, Vore M, Kindt E, et al. The effect of troglitazone biliary excretion on metabolite distribution and cholestasis in transporter-deficient rats. Drug Metab Dispos 2001;29(12):1561–1566.
277. Fattinger K, Funk C, Pantze M, et al. The endothelin antagonist bosentan inhibits the canalicular bile salt export pump: a potential mechanism for hepatic adverse reactions. Clin Pharmacol Ther 2001;69(4):223–231.
278. Bolder U, Trang NV, Hagey LR, et al. Sulindac is excreted into bile by a canalicular bile salt pump and undergoes a cholehepatic circulation in rats. Gastroenterology 1999;117(4):962–971.
279. Torok M, Gutmann H, Fricker G, Drewe J. Sister of P-glycoprotein expression in different tissues. Biochem Pharmacol 1999;57(7):833–835.
280. Wang E-J, Casciano CN, Clement RP, Johnson WW Fluorescent substrates of sister-P-glycoprotein (BSEP) evaluated as markers of active transport and inhibition: evidence for contingent unequal binding sites. Pharmaceutical Res 2003;20(4):537–544.
281. Smith AJ, van Helvoort A, van Meer G, et al. MDR3 P-glycoprotein, a phosphatidylcholine translocase, transports several cytotoxic drugs and directly interacts with drugs as judged by interference with nucleotide trapping. J Biol Chem 2000;275(31):23530–23539.

9

Human Intestinal Cellular Characteristics and Drug Permeability

James E. Polli, Anand Balakrishnan, and Paul R. Seo

1. Introduction

Oral drug administration is the most convenient method to administer drugs, particularly in the treatment of chronic diseases. This preference is generally the case, even when the site of drug action is beyond the gastrointestinal tract (GI), requiring the drug to be systemically available. For systemic drug availability, drug dosing via the oral route requires the drug to exhibit sufficient gastrointestinal absorption. Two components of drug absorption from solid oral-dosage forms (e.g., tablets, capsules) are: (1) drug dissolution from the dosage form, which results in the drug being in solution within the gut lumen, and (2) drug permeation across the GI wall. In this chapter, elements of GI drug permeability are discussed, with special emphasis on the enterocyte and the in vitro Caco-2 cell model. Included is a discussion of the lipid plasma membrane, in vitro drug permeability and its relation to in vivo permeability, and a discussion of future research directions.

2. Methods to Study Intestinal Drug Absorption

Prior to discussing the in vitro Caco-2 monolayer model as a model for intestinal drug permeability, it should be recognized that numerous methods are available to study intestinal drug absorption. In Table 1, methods can be categorized as in vivo, *in situ*, and in vitro. Selection of any method will depend on several considerations, including the question to be answered and the availability of resources. Methods range in degree of integration and complexity.

2.1. In Vivo Methods

In vivo systems, such as whole animal or human studies, to measure oral drug bioavailability can be considered the most complex experimental systems, because all potential barriers (e.g., dissolution, intestinal permeability, metabolism) are present to limit drug absorption. Consequently, mechanistic interpretations from such complex systems is often limited, if not impossible. Rather, the main advantage of in vivo studies is their ability to answer specific, often practical questions. In vivo studies are used to measure absolute oral drug bioavailability, determine bioequivalence of one product to another, and assess pharmacokinetic impact of a formulation change.

From: *Cellular Drug Delivery: Principles and Practice*
Edited by: D. R. Lu and S. Øie © Humana Press Inc., Totowa, NJ

Table 1
Hierarchy of Methods to Study Intestinal Drug Permeability

In vivo	Bioavailability; human bioequivalence; scintigraphy study; human intestinal perfusion; fistulated animals
In situ	Intestinal perfusion (various formats); chronically isolated intestinal loop
In vitro	Cell cultures (e.g., cell monolayers; cell suspensions), including transfected cells; excised tissue (e.g., tissue flap, everted sacs, intestinal rings); artificial lipid membranes (e.g., parallel artificial membrane, black lipid membrane); membrane vesicles (e.g., brush-border membrane vesicles, basolateral membrane vesicles).

In vitro systems, such as the Caco-2 cell monolayer system, benefit from relative simplicity in design, allowing for mechanistic interpretation. However, this lack of integration with other physiologic and biochemical components that can influence overall drug oral bioavailability, and naturally limits the scope of interpretation for such simple systems. A panel of multiple in vitro tests is typically required to estimate oral drug bioavailability, or delineate the major mechanism of absorption.

These in vivo designs can be supplemented with scintigraphy, to allow for imaging of the dosage form, in order to relate observed pharmacokinetics with the disposition of the dosage form during intestinal transit. For example, scintigraphy elucidated the role of small intestinal residence time as a contributing factor in oral saquinavir bioavailability *(1)*. In an in vivo study of saquinavir capsules in humans, saquinavir bioavailability was greatly enhanced when the drug was co-administered with food, relative to the drug without food. Scintigraphy data allowed for the relating of saquinavir pharmacokinetic profiles to the temporal disposition of the drug in the stomach and small intestine. Data indicated that that drug absorption was enhanced when drug transit to the colon was retarded. Food increased saquinavir retention in the stomach, in part by allowing more time for the drug to dissolve, and hence be absorbed. It should be noted that fasting data also indicated that drug absorption was enhanced when drug transit to the colon was retarded, since drug absorption was essentially zero after the drug reached the colon.

In addition to scintigraphy, another technique that can be noted as an in vivo method to study intestinal drug absorption is the human intestinal perfusion technique. Through the efforts of a few research groups, various human perfusion approaches to measure drug intestinal permeability have been developed, including the double-balloon method *(2)*. In the double-balloon method, human subjects are intubated with a multilumen tube (i.e., an outer tube containing several parallel tubes inside the outer tube). After a subject is intubated with the multilumen tube, pressurized air is introduced into two of the tube lumens, in order to inflate two balloons. The two balloons, which are positioned about 10 cm apart, create an isolated intestinal segment, which can subsequently be perfused with drug solution by pumping drug solution through a third lumen of the tube. The perfused drug solution can be collected from yet another tube lumen(s). Drug permeability across the intestinal mucosa can be calculated, by considering initial and final drug concentrations. The intestinal segment most frequently studied is the jejunum; the colon has also been studied. In the last decade, a range of drugs has been subjected to the human jejunal perfusion technique, providing in vivo drug-permeability values. In Subheading 5., permeability values from the Caco-2 method will be compared to values from human perfusion. Fistulated animals are also an in vivo method noted in Table 1. For example, duodenally fistulated dogs (i.e., dogs with an external access port to the duodenum) can be used to study absorption and/or stability within the duodenum.

2.2. In Vitro Methods

At the other end of the hierarchical spectrum of techniques, in vitro systems are preferentially employed to address mechanistic questions, as well as perform high-throughput screening, as described in Chapter 17. In vivo systems represent highly integrated systems and thus are inherently handicapped to address mechanistic questions, although they offer an excellent approach to obtain practical pharmacokinetic data. By virtue of their design, the strength of in vitro systems is their simplicity, which is not to suggest that such experiments are readily performed. Perhaps the most complex in vitro methods are excised tissues and cell cultures. In vitro platforms with greater simplicity are artificial lipid membranes and membrane vesicles.

Numerous cell-culture models have been applied to study and estimate drug permeability. In Chapter 7, several cell-culture models of carrier-mediated transport are described. Two common cell-culture systems to study drug permeability are Caco-2 cell monolayers and Madin-Darby canine kidney (MDCK) cell monolayers. Each system, particularly Caco-2 cell monolayers, is described in greater detail in Subheading 3. It should be noted that a main reason for the utility of Caco-2 and MDCK cells is that each is an adherent cell type that forms a "tight" monolayer. By virtue of being adherent cells (i.e., cell that preferentially grow on surfaces), each can be grown on porous supports. Because of the near-daily attention required to culture cells, excised tissue remains a perhaps more convenient alternative; tissue flap, everted sacs, and intestinal rings are three means to employ intestinal tissue. Artificial lipid membranes include the parallel artificial membrane method (*see* Chapter 17), where a series of lipid bilayers are formed within a filter support, which is subjected to drug flux. A black lipid membrane (BLM) is composed of a single lipid bilayer, and perhaps represents the most simplified system to study membrane permeability, although BLM studies are technically challenging to construct.

2.3. *In Situ* Methods

Latin for "in place," *in situ* lies between in vivo and in vitro, and is exemplified by methods where drug solution is perfused through the rat intestinal lumen. Several experimental variations of the rat intestinal perfusion method are practiced, such as the single-pass perfusion, the single-pass perfusion with venous collection, and the multiple-pass perfusion. By virtue of being an *in situ* method, the animal is alive during the course of the study, but must be sacrificed after the experiment.

Denoted here as an *in situ* method, the chronically isolated intestinal loop is similar to the perfusion methods mentioned previously because the intestine is suitably prepared and perfused to determine intestinal drug permeability. However, this method is unique because the animal is conscious and is not sacrificed at the end of the experiment. In the chronically isolated intestinal loop method, a segment of the intestine is isolated and externalized to the abdominal wall. The isolated segment is then connected to two cannulas, inserted through the skin and attached to the abdominal wall. The remaining intestine is rejoined by end-to-end anastomosis.

3. The Enterocyte and the Caco-2 Cell Model

3.1. Morphology

The GI tract is composed of the mouth, pharynx, esophagus, stomach, small intestine, large intestine, and rectum, and functions to absorb food, nutrients, and water. Arguably, the most important component of the GI tract is the small intestine, because the majority of food digestion and nutrient absorption (and drug absorption) occurs there. The small intestine is a tubular structure with an approximate radius of 1.5 cm. Drug is absorbed from the tube lumen, which can be considered external to the body. As a result, the small intestine is protected by an epithelial barrier, to guard the underlying serosal tissue and avoid systemic exposure of toxic xenobiotics. Epithelial cells of the small intestine are predominately composed of absorptive cells called enterocytes.

Figure 1 illustrates the small intestine at the organ level, the enterocyte level, and three levels in between *(3)*. In panel 1, circular folds and longitudinal and circular muscles facilitate absorption by enhancing surface area and mixing action, respectively. In panel 2, the surface area of the circular folds is enhanced by villi. In panels 3 and 4, the villus is covered by a layer of enterocytes; other specialized cells, which function in mucous secretion, endocrine secretion, and immunity, are also present. In the small intestine, cells at the base of the villi undergo mitosis. These cells subsequently differentiate, as they migrate to the top of the villi, replacing older cells, and release into the lumen. The center of each villus is occupied by lymphatic vessels, which combine with other lymphatic vessels from other villi to form a lymph duct. The lymphatic vessel within a villus is called a lacteal. Also within a villus is a microcirculation system (i.e., arterioles, capillaries, and venules), which is situated around the lacteal. From panel 4, enterocytes present the main resistive barrier between the gut lumen and the blood and lymph, and present a physical and biochemical barrier to permeation. The physical barrier is constituted by the cell membrane and the presence of tight junctions, which restrict transcellular and paracellular permeability of solutes, respectively. The presence of various metabolic enzymes (e.g., Cytochrome P450) and efflux proteins (e.g., P-glycoprotein [P-gp]) contribute toward the biochemical barrier.

The absorbed nutrients and drugs enter the blood capillaries, while absorbed fat enters the lacteals. The enterocyte is illustrated in panel 5. Cell structures and their functions are described in Chapter 1. Enterocytes are columnar and highly polarized, with apical and basolateral membranes that are separated by tight junctions. Microvilli on the apical membrane (or mucosal side) allow for enterocytes to function as absorptive cells. These finger-like projections enhance the absorptive surface area in a multiplicative fashion, as microvilli further amplify the effects of villi and circular folds.

Since the early 1990s, the Caco-2 cell monolayer model has been and continues to be a frequently employed model system to answer questions and test hypotheses concerning intestinal drug absorption. Caco-2 cells were derived from a primary colon tumor from a 72-yr-old man *(4)*. Like most tumor cell lines, Caco-2 cells possess a transformed phenotype, which has given rise to an immortalized cell line. Daughter cultures of a continuous cell line have similiar growth characteristics and genetic properties as the parent culture. The Caco-2 cell monolayer model initially became a model of interest because Caco-2 monolayers mimic enterocytes, in terms of cell morphology and biochemistry. Most notably, these cells can be cultured with ease in vitro and undergo spontaneous, enterocyte-like differentiation after reaching confluence *(5)*. In fact, Caco-2 cells are frequently used as a model to study cell differentiation. Efforts to

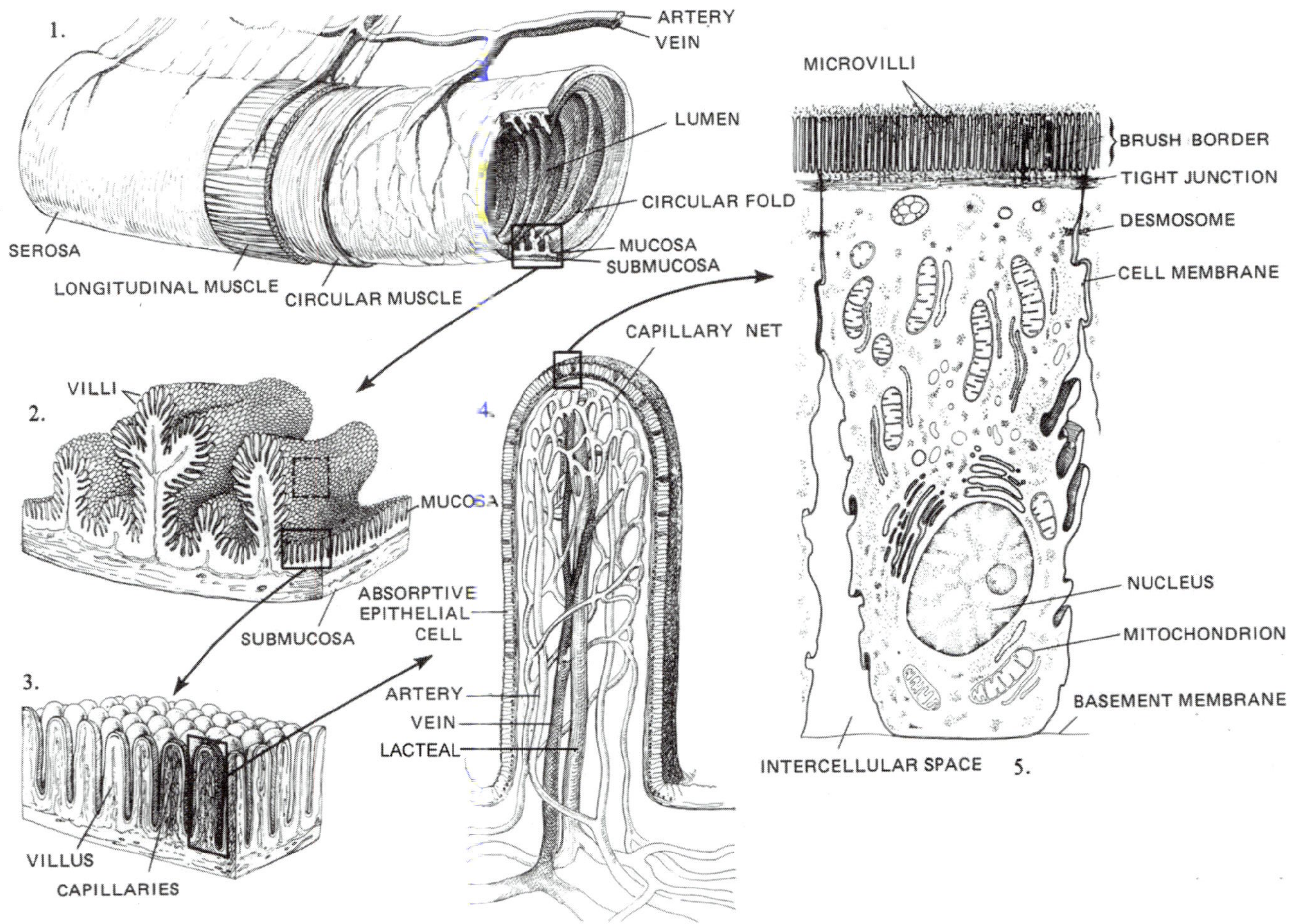

Fig. 1. Increasingly magnified views of the small intestine. Adapted with permission from ref. *3*.

culture noncancerous enterocytes is technically challenging, because cells dedifferentiate (i.e., lose their specialized properties in vitro). Although these cells are colorectal adenocarcinoma cells, Caco-2 cells are broadly representative of enterocytes (e.g., form a tight monolayer of polarized, columnar cells with microvilli, containing brush-border enzymes such as peptidases, disaccharidases, and alkaline phosphatase) *(6)*. These cells are an enterocyte model in many research areas, from cellular biochemistry and signaling to cellular physiology. These cells have also been used to study the transport of various nutrients and endogenous compounds, including amino acids, bile acids, and neurotransmitters *(7)*. These studies indicate the expression of a variety of active transport systems by these cells similar to those expressed in the small intestine, which help in the translocation of various nutrients. In addition, these cells also express various receptors involved in the endocytotic transport pathways, such as vitamin B_{12}.

However in spite of these similarities, there are significant differences between the in vivo intestinal epithelium and the Caco-2 system, because a simple tumor cell line is inherently unable to duplicate all attributes of a normal cell, much less a collection of cells that form a complex epithelium. Another inherent challenge is the maintenance of phenotypic stability of cultured cells, which varies with each passage *(8)*. Phenotypic stability is also affected by differences in culture conditions.

As described later, recent research efforts have focused on: (1) differences in physical barrier function between the Caco-2 system and the small intestine, (2) Caco-2 expression properties relevant to drug transport and metabolism, (3) methods to induce Caco-2 biochemical barrier function, and (4) cell-culture methods to facilitate phenotypic stability and rapid assay.

3.2. Differences in Physical Barrier Function Between the Caco-2 System and the Small Intestine

3.2.1. Relative Homogenous Nature of the Caco-2 Model and Absence of Mucus

The intestinal epithelium comprises of a heterogeneous population of cells, including enterocytes as well as goblet and vacuolated cells, tuft cells, microfold cells, theliolymphocytes, Paneth cells, and basally granulated cells. The enterocytes and goblet cells represent the two major phenotypes in the intestinal epithelium *(9)*. Mucus-producing goblet cells are of particular potential interest with regard to drug absorption, but are absent in the Caco-2 system. Absence of mucin is just one of the many imperfections of the Caco-2 system, in terms of its suitability as a comprehensive model for the in vivo epithelium. The role of mucus as a diffusion barrier to drug absorption is not clearly understood, owing to the limited number of studies and conflicting results observed so far *(10)*. HT-29 cells are a mucus-producing goblet-cell line that potentially may address this limitation. These cells form monolayers with a large proportion of mature goblet cells, which secrete mucin to form a layer of mucus at the apical-cell surface. A limited number of studies have also grown co-cultures of Caco-2 cells with HT-29 goblet-cell lines and apparently have shown limited success *(9)*.

3.2.2. Paracellular Permeability

Two major routes for passive drug transport across an epithelial monolayer are the paracellular and transcellular pathways. The transcellular pathway refers to the movement of solutes through the cell via the apical and basolateral plasma membranes. The paracellular transport refers to the passage of solute molecules through the intercellular spaces or pores between cells. Paracellular permeability is a function of the intercellular spaces, which are gated by tight junctions. This barrier function of the tight junctions is a significant factor in the transport of hydrophilic drugs across epithelial barriers. Caco-2 monolayers exhibit relatively low paracellular permeability (10–100 times lower) compared to human intestinal permeability *(11)*. This difference can result in underestimation of permeabilities for compounds with low transcellular permeability. Nevertheless, Caco-2 cells have been used to study qualitative and quantitative aspects of paracellular transport. There have also been efforts to develop alternate cell lines that exhibit paracellular permeabilities comparable to human intestine. A majority of these efforts focus on goblet-cell lines (e.g., HT29) or co-cultures *(9)*. Recent efforts use normal intestinal cells expressing a temperature-sensitive mutant of Simian Virus 40 (SV40) T oncogene (2/4/A1). These cells are based on a temperature-switch mechanism, where a change in culture temperature promotes cell differentiation, such that the cells exhibit high paracellular permeability.

3.3. Caco-2 Expression Properties Relevant to Drug Transport and Metabolism

Historically, the first models of drug permeability assume passive permeability, with little attention focused on active transport and the contribution of intestinal metabolism, largely because models for active transport and intestinal metabolism were lacking. Also lacking was an understanding about the interplay between the various components of the biochemical barrier to the absorption of xenobiotics. Cell-culture models now provide critical tools to uncover and understand mechanisms controlling this drug transport. Mechanisms include influx transporters, efflux transporters, and intestinal metabolism.

3.3.1. Influx Transporters

In concert with broad developments in gene-based advances, a significant amount of research has been performed in the last decade to identify and characterize previously unknown transporters that serve to enhance absorption (i.e., influx transporters). There are transport systems for amino acids, peptides, [illegible] acids, monocarboxylic acids, organic cations, organic anions, phosphates, nucleosides, and vitamins *(12,13)*. Table 2 lists several influx transporters in Caco-2 cells. Several transporters have been cloned and studied in detail and play a major role in determining the intestinal absorption of xenobiotics. With the commercial success of the anti-viral prodrug valacyclovir, researchers have made significant progress to understand the role of transport proteins in drugs absorption. For example, a large body of work has been reported on the peptide transporter and its various isoforms. Peptide transporters are known to play an important role in the oral absorption of various classes of drugs, including β-lactam antibiotics and ACE inhibitors. The peptide transporters have also been exploited in a prodrug fashion to improve the intestinal uptake of the otherwise poorly absorbed nucleoside analogs acyclovir and ziduvidine *(14)*. In contrast to nutrient transporters, which often exhibit narrow substrate specificity, other transporters exhibit broad substrate specificity, such as monocarboxylic acid transporters (MCTs) *(12)*.

Although the Caco-2 system expresses a significant proportion of the influx transporters, this system does not reflect the degree of expression of active transporters in vivo. Many studies indicate that the level of transporter expression in Caco-2 monolayers differs by more than fivefold *(15,16)* relative to their expression in vivo levels. The level of expression of influx transporters is typically low in vitro, resulting in an under-

Table 2
Representative Influx Active Transporters in Caco-2 Cells

Transporter	Natural substrate	Drug substrates
Large neutral amino acid system	Valine	Gabapentin, L-dopa, α-methyldopa
Pept1	Small peptides	β-lactam antibiotics, valacyclovir, ACE inhibitors
Glucose transport systems	Glucose	Dehydroascorbic acid
Monocarboxylic acid transport system	Lactic acid Short-chain fatty acids	Atorvastatin, Salicylic acid
Inorganic phosphate transport system	Inorganic phosphates	Foscarnet
Bile-acid transport system	Bile acids	Chlorambucil-bile acid conjugates (S2521)
Organic cation transport system	Carnitine	Cimetidine
Multivitamin transport system	Biotin, pantothenic acid	Tat9-K-biotin (bioactive peptide conjugated with biotin)

estimation of drug permeability. It should be noted that in vivo expression varies. For example, a significant correlation was observed between Pept1 expression and cephalexin permeability, indicating that variability in the intestinal expression of Pept1 is a major determinant of variable in vivo cephalexin absorption *(16)*. Moreover, in vitro transporter expression level can be dependent on passages and age of the culture *(8)*. Not surprisingly, considerable variability is observed between different labs. The transport of bile acids across Caco-2 cells varies almost 20-fold with regard to the contribution of the transporter, with one report showing a 60-fold enhance of bile-acid permeability owing to the transporter, whereas another report indicates only a 3.6-fold effect *(7,17,18)*. Similarly, the activity of the amino acid transporter differs by as much as fivefold between labs when comparing bi-directional transport ratios.

3.3.2. Efflux Transporters

The intestinal epithelium expresses a variety of efflux transporters and metabolic enzymes, which constitute a biochemical barrier to drug absorption. Efflux serves as a detoxification system in the small intestine and limits the absorption of toxic xenobiotics from the small intestine. Several efflux transporters are members of the adenosine 5'-triphosphate (ATP)-binding cassette (ABC) transporter family and are detailed in Chapter 8. Efflux proteins generally exhibit broad substrate specificity and high capacity. The low and variable bioavailability of drugs such as digoxin and HIV protease inhibitors is partly owing to efflux transporters. Owing to overlapping substrate specificity, various efflux proteins may operate in parallel to limit intestinal drug absorption in a complex manner. Caco-2 cells express several efflux transporters such as P-gp and multiple drug resistance proteins (MRPs). Expression levels vary significantly and can depend on age in culture, culture conditions, and previous drug exposure *(19)*.

A few studies correlate level of expression in the Caco-2 system with in vivo human tissue *(20,21)*. Taipalensuu et al. *(20)* compared the expression of 10 members of the ABC transporter family in Caco-2 cells to expression levels in human jejunum. Results indicate that the Caco-2 system expresses most of the genes responsible for efflux proteins at levels similar to those observed in human jejunum. In most cases, the Caco-2 values were within a factor of 2.5 of human jejunum values. The exception was BCRP, which was found to be expressed at very high levels in the jejunum compared to Caco-2. The in vivo level of expression of BCRP was even higher than the transcript for MDR1. This finding indicates that the role of BCRP may have a greater role in limiting drug absorption than previously thought. Indeed, studies with mice have indicated that BCRP has a major role in limiting the bioavailability of topotecan. Similar findings were reported by Anderle et al. *(15)*, who examined mRNA expression of various transporter and ion-channel genes using a microarray chip and compared the transcript levels between Caco-2 and human intestine. Their study indicated lack of significant differences in the expression levels of ABC transporters in Caco-2 cells and human intestine. Minor differences were seen in expression levels of Pept1, with a higher level in small intestine than Caco-2 cells. A more comprehensive study has been carried out by Sun et al. *(21)*, who compared the gene expression profiles of Caco-2 cells and human duodenum for various transporters, channels, and metabolic enzymes. They further related these results to drug permeability differences between human duodenum and the Caco-2 system. Their study indicates at least a 10-fold difference in influx transporter-expression levels between Caco-2 and small intestine. P-gp was expressed at a much higher level in the duodenum than Caco-2 cells; MRPs were expressed at comparable levels between the Caco-2 and small intestine.

3.3.3. Intestinal Metabolism

Metabolic enzymes comprise another biochemical barrier to drug absorption and can be broadly classified into two main groups: the phase I oxidative cytochrome P-450s (CYPs) and the phase II conjugative enzymes (e.g., UDP-glycosyltransferases [UGTs] and sulfotransferases [SULTs]). These enzymes can present a major barrier, particularly in coordination with efflux transporters.

CYP3A4 is the principal component of the oxidative enzymes in the small intestine. Intestinal CYP3A4 contributes significantly to the poor bioavailability of certain drugs. As both CYP3A4 and P-gp functions to protect the body from toxic xenobiotics, there is substantial over lap in substrate specificity between these two systems (e.g., cyclosporin, verapamil, etoposide). Likewise, CYP3A4 inhibitors, such as ketoconazole and erythromycin, significantly inhibit the activity of P-gp.

It has been proposed that P-gp not only functions as a transport barrier but also promotes intestinal metabolism by CYP3A4, resulting in a synergistic effect. Two mechanisms have been suggested to explain the enhancement of metabolism by P-gp. One entails the repeated cycling of the compound in the intestine, such that there is increased exposure of drug to intestinal CYP3A4. An alternate mechanism points to P-gp effluxing primary metabolites from the cell, thus avoiding product inhibition of the enzymes.

Glucouronidation and sulphate conjugation are two major phase II metabolic pathways in enterocytes and are mediated by UGTs and SULTs, respectively. Phase II conjugation enzymes usually work in conjunction with organic anion efflux mechanisms in restricting the absorption of drugs. Flavanoids are one class of compounds whose absorption is adversely affected by the activity of these enzymes. Moreover, these compounds are reported to promote their own metabolism by the inducing expression of these enzymes. Once flavanoids are conjugated inside the enterocytes, their metabolites are usually extruded from the enterocytes via organic anion transporters.

3.4. Methods to Induce Caco-2 Biochemical Barrier Function

In vitro cell culture has potential to predict drug interactions. Specific efforts have been directed towards understanding drug interactions involving drug absorption, with particular interest in the role of metabolic enzymes and efflux transporters. However, systems such as Caco-2 cells do not express significant amounts of CYP3A4 and phase II metabolic enzymes. Several approaches have been taken to circumvent this limitation. One method is to stably transfect Caco-2 cells or other cells with cDNA encoding for CYP3A4. Another approach is enzyme induction by exposing cells to inducing agents. Treatment of Caco-2 cells with 1-α,25-dihydroxy vitamin D_3 is used to induce CYP3A4 expression, as well as P-gp, in a dose dependent fashion *(22)*. UGT has been induced by the aryl hydrocarbon receptor agonist t-butylhydroquinone (TBHQ) and the anti-oxidant type inducer 2,3,7,8-tetrachlorodibenzo-*p*-dioxin (TCDD). This induction is isoform-specific, with UGT1A6 and UGT1A9 being inducible by TBHQ, and UGT1A6, 1A9, and 2B7 responding to TCDD.

Recent studies indicate the potential of drugs to induce or repress metabolic enzymes expression, which can result in modulated exposure of other co-administered drugs. For example, P-gp and CYP3A4 induction has been proposed as a mechanism to explain decreased verapamil absorption after repeated oral rifampin administration. Similarly, reduced HIV protease-inhibitor exposure after repeated administration has been ascribed to self-induction of intestinal P-gp and CYP3A4. Dietary components like grapefruit juice and other citrus fruits have been observed to increase oral drug bioavailability by downregulation of CYP3A4 and P-gp inhibition.

To study such drug interactions, subclones of Caco-2 (e.g., TC-7) and other cell lines (e.g., LS-180) have been used. These alternatives may prove useful to predict induction of the biochemical barrier by drugs and the resultant potential for drug interactions. In recent years, orphan nuclear receptors such as the pregnane X receptor (PXR), constitutive androgen receptor (CAR), arylhydrocarbon receptor (AhR), retinoic acid receptor (RAR), and farsenoid X receptor (FXR) have been established as xeno-sensors that regulate the expression of phase I and II enzymes and efflux transporters. This transcriptional regulation is achieved by the binding of these xenobiotic receptors to the nuclear-response elements found within the promoter regions of target genes. There is significant evidence indicating common regulatory mechanisms at the molecular level controlling the expression of metabolic enzymes and efflux proteins. A study of these controlling elements will be of significant importance in delineating the expression variability of efflux proteins and metabolic enzymes in different individuals. For example, the CYP3A4 and P-gp share a common regulatory-expression control through PXR, which partly explains why inducers of CYP3A4 expression increase P-gp *(23)*.

3.5. Cell-Culture Methods to Facilitate Phenotypic Stability and Rapid Assay

Traditional Caco-2 cell culture for permeability studies employs a rather generic cell-culture approach, resulting in a 3-wk period to attain functional, differentiated monolayers. Medium changes during this culture period are cumbersome and expensive. Moreover, the use of 10% fetal bovine serum (FBS), which varies in composition from lot to lot, contributes to phenotypic instability. Owing to these reasons, the traditional Caco-2 system is typically a low-through-put system that is susceptible to phenotypic drift. We developed a rapid culture system for growing Caco-2 cell monolayers that reduces the need for serum *(24)*. In this system, confluent and polarized Caco-2 cell monolayers are formed in 4 d. One salient feature of this system is the use of reduced serum (2% calf serum compared to 10% FBS), which results in reduced variability (usually associated with serum composition) and significantly lowers costs. Another advantage of this system is the reduction in transporter expression, which better allows the estimation of passive permeability. Liang et al. found this system to provide permeability values comparable to the 21-d system, with a low level of P-gp expression *(25)*. This model has been proposed as a rapid, accurate, and economical mean of classifying drugs candidates within the Biopharmaceutics Classification System (BCS).

The MDCK model is a nonintestinal epithelial cell line that also serves as a rapid approach to measuring passive permeability. A main advantage of MDCK cells is their adaptability to high-throughput screening. Native fast growers, MDCK cells may be used for permeability studies in as little as 3 d. MDCK monolayers are a model for studying tight-junction functioning and provide low paracellular permeability. Passive permeability values from 3-d MDCK cells are comparable to Caco-2 cells *(18,26)*. Additionally, MDCK monolayers express a subset of several major transporters, such as peptide transporter, P-gp and MCTs *(18)*.

MDCK monolayers have been used to stably overexpress specific transporters to develop simplistic models for the specific transporter in question. Relative to a system such as Caco-2 that expresses a variety of transporters, the advantage of a transfection approach is the resulting high expression of the single transporter, essentially without confounding effects of other transporters. MDCK cells transfected with the MDR1 gene (i.e., MDR1-MDCK cells) are often used to study the effect of P-gp on drug permeability. Although only one transporter is typically overexpressed, doubly transfected systems have been generated to study nucleoside transport.

4. Cell Plasma Membrane and Drug Permeability

4.1. The Lipid Bilayer

The apical plasma membrane is generally thought to be the main barrier for passive drug absorption, as the apical membrane defines the interface between self and the external environment. The plasma membrane is a bilayer, composed primarily of amphipathic phospholipids, as well as other components (e.g., proteins, sugars, cholesterol). Phospholipids contain a hydrophilic phosphate head and two hydrophobic fatty-

Table 3
Phospholipid Composition of Apical Membranes of the Small Intestine

Phospholipid	Percent of total phospholipid			
	Pig	Rat	Rabbit	Mouse
Phosphatidylethanolamine	34.0	38.3	30.3	49.1
Phosphatidylcholine	34.0	32.6	35.0	25.1
Phosphatidylserine	12.0	13.0	7.5	6.0
Phosphatidylinositol	10.0	8.7	8.3	11.6
Sphingomyelin	10.0	6.4	10.5	8.3
Lysophosphatidylethanolamine	—	—	3.	—
Lysophosphatidylcholine	—	1.0	3.1	—
Phosphatidic acid	—	—	1.2	—
Cardiolipin	—	—	—	—
Phosphatidylglycerol	—	—	—	—

Table 4
Fatty Acid Tail Composition of Apical Membranes in Rat Colon

Fatty acid	Composition (%w/w)
16:0	28.0
16:1	2.5
18:0	19.2
18:1	24.0
18:2	15.2
20:4	11.2

acid tails. The phosphate head group is polar, with the phosphate functional group being negatively charged. Table 3 lists several types of phospholipid-head groups found in the apical membranes of various animals *(27)*. Phosphatidylethanolamine and phosphatidylcholine are the two most common phospholipids, with each phospholipid possessing a net neutral charge. Phosphatidylserine carries a net negative charge. Phospholipids also differ in fatty-acid tail composition. Because there are two chains present per phospholipid, the tails can exhibit symmetry or nonsymmetry (i.e., identity or nonidentity in fatty-acid tail composition). Each fatty-acid tail is characterized by its carbon-chain length, as well as its degree of unsaturation. The vast majority of fatty-acid tails are 16, 18, or 20 carbons in length. Unsaturation generally can occur between carbon 9 and 10, and 12 and 13. In Table 4, two common phospholipid fatty-acid tails are 16:0 and 18:1 *(28)*. These designations denote a fatty-acid tail of length 16 carbons and no double bond, and fatty-acid tail of length 18 carbons and one double bond, respectively. Hence, a range of phospholipids, that differ in phosphate-head group and fatty-acid tails, exist.

The phosphate-head group and fatty-acid tails vary in chemical composition, such that phospholipid properties change, resulting in modulated membrane properties. Different head groups and tails alter lipid packing, membrane-surface charge, and fluidity or anisotropy. For example, the head group's bulkiness influences both phospholipid

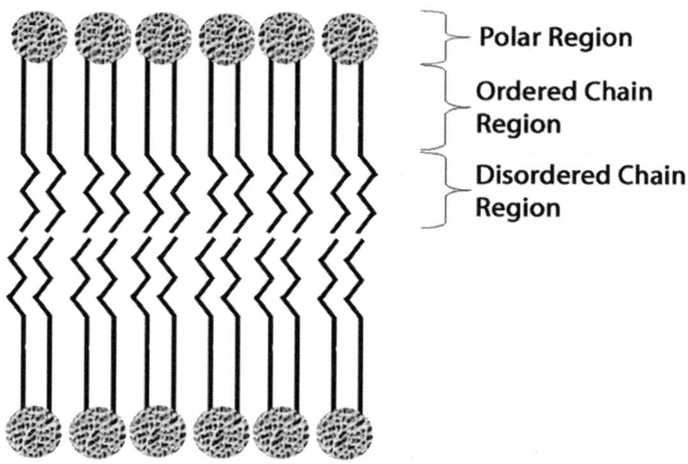

Fig. 2. Schematic of phospholipid bilayer and three regions in a bilayer leaflet.

packing density and nominal membrane charge. Also, fatty-acid tails composition impacts the lipophilic region of the lipid bilayer. Because fatty-acid double bonds are generally *cis* in conformation, unsaturation introduces sharp bends in the carbon chain, resulting in less membrane order and greater membrane fluidity.

Figure 2 is a schematic of phospholipid bilayer. Each lipid leaflet consists of three regions: the polar region, the highly ordered chain region, and the disordered chain region. The polar region consists of the hydrated phosphate-head group and the ester/ether linkages between the head groups and acyl chains. The highly ordered chain region is the fatty-acid chain region with no unsaturation. The disordered chain region includes the fatty-acid chain regions with unsaturation. Phospholipid composition changes impact these regions. For example, small phosphate-head group moieties and the addition of cholesterol each increase lipid-packing density, to result in a more rigid membrane. Of course, the charge of the phosphate-head group influences membrane charge. Also, fatty-acid saturation generally will reduce the degree of disorder in the disordered chain region.

4.2. Bilayer Properties and Passive Drug Permeability

A model that relates bilayer properties and solute permeability though the bilayer is the solubility-diffusion model. Both solute solubility in the bilayer and solute diffusivity across the bilayer dictate solute permeability across the bilayer. Solute solubility in the bilayer, as a determinant of permeability, has received greater attention. Typically, increased membrane lipophilicity—for example, by increasing fatty-acid chain length—enhances the permeability of lipophilic drugs. Meanwhile, reduced membrane lipophilicity—for example, by increasing membrane protein content—enhances the permeability of hydrophilic drugs. For various solutes, Overton's Rule suggests that solute permeability coefficients follow oil/water partition coefficients. The hydrocarbon domain of bilayers has been compared with various hydrocarbon solvents (e.g., hexane, dodecane, hexadecane, octanol, heptane). Additionally, the chain ordering in lipid bilayers, particularly in the ordered chain region, serves as a diffusional resistance and entropic barrier to partitioning. For example, after correcting for its hydrophobic effect, the incorporation of cholesterol into a membrane reduced solute permeability, owing to cholesterol's membrane-ordering effect *(29)*.

4.3. Bilayer Properties and Active Drug Transport

Plasma-membrane properties, such as membrane fluidity, affect not only passive drug permeability, but also the conformation and functioning of membrane-bound transporters. Because most of transport proteins are membrane-bound, the physical state of the cell membrane is generally important in transporter function. Hence, transporter-mediated permeability is modulated by changes in plasma bilayer. Several studies have investigated the influence of cholesterol content and cell-membrane fluidity on P-gp function. Fluidization usually resulted in decreased P-gp activity. Studies of the effect of rigidizing the cell membrane on P-gp activity have yielded contradictory results, with some reports suggesting an increase in P-gp activity, and others suggesting a decrease.

We previously studied the effect of three pharmaceutical surfactants on membrane fluidity and transporter function in Caco-2 cells *(30)*. Tween-80 and Cremophor EL increased fluidity in the hydrophobic region of Caco-2 plasma membranes. In contrast, vitamin E TPGS rigidized the hydrocarbon region. All three surfactants contained poly(ethylene oxide) and were effective in inhibiting P-gp, although Vitamin E TPGS showed a markedly different pattern of P-gp inhibition than Tween-80 and Cremophor EL. Tween-80 and Cremophor EL progressively increased AP-BL permeability of rhodamine 123 over a concentration range of 0–1 m*M*.

Vitamin E TPGS's effect was equally large, mainly owing to reduced BL-AP permeability of rhodamine 123. Moreover, its effect was already maximal at 0.025 m*M*. These P-gp inhibition effects would appear to be related to these excipients' ability to modulate membrane fluidity. Tween-80 and Cremophor EL fluidized lipid bilayers, whereas vitamin E TPGS rigidized lipid bilayers. However, among the three surfactants, only Tween-80 inhibited the peptide transporter. Likewise, only Cremophor EL inhibited the monocarboxylic acid transporter. Nevertheless, P-gp, the human intestinal peptide transporter, and the monocarboxylic acid transporter were each inhibited by at least one of these surfactants. A common functional feature of these surfactants was their ability to modulate fluidity, although the present studies indicate that strong membrane-fluidity modulation by itself was not sufficient to reduce transporter activity. The protein kinase C (PKC) inhibitors failed to inhibit transporter activity, suggesting inhibition of PKC was not the mechanism of surfactant-induced transporter inhibition. In total, these results suggest that surfactants can inhibit multiple transporters, but that changes in membrane fluidity itself are not sufficient to reduce transporter activity.

5. In Vitro Caco-2 Passive Drug Permeability and Its Relation to In Vivo

In addition to Caco-2 monolayers mimicking the gut epithelium, a second rationale for the use of Caco-2 monolayers is their ability to be correlated with in vivo oral absorption. Figure 3 exemplifies the relation between the oral percent dose absorbed in humans and Caco-2 permeability. Each data point is derived from measuring the Caco-2 permeability of a drug and plotting its permeability value against its oral percent dose absorbed from human pharmacokinetic studies. Each compound is known to be passively absorbed (i.e., not a substrate for active intestinal transport), not limited by dissolution, and not suspect to lumenal degradation. Similar plots appear throughout the literature *(24,31)*.

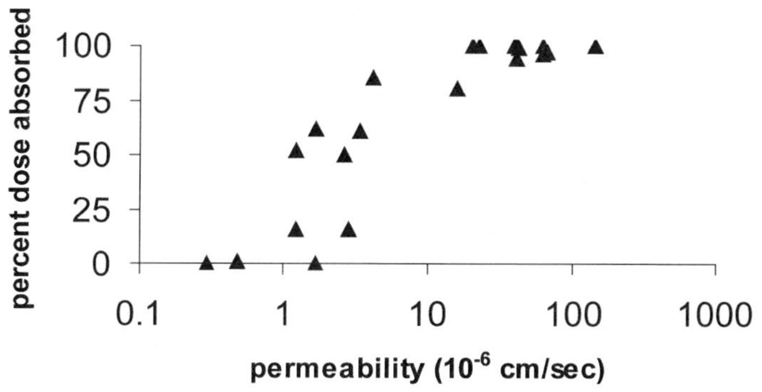

Fig. 3. Relationship between percent oral dose absorbed in humans and Caco-2 permeability.

Such plots are generally similar in permeability values and sigmoid curvature, with three general areas of the plot. On the left side of the plot with low-permeability values, a plateau persists with percent dose absorbed equal to about 0%; for a range of sufficiently small permeability values, such as permeability less than 1×10^{-6} cm/s, the percent dose absorbed is 0%. Likewise, a second area of the plot with sufficiently high permeability (e.g., permeability greater than 10×10^{-6} cm/s) exhibits a plateau with the percent dose absorbed equal to about 100%. In between these two areas, the percent dose absorbed increases approximately log-linearly with permeability (between 1×10^{-6} cm/s and 10×10^{-6} cm/s). As described in Subheading 3, even passive Caco-2 permeability values are subject to incompletely explained variation with culture conditions, time in culture, and passage number. Between laboratories, and also within a laboratory over the course of time, the greatest variation in plots such as Fig. 3 manifests in low-permeability values. For example, permeability value of the hydrophilic solute mannitol easily vary 10-fold between laboratories from 0.1×10^{-6} cm/s to 1×10^{-6} cm/s. This variability presumably is owing to variation in tight-junction formation and/or imperfection in monolayer integrity.

The two extremes of the sigmoid curve in Fig. 3 indicate that sufficiently low-permeability values result in essentially no drug absorption, and sufficiently high-permeability values result in essentially complete drug absorption. In the middle region of such plots, the general slope is about one, or even greater than one *(24,31)*. This characteristic indicates that Caco-2 monolayers mimic the relative passive permeability barrier function of the human intestine in a reasonable fashion. Absolute comparisons have been made between Caco-2 permeability and human jejunal permeability obtained from human perfusion *(2)*. For low-permeability compounds, Caco-2 values are typically about 50-fold less than human jejunal permeability values. This difference is attributed to a lower paracellular permeability and/or a larger surface area in vivo in humans. For high-permeability compounds, Caco-2 values are typically about threefold less than human jejunal permeability values.

It should be noted that plots such as Fig. 3 have intrinsic difficulty in showing a strong correlation (i.e., a tight degree of association) between Caco-2 permeability and percent dose absorbed. A main reason for a limiting correlation is the intrinsic variability is the percent dose absorbed parameter, which is obtained from human pharmacokinetic studies. For many drugs, particularly those with incomplete absorption, oral bioavailability is suspect to at least modest variability (e.g., %CV = 15%). Additionally, first-pass metabolism adds to the uncertainty of percent dose absorbed estimates. In contrast to the variability in y-axis values of Fig. 3, we have found the intra-day variability in Caco-2 permeability generally to be 10% or less *(30)*.

6. Future Directions

Many efforts have been directed to develop an ideal cell line to duplicate drug absorption across the small intestine. Of course, ideal characteristics depend on the application. Ideal characteristics in one circumstance may contrast desirable properties in other applications. For example, a general advantage of Caco-2 cells is the low native expression of metabolic enzymes, in order to measure permeability values without bias from metabolism. However, as described in Subheading 3., Caco-2 cells induced to express metabolic enzymes have been described as a model system for coupled transport and metabolism.

Given the explosion in our understanding of drug absorption across epithelia, particularly transporter-mediated absorption, the need for an even wider array of transport models continues, particularly at the molecular level for specific individual processes. Future approaches should include the development of singly and co-transfected cell lines to facilitate a better understanding of the interaction between the drugs and cell proteins (e.g., transporter proteins or metabolic enzymes), along with any co-factors. Practical applications of such cell lines with high selective expression will be the design and development of new therapeutic agents with acceptable pharmacokinetic properties. Also, reporter-gene assays are needed for drug-interaction studies that involve induction or inhibition of enzymes and transporters by drugs and dietary components. Additionally, the identification of nuclear receptors as xeno-sensors represents a major step forward in understanding the genetic mechanisms controlling the expression of drug-metabolizing enzymes. Such efforts may potentially explain intra-individual and inter-individual variability in drug bioavailability. There is tremendous potential for a cell-based screening system incorporating the role of nuclear receptors to mechanistically screen for potential drug candidates with low drug-interaction potential, thus facilitating the development of safer drugs. Such systems will further lead to the development of therapeutic options based on an individuals genetic profile.

Future progress is also needed in our understanding of passive permeability, including the role of the lipid bilayer. Although the solubility-diffusion model provides a basis for understanding the interacting roles of drug and lipid bilayer in determining drug permeability through a specific lipid bilayer, it is clear that much remains to be understood. Moreover, given the diversity of plasma-membrane composition in different organs, tissues, and even different parts of the same cell, it can be expected that future efforts will explore the effects of these differences on drug permeability and organ disposition, and their basis. Furthermore, a future direction is the elucidation of excipients on passive and active drug transport, with the aim to control and modulate drug bioavailability and disposition.

Finally, the need to better integrate our understanding of drug permeability with other arenas continues, including applied areas of drug design and development. For example, the FDA Biopharmaceutics Classification System (BCS) is widely viewed as conservative, with potential to extend biowaivers to highly soluble drugs with moderate (rather just simply high) permeability. One major limitation in this regulatory consideration is a comprehensive analysis of allowing biowaivers of moderately permeable drugs. A comprehensive analysis needs to consider the risk of formulation changes to result in bioinequivalence for moderately permeable drugs. Although permeability models have been related to pharmacokinetic measures such as percent dose absorbed, and although some studies have assessed the potential for excipients to modulate drug permeability, permeability is only infrequently considered in assessing the overall robustness of the drug product.

References

1. Kenyon CJ, Brown F, McClelland GR, Wilding IR. The use of pharmacoscintigraphy to elucidate food effects observed with a novel protease inhibitor (saquinavir). Pharm Res 1998;15:417–422.
2. Lennernas H. Human intestinal permeability. J Pharm Sci 1998;87:403–410.
3. Friedman MH. Principles and Models of Biological Transport. Berlin, Springer-Verlag, 1986:216.
4. Fogh J, Fogh JM, Orfeo T. One hundred and twenty-seven cultured human tumor cell lines producing tumors in nude mice. J Natl Cancer Inst 1977;59:221–226.
5. Pinto M, Robine-Leon S, Appay MD, et al. Enterocytic like differentiation and polarization of the human colon carcinoma cell line Caco-2 in culture. Biol Cell 1983;47:323–330.
6. Hidalgo IJ, Raub TJ, Borchardt RT. Characterization of the human colon carcinoma cell line (Caco-2) as a model system for intestinal epithelial permeability. Gastroenterology 1989;96:736–749.
7. Hidalgo IJ, Borchardt RT. Transport of bile acids in a human intestinal epithelial cell line, Caco-2. Biochim Biophys Acta 1990;1035:97–103.
8. Yu H, Cook TJ, Sinko PJ. Evidence for diminished functional expression of intestinal transporters in Caco-2 cell monolayers at high passages. Pharm Res 1997;14:757–762.
9. Walter E, Janich S, Roessler BJ, et al. HT29-MTX/Caco-2 cocultures as an in vitro model for the intestinal epithelium: in vitro-in vivo correlation with permeability data from rats and humans. J Pharm Sci 1996;85:1070–1076.
10. Pontier C, Pachot J, Botham R, Lenfant B, Arnaud P. HT29-MTX and Caco-2/TC7 monolayers as predictive models for human intestinal absorption: role of the mucus layer. J Pharm Sci 2001;90:1608–1619.
11. Artursson P, Ungell AL, Lofroth JE. Selective paracellular permeability in two models of intestinal absorption: cultured monolayers of human intestinal epithelial cells and rat intestinal segments. Pharm Res 1993;10:1123–1129.
12. Tsuji A, Tamai I. Carrier-mediated intestinal transport of drugs. Pharm Res 1996;13:963–977.
13. Yang C, Tirucherai GS, Mitra AK. Prodrug based optimal drug delivery via membrane transporter/receptor. Expert Opin Biol Ther 2001;1:159–175.
14. Han H, de Vrueh RL, Rhie JK, et al. 5'-Amino acid esters of antiviral nucleosides, acyclovir, and AZT are absorbed by the intestinal PEPT1 peptide transporter. Pharm Res 1998;15:1154–1159.
15. Anderle P, Rakhmanova V, Woodford K, et al. Messenger RNA expression of transporter and ion channel genes in undifferentiated and differentiated Caco-2 cells compared to human intestines. Pharm Res 2003;20:3–15.
16. Chu XY, Sanchez-Castano GP, Higaki K, et al. Correlation between epithelial cell permeability of cephalexin and expression of intestinal oligopeptide transporter. J Pharmacol Exp Ther 2001;299:575–582.
17. Chandler CE, Zaccaro LM, Moberly JB. Transepithelial transport of cholyltaurine by Caco-2 cell monolayers is sodium dependent. Am J Physiol 1993;264:G1118–G1125.
18. Putnam WS, Ramanathan S, Pan L, et al. Functional characterization of monocarboxylic acid, large neutral amino acid, bile acid and peptide transporters, and P-glycoprotein in MDCK and Caco-2 cells. J Pharm Sci 2002;91:2622–2635.
19. Anderle P, Niederer E, Rubas W, et al. P-Glycoprotein (P-gp) mediated efflux in Caco-2 cell monolayers: the influence of culturing conditions and drug exposure on P-gp expression levels. J Pharm Sci 1998;87:757–762.

20. Taipalensuu J, Tornblom H, Lindberg G, et al. Correlation of gene expression of ten drug efflux proteins of the ATP- binding cassette transporter family in normal human jejunum and in human intestinal epithelial Caco-2 cell monolayers. J Pharmacol Exp Ther 2001;299:164–170.
21. Sun D, Lennernas H, Welage LS, et al. Comparison of human duodenum and Caco-2 gene expression profiles for 12,000 gene sequences tags and correlation with permeability of 26 drugs. Pharm Res 2002;19:1400–1416.
22. Schmiedlin-Ren P, Thummel KE, Fisher JM, et al. Expression of enzymatically active CYP3A4 by Caco-2 cells grown on extracellular matrix-coated permeable supports in the presence of 1alpha,25-dihydroxyvitamin D3. Mol Pharmacol 1997;51:741–754.
23. Geick A, Eichelbaum M, Burk O. Nuclear receptor response elements mediate induction of intestinal MDR1 by rifampin. J Biol Chem 2001;276:14581–14587.
24. Lentz KA, Hayashi J, Lucisano LJ, Polli JE. Development of a more rapid, reduced serum culture system for Caco-2 monolayers and application to the biopharmaceutics classification system. Int J Pharm 2000;200:41–51.
25. Liang E, Chessic K, Yazdanian M. Evaluation of an accelerated Caco-2 cell permeability model. J Pharm Sci 2000;89:336–345.
26. Irvine JD, Takahashi L, Lockhart K, et al. MDCK (Madin-Darby canine kidney) cells: A tool for membrane permeability screening. J Pharm Sci 1999;88:28–33.
27. Proulx P. Structure-function relationships in intestinal brush border membranes. Biochim Biophys Acta 1991;1071:255–271.
28. Brasitus TA, Dudeja PK. Effect of hypothyroidism on the lipid composition and fluidity of rat colonic apical plasma membranes. Biochim Biophys Acta 1988;939:189–196.
29. Xiang TX, Chen J, Anderson BD. A quantitative model for the dependence of solute permeability on peptide and cholesterol content in biomembranes. J Membr Biol 2000;177:137–148.
30. Rege BD, Kao JPY, Polli JE. Effects of nonionic surfactants on membrane transporters in Caco-2 cell monolayers. Eur J Pharm Sci 2002;16:237–246.
31. Artursson P, Karlsson J. Correlation between oral drug absorption in humans and apparent drug permeability coefficients in human intestinal epithelial (Caco-2) cells. Biochem Biophys Res Commun 1991;175:880–885.

10

Targeted Drug Delivery to Tumor Cells Using Colloidal Carriers

Sushma Kommareddy and Mansoor Amiji

1. Introduction

Advances in biotechnology and medicine have provided an opportunity for the development of a number of carrier systems for tumor-targeted delivery of anticancer drugs and genes. Tumors are an uncontrolled mass of proliferations of a single malignant cell arising from mutations, which are either inherited or caused by environmental factors *(1)*. In order to reach the tumor cells, systemically administered drugs have to overcome a number of obstacles, which may include rapid metabolism and clearance of drugs from the body, physiological barriers in transportation of the drugs from the site of administration to the tumor cells, drug resistance, and toxicity of the anticancer drugs to normal cells *(2–4)*. The high interstitial pressure and absence of lymphatic drainage were also reported to cause problems in distribution of the drug in the solid tumor *(5)*.

Most of the solid tumors have increased levels of vascular permeability factors (VPF) *(6)*, vascular endothelial growth factor (VEGF) *(7)*, bradykinin *(8)*, nitric oxide (NO) *(9)*, and peroxy nitrite, which enhance the permeability of the tumor vasculature. These VEGFs secreted from the blood vessels promote angiogenesis, a complex process that involves endothelial-cell division and migration, and selective degradation of vascular basement membrane and extracellular matrix (ECM) *(10)*. The blood capillaries of the tumor become dense, disorganized, and tortuous. The resulting vasculature in the tumor was found to be heterogeneous leading to a nonuniform distribution of the drug. Based on the permeability studies by Jain and co-workers *(11,12)*, solid tumors were found to have a well-vascularized periphery, an intermediate semi-necrotic region, and a central necrotic region. The vascular permeability was found to vary from one tumor to the other, within the tumor in terms of both temporal and spatial distribution, and during tumor growth and regression *(13)*. Thus, the microenvironment of the tumor was found to play an important role in the delivery of drugs to the cancer cells. An improper supply of oxygen and nutrients was also observed owing to lack of normal vasculature. The hypoxia and production of lactic acid lead to acidic conditions within the tumor mass *(14)*.

Dvorak et al. *(15–17)* suggested a pathway, which is interconnected by vesiculovacuolar organelles (VVO). These structures are comprised of 12–20 vesicles or vacuoles that are interconnected with each other and with the endothelial cell by means of fenestrae

From: *Cellular Drug Delivery: Principles and Practice*
Edited by: D. R. Lu and S. Øie © Humana Press Inc., Totowa, NJ

that open and close by diaphragms. The size of the VVOs was found to be in the range of 50 nm to 415 nm in diameter *(18)*. The VVOs provide a pathway for the transportation of plasma proteins into tissues. In case of tumor cells the increase in secretion of VEGF, VPF, and other cytokines *(19,20)* was found to enhance the number of VVOs and fenestrae that in turn increase the permeability of tumor vasculature.

Targeting of drugs to the tumor can be achieved either by passive or active mechanisms. Specialized polymeric or colloidal carriers with target-specific ligands, such as antibodies, attached to their surface are used for active targeting. Passive tumor delivery of macromolecules and colloidal particles carrying drugs is achieved primarily by the enhanced permeability of the tumor vasculature and lack of lymphatic drainage. These processes are together termed as the "enhanced permeability and retention (EPR) effect." Using macromolecular conjugates of anticancer drugs, Maeda et al. *(21–23)* were the first to describe the EPR effect of solid tumor. The EPR effect for passive targeting to tumor was later extended to nanoparticles, liposomes, and other colloidal carriers *(24)*. For the EPR effect, the carrier systems used are not target specific and act only to transport the drug to the site of action. In this chapter, we will limit our discussion to the colloidal carrier systems used for passive targeting. These include polymeric conjugates, nanoparticles, liposomes, and micellar systems (Table 1).

2. Polymeric Conjugates

Polymeric conjugated systems are most widely used for the delivery of anticancer agents. Low-molecular-weight drugs penetrate readily through the cellular membrane and cause cytotoxic effects to healthy tissues *(25)*. In order to overcome this limitation, the drug is conjugated with a polymeric carrier. There was a significant inhibition of pinocytotic uptake when macromolecular conjugates of the drug were used as compared to the free drug *(26)*. The polymer-drug conjugates were found to accumulate passively in solid tumors by the EPR effect. The polymer conjugates are also used to enhance the solubility and half-life of the drug in the systemic circulation. Finally, multidrug resistance (MDR) can be overcome by using the conjugated systems as the macromolecular drugs enter the cells through a different pathway as compared to the small-molecule drugs *(27)*.

The polymeric carriers used for delivery of anticancer drugs are broadly divided into synthetic and natural systems. The synthetic polymers are nonimmunogenic, but many do not degrade in the biological environment. These include poly(*N*-[2-hydroxypropyl]-methacrylamide) (HPMA), poly(styrene-co-maleic anhydride), poly(divinylether-co-maleic anhydride) (DIVEMA), and poly(ethylene glycol) (PEG). Natural biodegradable polymers include poly(amino acids) and proteins, hyaluronic acid, dextran, and chitosan are the ones used most often for the preparation of polymer-drug conjugates.

2.1. HPMA Conjugates

HPMA is a biocompatible, synthetic polymer, which is used systemically to deliver anticancer drugs to tumor tissue. The polymer is synthesized in such a way that any drug containing an aliphatic amino group can be conjugated by aminolysis reaction *(28)*. Low-molecular-weight anticancer agents penetrate readily through the cellular membrane, but the HPMA copolymer conjugates are too big and enter the cell by nonspecific endocytotic uptake. This decreases the toxicity of the drug to the normal tissues and increases the targeting of the drug to the tumor by increasing the plasma residence time.

Table 1
Classification of Carriers Used in Tumor-Targeted Drug Delivery

Carrier System	Components
Polymer conjugates	*N*-(2-hydroxypropyl)-Methacrylamide
	Styrene-Maliec Anhydride-Neocarzinostatin (SMANCS)
	Poly(ethylene glycol) (PEG)
	Poly(L-glutamic acid)
	Hyaluronic acid
	Dextran
	Chitosan
Polymeric nanoparticles	Poly(D,L-lactide-co-glycolide)
	Polyalkylcyanoakrylates
	Chitosan
	Gelatin
Liposomes and related systems	Conventional liposomes
	Lipoplexes
	Stealth liposomes
	Niosomes
Polymeric micelles	Poly(ethylene glycol) or poly (ethylene oxide) block co polymers
	Thermoresponsive micelles

In addition, the biocompatibility and low immunogenicity of the HPMA copolymers makes it more useful as carriers of anticancer agents *(29)*. Table 2 gives the list of HPMA copolymer-drug conjugates that have entered Phase I/II clinical trials *(30–33)*.

When anticancer drugs are conjugated with HPMA copolymer through a glycylphenylalanylglycl (GFLG) oligopeptide spacer, the drug release occurs by hydrolytic degradation of the spacer in an acidic environment. These types of polymer conjugates are particularly attractive for tumor targeting, owing to low pH around the tumor (pH ~6.5), and were found to be more effective than those containing nondegradable GG spacer. The antitumor activity of doxorubicin conjugates was evaluated in mice containing EL-4T lymphoma. Free doxorubicin and free polymer precursor were also given along with the 0.15 *M* sodium chloride as controls. When compared to the free drug, the doxorubicin-polymer conjugates were found to have a greater reduction in tumor volume and an increase in survival time of more than 80 d in 40% of the treated animals *(34)*. HPMA-doxorubicin conjugate, called PK1, is currently under Phase II clinical trial in the United Kingdom. The Phase I clinical trial results showed that the circulation time of the drug conjugates increased when compared to the free doxorubicin. Reduced toxicity and increased antitumor activity in refractory tumors was also observed with the conjugates *(30)*. These polymer conjugates were found to decrease the elimination rate and increase the tumor residence time of the drug. When injected intravenously into rats bearing Walker sarcoma, daunomycin-HPMA conjugate was found to be four times more concentrated in the tumor than the free drug *(35)*. Duncan et al. *(36)* evaluated the activity of HPMA-GFLG-daunomycin conjugates in mice bearing L1210 leukemia. Results showed that the polymer conjugates containing

Table 2
HPMA Copolymer Drug Conjugates Under Clinical Trial

Polymer-drug conjugate	Reference
HPMA-doxorubicin	*(30)*
HPMA-paclitaxel	*(31)*
HPMA-camptothecin	*(32)*
HPMA-platinate	*(33)*

biodegradable GFLG linkage had a significant increase in the mean survival time relative to the free drug. In addition, HPMA has been conjugated with cisplatin and glendamycin *(33,37)*.

HPMA copolymer containing a degradable tetrapeptide GFLG was also used to deliver anticancer agents to multidrug resistant cells that expressed P-glycoprotein (P-gp) *(38)*. In order to examine the efficacy of the conjugated drug in resistant cells, internalization of the HPMA copolymer-doxorubicin conjugates was compared in both sensitive and resistant cells of ovarian carcinoma cell line (A2780). Upon measurement of the intracellular fluorescence signal, the results showed that there was an increase in the concentration of the polymer-bound doxorubicin in resistant cells as compared to the free drug. HPMA copolymer bearing GFLG was also used as carrier for anticancer agents like cathepsin B *(39)* and elliciptine *(40)*. The anticancer activity of these polymer drug conjugates was evaluated in mice bearing B16F10 melanoma. The results indicate an increased tumor accumulation of the drug (4.2-fold) and decreased cytotoxicity to healthy tissues. HPMA-GFLG copolymer conjugates of mesochlorin e_6 monoethylenediamine were used in the treatment of ovarian cancer *(41)*.

2.2. Poly(Styrene-Co-Maleic Anhydride) Conjugated With Neocarzinostatin (SMANCS)

Neocarzinostatin (NCS) is an anticancer drug consisting of two components—a protein component (apo-NCS) and a nonprotein component (NCS-chr)—that causes breaks in double-stranded DNA, leading to cell death *(42)*. NCS consists of 112 amino acids with two disulfide bridges and a prosthetic chromophore *(43,44)*. NCS is conjugated with styrene-co-maleic anhydride copolymer resulting in a total molecular mass of 15,000 Daltons. This further binds to albumin in the plasma resulting in a macromolecule of 83,000 Daltons. The macromolecular conjugate was found to achieve increased concentrations of the drug in tumor mass as compared with free NCS *(45)*. In addition, the blood circulation time in mice increased from 2 min with the free drug to 18 min with the macromolecular conjugate. It was observed that prolonged circulation of the drug resulted in an increased concentration of the drug in tumor tissues. Compared to the free drug, an eightfold increase in concentration of NCS was observed upon intravenous administration of the drug conjugate *(23,46)*.

Konno et al. *(47)* treated primary hepatoma patients with SMANCS. Of the 44 patients treated, 95% of them showed a decrease in tumor volume. The mean survival time increased to 18 mo in the polymer conjugate-treated group as compared with 3.7 mo for the control group. Nine patients showed 40–99% decrease in tumor mass over a period of 1–5 mo. In a similar study, the same authors also treated 24 patients with tumors other than primary hepatoma *(48)*. Of those patients who could be evaluated, regression in tumor growth was observed in six (of nine) with metastatic liver cancer, one (of three) unresectable gall bladder, and in all four with lung adenocarcinoma. In both trials, fever was the most common side effect observed. In addition, abdominal pain (in 20% cases) and moderate leukocytosis (in 65% cases) were also observed. No adverse effects were observed in lung, heart, or the brain.

NCS was also conjugated with DIVEMA in order to decrease the toxicity and achieve higher tumor concentrations. Hirano et al. *(49)* prepared conjugates of doxorubicin using DIVEMA by reacting amino group of the drug with anhydride group of the copolymer. When these conjugates were injected into mice bearing an intraperitoneal tumor, a significant increase in survival time (up to 60 d) was observed when compared with the free drug (less than 21 d).

2.3. Poly(Ethylene Glycol) Conjugates

Poly(ethylene glycol) (PEG) has a general structure of HO-$(CH_2CH_2O)_n$-CH_2CH_2-OH. The polymer has a polyether backbone that is chemically inert and terminal hydroxyl groups that can be activated for conjugation with different types of drugs. The high aqueous solubility, lack of toxicity, and ability to decrease immunogenicity makes PEG a very attractive carrier polymer for anticancer agents *(50)*. Paclitaxel is one of the most widely used anticancer drugs. The in vivo activity, solubility, and stability of paclitaxel are improved by linking the drug covalently to monomethoxy-PEG. The PEG derivatives of paclitaxel were found to have significant antitumor activity in mice bearing HA-29 (colon), A549 (lung), and SKOV3 (ovarian) tumors. Upon administration of the conjugates in vivo in mice, the total body clearance and uptake by liver and spleen were reduced *(51,52)*. PEG conjugation with camptothecin was found to enhance the toxicity of the drug in both sensitive and drug resistant human ovarian carcinoma cells. Compared to the nonbiotinylated conjugate, the biotinylated PEG-camptothecin conjugate was found to further enhance the activity of the drug. The increase in toxicity by as much as 60 times in sensitive cells and 30 times in resistant cells was attributed to increased uptake of these conjugates by sodium-dependent multivitamin transporter *(53)*. The presence of amino acid or peptide spacer was found to increase the stability of the PEG conjugates against alkaline degradation. The tripeptide spacer glycil-L-valil-L-phenylalanine was found to be more active than the phenylalanine and tryptophan derivatives. This resulted in decreased toxicity of the drug to healthy cells *(54)*.

Various proteins and enzyme conjugates that act as anticancer agents are also prepared using PEG. L-Asparginase is an enzyme that converts amino acid L-aspargine to L-aspartic acid. It acts as an anticancer agent by depleting the essential amino acid L-aspargine in tumor cells. Conjugation of this enzyme with PEG was found to prevent the proteolytic degradation and enhance the activity *(55)*. In a clinical study conducted in 31 patients with acute lymphoblastic leukemia (ALL), conjugation of L-asparginase with PEG was found to increase the plasma half-life from 20 h to 357 h. Absence of hematological toxicity and reduced hypersensitivity were also observed *(56)*. In another clinical trial with 21 patients having non-Hodgkin's lymphoma (NHL), severe nausea and vomiting were observed in 50% of the cases. However, the absence of hematological toxicity makes the PEG-L-asparginase conjugate very useful in combination therapy *(57)*. Similarly, PEG conjugates of arginine deaminase are used in the treatment of hepatoma (MH134), colon carcinoma (colon 26), sarcoma (S180), and melanoma (B16) *(58)*. Xanthine oxidase (XO) was found to exhibit anticancer activity by generating cytotoxic oxygen radicals. The use of XO was limited owing to its high binding capacity to blood vessels, which results in vascular damage. Conjugation of XO with PEG was found to decrease toxicity and enhance the anticancer activity *(59)*.

2.4. Poly(L-Glutamic Acid) Conjugates

Poly(L-glutamic acid) was widely used to decrease the toxicity and enhance the antitumor activity of the anticancer agents *(60)*. Singer et al. *(61)* used the poly(L-glutamic acid) in order to improve the solubility and anticancer activity of camptothecin. The polymer was conjugated with the drug through γ-carboxylic acid side chain. The conjugations of the drug to the polymer, with and without glycine linker, were evaluated and found to have similar antitumor activity against B-16 melanoma. An increase in the molecular weight of the polymer was found to increase the antitumor activity owing to increased residence time in the tumor and decreased renal clearance. When the activity of conjugated camptothecin was evaluated against athymic mice bearing human lung cancer tumors NCI-H460, a threefold decrease in the tumor volume was found relative to the control. The biodegradable conjugate of poly(L-glutamic acid) and paclitaxel, prepared under similar conditions, is under Phase-II clinical trials in the United States. Biodistribution studies carried out in mice bearing B16 melanoma and OCA-1 ovarian tumors show greater accumulation of the paclitaxel-polymer conjugate as compared with the free drug.

2.5. Hyaluronic Acid Conjugates

Hyaluronic acid (HA) is a nonsulfated glycosaminoglycan that is abundant in synovial fluid and ECM *(62)*. HA conjugated with polymeric carriers and anticancer drugs is used to target tumor cells expressing HA receptor. Prestwich and co-workers *(63)* prepared *N*-(2-hydroxypropyl) methacrylamide (HPMA)-HA conjugates for the delivery of the anticancer drug doxorubicin. The results of the in vitro cell-culture experiments show that the HPMA-HA-doxorubicin conjugates have greater cytotoxicity than HPMA-doxorubicin conjugates against HBL-100 (human breast cancer), SKOV-3 (ovarian cancer), and HCT-116 (colon cancer) cell lines expressing the HA receptor. This was attributed to a more efficient receptor-mediated uptake of the macromolecular drug by the tumor cells as compared to nonspecific uptake of HPMA-doxorubicin conjugates. Similar results were obtained using HA-paclitaxel conjugates with selective toxicity against tumor cells (breast, colon, and ovarian) expressing HA receptors and minimal toxicity against mouse fibroblast cell line, which does not express HA receptor *(64)*.

2.6. Dextran Conjugates

Dextran is a biodegradable polysaccharide widely used as a carrier of anticancer agents. Okuno et al. *(65)* prepared carboxymethyldextran-camptothecin conjugate with a triglycine spacer. When injected intravenously into rats bearing Walker sarcoma-256 tumor, the conjugate was found to have increased blood circulation, resulting in an increased accumulation of the drug in the tumor. These conjugates were used to treat mammary carcinoma (MX-1), lung carcinoma (LX-1), gastric (St-4), and colorectal (HT-29) tumors, which are refractory to camptothecin. Carboxymethyldextran conjugates with cisplatin are used in the treatment of human osteogenic sarcoma and human ovarian carcinoma in athymic mice and embryonic carcinoma in Balb/c mice *(66)*. Bernstein et al. *(67)* showed that doxorubicin can be conjugated with oxidized dextran, thus producing greater antitumor activity than the free drug.

2.7. Chitosan Conjugates

Chitosan is a biodegradable co-polymer of *N*-acetyl-D-glucosamine and D-glucosamine linked together by a β (1–4) glycosidic linkage. Chitosan, obtained by alkaline deacetylation of chitin, is usually greater than 70% deacetylated and has a molecular weight in the range of 50,000 to 2 million Daltons *(68)*. Sato et al. *(69)* synthesized *N*-succinyl chitosan and glycol chitosan conjugates of mitomycin C. They also evaluated the antitumor activity of these drug-polymer conjugates in vivo against P388 leukemia and sarcoma 180. The drug conjugates of glycol-chitosan were found to have a significant increase in the survival time against P388 leukemia. In case of sarcoma 180 inoculated subcutaneously, the drug conjugates of *N*-succinyl chitosan were found to have greater tumor-growth suppression. When compared with the free drug, both the polymer conjugates were found to exhibit increased capacity for suppression of tumor growth.

2.8. Other Polymeric Conjugates

The potential use of aspartic acid polymers and pyrrolo [2, 1-c] [1, 4] benzodiazepine-(PBD) polyamide conjugates as carriers of anticancer agents were evaluated in human cancer cell lines *(70,71)*. Ferrocene-containing anti-neoplastic agents that are covalently bound to aspartic-acid polymers were tested for their cytotoxicity against EMT-6 cancer cell lines. The PBD dimers were found to be more effective against colon, melanoma, renal, and breast cancer cell lines. In both the cases, an increase in the activity was observed with an increase in the chain length of the polymer.

3. Polymeric Nanoparticles

Nanoparticles are colloidal systems of submicron size (<1 μm). They may be classified as either nanospheres, in which the drug is dispersed throughout the system, or as nanocapsules, in which the drug is entrapped in a cavity surrounded by a polymer layer. Sometimes the drug may be also be adsorbed to the surface of the nanoparticles *(72,73)*. Because of their small size, nanoparticles extravasate into the tumor cells and are used as carriers for the delivery of anticancer agents. This increases the efficiency of drug targeting and reduces the cytotoxicity of the drug in healthy tissues. From the histological examinations, it was proved that the nanoparticles first reach the extravascular compartments and are then taken up by tumor cells by endocytosis *(74)*.

Polymeric nanoparticles can be prepared either by emulsion polymerization with monomers or by solvent precipitation of formed polymers. Emulsion polymerization is done in organic solutions of the monomers and the formed polymeric nanoparticles are precipitated in the presence of aqueous surfactant solutions with controlled stirring. This method does not involve high temperatures and can be used for encapsulation of heat-sensitive compounds. The solvent precipitation method involves gradual addition of a nonsolvent to the polymer solution under controlled stirring conditions in order to precipitate the nanoparticles.

The polymer used for nanoparticle preparation should be biocompatible and biodegradable, as accumulation of nonmetabolizable materials in vivo results in toxicity. The conventional nanoparticles, like many other colloidal systems, are engulfed predominantly by the cells of the reticuloendothelial system (RES) upon systemic administration. Surface modification of the conventional nanoparticles with PEG and other water-soluble polymers can result in long-circulating nanoparticles. The hydrophilic nature of these surface-modified particles reduces their binding capacity to proteins (opsonins), resulting in reduced uptake by the RES.

3.1. Poly(D, L-Lactide-Co-Glycolide) (PLGA) Nanoparticles

PLGA is a biocompatible and biodegradable polymer, which is used for the encapsulation of lipophilic drugs. The polymeric nanoparticles are used for sustained release of paclitaxel, doxorubicin, and cisplatin. Yoo et al. *(75)* prepared polymeric nanoparticles by conjugating the hydroxyl group of doxorubicin with carboxyl group of PLGA. The nanoparticles containing conjugated doxorubicin were found to exhibit sustained drug release. The in vitro studies using HepG2 cell line and in vivo antitumor activity showed that doxorubicin conjugated with PLGA nanoparticles has greater antitumor activity than free doxorubicin. Feng et al. *(76)* prepared PLGA-poly(L-lactic acid) (PLA) nanoparticles, which can act as carrier for administration of paclitaxel, a highly hydrophobic anticancer drug. In addition, PLGA-mPEG nanoparticles are used as carriers of cisplatin *(77)*. The nanoparticles were prepared by the double-emulsion method. The in vitro studies showed that the rate of degradation and the rate of release increased with increase in mPEG content. The in vivo studies conducted in BALB/C mice showed an increase in the circulation time of cisplatin-containing PLGA-mPEG nanoparticles. Biodegradable nanoparticles of PLA were prepared by emulsification-diffusion technique using propylene carbonate. A significant decrease in the size of the nanoparticles was observed when stabilizing agents, like poloxamers, are used in the concentration range of 0.5–5% (w/v). P^H of the external phase, internal phase to external phase volume ratio, and stirring speed were found to influence the particle size. These biodegradable nanoparticles were found to have potential for drug targeting to solid tumors *(78)*.

PLGA nanoparticles were also prepared to encapsulate plasmid DNA for the delivery of gene constructs to the tumors. Sustained release of DNA from the nanoparticles with high encapsulation efficiency was obtained. DNA is highly negatively charged and hydrophilic, hence it is combined with a cationic agent, such as dimethyldioctylammonium bromide (DDAB), and encapsulated in PLGA *(79)*. Cohen et al. *(80)* prepared PLGA nanoparticles that can protect the plasmid DNA from nuclease degradation. The DNA associated with the nanoparticles was found to be potent for a longer period of time than the liposomal DNA. However, the cell culture studies using NIH-3T3 cells showed that the transfection efficiency of DNA delivered in these particles was lower than with liposomes. Maruyama et al. *(81)* prepared graft co-polymers of poly(L-lysine) and PLA as carriers of DNA. PLA nanoparticles are negatively charged owing to the terminal carboxyl group. Association of the PLA with poly(L-lysine) resulted in positively charged particles, which could bind to the negatively charged DNA. The nanoparticles of the graft copolymer were obtained by solvent evaporation method or diafilteration method in which the nanoparticles are obtained by dialysis followed by centrifugation. The size of the nanoparticles and the adsorption capacity of DNA to the nanoparticles were found to vary with the concentration of the graft copolymer.

3.2. Poly(Alkylcyanoacrylates) Nanoparticles

Poly(alkylcyanoacrylates) are biodegradable polymers that are widely used for the delivery of anticancer drugs. One of the major advantages of using poly(alkylcyanoacrylates) is the ease of polymerization, because the monomers do not require any energy for polymerization *(82)*. The most widely used poly(alkylcyanoacrylates) are poly-

(butylcyanoacrylate) (PBCA), poly(hexylcyanoacrylate) (PHCA), poly(isohexylcyanoacrylate) (PIHCA), poly(isobutylcyanoacrylate) (PIBCA) and poly(methylcyanoacrylate) (PMCA). The molecular weight of these polymers, the residence time of the polymer in the plasma, as well as biodegradation kinetics depends on the alkyl chain length *(83)*. This was further proved by the experiments of Couvreur et al. *(82)* that the heavier polymer (PHCA) would be cleared more slowly than the lighter one (PIBCA). It was also observed that a large proportion of the nanoparticles were taken up by the Kupffer cells in the liver. However, in the presence of a lung tumor, these nanoparticles are localized preferentially in the lung tissue as compared with the liver. The increase in efficiency of the anticancer drugs loaded in the polymeric nanoparticles was explained by Brasseur et al. *(84)* conducted experiments that showed slow release of the drug from the nanoparticles that resulted in an increase in the bioavailability and cytotoxic activity of the drug at the tumor site.

Soma et al. *(85)* prepared doxorubicin-loaded PIBCA nanoparticles for tumor-targeted delivery. An increase in antitumor activity and a decrease in clearance were observed for the doxorubicin nanoparticles. There was also a decrease in the concentration of the drug in the heart, indicating a reduction in the cardiotoxic effects of the drug. Similar effects were observed by Verdun et al. *(86)* using doxorubicin-loaded PIHCA nanoparticles. The biodistribution of the free and bound dactinomycin to polymethyl (PMN), polyethyl (PEN), and polybutyl cyanoacrylate nanoparticles was compared by Couvreur et al. *(82)* It was observed that PBN significantly increased the concentrations of the drug in liver (64-fold), spleen (44-fold), and lungs (4.7-fold) when compared to the free drug. It was also found that dactinomycin-loaded PMN nanoparticles resulted in a significant reduction in tumor volume against S250 sarcoma in rats. Similar experiments conducted using vinblastine-loaded nanoparticles showed that there was an increased accumulation in the spleen (21-fold) followed by lungs (fourfold) and liver (1.7-fold). Simeonova et al. *(87)* prepared PBCA nanoparticles associated with anticancer drug vinblastine. Intraperitoneal injection of these nanoparticles into mice resulted in a significant reduction in leucopenia caused by vinblastine.

PIHCA nanoparticles were also used to overcome MDR, which is the main cause of failure in systemic cancer therapy. In order to overcome the problem of low drug accumulation, Cuvier et al. *(88)* evaluated the effect of nanoparticles loaded with doxorubicin in five different MDR cell lines. An increase in the cytotoxicity of the cells was explained by an increase in drug concentration and an enhancement of contact time with the tumor cells. The nanoparticles of doxorubicin were tested in the C6 rat glioblastoma cell lines with varying degrees of resistance *(89)*. It appeared that the reversal of resistance by the doxorubicin nanoparticles depended on the nature of the resistance.

The half-life and distribution of the nanoparticles was altered by coating them with surfactants like polysorbate 80, poloxamer, and poloxamine *(90–93)*. The surfactants were found to induce steric repulsion, which reduced the uptake of these nanoparticles by the mononuclear phagocyte system (MPS). Surface coating of the nanoparticles with polysorbate 80 (Tween® 80) significantly altered the biodistribution of the drug. Doxorubicin concentration decreased in the heart, whereas there was a 60-fold increase in the brain with the use of nanoparticles. The author concluded that these nanoparticles could be used for delivery of drugs to the tumors in the brain. When compared to liposomes, poly(alkylcyanoacrylate) nanoparticles coated with poloxamine were found to deliver the anticancer drug mitoxantrone more efficiently resulting in a greater reduction in tumor growth in mice bearing B16 melanoma.

Poly(alkylcyanoacrylate) nanoparticles were also used to deliver antisense oligonucelotides to the tumor cells *(94)*. These nanoparticles increased the bioavailability of the nucleotides by preventing their degradation by nucleases. The cyanoacrylates are negatively charged so they are combined with a cationic polymer (e.g., diethylaminoethyl dextran) or cationic detergent (e.g., hexadecyltrimethylammonium bromide [CTAB]), which facilitates the binding of the oligonucleotides by electrostatic interactions. From the in vitro studies using Vero, U937, and HBL100ras1 cell lines, the uptake of the oligonucleotide nanoparticles was found to be 50-fold higher with the nanoparticles than from the oligonucleotides solution *(95)*. The in vivo studies conducted in OF1 mice showed that there was an increase in the biological half-life of the nanoparticle-delivered oligonucleotides. The nanoparticles were distributed preferentially in the liver after systemic administration and this significantly limits their use as colloidal carriers *(96,97)*. Schwab et al. *(98)* used oligonucleotides adsorbed to poly(alkylcyanoacrylate) nanoparticles to inhibit tumor cells resulting from Ha-ras mediated mutations in nude mice. In mice implanted with HBL100ras1 cells, tumor-growth inhibition was achieved at the oligonucleotide concentrations that were approx 100-fold less than when free oligonucleotide were used.

Kattan et al. *(99)* conducted clinical studies of doxorubicin-loaded PIHCA nanoparticles in 21 patients with refractory solid tumors. Doppler-echocardiography showed complete absence of cardiotoxicity in patients receiving the nanoparticle formulations. However, both dose-limiting granulocytopenia as well as allergic reactions were observed when the drug was given by a 10-min intravenous infusion. Although the doxorubicin PIHCA nanoparticles had a higher therapeutic effect according to the authors as compared with the free drug, the effects could not be confirmed owing to the small number of patients and low cumulative dose.

3.3. Chitosan Nanoparticles

The positively charged chitosan nanoparticles were used as colloidal carriers of doxorubicin. The encapsulated efficiency of the cationic anticancer drug was improved by conjugating it with dextran sulfate prior to encapsulation in chitosan microspheres. The in vitro studies using human melanoma A375 cells and C26 murine colorectal carcinoma cells indicated that doxorubicin nanoparticles had cytotoxic activity equivalent to that of free doxorubicin *(100)*. Mitra et al. *(101)* prepared chitosan nanoparticles incorporated with dextran-doxorubicin conjugate. The in vivo antitumor activity of the dextran-conjugated doxorubicin nanoparticles was evaluated in BALB/C mice implanted with J774A.1 macrophage tumor cells. The results showed that the animals treated with nanoparticle formulation had an increased survival time and greater tumor regression as compared with the free drug.

Chitosan nanoparticles are also used for the delivery of DNA to the tumor cells. Kabbaj et al. *(102)* prepared *Mycobacterium pheli* DNA-incorporated chitosan nanoparticles for inhibition of cancerous-cell division. The nanoparticles prevented degradation of DNA by nucleases resulting in a 20-fold increase in the transfection efficiency as compared with that of naked DNA. Mao et al. *(103)* prepared chitosan-DNA nanoparticles of 100–250 nm diameter by the complex coacervation process. On conjugation with transferrin and KNOB protein (cloned from adenovirus-fiber protein), the nanoparticles were reported to have significantly higher transfection efficiency in HEK293 (fourfold) and HeLa cells (130-fold). The distribution profile of the nanoparticles in vivo was obtained following iv injection of the nanoparticle suspension through the tail vein of AKR/J mice. A significant fraction of the administered dose of the nanoparticles was taken up by the liver and the kidneys. PEG surface modification of these nanoparticles was found to enhance their circulation time.

3.4. Gelatin Nanoparticles

Gelatin is a biocompatible and biodegradable polymer obtained by hydrolysis of collagen. It is a polyampholite, which gels below 35–40°C. Because there are many different Bloom strengths of gelatin available, a wide range of nanoparticles can be made by salt-induced complex coacervation *(104)*. Troung-Le et al. *(105,106)* prepared DNA-containing gelatin nanoparticles by the complex coecervation process. The cellular uptake of these nanoparticles was enhanced by conjugation with transferrin, and incorporation of calcium or chloroquine. The transfection efficiency of the DNA-gelatin nanoparticles in 9HTEo cells (human tracheal epithelial cells) was proved by expression of the cystic fibrosis transport regulator (CFTR) protein in more than 50% of the cells. Intramuscular injection of these nanoparticles into mice indicates that transfection efficiency was higher with the nanoparticles as compared to DNA-complexed with lipofectamine. Coester et al. *(107)* prepared avidin-conjugated gelatin nanoparticles by the desolvation method as carriers of biotinylated antisense drug derivatives. Additionally, Kaul et al. *(108,109)* prepared 250–400 nm PEG-modified gelatin nanoparticles by solvent-displacement method for encapsulation and delivery of hydrophilic macromolecules (e.g., DNA) to solid tumors. The in vitro studies using BT-20 cells (human breastcancer cell line) show that these nanoparticles can be used as carriers for gene delivery.

3.5. Other Polymeric Nanoparticles

Sharma et al. *(110)* prepared paclitaxel-containing poly(vinylpyrrolidine) (PVP) nanoparticles by reverse micro-emulsion method. The in vivo anti-tumor activity of these nanoparticles was evaluated in C57B1/6 mice bearing murine melanoma (B16F10). When compared to the free paclitaxel, these nanoparticles had increased survival time and regression in tumor volume. Kim et al. *(111)* prepared methoxyPEG and poly(ε-caprolactone) (MePEG/PCL) nanoparticles loaded with the anticancer drug paclitaxel. The size of the optimized nanoparticles was less than 100 nm and had a paclitaxel load of 20%. The in vivo studies conducted in ICR mice showed that the polymer used had very low toxicity. The authors concluded that these nanoparticles have significant potential as carriers of anticancer drugs.

Additionally, poly(γ-benzyl-L-glutamate)/PEO (PBLG/PEO) block copolymer nanoparticles were prepared for targeting of the anticancer drug doxorubicin *(112)*. The drug release from the PBLG/PEO nanoparticles was dependent on the molecular weight of the hydrophobic component of the block copolymer. The circulation time of the adriamycin nanoparticles was found to be threefold higher than that of the free drug. Potineni et al. *(113)* recently developed poly(ethylene oxide) (PEO)-modified poly(β-amino ester) nanoparticles (100–150 nm in diameter) for encapsulation of hydrophobic anticancer drugs like paclitaxel. The cell-uptake studies using PEO-modified poly(β-amino ester) nanoparticles encapsulated with fluorescent dye (rhodamine-123) indicate that the particles were taken up by the tumor cells (BT-20) by nonspecific endocytosis. Being pH-sensitive, the poly(β-amino ester) nanoparticles are expected to spontaneously release the encapsulated payload in the low pH environment of the tumor or the endosome/lysosome complex. Thunemann et al. *(114)* developed retinoic acid and poly(ethylene imine) nanoparticles using different molecular weight of the polymer. The size of the nanoparticles was found to decrease with increasing molecular weight of PEI.

4. Liposomes and Related Systems

4.1. Conventional Liposomes

First described by Bangham et al. *(115)*, liposomes are phospholipid vesicles composed of a bilayered lipid membrane. These vesicles are extensively used as drug carriers because of their ability to carry hydrophilic drugs in their aqueous compartment and lipophilic drugs in the bilayered membrane. Depending on the method of preparation and the lipids used the size of the liposomes range from 0.05 to 5.0 μm *(116)*. Liposomes used for drug-delivery applications are usually made from natural lipids such as phosphatidylcholine, phosphatidylglycerol, and cholesterol.

There are three major types of liposomes. The multi-lamellar vesicles (MLV) consist of several concentric bilayers of phospholipids. MLV range in size from 1.0–5.0 μm and are prepared by hydrating the lipid film with an aqueous solution. MLV can encapsulate up to 5% efficiency of the hydrophilic drug and up to 10% efficiency of the hydrophobic drug *(117)*. The small unilamellar vesicles (SUV) range from 30 to 50 nm and could not be used for encapsulation of hydrophilic drugs because of low encapsulation efficiency (~1%) *(118)*. The large unilamellar vesicles (LUV) range from 50 to 100 nm and are used to encapsulate hydrophilic drugs. Drugs are incorporated into liposomes by encapsulation, partitioning, or reverse loading *(116)*.

Liposomes have been used as carriers of a number of anticancer drugs. The drug-carrying vesicles reach the tumor tissue through the leaky vasculature and are retained in the interstitial space by passive targeting mechanism. The colloidal drug-carrying vesicles are recognized as foreign particles by MPS and are trapped by the cells of the RES *(119)*. Hence, these liposomes can be used to target tumors present in liver, spleen, and the bone marrow. The liposomes of different compositions and the results of their therapeutic activity in various animal tumor models are presented in Table 3 *(120–130)*.

Liposome formulations composed of distearoylphosphatidylcholine and cholesterol have unique advantages over those containing egg phosphatidylcholines because they are more resistant to leakage of entrapped drug. This is owing to the presence of lipids, whose phase-transition temperature is much higher than the physiological temperature *(131)*. Daunorubicin encapsulated in liposomes, now marketed as DaunoXome®, was the first drug to enter into clinical trials. The liposomal formulation is composed of distearoylphosphatidylcholine and cholesterol in a 2:1 molar ratio. In one study, comparison of daunorubicin associated with the lipid bilayer and the citrate salt of the drug entrapped in the aqueous core, showed that the core-containing drug had higher stability, better biodistribution profile, and was more effective than the drug associated with the bilayers *(132)*.

In the preclinical studies conducted in mice bearing P1798 lymphosarcoma, it was found that DaunoXome® resulted in greater mean survival time (18.5 d) than free daunorubicin (13 d) at a dose of 30 mg/kg *(133)*. Money-Kyrle et al. *(134)* conducted Phase I/II trials in 25 patients with advanced AIDS-associated Kaposi's sarcoma (KS). In this study, it was found that DaunoXome®, given at 40 mg/m^2 dose, resulted in a partial response of 40%. In another study with patients who had not received previous treatment, a 57% response rate was observed *(135)*. When compared to the free

Table 3
Therapeutic Efficacy of the Conventional Liposome-Associated Anticancer Drugs in Animal Tumor Models

Drug	Liposomes	Tumor	Host	Dose	Results	Reference
Actinomycin D	PC:C:PA 7:2:1	AKR-A ascites 1 × 10^6 cells	AKR mice	Ip 1.0 µg from d 2–6	Prolonged survival compared with drug	*(120)*
	PC:C:SA 18:4:5	Ridgeway osteosarcoma	AKR mice	Iv 0.8 mg/kg kg on d 1	Less effective than drug	*(121)*
Adriamycin	PC:PS:C 10:4:1	P388 ip 1 × 10^5 cells	BD/F_1 mice	Ip 4 mg/ kg on d 1	As effective as drug and less cardiotoxic	*(122)*
	PC:C:SA 10:4:3	Lewis lung sc 1 × 10^5 cells	CD/F_1 mice	Iv 4 mg/kg on d 8, 10, 12	As effective as drug	*(122)*
	PC:C:DCP 7:2:1 PC:C:SA 7:2:1	Ehrlich ascites sc 2 × 10^6 cells	ICR mice	Ip 1.25 mg/kg on d 7, 8, 9	Reduction in tumor weight compared to control	*(123)*
	PS:PC:C 3:7:10	J6456 lymphoma iv 10^6 cells	BALB/c mice	iv 6 mg/kg on d 3, 7, 10, 24	Increase in mean survival time	*(124)*
Cisplatin	DSPC:DPPC C 7:1	Sarcoma 180 sc 5 × 10^5 cells	MMR1 mice	Iv 0.7–1.2 mg/kg on, d 7 followed by 1 hr hyperthermi	Delay in tumor growth	*(125)*
Daunomycin	PC:C:PA 7:2:1	L1210 iv 10^5 cells	B-60/F_1 mice	Iv 5-80 mg/ kg on d 1	Increase in mean survival time at high doses	*(126)*
Methotrexate	PC:C:SA 3.25:2.25:2	L1210 ip 10^5 cells	DBA/2 mice	Ip or iv 25–50 mg/kg on d 1	As effective as free drug	*(127)*
	PC:C:SA 4:3:1	P-1798 lymphosarcoma	CD/F_1 mice	Ip 1 mg/kg on d 10–13	Reduction in tumor weight	*(128)*
	DPPC:DSPC C 7:3	L1210 sc 0.5-20 × 10^6 cells	B6D2/F_1 mice	Iv 3 mg/kg on d 1, 2, 3	Delay in tumor growth	*(129)*
	DPPC:C:SA A 34:23:10	Hepatoma 129 ip 10^6 cells	C3H mice	Iv or Ip 2–3 mg/kg on d 1	Iv treatment was ineffective, ip treatment increases the mean survival time	*130*

C, cholesterol; DCP, dicetylphosphate; DSPC, disteroylphosphatidylcholine; PC, phosphatidylcholine; PA, phosphatidic acid; PS, phosphatidylserine; SA, steroylamine; ip, intraperitoneal; iv, intravenous; sc, subcutaneous.

daunorubicin, DaunoXome® was also found to result in a 35-fold increase in the area under the curve (AUC) in plasma concentration vs time profile. Although these studies showed an absence of cardiotoxicity with DaunoXome, moderate myelosuppression was observed.

Gill et al. *(136)* conducted a Phase III trial in 232 patients with AIDS-related KS and compared the therapeutic efficacy of DaunoXome (40 mg/m^2) with a standard ABV regimen of doxorubicin (10 mg/m^2), bleomycin (15 U), and vincristine (1 mg). A 25% response rate (3 complete and 26 partial of 116 treated) was observed for DaunoXome and 28 % (1 complete and 30 partial of 111 treated) with ABV. No significant difference in the response rates and survival times were observed in the treatments. A higher incidence of alopecia and neuropathy were observed in patients treated with ABV and grade 4 neutropenia was observed in patients treated with DaunoXome. The aforementioned preclinical and clinical studies indicate the efficacy of DaunoXome, which is now approved for treatment of KS in HIV-positive patients in several countries, including the United States.

To investigate the effect of cationic charge on tumor targeting potential, Campbell et al. *(137)* prepared liposomes using 1, 2-dioleoyl-3-trimethylammonium propane (DOTAP), a cationic lipid, for the delivery of paclitaxel, a poorly soluble anticancer drug. The cationic lipid was combined with varying ratios of dimyristylphosphatidylcholine (DMPC), dipalmitoylphosphatidylcholine (DPPC), and distearoylphosphatidylcholine (DSPC). It was observed that the stability of the formulations increased with an increase in the concentration of DOTAP, reaching a maximum when the lipid concentration was 30–50 mol%.

Liposomes-encapsulated drugs can also be used in the treatment of MDR tumors. The drug resistance is caused mainly by efflux of drugs in tumor cells by the P-gp pump. Rahman and co-workers *(138,139)* prepared doxorubicin-entrapped liposomes containing cardiolipin, which interacts with the P-gp pump and prevents the drug efflux. The major disadvantage of the formulation was that the phagocytes of the RES recognize cardiolipin and clear the vesicles from blood circulation in a relatively short period of time.

Even though the small-sized liposomes have the advantage of longer circulation than large ones, they are not capable of carrying large quantities of the drug *(140)*. The large multilamellar liposomes have the problem of aggregation in the lungs *(141)*. In addition, these conventional liposomes are internalized rapidly by the macrophage-rich RES and cleared from the systemic circulation *(142,143)*. In order to overcome the rapid clearance problem, specialized liposomes like the PEG-coated, long-circulating, and cationic liposomes are being used to achieve better passive targeting to the tumor tissues.

4.2. Lipoplexes

Lipoplexes are cationic lipid-DNA complexes that are most widely used in gene delivery to tumor cells. They are formed by electrostatic interaction between the positively charged lipid and negatively charged DNA *(144)*. Lipofectin and other cationic lipids are under clinical trials. Lipofectin is a cationic lipid composed of 1:1 (w/w) ratio of *N*-[1-(2, 3-dioleyloxy) propyl]-*N,N,N*-trimethylammonium chloride (DOTMA) and dioleylphosphatidylethanolamine (DOPE). Other cationic lipids commonly used are 2,3-dioleyloxy-*N*-[2-(sperminecarboxamido)ethyl]-*N,N*-dimethyl-1-propanaminium trifluoroacetate (DOSPA), dioctamido-decylamidoglycyl spermine (DOGS), 1,2-

dimyristyloxypropyl-3-dimethylhydroxyethyl ammonium bromide (DMRIE), 1,2-(dioleoyl-3-trimethylammonium) propane (DOTAP), dimethyl dioctadecylammonium bromide (DDAB), and 3β-(*N,N*-dimethylaminoethane)carbamoyl cholesterol (DC-chol) *(145–148)*.

Reimer et al. *(149)* prepared cationic liposomes using dioleoyldimethylammonium chloride and dioleoylphosphatidylethanolamine in a 1:1 ratio in order to deliver chloramphenicol acetyl transferase (CAT) gene to tumor cells. Upon ip administration of these liposomal formulations to a murine model bearing B16/B16 tumor, a highly variable gene expression was observed ranging from 100 mU/g to 2000 mU/g tumor. Small tumors did show much higher transfection efficiency relative to larger tumors. In order to minimize the toxic effects of lipoplexes in the systemic circulation, intratumoral delivery has been suggested *(150)*.

4.3. Long-Circulating (Stealth®) Liposomes

Modification of conventional liposomes by hydrophilic polymers, like PEG, enhances the repulsive interactions between the colloidal particles and prevents opsonization by the macrophages of the RES. This results in long circulating liposomes, also called Stealth® liposomes *(151)*, sterically stabilized liposomes *(152)*, or cryptosomes *(153)*. The mechanism of long circulation of the polymer-coated liposomes was explained by Torchillin and co-workers *(154–156)*. The authors suggested that the coated polymer is in constant motion and exists as a cloud above the liposome surface, thus protecting the liposome from opsonization and recognition by RES cells. The high hydrophilicity, flexibility, and low immunogenicity were found to make PEG most acceptable for surface modification of the liposomes *(157–159)*. Other synthetic polymers that can serve as protective coating on liposomes include poly(acrylamide) and poly(*N*-vinyl pyrrolidine) *(160,161)*. These long circulating liposomes were found to deliver anticancer drugs to tumor tissues by passive targeting due to the EPR effect.

The accumulation of long-circulating liposomes in tumors was studied by Ning et al. *(162)* by injecting fluorescently labeled, long-circulating liposomes into rats with dorsal skin-flap window chambers and having mammary adenocarcinoma. Long-circulating liposomes made from egg phosphatidylcholine:cholesterol:disteroylphosphatidylethanolamine-PEG 2000: *N*-(6-tetramethylrhodamine-thiocarbamoyl)-1, 2-dihexadecanoyl-*sn*-glycero-3-phosphoethanolamine (TRITC-DHPE, a fluorescent label) in a weight ratio of 60:33:5:2 were used in the study. Within 90 min after injection, the long-circulating liposomes accumulated at three- to fourfold greater concentration than the conventional liposomes in the tumor mass. In addition, after 90 min, about 60% of the long-circulating liposomes were found to remain in systemic circulation, as compared with only 20% of the conventional liposomes.

Vincristine is an anticancer drug that acts by binding to the microtubules *(163)*. Conventional liposomal formulations of the vincristine were found to be no more effective than the free drug (0.9 mg/kg). At higher doses, the liposomal formulation was found to be toxic owing to the high lipid content *(164)*. In order to enhance the tumor targeting and anticancer activity of the drug, vincristine was encapsulated in PEG-coated liposomes. In vivo studies of the drug entrapped long-circulating liposomes containing methoxyPEG-disteroylphosphatidylethanolamine, hydrogenated soybean phosphatidylcholine, and cholesterol in a weight ratio of 5:56:39 were conducted in various tumor models *(165–167)*. The results in Table 4 show an increased mean survival time in leukemia models and reduction in tumor growth in colon carcinoma and mammary carcinoma models.

Doxorubicin, an anthracycline antineoplastic drug, is most widely used in the treatment of many different forms of human cancers. The clinical use of this drug is limited by a dose-dependent cardiotoxicity, myelosuppression, and stomatitis. The early studies using the conventional liposomes containing doxorubicin showed the problems of drug targeting to the tumor tissue and rapid clearance by RES. These drawbacks can be overcome by using long-circulating liposomes.

Doxil® is the long-circulating liposomal formulation of doxorubicin marketed by Liposome Technology, Inc. (Menlo Park, CA). The lipid bilayer of Doxil is made of hydrogenated soybean phosphatidylcholine, cholesterol, and PEG (molecular-weight 1900–2000)-modified disteroylphosphatidylethanolamine in a weight ratio of 56:5:39. The formulation has doxorubicin at a concentration of 2 mg/mL and the lipids at 16 mg/mL *(168)*. The published results of preclinical studies conducted in animals show the effectiveness of Doxil over free doxorubicin *(169–171)*. The results, summarized in Table 5, prove that Doxil has longer circulation time and greater inhibition of tumor growth than free doxorubicin.

Clinical trials of the Doxil formulation were conducted in patients with different types of cancer. In a Phase III clinical trial of patients with AIDS-related KS, Doxil was compared with a combination therapy of doxorubicin, bleomycin, and vincristine (ABV) *(172)*. The drugs were administered every 14 d in 6 cycles. There was no significant difference in the survival times of the patients. However, a higher incidence of alopecia, nausea, and vomiting was observed in patients treated with ABV regimen as compared with Doxil. The reduction in toxicity was found to be a major advantage of Doxil over the conventional ABV regimen.

Lyass et al. *(173)* conducted clinical trials of Doxil in breast-cancer patients who were previously treated with doxorubicin at a dose of less than 400 mg/m^2. The results indicate a high incidence of mucositis and palmar-plantar erythrodysaesthesia. No significant cardiotoxicity was observed in these trials, except for the occurrence of congestive heart failure in one patient who had prior mitoxantrone and radiotherapy. Partial response (five patients) and complete response (two patients) was observed in patients who were present in different treatment groups. In all the preclinical and clinical trials of Doxil, a high incidence of stomatitis was observed. Stomatitis is also a dose-limiting toxicity of continuous doxorubicin infusion.

The sterically stabilized liposomes are also used to encapsulate other anticancer drugs including cisplatin. SPI-077, a liposomal formulation of cisplatin, with a lipid composition hydrogenated soybean phosphatidylcholine: cholesterol: PEG (2000)-modified disteroylphosphatidylethanolamine in a weight ratio 51:5:44 is currently in clinical trials. The preclinical studies indicate inhibition of tumor growth in colon C26 carcinoma and Lewis lung carcinoma models when compared with the delayed tumor growth observed in the treatment with cisplatin and carboplatin *(174,175)*.

Long-circulating liposome combined with local hyperthermia is used to maximize tumor targeting of the anticancer drugs *(176)*. The combination enhances the cytotoxic effects of the drug against tumor cells. Liposomes with the aforementioned properties were prepared by using DPPC, DSPC, distearylphosphatidylethanolamine-monomethoxypolyethylene glycol succinimidyl succinate of molecular weight 2000

Table 4
Therapeutic Efficacy of Long-Circulating Liposomes Loaded With Vincristine

Tumor	Host	Dose	Results	Reference
P388 Lymphocytic leukemia 10^6 cells iv or ip	B6D2F1 mice	Single dose of 1.3 mg/kg or 2.0 mg/kg iv or ip after 24 h	Increase in mean survival time (MST); MST of ip tumor is greater than MST of iv tumor	*(165)*
L1210 leukemia 10^6 cells ip	DBA/2J mice	Single dose 0.5–3 mg/kg after 24 h	Increase in survival time 17–117% for the given dose range	*(166)*
C26 colon carcinoma 4×10^5 cells sc	BALB/c mice	Three weekly doses of 1.3 or 2 mg/kg starting on d 6 or 10	Reduction in tumor growth with stealth liposomes when compared with the delay in tumor growth of the free drug	*(165)*
MC2B mammary carcinoma 1 mm^3 piece implanted sc	C3H/HE mice	Three weekly injections of 1.0 or 1.3 mg/kg starting on d 3	Reduction in tumor growth with 1.3 mg/kg dose of the liposome formulation being more effective	*(167)*

ip, intraperitoneal; iv, intravenous; sc, subcutaneous.

Table 5
Therapeutic Efficacy of Doxorubicin-Encapsulated in Long-Circulating Liposomes (Doxil®) in Athymic Mice Xenografted With Human Tumors

Tumor	Dose	Results	Reference
Human ovarian carcinoma HEY sc or iv	6 mg/kg on day 1, 8 and 15 iv or ip	Inhibition of tumor growth with 15 out of 18 treated mice being tumor-free when compared with 5 out of 18 in case of free adriamycin	*(169)*
Pancreatic adenocarcinoma AsPC-1 sc	3 mg/kg on d 1, 8, 15, 22, and 29 iv	Inhibition of tumor growth and more effective than free adriamycin (7/20 Vs 1/20). The doxorubicin could be detected in the tumor long after treatment (168 h) when compared with free adriamycin (24 h).	*(170)*
Prostatic carcinoma PC-3	6 mg/kg on d 1, 8, 15, and 22 iv	More effective than free adriamycin in inhibiting tumor growth	*(171)*

ip, intraperitoneal; iv, intravenous; sc, subcutaneous.

(DSPE-PEG-Osu 2000), and cholesterol. The activity of the thermosensitive liposomes with entrapped doxorubicin was studied in mice bearing colon 26 tumor. When compared to conventional thermosensitive liposomes, the long-circulating thermosensitive liposomes resulted in 2–2.5 times higher drug levels in the tumor tissue upon application of heat (42°C).

Tumor targeting is also enhanced by combining the unique properties of long-circulation and pH sensitivity. Slepushkin et al. *(177)* have prepared sterically stabilized, pH-sensitive liposomes using cholesteryl hemisuccinate (CHEMS), DOPE, and PEG-derivatized phosphatidylethanolamine (PEG-PE). The pharmacokinetics of the ^{131}In-labeled liposomes was studied by injecting them iv in rats. The sterically stabilized, pH-sensitive liposomes were found to have enhanced tumor accumulation (100-fold) and longer circulation time ($t_{1/2}$ = 11.1 h) when compared with the pH-sensitive liposomes, which are rapidly cleared within the first 30 min.

4.4. Niosomes

Niosomes are nonionic surfactant vesicles (NSV) formed by hydration of a single alkyl chain nonionic surfactant with cholesterol or a steroid *(178)*. These vesicles are considered to be an alternative to liposomes and are capable of entrapping water-soluble drugs, altering their distribution and metabolic stability. Niosomes are used to encapsulate a wide range of anticancer drugs such as doxorubicin, vincristine, methotrexate, pentoxyfiline, and 5-fluorouracil (5-FU).

Uchegbu et al. *(179)* prepared doxorubicin-loaded niosomes using sorbitan monostearate (Span® 60), cholesterol and choleth-24 (24 oxyethylene cholesteryl ether) in a weight ratio of 45:45:10. The niosomal formulation, at doxorubicin dose of 5 mg/kg and 10 mg/kg, was injected iv into female NMRI mice bearing subcutaneously (sc) implanted MAC15A tumor. The results showed an increase in the area under the plasma concentration-time curve (sixfold) and tumor level-time curve (50%) as compared to the control doxorubicin administered as an aqueous solution. However, an increase in drug metabolites was observed in the liver with no significant difference in drug concentration in the heart.

Parthasarathi et al. *(180)* prepared niosomal encapsulated vincristine using nonionic surfactant (Span® 40) and cholesterol in a 1:1 ratio. The antitumor activity was evaluated by injecting the formulation containing 0.5 mg/kg and 1.0 mg/kg of vincristine into BALB/c mice bearing Sarcoma-180 and Ehrlich ascites. The results indicate 110% increase in life span of the animals receiving the niosomal formulation relative to the control. Diarrhea, severe weight loss, and muscular weakness, which are characteristic of vincristine, were not observed in animals receiving the niosomal formulation of the drug.

Hao et al. *(181)* evaluated the encapsulation efficiency of niosomes prepared by evaporation sonication method using different kinds of Spans (20, 40, 60, and 80). Span 60 was found to have the highest encapsulation efficiency. This study also showed that the encapsulated formulation provided prolonged release of colchicine and 5-FU over a 24-h period. Niosomes were also found to alter the metabolic profile of methotrexate by preventing the rapid formation of 7-hydroxy methotrexate.

5. Micellar Systems

Micelles are defined as colloidal dispersions with a size ranging from 5 to 50 nm. Micelles are made from amphipathic polymer molecules that have hydrophilic and hydrophobic regions or segments *(182)*. At a low concentration, these compounds exist as single molecules or unimers. As the concentration in aqueous solution increases, the amphipathic polymer chains aggregate to form three-dimensional (3D) structures or micelles. The concentration of the polymer at which the micelles are formed is known as the critical micelle concentration (CMC). Polymeric micelles may be formed from diblock, triblock, or graft copolymers *(183–185)*. In aqueous systems, these micelles consist of an inner hydrophobic core and an outer hydrophilic shell. An increase in the length hydrophobic segment was found to be associated with a decrease in the CMC value *(186)*. Hydrophobic anticancer drugs can be entrapped into the core of the micelle and can result in an improvement in the solubility and bioavailability of these drugs. For drug-delivery applications, it is important that the CMC value of the polymeric molecules is low enough so as to preserve the micellar structure even under high dilutions, such as after systemic administration. For this reason, amphiphilic-block copolymers that form micellar structures at low concentrations are used for drug-delivery applications.

Drug delivery to the tumor tissue using micellar carriers is achieved by passive targeting based on the EPR effect. Passive targeting followed by an increase in temperature, acidic conditions, or ultrasound may also enhance the release of anticancer drugs at the tumor site *(182)*. One of the major advantages of micellar delivery systems is the presence of the hydrophobic core that can entrap and retain hydrophobic drugs. In addition, relative to liposomes and nanoparticles, micellar carriers usually have much smaller diameter and thus provide a far more efficient tumor uptake. When compared to the free drug, an increase in circulation time was achieved by using the polymeric micelles. The increased circulation time allows for the drug to accumulate in target tissue. In contrast to many surface-active agents, polymer micelles are preferred as carriers of anticancer drugs owing to their high loading capacity and long circulating properties *(187,188)*. However, the circulation time of micellar structures is usually shorter than that of long-circulating liposomes. This is because micelles are smaller than long-circulating liposomes and are eliminated at a faster rate. Micellar systems also reduce the toxicity and enhance the antitumor activity of the drugs *(189)*.

5.1. Poly(Ethylene Glycol) Block Copolymer Micelles

PEG and PEO are water-soluble polymers used as hydrophilic blocks in the preparation of diblock and triblock copolymers. The block copolymers are synthesized by anionic polymerization or by ring-opening polymerization. The outer shell of the block copolymers is formed by the hydrophilic polymers (PEG or PEO) and the inner core is formed by the hydrophobic polymers such as poly(L-aspartic acid) (PAsp), poly(β-benzyl L-aspartate) (PBLA), Poly(γ-benzyl-L-glutamate) (PBLG), poly(D-lactic acid) (PLA), poly(propylene oxide) (PPO), and distearoylphosphatidylethanolamine (DSPE) (Table 6). The hydrophilic nature of the outer shell prevents the interaction of the micelles with the MPS and, thus increases the circulation time after systemic administration.

PEG or PEO is a nonionic homopolymer of ethylene oxide that is distinguished based on the molecular weight (i.e., <20,000 Daltons are PEG). The total molecular weight of these polymers depends on the average number of oxyethylene groups present in the molecule. The chemical composition and structures the polymers PEG and PEO are the same except for the presence of an extra hydroxyl group in PEG. Hence PEG and PEO are discussed together.

Table 6
Some Examples of Block Copolymers of Poly(Ethylene Glycol) or Poly(Ethylene Oxide) Using for Tumor-Targeted Drug Delivery

Block Copolymer	Reference
Poly(ethylene glycol)-poly(L-aspartic acid)	*(190)*
Poly(ethylene glycol)-poly(β-benzyl L-aspartate)	*(192)*
Poly(ethylene glycol)-poly(D,L-lactic-co-glycolic acid)	*(209)*
Methoxypoly(ethylene glycol)-poly(D,L-lactide)	*(210)*
Poly(ethylene glycol)-phosphatidyl ethanolamine	*(214)*
Poly(ethylene oxide)-poly(aspartic acid)	*(197)*
Poly(ethylene oxide)-poly(β-benzyl L-aspartate)	*(200)*
Poly(ethylene oxide)-poly(γ-benzyl glutamate)	*(213)*
Poly(ethylene oxide)-poly(propylene oxide)	*(203)*

5.1.1. PEG-Poly(L-Aspartic Acid) or PEO-Poly(L-Aspartic Acid)

Yokoyama et al. *(190–193)* prepared polymeric micelles from PEG-poly(L-aspartic acid) (PEG-PAsp) block copolymer. The anticancer drug doxorubicin was incorporated into the micelles by chemical conjugation and physical entrapment. Physical entrapment of the drug resulted in the formation of a doxorubicin dimer, which contributed to the stability and slow release of the drug. The incorporation of the drug in the micelles also resulted in an increase in the maximum tolerated dose. This caused a significant regression in tumor volume upon systemic administration. The antitumor activity was evaluated in mice bearing different solid tumors (C26, P388, C38, M 5076, MKN-45, and MX-1). A significant reduction in weight loss and regression in tumor volume were observed in all of the tumor models. It was also observed that the physically entrapped drug contributed positively towards the antitumor activity as compared to the covalently attached drug. The increased half-life of the doxorubicin and the small size of the micelles (50 nm) were found to further increase the potential of these systems as carriers of anticancer drugs. Cisplatin-containing micelles of PEG-PAsp block copolymer were prepared by ligand-substitution reaction. The size of the micelles (20–100 nm) was found to increase with an increase in size of PAsp segment. These micelles were found to be highly stable upon dilution *(194)*. When compared to the free cisplatin, the micelles were found to have an increased antitumor activity and reduced nephrotoxic acitivity in nude mice bearing MKN45, a human gastric cancer xenograft tumor *(195)*.

Nakanishi et al. *(196)* prepared a micellar carrier NK911, which is a block copolymer of PEG-PAsp. The conjugated doxorubicin was found to increase the hydrophobicity of the core and the entrapped doxorubicin contributed to the anticancer activity. High-performance liquid chromatography (HPLC) analysis of the doxorubicin concentrations in the plasma and tumor tissue, after systemic administration, showed that there was a significant enhancement of antitumor activity with the NK911 micelles (40 nm). The authors concluded that these micelles were passively targeted to the tumor mass by the EPR effect. The antitumor activity of the doxorubicin-loaded NK911 and free doxorubicin was also determined by the in vivo experiments conducted on CDF1 mice inoculated with four sc tumors (C26, M5076, Lu-24, MX-1) and one intravenous tumor (P388). The results of the in vivo experiments given in Table 7 and Table 8, show that

Table 7
Antitumor Activity of NK911 and Doxorubicin in Murine Tumor Models

Dose (mg/kg)	Mouse colon 26 carcinoma		Mouse M5076 sarcoma		Mouse P388 leukemia	
	NK911 T/C (%)[a]	DOX T/C (%)	NK911 T/C (%)[b]	DOX T/C (%)	NK911 ILS (%)[c]	DOX ILS (%)
30.0	2.4	10.3	19.2	—	379	199
24.0	14.4	21.2	30.5	26.8	390	290
19.2	31.6	33.7	35.4	54.1	203	73
15.4	50.6	48.5	34.4	72.5	68	58

[a]T/C (%) - (Tumor volume of treated mouse / Tumor volume of control mouse) × 100 on d 13.
[b]T/C (%) - (Tumor volume of treated mouse / Tumor volume of control mouse) × 100 on d 14.
[c]ILS (%) - Increased life span: [(mean survival time of treated mice / mean survival time of control mice)-1] × 100 on d 60.
Adapted from ref. *196*.

doxorubicin in NK911 micelles provides better antitumor activity as compared with the free drug. The micellar systems presently are undergoing clinical trials in Japan.

Yokoyama and co-workers *(197,198)* developed doxorubicin-conjugated PEO-PAsp micelles and also entrapped additional doxorubicin by physical means. Any unconjugated doxorubicin remaining in the micelles was removed by dialysis and a known amount of the drug was entrapped. Micelles prepared by this method were found to have 100 times lower in vitro cytotoxic activity than the free doxorubicin against $P388D_1$ cells. It was also observed that PEO-PAsp micelles resulted in increased survival time and reduced cytotoxicity when given intraperitoneally to mice bearing P388 leukemia.

5.1.2. PEG-Poly(β-Benzyl L-Aspartate) or PEO-Poly(β-Benzyl L-Aspartate)

Yokoyama et al. *(192)* evaluated the potential of PEG-poly(β-benzyl L-aspartate) (PEG-PBLA) block copolymer micelles and its derivatives as carriers of an investigational anticancer drug KRN 5500. The polymers were synthesized by ring-opening polymerization and were formed by partial hydrolysis and cetyl ester substitution of the β-benzyl L-aspartate units. KRN 5500 was incorporated by dialysis, followed by sonication, which resulted in drug-containing micelles (71 nm in diameter). Kataoka et al. *(199)* prepared doxorubicin-loaded micelles of PEG-PBLA by the emulsion method to enhance the stability of the drug and provide sustained release. The dimer derivatives of doxorubicin thus formed enhanced the stability and sustained release. The accelerated release of doxorubicin at pH 5.0 showed that the micelles were pH-sensitive and would be beneficial for tumor-targeting, because the pH around the tumor is much lower than the systemic circulation. Intravenous injection of these micelles into mice with C26 tumor indicated higher anticancer activity and long circulation as compared with free doxorubicin.

Micelles of PEO-PBLA have also been used as carriers of anticancer drugs *(200–202)*. These micelles provide an opportunity for drug targeting and decrease toxicity. For instance, in the case of doxorubicin-loaded PEO-PBLA micelles (approx 40 nm in diameter), entrapment of the drug was found to enhance the stability of the drug.

Table 8
Anti-Tumor Activity of NK911 and Doxorubic Against Subcutaneously Implanted Tumors

Dose (mg/kg)	Human Lu-24 lung cancer NK911 T/C (%)[a]	Human Lu-24 lung cancer DOX T/C (%)	Human MX-1 breast cancer NK911 T/C (%)	Human MX-1 breast cancer DOX T/C (%)
18.0	23.5	—	30.7	—
14.4	35.5	28.3	42.2	48.3
11.5	54.7	48.1	67.6	63.1

[a]T/C (%) - (Tumor volume of treated mouse / Tumor volume of control mouse) × 100 on d 14.
Adapted from ref. *196*.

5.1.3. PEO-PPO-PEO Triblock Copolymers (Pluronics® or Poloxamers)

The PEO-PPO-PEO triblock copolymers are commercially available as Pluronics® or poloxamers from BASF Corporation (Parsipanny, NJ). The PPO segment forms the inner hydrophobic core and the two PEO chains forms the outer hydrophilic shell *(203)*. There are more than 30 different types of PEO-PPO-PEO triblock copolymers available with varying PPO and PEO chain lengths. The CMC of these copolymers also depends on the chemical composition. These triblock copolymers form micelles that are highly hydrophilic and escape the RES, which results in long circulation time. Hence these polymeric micelles were used to increase the efficiency of anticancer drugs like camptothecin *(187)* and doxorubicin *(204)*. However, a decrease in efficiency of the Pluronic systems was observed at concentrations above the CMC (44–46). Drug release from the micelles was enhanced by the application of ultrasound at the tumor site *(205)*.

It was also observed that the Pluronic micelles could be used to deliver anticancer agents to MDR tumors. The Pluronic effects on the drug resistance mechanisms include inhibition of drug efflux transporters and glutathione/glutathione S-transferase detoxification system. These energy-dependent mechanisms were inhibited by ATP depletion induced by Pluronic block copolymer *(206,207)*. Batrakova et al. *(208)* evaluated the anticancer activity of Pluronic L-61, P-85, and F-108 micelles carrying epirubicin (EPI) and doxorubicin (DOX) in animals bearing murine leukemia (P388) and daunorubicin-sensitive (Sp2/0) and resistant (Sp2/0[DNR]) myeloma tumors grown sc. This study has shown that the Pluronic micelles are associated with a greater than 150% increase in survival time between treatment and control groups. The formulations EPI/P85 and DOX/L61 were found to inhibit tumor growth in more than 90% of the animals and complete disappearance of tumor growth in 33–50% of animals bearing Sp2/0 tumors. An increase in the therapeutic activity of the micellar drug was observed with an increase in the hydrophobicity of the copolymer as a result of increasing the PPO segment length.

5.1.4. Miscellaneous Diblock Copolymers of PEG and PEO

Yoo et al. *(209)* prepared biodegradable polymeric micelles from the diblock copolymer of PEG and PLGA. Doxorubicin was incorporated into the micelles by conjugating the amino group of the drug with the hydroxyl group of PLGA. The micelles containing conjugated doxorubicin were found to exhibit sustained release to a greater extent than the micelles with entrapped drug. The cytotoxic activity of the micelles against HepG2 cells was found to be greater than the free drug. Micelles of the diblock copolymer of methoxyPEG and poly(D,L-lactide) (MPEG-PLA) are used as carriers of the anticancer drug paclitaxel *(210)*. The diblock copolymer is formed by ring-opening polymerization of the monomers. The antitumor activity of the micelles was tested in vitro in different cell lines (OVCAR-3, MCF7, Hs578T, SKMES, and HT-29) *(211,212)*. The biodistribution studies of paclitaxel in mice bearing B16 melanoma showed that the concentration of micellar paclitaxel is two to three times higher than the free drug in tissues, including liver, spleen, kidneys, lungs, heart, and tumor. Micellar paclitaxel administered in mice bearing SKOV-3 tumor (human ovarian cancer) and P388 leukemia showed a significantly enhanced activity than did the free drug.

Doxorubicin-loaded polymeric micelles of PEO-PBLG were prepared by dialysis and ultrafiltration (diafiltration) method. The size of the micelles formed was found to increase with an increase in PBLG content (20 nm to 70 nm). The release rate of the drug was reduced with an increase in PBLG chain length and the amount of incorporated doxorubicin *(213)*.

5.1.5. PEG-Diacyllipids

PEG-phosphatidylethanolamine conjugates are used to prepare highly stable micelles, which can act as carriers for the delivery of different hydrophobic anticancer drugs *(214–216)*. The phosphatidylethanolamine used is highly hydrophobic owing to the presence of two long-chain fatty acyl groups and PEG is highly hydrophilic. These PEG-lipid conjugates can form a micellar system that are very stable and have a very low CMC value in the 10^{-5} molar range. PEG of varying chain lengths was used to prepare micelles 7–35 nm in diameter. An increase in the circulation time and selective tumor accumulation was observed upon iv administration of the micellar system into mice bearing sc Lewis lung carcinoma and EL4 lymphoma *(217)*. Weissig et al. *(216)* have compared the biodistribution of the PEG-distearoylphosphatidylethanolamine (PEG-DSPE) micelles with PEG-modified liposomes using mice bearing subcutaneous Lewis lung carcinoma. It was observed that the micelles containing the model protein, soybean trypsin inhibitor, accumulated in the tumor to a greater extent than the long-circulating liposomes. This could be explained by the fact that the cut-off size of the vascular pores in Lewis lung carcinoma is small and permits micelles (5–50 nm) more efficiently than long-circulating liposomes (>100 nm) by passive targeting.

5.2. Thermoresponsive Micelles

Block copolymers of poly(*N*-isopropylacrylamide) (NIPAAm) are used in the preparation of thermoresponsive micelles along with poly(D,L-lactide) *(218)*, polystyrene *(219)*, poly(polybutylmethacrylate) *(220)*, poly(dimethylacrylamide) *(221)*, and alkylterminated PIPAAm *(222)*. NIPAAm is a water-soluble polymer, which undergoes a phase-transition from hydrated to dehydrated state (cloud point) with a slight change in solution temperature in the range of 35°–38°C. Diblock copolymers of PIPAAm, obtained by ring-opening polymerization, were used in the preparation of thermoresponsive micelles, which act as carriers of anticancer drugs. The anticancer drug doxorobicin was incorporated into the micelles by the dialysis method. The swollen core formed owing to the presence of water was found to increase the doxorubicin-loading capacity *(221)*. Passive targeting by the EPR effect followed by an increase in temperature at the tumor site would enhance the release of the drug from the micelles. These micelles were associated with a reduction in the cytotoxicity and an increase in therapeutic efficiency of the anticancer drugs.

5.3. Other Polymeric Micelles

Miwa et al. *(223)* synthesized *N*-lauryl-carboxymethyl-chitosan (LCC), which is used as a carrier of the anticancer drug paclitaxel. Chitosan, a hydrophilic polymer of D-glucosamine, cannot form micelles and so LCC was synthesized by attaching lauryl groups to the primary amino groups of D-glucosamine in carboxymethyl chitosan. An increase in the solubility of paclitaxel was observed upon incorporation in the micellar structure. The in vitro studies against KB cells indicate an increase in the anticancer activity of the entrapped paclitaxel relative to the free drug.

6. Conclusions

Passive targeting of anticancer drugs to solid tumors using colloidal-carriers systems provides a unique opportunity to develop a newer generation of safer drugs. Preclinical and clinical results presented in this chapter show that the different types of colloidal carriers can improve the therapeutic benefits of anticancer drugs when directed to its intended target. A very promising observation in colloidal delivery of anticancer agents is the opportunity to overcome MDR as the transport of the macromolecules or particulate carrier into the cells bypasses the P-gp efflux pump present in the cell membrane. Biocompatibility of the carrier system is the most important issue that needs to be considered in developing these systems for tumor targeting.

References

1. Seymour WL. Systemic cancer therapy using polymer-based prodrugs and progenes. In: Dumitriu S, ed. Polymeric Biomaterials, 2nd edition, New York, NY, Marcel-Dekker, Inc., 2001, pp. 843–850.
2. Jain RK. Normalizing tumor vasculature with anti-angiogenic therapy: a new paradigm for combination therapy. Nature 2001;7:987–989.
3. Teicher BA. Drug Resistance in Oncology. New York, NY, Marcel Dekker, 1994.
4. Yuan F. Transvascular drug delivery in solid tumors. Semin Radiat Oncol 1998;8:164–175.
5. Jain RK. Delivery of molecular and cellular medicine to solid tumors. Microcirculation 1997;4:1–23.
6. Senger DR, Galli SJ, Dvorak AM, et al. Tumor cells secrete a vascular permeability factor that promotes accumulation of ascites fluid. Science 1983;219:983–985.
7. Leung DW, Cachianes G, Kuang WJ, et al. Vascular endothelial growth factor is a secreted angiogenic mitogen. Science 1989;246:1306–1309.
8. Maeda H, Matsumura Y, Kato H. Purification and identification of (hydroxypropyl) bradykinin in ascitic fluid from a patient with gastric cancer. J Biol Chem 1988;263:16051–16054.
9. Maeda H, Noguchi Y, Sato K, Akaike T. Enhanced vascular permeability in solid tumor is mediated by nitric oxide and inhibited by both new nitric oxide scavenger and nitric oxide synthase inhibitor. J Cancer Res 1994;85:331–334.
10. Folkman J, Shing Y. Angiogenesis. J Biol Chem 1992;267:10931–10934.
11. Munn LL, Koenig GC, Jain RK, Melder RJ. Kinetics of adhesion molecule expression and spatial organization using targeted sampling fluorimetry. Biotechniques 1995;19:622–631.
12. Jain RK. Transport of molecules across tumor vasculature. Cancer Metas Rev 1987;6:559–593.
13. Jain RK, Safabakhsh N, Sckell Y, et al. Endothelial cell death angiogenesis, and micro vascular function following castration in an androgen-dependent tumor: role of VEGF. Proc Natl Acad Sci USA 1998;95:10820–10825.
14. Tannock IF, Rotin D. Acid pH in tumors and its potential for therapeutic exploitation. Cancer Res 1989;49:4373–4384.
15. Dvorak HF, Brown LF, Detmar M, Dvorak AM. Vascular permeability factor/vascular endothelial growth factor, micro vascular hyperpermeability and angiogenesis. Am J Pathol 1995;146:1029–1039.

16. Kohn S, Nagy JA, Dvorak AM. Pathways of macromolecular tracer transport across venules and small veins: structural basis for hyperpermeability of tumor blood vessels. Lab Invest 1992;67:596–607.
17. Dvorak AM, Kohn S, Morgan ES. The vesiculo-vacuolar organelles (VVO): a distinct endothelial cell structure that provides transcellular pathway for macromolecular extravasation. J Leukoc Biol 1996;59:100–115.
18. Feng D, Nagy JA, Hipp J. Vesiculo-vacuolar organelles and the regulation of venule permeability to macromolecules by vascular permeability factor, histamine and seratonin. J Exp Med 1996;183:1981–1986.
19. Roberts WG, Palade GE. Increased vascular permeability and endothelial fenestration induced by vascular endothelial growth factor. J Cell Sci 1995;108:2369–2379.
20. Roberts WG, Palade GE. Neovasculature induced by vascular endothelial growth factor is fenestrated. Cancer Res 1997;57:765–772.
21. Maeda H, Matsumura Y. Tumoritropic and lymphotropic principles of macromolecular drugs. Crit Rev Ther Drug Carrier Syst 1989;6:193–210.
22. Maeda H. SMANCS and polymer-conjugated macromolecular drugs: advantages in cancer chemotherapy. Adv Drug Delivery Rev 1991;6:181–202.
23. Matsumura Y, Maeda H. A new concept for macromolecular therapeutics in cancer chemotherapy: mechanism of tumorotropic accumulation of proteins and the antitumor agent SMANCS. Cancer Res 1986;46:6387–6392.
24. Muggia F. Doxorubicin-polymer conjugates: further demonstration of the concept of enhanced permeability and retention. Clin Cancer Res 1999;5:7–8.
25. Duncan R. Drug-polymer conjugates: potential for improved chemotherapy. Anticancer Drugs 1992;3:175–210.
26. Duncan, R. Selective endocytosis. In: Robinson JR, ed. Sustained and Controlled Drug Delivery Systems. New York, Marcel Dekker, 1987, pp. 581–621.
27. Luo Y, Prestwich GD. Cancer-targeted polymeric drugs. Curr Cancer Drug Targets 2002;2:209–226.
28. Kopecek J, Bazilova H. Poly[N-(2-hydroxypropyl) methacrylamide] 1. Radical polymerization and copolymerization. Eur Polym 1973;9:7–14.
29. Seymour WL, Duncan R, Kopecekova P, Kopecek J. Daunomycin and adriamycin-N-(2-hydroxypropyl) methacrylamide copolymer conjugates: toxicity reduction by improved drug-delivery. Cancer Treatment Rev 1987;14:319–327.
30. Vasey PA, Kaye SB, Morrison R, et al. Phase I clinical and pharmacokinetic study of PK1 [N-(2-hydroxypropyl)methacrylamide copolymer doxorubicin]: first member of a new class of chemotherapeutic agents-drug-polymer conjugates. Clin Cancer Res 1999;5:83–94.
31. Merum Terwogt JM, Ten Bokkel Huinink WW, Shellens JH, et al. Phase I clinical and pharmacokinetic study of PNU166945, a novel polymer-conjugated prodrug of paclitaxel. Anticancer Drugs 2001;12:315–323.
32. Caiolfa VR, Zamal M, Fiorini A, et al. Polymer bound camptothecin: initial biodistribution and antitumor activity studies. J Control Release 2000;65:105–120.
33. Gianasi, E., Wasil, M., Evagorou, E.G., Keddle, A., Wilson G, Duncan R. HPMA copolymer platinates as novel antitumor agents: *in vitro* properties, pharmacokinetics and antitumor activity. Eur J Cancer 1999;3:994–1002.
34. Etrych T, Jelinkova M, Rihova B, Ulbrich K. New HPMA copolymers containing doxorubicin bound via pH-sensitive linkage: synthesis and preliminary *in vitro* and *in vivo* biological properties. J Control Release 2001);73:89–102.
35. Cassidy J, Duncan R, Morrison GJ, et al. Activity of N-(2-hydroxypropyl) methacrylamide copolymers containing daunomycin against rat tumor model. Biochem Pharmacol 1989;38:875–879.
36. Duncan R, Kopecekova P, Strohalm J, et al. Anticancer agents coupled to N-(2-hydroxypropyl) methacrylamide copolymers. II. Evaluation of daunomycin conjugates in vivo against L1210 leukemia. Br J Cancer 1988;57:147–156.
37. Kasuya Y, Lu ZR, Kopecekova P, et al. Synthesis and characterization of HPMA copolymer-aminopropylgeldanamycin conjugates. J Control Release 2001;74:203–211.

38. Omelyanenko V, Kopecekova P, Gentry C, Kopecek, J. Targetable HPMA copolymer-adriamycin conjugates. Recognition, internalization and subcellular fate. J Control Release 1998;53:25–37.
39. Satchi R, Connors TA, Duncan R. PDEPT: polymer-directed enzyme prodrug therapy. ;I.HPMA copolymer-cathepsin B and PK1 as a model combination. Br J Cancer 2001;85:1070–1076.
40. Searle F, Gac-Breton S, Keane R, et al. N-(2-hydroxypropyl) methacrylamide copolymer-6-(3-aminopropyl)-elliciptine conjugates. Synthesis, in vitro, and preliminary in vivo evaluation. Bioconjug Chem 2001;12:711–718.
41. Tijerina M, Kopecekova J, Kopecek J. The effects of subcellular localization of N-(2-hydroxypropyl) methacrylamide copolymer-Mce6 conjugates in a human ovarian carcinoma. J Control Release 2001;74:269–273.
42. Seymour LW. Passive tumor targeting of soluble macromolecules and drug conjugates. Crit Rev Ther Drug Carrier Syst 1992;9:135–187.
43. Kuromizu K, Tsunasawa S, Maeda H, et al. Reexamination of the primary structure of an antitumor protein, neocarzinostatin. Arch Biochem Biophys 1986;246:199–205.
44. Kuromizu K, Abe O, Maeda H. Location of the disulfide bonds in the antitumor protein, neocarzinostatin. Arch Biochem Biophys 1991;286:569–573.
45. Maeda H, Takeshita J, Yamashita A. Lymphotropic accumulation of an antitumor antibiotic protein neocarzinostatin. Eur J Cancer 1980;16:723–731.
46. Maeda H, Matsumura Y. New tactics and basic mechanisms of targeting chemotherapy in solid tumors. In: Kimura K, Carter SK, Ota K, Pinedo HM, eds. Cancer Chemotherapy: Challenges for the Future. Tokyo, Excerpta Medica, 1989, pp. 239–260.
47. Konno T, Maeda H. Targeting chemotherapy of hepatocellular carcinoma: arterial administration of SMANCS/Lipiodol. In: Okada K, Ishak KG, eds. Neoplasms of the Liver. New York, NY, Springer-Verlag, 1987, pp. 276–291.
48. Konno T. Targeting anticancer chemotherapy for primary and secondary liver cancer using arterially administered oily anticancer agents. In: Kimura K, ed. Cancer Chemotherapy: Challenges for the Future. Tokyo, Excerpta Medica, 1987, pp. 299–311.
49. Hirano T, Ohashi S, Morimoto S, et al. Synthesis of antitumor- active conjugates of adriamycin or daunomycin with the copolymer of divinyl ether and maleic anhydride. Makromol Chem 1986;187:2815–2824.
50. Zalipsky S. Functionalized poly(ethylene glycol) for preparation of biologically relevant conjugates. Bioconjug Chem 1995;6:150–165.
51. Greenwald RB, Gilbert CW, Pendri A, et al. Drug delivery systems: water soluble taxol 2'-poly(ethylene glycol) ester prodrugs: design and in vivo effectiveness. J Med Chem 1996;39:424–431.
52. Pendri A, Conover CD, Greenwald RB. Antitumor activity of paclitaxel-2'-glycinate conjugated to poly(ethylene glycol): a water-soluble prodrug. Anti-Cancer Drug Design 1998;13:387–395.
53. Minko T, Paranjpe PV, Qiu B, et al. Enhancing the anticancer efficacy of camptothecin using biotinylated poly(ethylene glycol) conjugates in sensitive and multidrug-resistant human ovarian carcinoma cells. Cancer Chemother Pharmacol 2002;50:143–150.
54. Calceti P, Monfardini C, Sartore L, et al. Preparation and properties of monomethoxy poly(ethylene glycol) doxorubicin conjugates linked by an amino acid or a peptide spacer. Farmaco 1993;48:919–932.
55. Keating MJ, Holmes R, Lermer S, Ho DH. Asparginase and PEG asparginase-past, present and future. Leuk Lymphoma 1993;10:153–157.
56. Ho DH, Brown NS, Yen A, et al. Clinical pharmacology of polyethylene glycol-asparginase. Drug Metab Disposit 1986;14:349–352.
57. Kurtzberg J, Moore JO, Scudiery D, Franklin A. A phase II study of polyethylene glycol (PEG) conjugated L-asparginase in patients with refractory acute leukemia's. Proc AACR 1988;29:213.
58. Duncan R, Spreafico F. Polymer conjugates. Pharmacokinetic considerations for design and development. Clin Pharmacokinet 1994;27:290–306.

59. Sawa T, Wu J, Akaike T, Maeda H. Tumor-targeting chemotherapy by a xanthine oxidase-polymerconjugate that generates oxygen-free radicals in tumor tissue. Cancer Res 2000;60:666–671.
60. Li C. Poly(L-glutamic acid)-anticancer drug conjugates. Adv Drug Delivery Rev 2002;54(5):695.
61. Singer JW, Bhatt R, Tulinsky J, et al. Water-soluble poly(L-glutamic acid)-gly-camptothecin conjugates enhance camptothecin stability and efficacy in vivo. J Control Rel 2001;74:243–247.
62. Luo Y, Prestwich GD. Hyaluronic acid-N-Hydroxysuccinate: a useful intermediate for bioconjugation. Bioconjug Chem 2001;12:1085–1088.
63. Luo Y, Bernshaw NJ, Lu ZR, et al. Targeted delivery of doxorubicin by HPMA copolymer-hyaluronan bioconjugates. Pharm Res 2002;19:396–402.
64. Luo Y, Prestwich GD. Synthesis and selective cytotoxicity of a hyaluronic acid-antitumor bioconjugate. Bioconjug Chem 1999;10:755–763.
65. Okuno S, Harada M, Yano S, et al. Complete regression of xenografted human carcinomas by camptothecin analogue-carboxymethyl dextran conjugate. Cancer Res 2000;60:2988–2995.
66. Avichezer D, Schechter B, Arnon R. Functional polymers in drug delivery: carrier-supported CDDP (cisplatin) complexes of polycarboxylates-effect on human ovarian carcinoma. React Funct Polym 1998;36:59–69.
67. Bernstein A, Hurwitz E, Maron R, et al. Higher antitumor efficacy of daunomycin when linked to dextran: in vivo and in vitro studies. J Natl Cancer Inst 1978;60:379–384.
68. Hejazi R, Amiji MM. Chitosan based delivery systems: physicochemical properties and pharmaceutical applications. In: Dumitriu S, ed. Biomaterials. Qubec, Marcel Dekker, 2001, pp. 213–237.
69. Sato M, Onishi H, Takahara J, et al. In vivo drug release and antitumor characteristics of water soluble conjugates of mitomycin C with glycol-chitosan and N-succinyl-chitosan. Biol Pharm Bull 1996;19(9):1170–1177.
70. Swarts JC, Swarts DM, Neuse EW, et al. Polyaspartamides as water-soluble drug carriers. part 1: Antineoplastic activity of ferrocene-containing polyaspartamide conjugates. Anticancer Res 2001;21:2033–2037.
71. Reddy BS, Damayanthi Y, Lown JW. Design synthesis and in vitro cytotoxicity studies of novel pyrrolo [2,1-c] [1,4] benzodiazepine (PBD)-polyamide conjugates and 2, 2'-PBD dimers. Anti-Cancer Drug Design 2000;15(3):225–238.
72. Kreuter J. Drug targeting with nanoparticles. Eur J Drug Metab Pharmacokinet 1994;3:253–256.
73. Brigger I, Dubernet C, Couvreur P. Nanoparticles in cancer therapy and diagnosis. Adv Drug Delivery Rev 2002l54:631–651.
74. Couvreur P, Dubernet C, Puisieux F. Controlled drug delivery with nanoparticles: current possibilities and future trends. Eur J Pharm Biopharm 1995;41:2–13.
75. Yoo HS, Lee KH, Oh JE, Park TG. In vitro and in vivo anti-tumor activities of nanoparticles based on doxorubicin-PLGA conjugates. J Control Release 2000;68:419–431.
76. Feng SS, Huang GF, Mu L. Nanospheres of biodegradable polymers: a system for clinical administration of an anticancer drug paclitaxel (Taxol). Ann Acad Med Singapore 2000;29:633–639.
77. Avgoustakis K, Beletsi A, Panagi Z, et al. PLGA-mPEG nanoparticles of cisplatin: in vitro nanoparticle degradation, in vitro drug release and in vivo drug residence in blood properties. J Control Release 2002;79:123–135.
78. Quintanar-Guerrero D, Fessi H, Allemann E, Dolker. Influence of stabilizing agents and preparative variables on the formation of poly(D, L-lactic acid) nanoparticles by an emulsification-diffusion technique. Int J Pharm 1996;143:133–141.
79. Hirosue S, Muller BG, Mulligan RC, Langer R. Plasmid DNA encapsulation and release from solvent diffusion nanospheres. J Control Release 2001;70:231–242.
80. Cohen H, Levy RJ, Gao J, et al. Sustained delivery and expression of DNA encapsulated in polymeric nanoparticles. Gene Ther 2000;7:1896–1905.
81. Maruyama A, Ishihara T, Kim JS, et al. Nanoparticles DNA carrier with poly(L-lysine) grafted polysaccharide copolymer and poly(D,L-lactic acid). Bioconjug Chem 1997;8:735–742.

82. Couvreur P, Grislain L, Lenaerts V, et al. Biodegradable polymeric nanoparticles as drug carrier for antitumor agents. In: Guiot P, Couvreur P. eds. Polymeric Nanoparticles and Microspheres. Boca Raton, FL, CRC Press, 1986, pp. 27–93.
83. Couvreur P, Kante, B, Roland M, Speiser P. Adsorption of antineoplastic drugs to polyalkylcyanoacrylate nanoparticles and their release characteristics in a calf serum medium. J Pharm Sci 1979;68:1521.
84. Brasseur F, Couvreur P, Kante B, et al. Actinomycin-D adsorbed on polymethylcyanoacrylate nanoparticles: increased efficiency against experimental tumor. Eur J Cancer 1980;16:1441.
85. Soma CE, Dubernet C, Barratt G, et al. Ability of doxorubicin-loaded nanoparticles to overcome multidrug resistance of tumor cells after their capture by macrophages. Pharm Res 1999;16:1710–1716.
86. Verdun C, Brasseur F, Vrancks H, et al. Tissue distribution of doxorubicin associated with polyisohexylcyanoacrylate nanoparticles. Cancer Chemother Pharmacol 1990;26:13–18.
87. Simeonova M, Ilarionova M, Ivanova T, et al. Nanoparticles as drug carriers for vinblastine. Acute toxicity of vinblastine in a free form and associated to polybutylcyanoacrylate nanoparticles. Bulgarian Acd Sci Acta Phy et Phram Bulg 1991;17:43–48.
88. Cuvier C, Roblot-Treupel L, Millot JM, et al. Doxorubicin-loaded nanospheres bypass tumor cell multidrug resistance. Biochem Pharmacol 1992;44:509–517.
89. Bennis S, Chapey C, Couvreur P, Roberts J. Enhanced cytotoxicity of doxorubicin encapsulated in poly-isohexylcyanoacrylate nanospheres against multidrug resistant tumor cells in culture. Eur J Cancer 1994;30A:89–93.
90. Gulyaev AE, Gelperina SE, Skidan IN, et al. Significant transport of doxorubicin into brain with polysorbate 80-coated nanoparticles. Pharm Res 1999;16:1564–1569.
91. Illum l, Davis SS. The organ uptake of intravenously administered colloidal particles can be altered using a non-ionic surfactant (poloxamer 338). FEBS Lett 1984;167:79–82.
92. Reszka R, Beck P, Fichtner I, et al. Body distribution of free, liposomal and nanoparticle-associated mitoxantrone in B16-melanoma-bearing mice. J Pharmacol Exp Ther 1997;280:232–237.
93. Beck P, Kreuter J, Reszka R, Fichtner I. Influence of polybutylcyanoacrylate nanoparticles and liposomes on the efficacy and toxicity of the anticancer drug mitoxantrone in murine tumor models. J Microencapsul 1993;10:101–114.
94. Zobel HP, Junghans M, Maienschein V, et al. Enhanced antisense efficacy of oligonucleotides adsorbed to monomethylaminoethylmethacrylate methylmethacrylate copolymer nanoparticles. Eur J Pharm Biopharm 2000;49:203–210.
95. Zimmer A. Antisense oligonucleotide delivery with polyhexylcyanoacrylate nanoparticles as carriers. Methods: Comp Meth Enzymol 1999;18:286–295.
96. Nakada Y, Fattal E, Foulquier M, Couvreur P. Pharmacokinetics and biodistribution of oligonucleotide adsorbed onto poly(isobutylcyanoacrylate) nanoparticles after intravenous administration in mice. Pharm Res 1996;13:38–43.
97. Fattal E, Vauthier C, Aynie I, et al. Biodegradable polyalkylcyanoacrylate nanoparticles for the delivery of oligonucleotides. J Control Release 1998;53:137–143.
98. Schwab G, Chavany C, Duroux I, et al. Antisense oligonucleotides adsorbed to polyalkylcyanoacrylate nanoparticles specifically inhibit mutated Ha-ras-mediated cell proliferation and tumorigenicity in nude mice. Proc Natl Acad Sci USA 1994;91:10460–10464.
99. Kattan J, Droz JP, Couvreur P, et al. Phase I clinical trial and pharmacokinetic evaluation of doxorubicin carried by polyisohexylcyanoacrylate nanoparticles. Invest New Drugs 1992;10:191–199.
100. Janes KA, Fresneau MP, Marazuela A, Fabra A. chitosan nanoparticles as delivery systems for doxorubicin. J Control Release 2001;73:255–267.
101. Mitra S, Gaur U, Ghosh PC, Maitra AN. Tumor targeted delivery of encapsulated dextran-doxorubicin conjugate using chitosan nanoparticles as carrier. J Control Release 2001;74:317–323.
102. Kabbaj M, Phillips NC. Anticancer activity of mycobacterial DNA: effect of formulation as chitosan nanoparticles. J Drug Target 2001;9:317–328.

103. Mao H, Roy K, Troung-Le VL, et al. Chitosan-DNA nanoparticles as gene carriers: synthesis characterization and transfection efficiency. J Control Rel 2001;70:399–421.
104. Farrugia CA, Groves JM. Gelatin behaviour in dilute aqueous solution: designing a nanoparticulate formulation. J Pharm Pharmacol 1999;51:643–649.
105. Truong-Le VL, Walsh SM, Schweibert E, et al. Gene transfer by DNA-gelatin nanospheres. Arch Biochem Biophys 1999;361:47–56.
106. Truong-Le VL, August JT, Leong KW. Controlled gene delivery by DNA-gelatin nanospheres. Hum Gene Ther 1998;9:1709–1717.
107. Coester C, Kreuter J, Briesen HV, Langer K. Preparation of avidin-labelled gelatin nanoparticles as carriers for biotinylated peptide nucleic acid (PNA). Int J Pharm 2000;196:147–149.
108. Kaul G, Amiji MM. Long-circulating poly(ethylene glycol)-modified gelatin nanoparticles for intracellular delivery. Pharm Res 2002;19:1062–1068.
109. Kaul G, Potineni A, Lynn DM, et al. Surface-modified polymeric nanoparticles for tumor-targeted delivery. surFacts Biomaterials 2002;7:1–5.
110. Sharma D, Chelvi TP, Kaur J, et al. Novel taxol formulation: polyvinylpyrrolidine nanoparticle-encapsulated taxol for drug delivery in cancer therapy. Oncol Res 1996;8:281–286.
111. Kim SY, Lee YM. Taxol-loaded block copolymer nanospheres composed of methoxy poly(ethylene glycol) and poly(epsilon-caprolactone) as novel anticancer drug carriers. Biomaterials 2001;22:1697–1704.
112. Oh I, Lee K Kwon HY, Lee YB, et al. Release of adriamycin from poly(gamma-benzyl-L-glutamate)/poly(ethylene oxide) nanoparticles. Int J Pharm 1999;181:107–115.
113. Potineni A, Lynn DM, Langer R, Amiji MM. Poly(ethylene oxide)-modified poly(β-amino ester) nanoparticles as a pH-sensitive biodegradable system for paclitaxel delivery. J Control Release 2003;86:223–234.
114. Thunemann AF. Polyethyleneimine complexes with retinoic acid: structure, release profile and nanoparticles. Macromolecules 2000;33:6878–6885.
115. Bangham AD, Standish MM, Warkins JC. Diffusion of univalent ions across the lamellae of swollen phospholipids. J Mol Biol 1965;13:238–252.
116. Chrai SS, Ahmad I. Liposomes (a review). Part one: manufacturing issues. BioPharm (2001);Nov:10–13.
117. Juliano RL, Stamp D. Interactions of drugs with lipid membranes. Characteristics of liposomes containing polar or non-polar antitumor drugs. Biochim Biophys Acta 1979;586:137–145.
118. Robert KY, Zee-Cheng, Cheng CC. Delivery of Anticancer drugs. Meth Find Exp Clin Pharmacol 1989;11:439–529.
119. Oku N, Tokudome Y, Asai T, Tsukada H. Evaluation of drug targeting strategies and liposomal trafficking. Curr Pharmaceut Design 2000;6:1669–1691.
120. Gregoriadis G, Neerunjun ED. Treatment of tumor bearing mice with liposome-entrapped actinomycin D prolongs their survival. Res Commun Chem Pathol Pharmacol 1975;10:351–362.
121. Kaye SB, Ryman BE. The fate of liposome-entrapped actinomycin-D in vivo and its therapeutic effect in a solid murine tumor. Biochem Soc Trans 1980;8:107–108.
122. Rahman A, Kessler A, More N. Liposomal protection of adriamycin induced cardiotoxicity in mice. Cancer Res 1980;40:1532–1537.
123. Shinozawa S, Maki Y, Oda T. Tissue distribution and antitumor effect of liposome entrapped doxorubicin (adriamycin) in Ehrlich ascites solid tumor bearing mice. Acta Med Okayama 1981;35:395–405.
124. Gabizon A, Dagan A, Goren D, et al. Liposomes as in vivo carriers of adriamycin: Reduced cardiac uptake and preserved antitumor activity in mice. Cancer Res 1982;42:4734–4739.
125. Yatvin MB, Muhlensiepen H, Porschen W, et al. Selective delivery of liposome associated cis-dichloro-diammineplatinum (II) by heat and its influence on tumor uptake and growth. Cancer Res 1981;41:1602–1607.
126. Fitchner I, Reszka R, Elbe B, Arndt D. Therapeutic evaluation of liposome encapsulated daunoblastin in murine tumor models. Neoplasma 1981;28:141–149.
127. Kimelberg HK, Atchison ML. Effects of entrapment of liposomes on the distribution, degradation and effectiveness of methotrexate on its chemotherapeutic efficacy in solid rodent tumors. Ann NY Acad Sci 1978;308:395–409.

128. Kosloski MJ, Rosen F, Miliholland RJ, Papahadjopoulos D. Effect of lipid vesicle (liposome) encapsulation of methotrexate on its chemotherapeutic efficacy in solid rodent tumors. Cancer Res 1978;38:2848–2853.
129. Weinstein JN, Magin RL, Cysyk RL, Zaharko DS. Treatment of solid L1210 murine tumors with local hyperthermia and temperature sensitive liposomes containing methotrexate. Cancer Res 1980;40:1388–1395.
130. Patel KR, Jonah MM, Rahman YE. In vitro uptake and therapeutic application of liposome-encapsulated methotrexate in mouse hepatoma. Eur J Cancer Clin Oncol 1982;18:833–843.
131. Forssen EA, Proffitt RT. Design and development of long circulating liposomal daunorubicin for in vivo targeting of solid tumors: DaunoXome®. In: Woodle M, Storm G. eds. Long Circulating Liposomes: Old Drugs, New Therapeutics. New York, NY, Springer-Verlag and Landes Bioscience, 1998, pp. 74–96.
132. Forssen EA, Coulter DM, Proffitt RT. Selective in vivo localization of daunorubicin small unilamellar vesicles in solid tumors. Cancer Res 1992;52:3255–3261.
133. Forssen EA. Ross ME. DaunoXome treatment of solid tumors: preclinical and clinical investigations. J Liposome Res 1994;4:481–512.
134. Money-Kyrle JF, Bates F, Ready J, et al. Liposomal daunorubicin in advanced Kaposi's sarcoma: a phase II study. Clin Oncol 1993;5:367–371.
135. Presant CA, Scolaro M, Kennedy P, et al. Liposomal daunorubicin treatment of HIV-associated Kaposi's sarcoma. Lancet 1993;341:1242–1243.
136. Gill PS, Wernz J, Scadden DT, et al. Randomized phase III trial of liposomal daunorubicin (DaunoXome) versus doxorubicin, bleomycin, vincristine, (ABV) in AIDS-related Kaposi's sarcoma. J Clin Oncol 1996;14:2353–2364.
137. Campbell R, Balasubramanian SV, Straubinger RM. Influence of cationic lipids on the stability and membrane properties of paclitaxel-containing liposomes. J Pharm Sci 2001;90:1091–1105.
138. Thierry AR, Rahman R, Dritschilo A. A new procedure for the preparation of liposomal doxorubicin: biological activity in multidrug-resistant tumor cells. Cancer Chemother Pharmacol 1994;35:84–88.
139. Gokhale PC, Radhakrishnan B, Husain SR, et al. An improved method of encapsulation of doxorubicin in liposomes: pharmacological, toxicological and therapeutic evaluation. Br J Cancer 1996;74:43–48.
140. Freeman AI, Mayhew E. Targeted drug delivery. Cancer 1986;58:573–583.
141. Rustum Y, Dave C, Mayhew E, Papahadjopoulos D. Anti-tumor effects of liposome-entrapped cytosine arabinoside against mouse L1210 leukemia: Role of liposome type and route of administration. Cancer Res 1979;39:1390–1395.
142. Gabizon A. Liposome circulation time and tumor targeting: implications for cancer chemotherapy. Adv Drug Deliv Rev 1995;16:285–294.
143. Dass CR, Walker TL, Burton MA, Decruz EE. Enhanced anticancer therapy mediated by specialized liposomes. J Pharm Pharmacol 1997;49:972–975.
144. Furgeson DY, Cohen RN, Mahato RL, Kim SW. Novel water insoluble lipoparticulates for gene delivery. Pharm Res 2002;19:382–390.
145. Anwer K, Kao G, Proctor B, Rolland A, Sullivan S. Optimization of cationic lipid/DNA complexes for systemic gene transfer to tumor lesions. J Drug Target 2000;8:125–135.
146. Meyer O, Kirpotin D, Hong K, et al. Cationic liposomes coated with polyethylene glycol as carriers for oligonucleotides. J Biol Chem 1998;273:15621–15627.
147. Birchall JC, Kellaway IW, Millis SN. Physico-chemical characterization and transfection efficiency of lipid-based gene delivery complexes. Int J Pharm 1999;183:195–207.
148. Whitmore M, Li S, Huang L. LPD lipoplex initiates a potent cytokine response and inhibits tumor growth. Gene Ther 1999;6:1867–1875.
149. Reimer DL, Kong S, Monck M, et al. Liposomal lipid and plasmid DNA delivery to B16/BL6 tumors after intraperitoneal administration of cationic liposome DNA aggregates. J Pharmacol Exp Ther 1999;289(2):807–815.
150. Dass CR. Biochemical and biophysical characteristics of lipoplexes pertinent to solid tumor gene therapy. Int J Pharm 2002;241:1–25.

151. Allen TM. Stealth™ liposomes: Avoiding reticuloendothelial uptake in liposomes. In: Lopez Berestain G, Fidler IJ, eds. Liposomes in the Therapy of Infectious Diseases and Cancer. New York, NY, Allen R. Liss, 1989, pp. 405–441.
152. Papahadjopoulos D, Allen TM, Gabizon A, et al. Sterically stabilized liposomes-improvements in pharmacokinetics and antitumor therapeutic efficacy. Proc Natl Acad Sci USA 1991;88:11460.
153. Blume G, Ceve G. Liposomes for the sustained release in vivo. Biochim Biophys Acta 1990;1029:91.
154. Torchilin VP, Papisov MI. Why do polyethylene glycol-coated liposomes circulate so long? J Liposome Res 1994;4:725–739.
155. Torchilin VP. How do polymers prolong circulation time of liposomes. J Liposome Res 1996;6:99–116.
156. Torchilin VP, Omelyanenko V, Papisov MI, et al. Poly(ethylene glycol) on the liposome surface: on the mechanism of polymer-coated liposomes longevity. Bichem et Biophys Acta 1994;1195:11–20.
157. Pang SNJ. Final report on the safety assessment of polyethylene glycols (PEGs)-6, -8, -32, -75, -150, -14M, -20M. J. Am College Toxicol 1992;12:429–456.
158. Yamaoka T, Tabata T, Ikada Y. Distribution and tissue uptake of poly(ethylene glycol) with different molecular weights after intravenous administration in mice. J Pharm Sci 1994;83:601–606.
159. Zalipsky S. Chemistry of polyethylene conjugates with biologically active molecules. Adv Drug Deliv Rev 1995;16:157–182.
160. Klibanov AL, Maruyama A, Torchilin VP, Huang L. Ampjipathic polyethylene glycols effectively prolong the circulation time of liposomes. FEBS Lett 1990;268:235–237.
161. Torchilin VP, Shtilman M, Trubetskoy VS, et al. Ampiphilic vinyl polymers effectively prolong liposome circulation time in vivo. Bichem Biophys Acta 1994;1195:181–184.
162. Ning ZW, Daphane D, Rudoll TL, et al. Increased microvascular permeability contributes to prefrential accumulation of stealth liposomes in tumor tissue. Cancer Res 1993;53:3765–3770.
163. Zhou XJ, Rahman R. Preclinical and clinical pharmacology of vinca alkaloids. Drugs 1992;44:1.
164. Layton D, Trouet A. A comparison of the therapeutic effect of free and liposomally encapsulated vincristine in leukemic mice. Eur J Cancer 1980;16:945.
165. Allen TM, Newman MS, Woodle M, et al. Pharmacokinetics and antitumor activity of vincristine encapsulated in sterically stabilized liposomes. Int J Cancer 1995;62:199–204.
166. Mayer LD, Bally MB, Laughery H, et al. Liposomal vincristine preparations which exhibit decreased drug toxicity and increased activity against murine L1210 leukemia. Cancer Res 1990;50:575.
167. Vaage J, Donovon D, Mayhew E, et al. Therapy of mouse mammary carcinomas with vincristine and doxorubicin encapsulated in sterically stabilized liposomes. Int J Cancer 1993;54:959.
168. Ceh B, Winterhalter M, Fredrik P, et al. stealth liposomes: from theory to product. Adv Drug Deliv Rev 1997;24:165–177.
169. Vaage J, Donovon D, Mayhew E. Therapy of human ovarian carcinoma xenografts using doxorubicin encapsulated in sterically stabilized liposomes. Cancer 1993;72:3671–3675.
170. Vaage J, Donovon D, Uster P. Tumor uptake of doxorubicin in polyethylene glycol-coated liposomes and therapeutic effect against a xenografted human pancreatic carcinoma. Br J Cancer 1997;75:482–486.
171. Vaage J, Barbera-Guillem E, Abra R. Tissue distribution and therapeutic effect of intravenous free or encapsulated liposomal doxorubicin on human prostate carcinoma xenografts. Cancer 1994;3:1478–1484.
172. Northfelt DW, Dezube BJ, Thommes JA. PEGylated-liposomal doxorubicin versus doxorubicin, bleomycin and vincristine in the treatment of AIDS-related Kaposi's sarcoma: results of a randomized phase III clinical trial. J Clin Oncol 1998;16:2445–2451.
173. Lyass O, Uziely B, Ben-Yosef R. Correlation of toxicity with pharmacokinetics of PEGylated liposomal doxorubicin (Doxil) in metastatic breast carcinoma. Cancer 2000;89(5):1037–1047.

174. Working P. Preclinical studies of lipid complexes and liposomal drugs. AMPHOTECTM DOXILTM and SPI-077. In: Lasic DD, Papahadjopoulos D, eds. Medical Applications of Liposomes. Amsterdam, Elsevier, 1998, pp. 605–624.
175. Newman MS, Colbem GT, Working P, et al. Comparitive pharmacokinetics, tissue distribution and therapeutic efficiency of cisplatin encapsulated in long circulating pegylated liposomes (SPI-077) in tumor bearing mice. Cancer Chemother Pharmacol 1999;46:155–165.
176. Maruyama K. Enhancement of doxorubicin by encapsulating in long circulating thermosensitive liposomes combined with local hyperthermia. In: Woodle MD, Storm G. eds. Long Circulating Liposomes: Old Drugs, New Therapeutics. New York, Springer-Verlag and Landes Bioscience, 1998; pp. 97–109.
177. Slepushkin VA, Simoes S, Dazin P, et al. Sterically stabilized pH-sensitive liposomes. Intracellular delivery of aqueous contents and prolonged circulation in vivo. J Biol Chem 1996;272:2382–2388.
178. Baillie AJ, Florence AT, Hume LR, et al. The preparation and properties of niosomes-nonionic surfactant vesicles. J Pharm Pharmacol 1985;37(12):863–868.
179. Uchegbu IF, Double JA, Turton JA, Florence AT. Distribution, metabolism and tumoricidal activity of doxorubicin administered in sorbitan monostearate (Span 60) niosomes in the mouse. Pharm Res 1995;12(7):1019–1024.
180. Parthasarathi G, Udupa N, Umadevi P, Pillai GK. Niosome encapsulated Vincristine sulfate: improved anticancer activity with reduced toxicity in mice. J Drug Target 1994;2:173–182.
181. Hao Y, Zhao F, Li N, et al. Studies on high encapsulation of colchicine by a niosome system. Int J Pharm 2002;244:73–80.
182. Torchilin VP. Structure and design of polymeric surfactant-based drug delivery systems. J Control Release 2001;73:137–172.
183. Yokoyama M, Miyauchi M, Yamada N, et al. Polymer micelles as novel drug carrier: Adriamycin-conjugated poly(ethylene glycol)-poly(aspartic acid) block copolymer. J Control Release 1990;11:269–278.
184. Rapoport N, Herron JN, Pitt WG, Pitina L. Micellar delivery of doxorubicin and its paramagnetic analogue ruboxyl, to HL-60 cells: effect of micelle structure and ultrasound on the intracellular drug uptake. J Control Release 1999;58:153–162.
185. Winnik FM, Davidson AR, Hamer GK, Kitano H. Amphiphilic poly(N-isopropylacrylamide) prepared by using a lipophilic radical initiator: synthesis and solution properties in water. Macromolecules 1992;25:1876–1880.
186. Kwon GS, Yokoyama M, Okano T, et al. Biodistribution of micelle-forming polymer-drug conjugates. Pharm Res 199310:970–974.
187. Cortesi R, Esposito E, Maietti A, et al. Formulation study for the antitumor drug camptothecin: liposomes, micellar solutions and microemulsion. Int J Pharm 1997;159:95–105.
188. Singla AK, Garg A, Aggarwal D. Paclitaxel and its formulations. Int J Pharm 2002;235:179–192.
189. Kataoka K, Harada A, Nagasaki Y. Block copolymer micelles for drug delivery: design, characterization and biological significance. Adv Drug Deliv Rev 2001;47(1):113–131.
190. Yokoyama M, Fukushima S, Uehara R, et al. Characterization of physical entrapment and chemical conjugation of adriamycin in polymeric micelles and their design for in vivo delivery to a solid tumor. J Control Release 1998;50:79–92.
191. Yokoyama M, Miyauchi M, Yamada N, et al. Characterization and anticancer activity of the micelle forming polymeric anticancer drug adriamycin-conjugated poly(ethylene glycol)-poly(aspartic acid) block copolymer. Cancer Res 1990;50:1693–1700.
192. Yokoyama M, Satoh A, Sakurai Y, et al. Incorporation of water-insoluble anticancer drug into polymeric micelles and control of their particle size. J Control Release 1998;55:219–229.
193. Yokoyama M, Okano T, Sakurai Y, et al. Toxicity and antitumor activity against solid tumors of micelle-forming polymeric anticancer drug and its extremely long circulation in blood. Cancer Res 1991;51:3229–3236.
194. Nishiyama N, Kataoka K. Preparation and characterization of size-controlled polymeric micelle containing cis-dichlorodiammineplatinum (II) in the core. J Control Release 2001;74:83–94.

195. Mizumura Y, Matsumura Y, Hamaguchi T, et al. Cisplatin incorporated polymeric micelles eliminate nephrotoxicity, while maintaining antitumor activity. Jpn J Cancer Res 2001;92(3):328–336.
196. Nakanishi T, Fukushima S, Okamoto K, et al. Development of polymer micelle carrier system for doxorubicin. J Control Release 2001;74:295–302.
197. Yokoyama M, Okano T, Sakurai Y, Kataoka K. Improved synthesis of adriamycin-conjugated poly(ethylene oxide)-poly(aspartic acid) block copolymer and formation of unimodal micellar structure with controlled amount of physically entrapped adriamycin. J Control Release 1994;32:269–277.
198. Kataoka K, Kwon GS, Yokoyama M, et al. Block copolymer micelles as vehicles for drug delivery. J Control Release 1993;24:119–132.
199. Kataoka K, Matsumoto T, Yokoyama M, et al. Doxorubicin-loaded poly(ethylene glycol)-poly(beta-benzyl-L-aspartate) copolymer micelles: their pharmaceutical characteristics and biological significance. J Control Release 2000;64:143–153.
200. Kwon GS, Naito M, Yokoyama M, et al. Block copolymer micelles for drug delivery: loading and release of doxorubicin. J Control Release 1997;48:195–201.
201. Kwon GS, Naito M, Yokoyama M, et al. Physical entrapment of adriamycin in AB block copolymer micelles. Pharm Res 1995;12(2):192–195.
202. Kwon GS, Naito M, Yokoyama M, et al. Micelles based on AB block copolymers of poly(ethylene oxide) and poly(β-benzyl L-aspartate). Langmuir 1993;9:945–949.
203. Kabanov AV, Batrakova EV, Melik-Nubarov NS, et al. A new class of drug carriers: micelles of poly(oxyethylene)-poly(oxypropylene) block copolymers as micro containers for drug targeting from blood in brain. J Control Release 1992;22:141–158.
204. Venne A, Li S, Mandeville R, et al. Hyper sensitizing effect of pluoronic L61 on cytotoxic activity, transport and subcellular distribution of doxorubicin in multi-drug resistant cells. Cancer Res 1996;56:3626–3629.
205. Rapoport N, Munshi N, Pitina L, Pitt WG. Pluronic micelles as vehicles for tumor-specific delivery of two anticancer drugs to HL-60 cells using acoustic activation. Polymer Reprints 1997;38:620–621.
206. Kabanov AV, Batrakova EV, Alakhov VY. Pluronic block copolymers for overcoming drug resistance in cancer. Adv Drug Deliv Rev 2002;54(5):759–779.
207. Kabanov AV, Batrakova EV, Alakhov VY. Pluronic block copolymers as novel polymer therapeutics for drug and gene delivery. J Control Release 2002;82:189–212.
208. Batrakova EV, Dorodnych TY, Klinskii EY, et al. Anthracycline antibiotics non-covalently incorporated into the block copolymer micelles: *in vivo* evaluation of anti-cancer activity. Br J Cancer 1996;74:1545–1552.
209. Yoo HS, Park TG. Biodegradable polymeric micelles composed of doxorubicin conjugated PLGA-PEG block copolymer. J Control Release 2001;70(1–2):63–70.
210. Zhang X, Jackson JK, Burt HM. Development of amphiphilic diblock copolymers as micellar carriers of taxol. Int J Pharm 1996;132:195–206.
211. Kim SC, Kim DW, Shim YH, et al. In vivo evaluation of polymeric micellar paclitaxel formulation: toxicity and efficacy. J Control Release 2001;72:191–202.
212. Zhang X, Burt HM, Von Hoff D, et al. An investigation of the antitumor activty and biodistribution of polymeric micellar paclitaxel. Cancer Chemother Pharmacol 1997;40:81–86.
213. Jeong YI, Nah JW, Lee HC, et al. Adriamycin release from flower-type polymeric micelle based on star-block copolymer composed of poly(gamma-benzyl L-glutamate) as the hydrophobic part and poly(ethylene oxide) as the hydrophilic part. Int J Pharm 1999;188:49–58.
214. Gao Z, Lukyanov AN, Singhal A, Torchillin VP. Diacyllipid-polymer micelles as nanocarriers for poorly soluble anticancer drugs. Nano Lett 2002;2(9):979–982.
215. Trubetskoy VS, Torchilin VP. Use of polyoxyethylene-lipid conjugates as long circulating carriers for delivery of therapeutic and diagnostic agents. Adv Drug Deliv Rev 1995;16:311–320.
216. Weissig V, Whiteman KR, Torchilin VP. Accumulation of protein-loaded long-circulating micelles and liposomes in subcutaneous Lewis lung carcinoma in mice. Pharm Res 1998;15:1552–1556.

217. Lukyanov AN, Gao Z, Mazzola L, Torchilin VP. Polyethylene glycol-diacyllipid micelles demonstrate increased accumulation in subcutaneous tumors in mice. Pharm Res 2002;19:1424–1429.
218. Kohori F, Sakai K, Aoyagi T, et al. Control of adriamycin cytotoxic activity using thermally responsive polymeric micelles composed of poly(N-isopropylacrylamide-co-N,N-dimethylacrylamide)-b-poly(D,L-lactide). Colloids Surfaces B: Biointerfaces 1999;16:195–205.
219. Cammas S, Suzuki K, Sone Y, et al. Thermo-responsive polymer nanoparticles with a core shell micelle structure as site-specific drug carriers. J Control Release 1997;48:157–164.
220. Chung JE, Yokoyama M, Yamato M, et al. Thermo-responsive drug delivery from polymeric micelles constructed using block copolymers of poly(N-isopropylacrylamide) and poly-(butylmethacrylate). J Control Release 1999;62:115–127.
221. Kohori F, Yokoyama M, Sakai K, Okano T. Process design for efficient and controlled drug incorporation into polymeric micelle carrier systems. J Control Release 2002;78:155–163.
222. Chung JE, Yokoyama M, Suzuki K, et al. Reversibly thermo-responsive alkyl terminated poly(N-isopropylacrylamide) core shell micellar structures. Colloids Surfaces B: Biointerfaces 1997;9:37–48.
223. Miwa A, Ishibe A, Nakano M, et al. Development of novel chitosan derivatives as micellar carriers of taxol. Pharm Res 1998;15:1844–1850.

11

Biological Protein Nanostructures and Targeted Drug Delivery

Guangliang Pan, Svein Øie, and D. Robert Lu

1. Introduction

Targeted drug delivery refers to the site-specific drug delivery that directs drugs mainly to certain cell types within a tissue and to certain molecular complexes or organelles within a cell while avoiding drug loading in nontargeted cells. Targeted delivery of drugs to specific cells involves the specific interactions between drugs or drug carriers and the cell-surface proteins through ligand–receptor interactions or antigen–antibody interactions. Targeted drug delivery to specific molecular complexes or organelles within a cell requires the specific interactions of drug with the targeted complexes to lead to the therapeutic effect. In the biological systems, these interactions generally occur on various types of biological nanostructures of protein origin. Understanding and utilization of these biological nanostructures could lead to significant improvement in drug targeting and drug carriers.

The biological protein nanostructures primarily include protein–lipid, protein–protein, protein–carbohydrate, and protein–nucleic acid complexes. Proteins, one group of the most important biological macromolecules in cells, are smaller nanoscale molecules with typical size range between 1 and 20 nm *(1)*. Through sophisticated interactions with other biomolecules, these protein nanostructures are formed and widely distributed in human body. For example, low-density lipoproteins (LDL), with a diameter of 25–28 nm, are protein–lipid complexes. They are the major circulatory [illegible] in the blood. When used as a drug carrier, these protein–lipid complexes offer a certain advantage of being endogenous nanostructures that do not trigger immunological response. They can also escape the recognition and elimination by the reticuloendothelial system (RES). On the other hand, glycoproteins, i.e., protein-carbohydrated complexes, are vital structural and regulatory proteins in viruses and can serve as important therapeutical targets for anti-viral drug development. Telomerase, a protein nanostructure formed from protein and nucleic acid, is activated only in cell immortalization and cancer progression. Thus telomerase is an ideal therapeutic target for anticancer therapy. Because protein nanostructures are so critical to various biological and physiopathological activities, they have received wide attentions in recent years in the development of drug-targeting strategies, either as drug carries or as therapeutic tar-

From: *Cellular Drug Delivery: Principles and Practice*
Edited by: D. R. Lu and S. Øie © Humana Press Inc., Totowa, NJ

gets. This chapter will focus on two aspects of biological protein nanostructures regarding their involvement in targeted drug delivery: (1) biological protein nanostructures as targeting drug carriers and (2) biological protein nanostructures as therapeutical targets for new drug development.

2. Protein–Lipid Nanocomplexes (Lipoproteins)

Lipoproteins are biological protein–lipid complexes in the nanoscale range. They have spherical shape consisting of a hydrophobic core and a polar shell that is incorporated with receptor-active proteins. The hydrophobic core contains triglycerides and cholesteryl esters, whereas the polar shell contains phospholipids, unesterified cholesterol, and one or several apolipoproteins. A schematic cross-section diagram of lipoprotein is shown in Fig. 1. Lipoproteins are commonly classified based on their densities, which can be determined through gradient ultracentrifugation. The classification thus is related to the respective amounts of lipid and protein in the complex. In an increasing order by density, lipoproteins include chylomicron, very low-density lipoprotein (VLDL), intermediate-density lipoprotein (IDL), LDL, and high-density lipoprotein (HDL).

Because lipoproteins are taken up in varying amounts through ligand-receptor interactions by different type of cells, they may be utilized as biocompatible nanoscale carriers for targeted drug delivery. For example, in hepatocytes, remnant receptor and asialoglycoprotein receptor can recognize chylomicrons and lactosylated HDL, respectively, in a molecular-specific manner. LDL receptor on the cell surface can specifically recognize LDL and its expression can be upregulated or downregulated depending on the type and state of these cells. By incorporating bioactive molecules into lipoproteins or modified lipoproteins, targeted drug delivery may be achieved, resulting in more bioactive molecules taken up by a select type of cells, e.g., cancer cells. Each class of lipoproteins has its unique biological property and thus can be utilized individually for targeted drug delivery.

2.1. Chylomicron for Drug Targeting

Chylomicrons are the largest lipoprotein complexes (80–500 nm) in the human body. Their main function is to transport dietary lipids from the intestine to the liver and adipose tissue. Assembled in the intestine from the absorbed dietary lipids and the apolipoproteins synthesized by the intestinal epithelium, they are transported out of the epithelial cells to the tissue fluid and further carried by lymphatic system for general circulation. When they enter the blood stream, their compositions of phospholipids and proteins are changed greatly through the hydrolysis of triglycerides and the component exchange with other lipoproteins in the plasma to form chylomicron remnants. Chylomicron remnants mainly are taken up by parenchymal cells in liver *(2)*. When they are oxidized, chylomicrons can be taken up by liver endothelial cells and Kupffer cells *(3)*. The uptake of chylomicron remnant by various cells is LDL receptor-mediated, which requires apoE protein as the ligand on the chylomicron-remnant particles *(3,4)*.

When associated with chylomicrons, many lipophilic drugs and xenobiotics can be absorbed via the intestinal lymphatic system *(5)*. This route can circumvent the first-pass effect in the intestine and, more importantly, can be used for drug targeting to liver cells because the liver is the destination of chylomicrons. Targeted drug delivery to the liver can help treat many critical diseases such as alcohol-induced liver disorders,

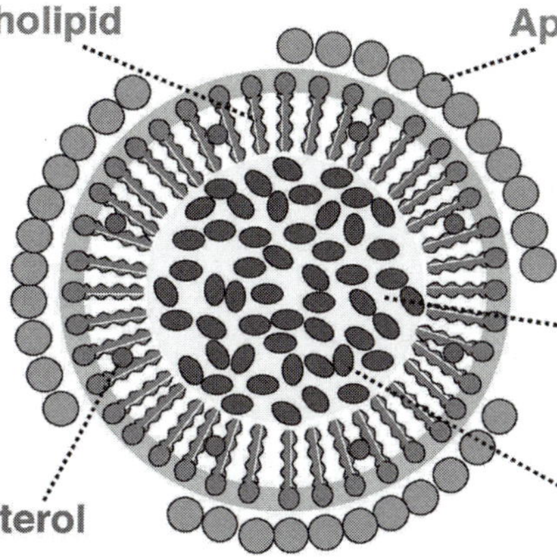

Fig. 1. Structure of lipoprotein.

chronic liver diseases and cirrhosis, virus-induced liver diseases (heptatitis), liver tumor, familial hypercholesterolemia, and type III hyperlipoproteinemia. To effectively treat viral infection, antiviral drugs or prodrugs can be incorporated into chylomicron and targeted to the liver. For example, by incorporating a nucleoside analog iododeoxyuridine into recombinant chylomicrons, the anti-viral drug was delivered selectively to liver parenchymal cells *(6)*. Liver is also an excellent target for gene therapy for the diseases caused by metabolic defect. Gene therapy involving viral vector is in general limited by the high immunogenicity and poor safety profile. Utilizing the hydrophobic core of the chylomicrons, Hara et al. incorporated a hydrophobic DNA complex into reconstituted chylomicron remnants and the DNA was successfully delivered to liver cells *(7,8)*.

2.2. VLDL for Drug Targeting

VLDL particles have a size range of 30–80 nm. They are assembled in the endoplasmic reticulum (ER) and matured in Golgi apparatus of hepatocytes before secretion *(9)*. After entering into the plasma, VLDL particles are catabolized by a series of biochemical actions including apolipoprotein exchange with apoC-I, apoC-II, apoC-III, and apoE; lipolysis by triglyceride lipase; and cell-surface receptor-mediated uptake by cells. As lipolysis proceeds, VLDL particles become smaller and smaller and eventually are converted to IDL. Some of the IDL particles are rapidly taken up by hepatocytes via a receptor-mediated mechanism and others undergo further hydrolysis before being converted to LDL.

The catabolism cascade of VLDL suggests the possibility of using VLDL as a drug carrier for targeted delivery. Because some cancer cells overexpress the receptors for apoE, a protein ligand present on the surface of VLDL, VLDL can potentially serve as an antineoplastic drug carrier. In vitro experiments demonstrated that VLDL could effectively incorporate cytotoxic drugs, 5-fluorouracil (5-FU), 5-iododeoxyuridine (IudR), doxorubicin (Dox), and vindesine, and the resultant complex showed effective cytotoxicity to human carcinoma cells *(10)*. By mimicking the composition and structure of VLDL, Shawer et al. developed a VLDL-resembling phospholipid nanoemulsion system that could carry a new antitumor boron compound for targeted delivery to cancer cells *(11)*.

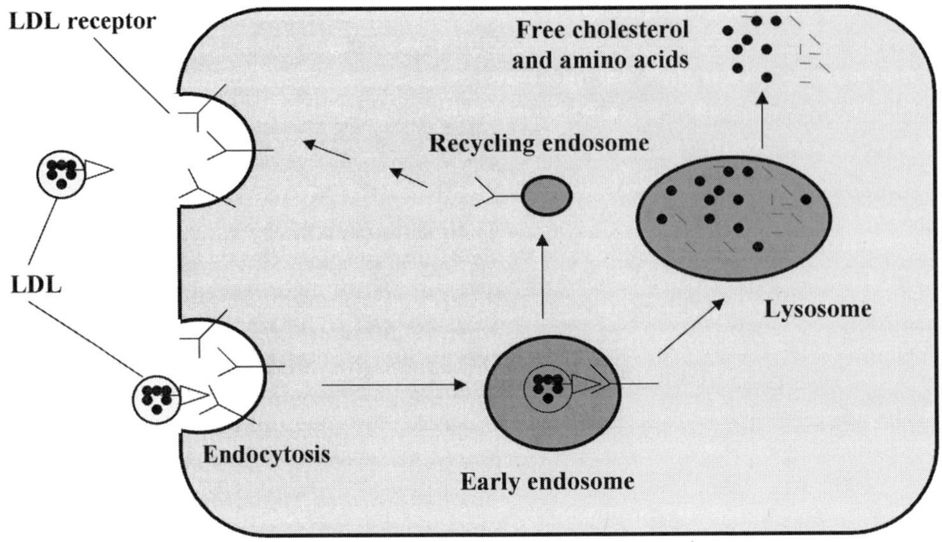

Fig. 2. LDL receptor-mediated endocytosis.

2.3. LDL for Drug Targeting

LDL (25–30 nm) is not directly synthesized in human body. Instead, most of them are formed through the VLDL pathway. LDL is the major circulatory lipoprotein for the transport of cholesterol and cholesteryl esters. Cholesterol required for cell-membrane construction is mainly obtained from LDL. LDL can be internalized by cells via LDL receptor-mediated endocytosis, a process that is determined by the availability of LDL receptors *(12)*. Apolipoprotein apoB-100 is the major ligand in LDL for recognition and binding by LDL receptor *(13)*. Once it is internalized, LDL is distributed to lysosomes in which cholesteryl esters are hydrolyzed (Fig. 2). Because cholesterol is required for cell growth and LDL is the main carrier for cholesterol in blood circulation, most cells can take up LDL through a receptor-mediated mechanism. It is estimated that 60–80% of LDL can be cleared from plasma by LDL receptor-mediated pathways *(14,15)*. As compared with chylomicron, VLDL, and IDL, LDL has a longer serum half-life of 2–4 d *(16)*. Thus, among various lipoproteins, LDL has a distinctive advantage to be used as drug carrier for targeted delivery and has been widely studied.

2.3.1. LDL for Anticancer Drug Targeting

It has been demonstrated that many tumor cells overexpress LDL receptors for the uptake of LDL particles to meet their increased requirement for cholesterol in cell-membrane construction *(17–22)*. Subsequently, a significant amount of work has been carried out to examine LDL as a candidate for antitumor drug carriers. Many lipophilic anticancer drugs have been incorporated into LDL particles for the purpose of drug targeting to various tumors *(10,23)*. When the antineoplastic drugs, methotrexate and floxuridine (FdUrd), were oleyl-derivatized and incorporated into LDL particles, they were effectively delivered into the hepatocellular carcinoma cell line Hep G2 *(24)*. The serum half-life of these drugs carried by LDL particles was considerably prolonged as compared to the free drug. Photodynamic therapy (PDT) of tumors is a recently developed therapeutic approach. It is based on the generation of highly cytotoxic oxygen species through the irradiation of photosensitizer such as porphyrins, chlorins, and phthalocyanines at selected wavelength. The efficacy of this therapy is dependent on the specific uptake of these photosensitizers by tumor cells. Using LDL as a carrier, photosensitizers were successfully targeted to tumor cells *(25–27)*.

2.3.2. LDL for Brain Drug Targeting

The blood–brain barrier (BBB) is a semi-permeable barrier that allows certain types of molecules to pass through but not others, depending on the lipophilicity, molecular size, and electric charge. It is a significant barrier for many drugs such as antibiotics, neuropeptides, and antineoplastic agents. In order to overcome this barrier, a number of methods have been employed including the use of prodrugs, antibody and drug-carrier systems such as liposomes *(28–30)*. Because the brain involves a variety of receptor-mediated transport systems to control the entry and exit of hydrophilic molecules and macromolecules, such systems can be utilized for brain drug targeting and transport. It is known that LDL receptors exist on endothelial cells of brain capillaries for LDL endocytosis *(31–33)*. Thus LDL potentially can be used as carriers for those drugs that are unable to pass through BBB freely.

2.3.3. Acetylated LDL for Drug Targeting

When chemically altered lipoproteins appear in the circulation of human body, the RES system is activated to remove these altered lipoproteins if they are recognized as foreign substances. The process involves the scavenger receptors on the cell surface of human macrophage *(34)*. Unlike T4 lymphocytes, which lead to collapse of the immune system when they are infected by human immunodeficiency virus (HIV), HIV-infected macrophages allow HIV to replicate for a long period of time. Macrophages play a very important role in HIV dissemination to various organs and to other parts of the immune system *(35,36)*. Experiments have shown that when antiviral drugs, e.g., AZT, are incorporated into chemically altered LDL, such as acetylated LDL, the HIV-infected macrophages can be targeted *(37)*.

2.3.4. Lactosylated LDL for Drug Targeting

In liver, Kupffer cells play a critical role during inflammation through enhanced expression of adhesion molecules, often resulting in the harmful infiltration of neutrophils into the liver. In addition, the production of inflammation mediators, such as interleukins (IL) and tumor necrosis factors (TNF) by Kupffer cells, causes a cascade of events that are related to serious physiological problems *(38)*. Because only Kupffer cells express galactose particle receptors in the liver, lactosylated LDL became a good candidate for drug targeting to Kupffer cells *(39)*. A cholesterol-conjugated oligonucleotide, which is a potent inhibitor to the gene expression of intercellular adhesion molecule-1, was associated with lactosylated LDL and the antisense oligonucleotide was efficiently delivered into Kupffer cells *(40)*, indicating the specific uptake of the encapsulated content by Kupffer cells.

2.3.5. Oxidized LDL for Drug Targeting

Atherosclerosis is responsible for more deaths than any other disease in Western countries. One important hallmark of this disease is the appearance of lipid-loaded macrophages in the vessel wall. Currently available therapies such as percutaneous angioplasty and bypass surgery are limited by recurrence or worsening of the atherosclerotic process. Photodynamic therapy involving various photosensitizers was considered to be a promising new therapy in recent years *(41)*. One obstacle to this therapy is how to efficiently deliver photosensitizers into macrophage cells. It is known that a high level of scavenger receptors are expressed on the cell surface of macrophages with the atheroscelosis plaque *(34)*. These scavenger receptors can be good candidates for targeted delivery. It has been shown that photosensitizer aluminum phthalocyanine chloride associated with oxidatively modified LDL (OxLDL) was delivered selectively to macrophages *(42)*.

2.3.6. Surface-Modified LDL for Gene Delivery

The success of gene-therapy is dependent on a safe and efficient gene-delivery system. Most of the current gene-therapy protocols are based on viral gene-delivery vectors, which may cause long term safety problems *(43)*. Although many nonviral gene-delivery vectors have been widely investigated, most of them were limited by low transfection efficiency. Lipoprotein has been used to construct a new gene-delivery system to increase safety and efficiency. Kim et al. developed a Terplex system, which had a diameter about 100 nm. The Terplex system was formed through the balanced hydrophobic and electrostatic interactions among LDL, lipidized poly(L-lysine), and plasmid DNA *(44,45)*. This system has demonstrated its efficiency by delivering both plasmid DNA and antisense oligonucleotide to smooth muscle-cells and lung fibroblasts. As an endogenous nanoparticle, LDL played a key role in the internalization of the Terplex system into the target cells via LDL receptor-mediated endocytosis.

2.4. HDL for Drug Targeting

HDL is the smallest lipoprotein with a diameter of 7–11 nm. It shares common structural characteristics with other lipoproteins. However, its polar shell contributes more than 80% of the total mass. Newly synthesized HDL hardly contain any cholesteryl ester molecules. Cholesteryl esters are gradually added to the particles via enzymatic reaction by lecithin:cholesterol acyltransferase (LCAT), a 59-kD glycoprotein associated with HDLs. The cholesteryl esters in HDL can also be transferred to VLDL and LDL via another associated protein, cholesteryl ester transfer protein. The uptake of HDL into cells appears to occur in a similar way to that of LDL. However, cholesterol uptake from HDL also involves more selective means than wholesale uptake because its cholesteryl esters can be transferred into the cells *(46)*. Although the function of HDL in the human body is not well-defined, in general it transports excess cholesterol and cholesteryl esters from various tissue cells to the liver. The major advantage of utilizing HDL for drug delivery and targeting is its small size and fast internalization by tumor cells. Among various lipoproteins, HDL has the smallest size. This makes it easier to pass through the vascular pores to reach the target tissue and quicker to be internalized by the cells. HDL has mainly been used for the incorporation of water-insoluble anticancer drugs for targeting *(47,48)*. When the anticancer drug Taxol was incorporated into HDL, stable complexes were formed for cancer-cell targeting *(48)*.

2.5. Artificial Lipoprotein for Drug Delivery and Targeting

Endogenous lipoprotein for drug delivery has usually been purified from plasma by gradient ultracentrifugation. These lipoproteins are limited in availability and loading additional transport or gene-transfection enhancers has been problematic. In order to overcome such limitations, the concept of artificial lipoprotein can be utilized. Previously, several research groups have attempted to develop artificial lipoproteins *(49–*

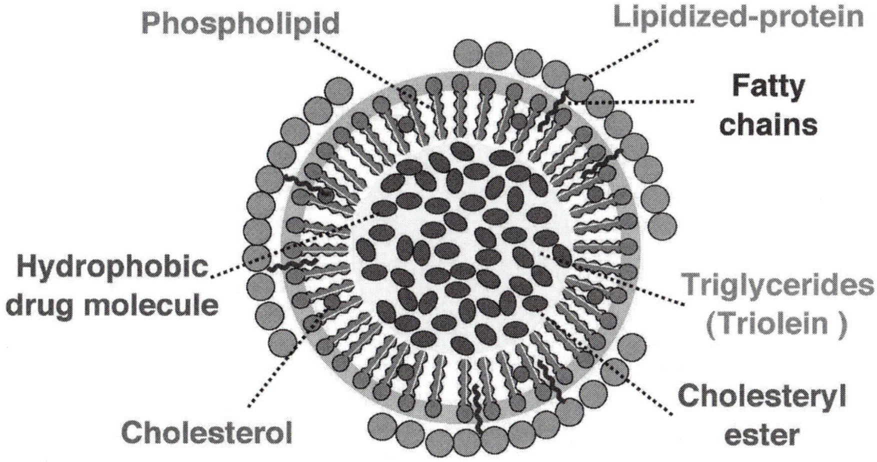

Fig. 3. Structure of artificial lipoprotein.

52). However, these studies have primarily been focused on the incorporation of natural apoB protein into lipid microemulsion for biochemical and metabolism research, and few of them on drug delivery and targeting.

By mimicking the structure of natural lipoproteins, artificial lipoproteins have been extensively investigated in our laboratory to incorporate different lipidized proteins or peptides for a diversified drug-delivery and drug-targeting strategy. The artificial lipoproteins consist of two structural portions, a hydrophobic core and a polar shell, containing surface proteins. The hydrophobic core is mainly composed of triolein and cholesterol oleate. The polar shell is composed of egg phosphatidylcholine, lysophosphatidylcholine, cholesterol, and lipidized protein or peptide. The fatty chains on the lipidized proteins and peptides serve as an anchor to interact with the phospholipid chains and to form stable protein–lipid nanocomplexes (Fig. 3). An early trial of such a system for drug delivery and targeting was through the constitution of a lipidized poly-L-lysine onto phospholipid nanoemulsion particles for gene delivery to tumor cells *(53)*. The incorporation of sufficient amount of palmitoyl poly-L-lysine (p-PLL) molecules onto the nanoemulsion particles led to positively charged complexes that were able to interact electrostatically with negatively charged DNA molecules. As demonstrated by Fig. 4, plasmid DNA migrated toward the positive anode because they were negatively charged (Lane 1). When plasmid DNA was incubated with p-PLL (Lane 2), no DNA migration band was observed. The binding of DNA molecules by p-PLL could block the intercalation of ethidium bromide molecules into the DNA molecules and thus no fluorescence emission occurred. When different ratios of p-PLL to nanoemulsion (i.e., the p-PLL to triolein ratio) were incubated, they demonstrated different DNA carrying capability (Lane 3 to Lane 7). A high ratio of p-PLL to nanoemulsion could tightly bind all the DNA molecules and no free DNA migration band appeared (Lane 3 to Lane 6). When the ratio of p-PLL to nanoemulsion became sufficiently low (0.0625:1 as the p-PLL to triolein ratio), plasmid DNA started to escape from the complex and free DNA bands (Lane 7) appeared on the agarose gel. Because the cell surface is normally negatively charged, the uptake of exogenous particles is affected largely by the surface charge of the particles. Positively charge particles appeared to be required

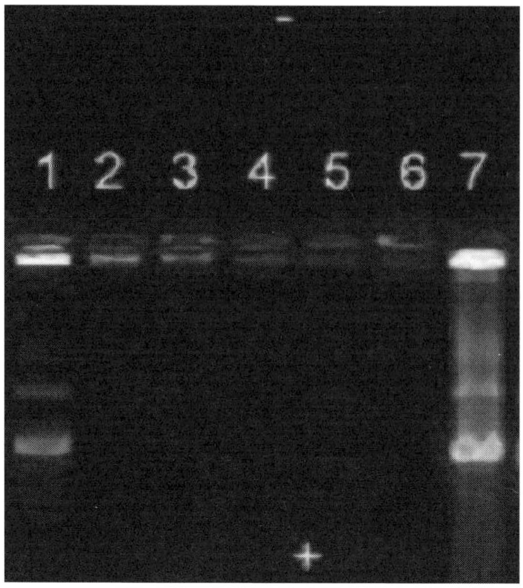

Fig. 4. 0.4% Agarose gel electrophoresis of plasmid DNA and its complexes with nanoemulsion and p-PLL stained with ethidium bromide. Lane 1: Pure DNA; Lane 2: DNA/p-PLL; Lane 3 to Lane 7 were complexes of nanoemulsion with different amount of p-PLL and DNA. The ratio of p-PLL to triolein was 1:1, 1:0.5, 1:0.25, 1:0.125, and 1:0.0625, respectively.

for the cellular uptake of the particles. However, excess positive charge on the particles could induce cellular toxicity and limit its use as DNA carrier. Thus particles with a properly balanced charge are required for the cellular uptake. The surface charge of the nanoemulsion/p-PLL/DNA complexes were measured and their zeta potentials are shown in Fig. 5. The zeta potentials of the particles increased with the increase of the amount of p-PLL when a fixed amount of DNA was used. Among the nanoemulsion/p-PLL/DNA complexes examined for the gene transfection, the complex with the zeta potential of 8.47 ± 1.85 mV resulted in the highest transfection efficiency. Such complexes demonstrated similar transfection efficiency as the Lipofectamine®, a commercial gene-transfection product. The artificial lipoprotein complex, however, had much lower toxicity (Fig. 6).

In recent years, cancer drug delivery and targeting have become a very active research area. Many cancer cells overexpress specific receptor proteins or peptides, which can recognize and bind specific ligands. For example, during the development of a tumor, angiogenic endothelial cells overexpress α_v integrin, which can specifically recognize cyclic peptides containing Arg-Gly-Asp motif *(54)*. Many other specific ligand-receptor binding has also been found in tumor cells *(55–57)*. In order to direct anticancer drug to the specific cellular site, an effective delivery system becomes critically important. The nanoscale size of the artificial lipoproteins and their capability of incorporating different recognition protein ligands may present a practical solution for the anticancer drug targeting and effective gene delivery.

3. Protein–Protein Complexes and Drug Targeting

Protein–protein interactions are critical events for a wide range of physiological and pathological processes. The biological formation of protein nanostructures through pro-

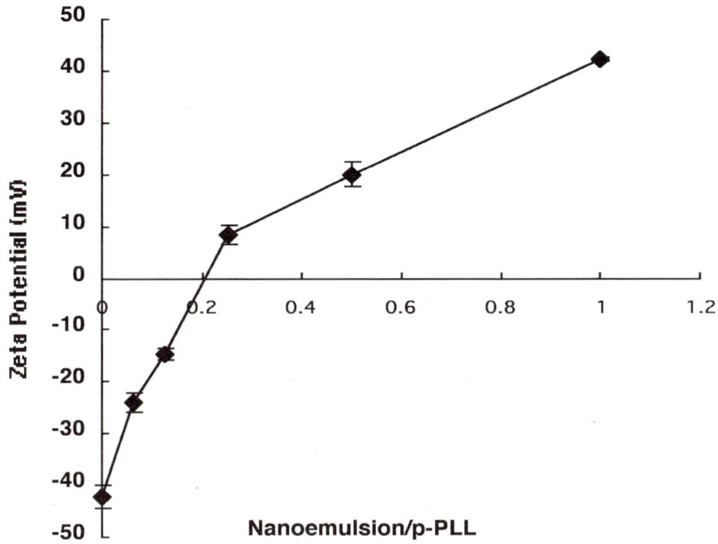

Fig. 5. Zeta potential of the nanoemulsion particles and their complexes with different amount of p-PLL and DNA (2 μg).

tein–protein interactions must be controlled in a precise manner in order to function properly. In viruses, virtually all the cellular processes, including the formation of protein nanostructures during replication and assembly, involve protein–protein interactions *(58)*. In the human immune system, the proper interaction between CD4 and CD8, the cell surface proteins expressed on T cells, with T-cell receptor (TCR) and major histocompatibility complex Class I (MHC-I) or II (MHC-II) is required to activate T cells *(59)*. The importance of such protein–protein interactions, as in the examples of viral replication and assembly and immune activation in human body, makes these nanocomplexes to be ideal therapeutic targets for new drug developments.

There are many types of nanoscale protein–protein complexes, including homodimers, heterodimers, antigen–antibody complexes, enzyme-inhibitor complexes, and multicomponents such as viral-coat protein and ribosomes *(60)*. The formation of these protein nanostructures via protein–protein interactions generally involves large and relatively flat surface areas with numerous contact sites, making it difficult to use small drug molecules to block such processes. However, these interactions require precise control in order to form pathophysiologically functional complexes. This presents an opportunity for therapeutic drug design and development. In herpesvirus (HSV), HSV ribonucleotide reductase (RR) is a tetramer ($\alpha_2\beta_2$) that consists of two large R1 subunits and two small R2 subunits *(61,62)*. The formation of an intact tetramer through proper interactions is important for the survival of HSV. When a synthetic peptide YAGAVVNDL was introduced into cells, the activity of RR was inhibited without causing significant side effect in the host cells *(63,64)*. The proper formation of other protein nanocomplexes in HSV, such as DNA polymerase (heterodimer) and helicase-primase complex, are also essential for the virus. Therefore, these nanocomplexes are also being considered as potential therapeutic targets.

In HIV, protease, intergrase, and reverse transcriptase are all homodimer nanostructures formed by protein–protein interactions. Protease has been one of the primary therapeu-

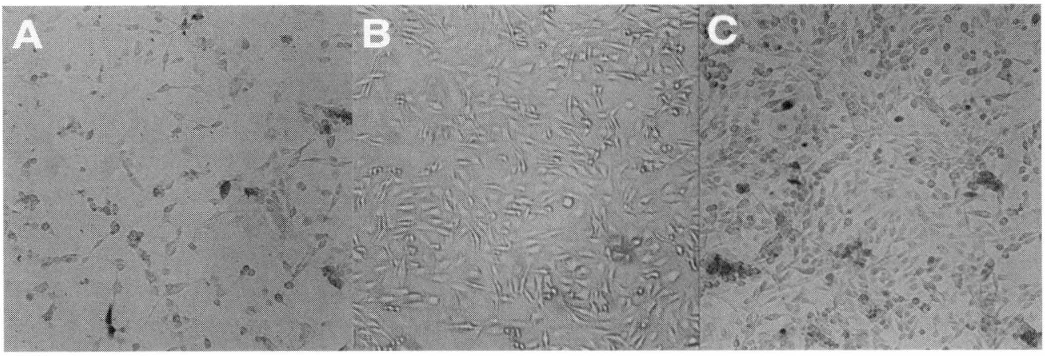

Fig. 6. X-Gal staining of glioma cells. **(A)** Cells transfected using Lipofectamine™ Reagent. **(B)** Control. **(C)** Cells transfected using nanoemulsion/p-PLL/DNA complex.

tic targets for AIDS chemotherapy, being critical for viral maturation. All successful inhibitors of HIV-1 protease to date are peptide mimetics that bind to the active site of the protease. Because of specific mutation within the HIV-1 genome *(65)*, drug-resistant proteases appeared in many HIV-1 strains. A good alternative to this therapeutic strategy would be agents that can block dimerization of protease *(66,67)*. The assembly of other important protein nanostructures, such as intergrase and reverse transcriptase, can also be inhibited by many peptides at the dimeric interface *(68,69)*. These types of protein-protein interactions are also widely observed in other types of viruses and have been considered as the therapeutic target for drug development *(70,71)*.

Although the formation of protein nanostructures through protein–protein interactions is critical to many physiopathological processes, it has been difficult to develop effective drug compounds to inhibit these processes. Recent design and screening strategies include rational structure-based drug design, peptide display technology, and in vivo genetic-selection systems *(72–74)*. However, many issues relating to drug delivery, such as cell permeability, intracellular localization, and physicochemical stability, must be resolved. Subsequently, various systems including scrape loading, electroporation, and delivery systems have been investigated. Among the delivery systems, liposomes, polycationic peptides, viruses, and proteins of eukaryotic, bacterial, or viral origin have been studied *(75–78)*. A good example of utilizing protein as delivery vehicle is the B-subunit of *Escherichia coli* heat-labile enterotoxin, which is able to deliver bioactive peptide to the cells to disrupt viral protein–protein complex *(79)*.

4. Protein–Carbohydrate Complexes and Drug Targeting

As one of the major groups of biological molecules, carbohydrates are unique in that they can have many branches and their monomeric units can connect to each other in different linkages in contrast to proteins and nucleic acids, which are exclusively linear and have only one type of linkage (amide linkage in proteins and 3'-5' phosphodiester linkage in nucleic acids). Most carbohydrates exist as nanoscale complexes with proteins (glycoproteins) or lipids (glycolipids). The complex sugar chains of glycoproteins and glycolipids play very important roles in the control of cellular functions and cell–cell recognitions, and therefore extensive investigations into the assembly of carbohydrate complexes may yield important information for drug-targeting development.

Glycoproteins are one of the major components in the outer surface of mammalian cells. They play critical roles in many important biological processes such as cell growth, fertilization, cell adhesion, immune responses, bacterial and viral infections, degradation of blood clot, and inflammation. Majority of glycoproteins are formed by the covalent attachment of carbohydrates to nitrogen atom (provided by asparagines residue) or oxygen atom (provided by serine or threonine residue) in proteins. The proteins are glycosylated as they move through the lumen of endoplasmic reticulum (ER) and Golgi apparatus in the cells, mostly by glycosidase and glycotransferase. The type and extent of glycosylation is dependent on the type and nature of proteins, cells, and tissues *(80)*.

The fusion of HIV envelope with host cell membranes is a critical step for HIV to enter the cells. The envelope glycoproteins of HIV are highly glycosylated. HIV-1 gp120 contains 20–25 glycosylation sites and the carbohydrates contribute about 50% of the apparent molecular weight. Blocking of the protein glycosylation can interfere significantly with the normal life-cycle of HIV *(81)*. Many sugar analogs have been screened for the anti-HIV activity in vitro. One of these analogs, *N*-butyldeoxygalactonojirimycin (NB-DNJ) has been shown to be a potent inhibitor of infection with minimal cytotoxicity. In hepatitis B virus, although there are only two glycosylation sites on the glycoprotein, the viral replication and assembly was inhibited by the treatment of NB-DNJ. Gp41, another HIV envelope glycoprotein, also plays an important role in the fusion of HIV envelope with host-cell membranes. Corresponding peptide and nonpeptide inhibitors to gp41 have been developed *(82)*. Knowledge about the HIV-1 envelope glycoprotein has provided insight into the possibilities for design of novel HIV vaccines *(83)*. Protein–carbohydrate nanostructures in HIV-1 currently have become the most important therapeutic targets for the development for anti-HIV drugs.

The molecular targets for new anticancer agents include inducers of cell differentiation, cell-cycle arrest, apoptosis, and signaling pathways for growth factors and cytokines. Because the protein glycosylation pathways are ubiquitous in cancer cells, they provide excellent opportunities for anticancer drug targeting. For example, alkaloid swainsonine, a Golgi α-mannosidase II inhibitor, is the first inhibitor to be selected for clinical test *(84)*. Because p-glycoproteins (P-gp) are multidrug transporters that result in multidrug resistance (MDR) in cancer chemotherapy, inhibitors targeting this protein have been developed *(85)*. Protein-glycosylation pathways are not only the ideal targets for drug development to treat cancers, but are also excellent targets in other diseases. For example, platelet plays an important role in the pathophysiology of certain diseases such as acute myocardial infarction and diabetes mellitus. The platelet activation and aggregation is caused by the activation of the glycoprotein IIb/IIIa receptor. Thus, glycoprotein IIb/IIIa has been considered to be a therapeutic target in these diseases *(86,87)*.

5. Protein–Nucleic Acid Complexes and Drug Targeting

Nucleic acids (DNA and RNA) are linear polymers of nucleotides with linkages of 3' to 5' by phosphodiester bridges. The genetic information for making all functional macromolecules are stored by the cellular DNA and accessed through the transcription of information into RNA. A typical DNA double helix has a diameter of 2 nm with varying length, depending on the organism. RNA occurs in various forms with different important biological functions such as messenger RNA (mRNA), ribosomal RNA (rRNA), transfer RNA (tRNA), and small nuclear RNA (snRNA).

The protein–nucleic acid nanostructures resulting from the interactions between protein and nucleic acids are critical to almost all aspects of genetic activity within an organism including DNA replication, transcription, packaging, rearrangement, and repair *(88)*. Based on the structural motif for DNA binding, there are four types of major protein–DNA complexes, i.e., helix-turn-helix (HTH), zinc finger (ZF), basic leucine zipper (B-Zip), and basic helix-loop-helix (B-HLH). Although the proteins in protein–DNA complexes are very diversified, the basic goal is to achieve a precise complementarity of the molecular shapes. This requires specific chemical recognition between proteins and their particular DNA targets. Thus, it has been proven possible to design proteins with novel recognition specificities for the purpose of breaking the normal protein–DNA binding *(89)*. On the other hand, a specific DNA or DNA complex can also be designed to bind the protein. For example, peptide nucleic acid (PNA), a DNA mimic in which the nucleobases are attached to a pseudopeptide backbone, was used to arrest transcription within a gene sequence and to provide an artificial open complex to promote transcription *(90)*.

Certain potent drug molecules interfere with DNA transcription by binding to the transcription factors and thus obstructing the specific DNA binding *(91)*. For example, doxorubicin can bind to NIL2A, a basic leucine-zipper transcription factor, to inhibit the formation of NIL2A/DNA complex. Echinomycin can bind to EGR1, a Cys2His2 ZF transcription factor, and inhibit the formation of EGR1/DNA complex. Intercalators such as nogalamycin and hedamycin, G/C-rich minor groove-binding drugs such as chromomycin A3, and A/T-specific drugs such as pluramycin can effectively inhibit the transcription factor-DNA complex formation *(91,92)*.

Interaction between protein and RNA plays a very important role in posttranscriptional RNA processing and protein biosynthesis. For example, spliceosomes, complexes of small nuclear RNAs and different proteins, are responsible for the precise formation of mRNAs. Ribosomes, the complexes of proteins with RNAs, are the agents for protein synthesis. For these complexes of proteins and RNAs, either component can be the target for chemotherapy. For example, PNA can be targeted to mRNA to block protein synthesis in an antisense strategy. PNA can also be targeted to the RNA components of ribonucleoproteins (RNPs) to inhibit their enzymatic activities *(90)*.

One of the most interesting protein-RNA complexes for drug targeting is telomerase, a protein-RNA complex that elongates telomeric DNA and appears to play an important role in cellular immortalization *(93)*. Telomeres are nucleoprotein structures at the end of human chromosomes. They play a fundamental role in the regulation of cellular lifespan *(94)*. The tandemly repeated DNA sequence of telomeres is specified and controlled by telomerase, which is repressed tightly in the vast majority of normal cells but becomes activated during cell immortalization and in cancers *(94,95)*. Telomerase has received much attention as a novel and potentially highly specific target for the development of new anticancer therapeutics *(96,97)*.

6. Concluding Remarks

Nanoscale protein complexes in the biological system, including protein–lipid, protein–protein, protein–carbohydrate, and protein–nucleic acid complexes, are ubiquitous in living systems. They play essential roles in various biochemical and genetic activities in cells and viruses. As a result, they become very important objects in pharmaceutical and biomedical research and development, especially in the development of targeted drug-delivery systems. The understanding of the formation, structure, and function of these protein nanostructures are essential for the development of targeted therapeutic delivery systems, either using these nanostructures as drug carriers or treating these nanostructures as the therapeutic targets. With the rapid advancement in life sciences, pharmaceutical and biomedical sciences, and nanoscience and nanotechnology, more and more efficient targeted drug-delivery strategies based on protein nanostructues can be developed.

References

1. Roco MC. Nanotechnology: a frontier for engineering education. Int J Eng Edu 2002;18(5):488–497.
2. Jones AL, Hradek GT, Hornick C, Renaud G, Windler EET, Havel RJ. Uptake and processing of remnants of chlomicrons and very low density lipoproteins by rat liver. J Lipid Res 1984;25:1151–1158.
3. Umeda Y, Redgrave TG, Mortimer B-C, Mamo JCL. Kinetics and uptake oxidatively modified lymph chylomicrons. Am J Physiol 1995;268:G709–G716.
4. Soued M, Mansbach II CM. Chylomicron remnant uptake by enterocytes is receptor dependent. Am J Physiol 270 1996;(1 Pt 1):G203–G212.
5. Porter CJH, Charman WH. Uptake of drugs into the intestinal lymphatic after oral administration. Adv Drug Deliv Rev 1997;25:71–89.
6. Rensen PCN, De Vrueh RLA, van Berkel TJC. Targeting hepatitis B therapy to the liver: clinical pharmacokinetic considerations. Clin Pharmacokinet 1996;31:131–155.
7. Hara T, Tan Y, Huang L. In vivo gene delivery to the liver using reconstituted chylomicron remnants as a novel nonviral vector. Proc Natl Acad Sci USA 1997;94:14547–14552.
8. Hara T, Liu F, Liu DX, Huang L. Emulsion formulations as a vector for gene delivery in vitro and in vivo. Adv Drug Deliv Rev 1997;24:265–271.
9. Olofsson, SO, Bjursell G, Bostrom K, et al. Apolipoprotein B: structure, biosynthesis and role in the lipoprotein assembly process. Atherosclerosis 1987;68:1–17.
10. Kader A, Pater A. Loading anticancer drugs into HDL as well as LDL has little affect on properties of complexes and enhances cytotoxicity to human carcinoma cells. J Control Rel 2002;80:29–44.
11. Shawer M, Greenspan P, Øie S, DR Lu. VLDL-resembling phospholipid-submicron emulsion for cholesterol-based drug targeting. J Pharm Sci 2002;91:1405–1413.
12. Brown MS, Goldstein JL. How LDL receptors influence cholesterol and atherosclerosis. Scientific American 1984;251:58–66.
13. Brown, MS, Goldstein JL. A receptor-mediated pathway for cholesterol homeostasis. Science 1986;232:34–47.
14. Shepherd J, Packard CJ, Bicker S, Lawrie TD, Morgan HG. Cholestyramine promotes receptor-mediated low-density-lipoprotein catabolism. New Engl J Med 1980;302:1219–1222.
15. Kesaniemi YA, Witztum JL, Steibrecher UP. Receptor-mediated catabolism of low density lipoprotein in man: Quantitation using glucosylated low density lipoprotein. J Clin Invest 1983;71:950–959.
16. Kader A, Davis PJ, Kara M, Liu H. Drug targeting using low density lipoprotein (LDL): physicochemical factors affecting drug loading into LDL particles. J Control Rel 1998;55:231–243.
17. Vitols S, Gahrton G, Ost A, Peterson C. Elevated low density lipoprotein receptor activity in leukemic cells with monocytic differentiation. Blood 1984;63:1186–1193.
18. Vitols S, Gahrton G, Bjorkholm M, Peterson C. Hypocholesterolaemia in malignancy due to elevated low-density-lipoprotein-receptor activity in tumor cells: evidence from studies in patients with leukemia. Lancet 1985;2:1150–1153.
19. Vitols S, Peterson C, Larsson O, Holm P, Aberg B. Elevated uptake of low density lipoprotein by human lung cancer tissue in vivo. Cancer Res 1992;52:6244–6247.

20. Rudling MJ, Angelin B, Peterson CO, Collins VP. Low density lipoprotein receptor activity in human intracranial tumors and its relation to the cholesterol requirement. Cancer Res 1990;50:483–487.
21. Jung-Testas I, Weintraub H, Dupuis D, et al. Low density lipoprotein-receptor in primary cultures of rat glial cells. J. Steroid Biochem. Mol Biol 1992;42:597–605.
22. Maletinska L, Blakely EA, Bjornstad KA, et al. Human glioblastoma cell lines: levels of low-density lipoprotein receptor and low-density lipoprotein receptor-related protein. Cancer Res 2000;60:2300–2303.
23. Firestone RA. Low-density lipoproteins as a vehicle for targeting antitumor compounds to cancer cells. Bioconjug Chem 1994;5:105–113.
24. de Smidt PC, van Berkel TJ. Prolonged serum half-life of antineoplastic drugs by incorporation into the low density lipoprotein. Cancer Res 1990;50:7476–7482.
25. Allison BA, Pritchard PH, Levy JG. Evidence for low-density lipoprotein receptor-mediated uptake of benzoporphyrin derivatives. Br J Cancer 1994;69:833–839.
26. Soncin M, Polo L, Reddi E, et al. Effect of the delivery system on the biodistribution of Ge(IV)octabutoxy-phthalocyanines in tumor-bearing mice. Cancer Lett 1995;89:101–106.
27. Jori G, Reddi E. The role of lipoproteins in the delivery of tumor-targeting photosensitizers. Int J Biochem 1993;25:1369–1375.
28. Pardridge WM, Buciak JL, Friden PM. Selective transport of an anti-transferrin receptor antibody through the blood-brain barrier in vivo. J Pharmacol Exp Ther 1991;259:66–70.
29. Chen D, Lee KH. Biodistribution of calcitonin encapsulated in liposomes in mice with particular reference to the central nerve system. Biochem Biophys Acta 1993;1158:244–250.
30. Huwyler J, Wu D, Pardridge WM. Brain drug delivery of small molecules using immunoliposomes. Proc Natl Acad Sci USA 1996;93:14164–14169.
31. Meresse S, Delbart C, Fruchart JC, Cecchelli R. Low-density lipoprotein receptor on endothelium of brain capillaries. J Neurochem 1989;53(2):340–345.
32. Lucarelli M, Gennarelli M, Cardelli P, et al. Expression of receptors for native and chemically modified low-density lipoprotein in brain microvessels. FEBS Lett 1997;401(1):53–58.
33. Dehouck B, Fenart L, Dehouck M, et al. A new function for the LDL receptor: transcytosis of LDL across the blood brain barrier. J Cell Biol 1997;138(4):877–889.
34. Matsumoto A, Naito M, Itakura H, et al. Human macrophage scavenger receptors: primary structure, expression, and localization in atherosclerotic lesions. Proc Natl Acad Sci USA 1990;87:9133–9137.
35. Meltzer MS, Gendelman HE. Mononuclear phagocytes as targets, tissue reservoirs, and immunoregulatory cells in human immunodeficiency virus disease. Curr Top Microbiol Immunol 1992;181:239–263.
36. Gendelman HE, Orenstein JM, Baca LM, et al. The macrophage in the persistence and pathogenesis of HIV infection. AIDS 1989;3(8):475–495.
37. Hu J, Liu H, Wang L. Enhanced delivery of AZT to macrophages via acetylated LDL. J Control Rel 2000;69:327–335.
38. Jaeschke H, Smith CW, Clemens MG, Ganey PE, Roth RA. Mechanisms of inflammatory liver injury: adhesion molecules and cytotoxicity of neutrophiles. Toxicol Appl Pharmacol 1996;139:213–226.
39. Bijsterbosch MK, Ziere GJ, van Berkel TJ. Lactosylated low density lipoprotein: a potential carrier for the site specific delivery of drugs to Kupffer cells. Mol Pharmacol 1989;36:484–489.
40. Bijsterbosch MK, Manoharan M, Dorland R, et al. Delivery of cholesteryl-conjugated phosphorothioate oligodeoxynucleotides to Kupffer cells by lactosylated low-density lipoprotein. Biochem Pharmacol 2001;62:627–633.
41. Nyamekye I, Buonaccorsi G, McEwan J, et al. Inhibition of intimal hyperplasia in balloon injured arteries with adjunctive phthalocyanine sensitized photodynamic therapy. Eur J Vasc Endovasc Surg 1996;11:19–28.
42. de Varies HE, Moor ACE, Dubbelman TMAR, et al. Oxidized low-density lipoprotein as a delivery system for photosensitizers: Implications for photodynamic therapy of atheroscleroris. J Pharmacol Exper Ther 1999;289:528–534.

43. Phillips AJ. The challenge of gene therapy and DNA delivery. J Pharm Pharmacol 2001;53:1169–1174.
44. Kim J-S, Maruyama A, Akaike T, Kim SW. In vitro gene expression on smooth muscle cells using a terplex delivery system. J Control Rel 1997;47:51-59.
45. Kim J-S, Kim B-I, Maruyame A, et al. A new non-viral DNA delivery vector: the terplex system. J Control Rel 1998;53:175–182.
46. Steinberg D. A docking receptor for HDL cholesterol esters. Science 1996;271:460–461.
47. Rensen PC, de Vrueh RL, Kuiper J, et al. Recombinant lipoproteins: lipoprotein-like lipid particles for drug targeting. Adv Drug Deliv Rev 2001;47(2–3):251–276.
48. Lacko AG, Nair M, Paranjape S, et al. High density lipoprotein complexes as delivery vehicles for anticancer drugs. Anticancer Res 2002;22:2045–2049.
49. Reisinger RE, Atkinson D. Phospholipid/cholesteryl ester microemulsion containing unesterified cholesterol: model systems for low density lipoproteins. J Lipid Res 1990;31:849–858.
50. Chun PW, Brumbauge EE, Shiremann RB. Interaction of human low density lipoprotein and apolipoprotein B with ternary lipid microemulsion. Biophys Chem 1986;25:223–241.
51. Hirata RDC, Hirata MH, Mesquita CH, et al. Effects of apolipoprotein B-100 on the metabolism of a lipid microemulsion model in rats. Biochim Biophy Acta 1999;1437:53–62.
52. Maranhao RC, Cesar TB, Pedross-Mariani SR, et al. Metabolic behavior in rats of a nonprotein microemulsion resembling low-density lipoprotein. Lipids 1993;28:691–695.
53. Pan G, Shawer M, Oie S, Lu RD. In vitro gene transfection in human glioma cells using a novel and less cytotoxic artificial lipoprotein delivery system. Pharm Res 2003;20(5):738–733.
54. Arap W, Pasqualini R, Ruoslahti E. Cancer treatment by targeted drug delivery to tumor vasculature in a mouse model. Science 1998;279:377–380.
55. Croix BS, Rago C, Velculescu V, et al. Genes expressed in human tumor endothelium. Science 2000;289:1197–1202.
56. Hong FD, Clayman GL. Isolation of a peptide for targeted drug delivery into human head and neck solid tumors. Cancer Res 2000;60:6551–6556.
57. Molema G. Tumor vasculature directed drug targeting: applying new technologies and knowledge to the development of clinically relevant therapies. Pharm Res 2002;19(9):1251–1258.
58. Loregian A, Marsden HS, Palu G. Protein-protein interactions as targets for antiviral chemotherapy. Rev Med Virol 2002;12:239–262.
59. Huang Z. Structural chemistry and therapeutic intervention of protein-protein interactions in immune response, human immunodeficiency virus entry, and apoptosis. Pharmacol Ther 2000;86:201–215.
60. Jones S, Thornton JM. Principles of protein-protein interactions. Proc Natl Acad Sci USA 1996;93:13–20.
61. Cohen GH. Ribonucleotide reductase activity of synchronized KB cells infected with herpes simplex virus. J Virol 1972;9:408–418.
62. Dutia BM. Ribonucleotide reductase induced by herpes simplex virus has a virus-specified constituent. J Gen Virol 1983;64:513–521.
63. Paradis H, Gaudreau P, Brazeau P, Langelier Y. Mechanism of inhibition of herpes simplex virus (HSV) ribonucleotide reductase by a nonapeptide corresponding to the carboxy-terminus of its subunits. J Biol Chem 1988;263:16045–16050.
64. McClements W, Yamanaka G, Garsky V. Oligonucleotides inhibit the ribonucleotide reductase of herpes simplex virus by causing subunit separation. Viology 1988;162:270–273.
65. Pillay D, Taylor S, Richman DD. Incidence and impact of resistance against approved antiretroviral drugs. Rev Med Virol 2000;10:231–253.
66. Zutshi R, Franciskovich J, Slultz M, et al. Targeting the dimerization interface of HIV-1 protease: inhibition with cross-linked interfacial peptides. J Am Chem Soc 1997;119:4841–4845.
67. Bouras A, Boggetto N, Benatalah Z, et al. Design, synthesis, and evaluation of conformationally constrained tongs, new inhibitors of HIV-1 protease dimerization. J Med Chem 1999;42:957–962.

68. Morris MC, Robert-Hebmann V, Chaloin L, et al. A new potent HIV-1 reverse transcriptase inhibitor. J Biol Chem 1999;274:24941–24946.
69. Maroun RG, Krebs D, Roshani M, et al. Conformational aspects of HIV-1 integrase inhibition by a peptide derived from the enzyme central domain and by antibodies raised against this peptide. Eur J Biochem 1999;260:145–155.
70. Kasukawa H, Howley PM, Benson JD. A fifteen-amino-acid peptide inhibits human papilloma E1-E2 interaction and human papillomavirus DNA replication in vitro. J Virol 1998;72:8166–8173.
71. Huang H, Chopra R, Verdine GL, Harrision SC. Structure of a covalently trapped catalytic complex of HIV-1 reverse transcriptase: implication for drug resistance. Science 1998;282:1669–1675.
72. Bundell TL. Structure-based drug design. Nature 1996;384:23–26.
73. Johnsson K, Ge L. Phage display of combinatorial peptide and protein libraries and their application in biology and chemistry. Curr Top Microbiol Immunol 1999;243:87–105.
74. Shih HM, Goldman PS, Demaggio AJ, et al. A positive genetic selection for disrupting protein-protein interactions: identification of CREB mutations that prevent association with the coactivator CBP. Proc Natl Acad Sci USA 1997;94:13396–13401.
75. Weiner AL. Liposomes for protein delivery selecting manufacture and development processes. Immunomethods 1994;4:201–209.
76. Blanke SR, Milne JC, Benson EL, Collier RJ. Fused polycationic peptide mediated delivery of diphtheria toxin A chain to the cytosol in the presence of anthrax protective antigen. Proc Natl Acad Sci USA 1996;93:8437–8442.
77. Elliot G, O'Hare P. Intracellular trafficking and protein delivery by a herpes structural protein. Cell 1997;88:223–233.
78. Fawell S, Seery J, Daikh Y, et al. Tat-mediated delivery of heterologous protein into cells. Proc Natl Acad Sci USA 1994;91:664–668.
79. Loregian A, D'Acunto MR, Battistutta R, et al. Crystal structure of the B subunit of Eshcherichia coli heat-labile enterotoxin carrying peptides with anti-herpes simplex virus type 1 activity. J Biol Chem 1999;274:8764–8769.
80. Dwek R. Glycobiology: toward understanding the function of sugars. Chem Rev 1996;96:683–720.
81. Karpas A, Fleet GWJ, Dwek RA, et al. Aminosugar derivatives as potential anti-human immunodeficiency virus agents. Proc Natl Acad Sci USA 1988;85:9229–9233.
82. Jiang S, Zhao Q, Debnath AK. Peptide and non-peptide HIV fusion inhibitors. Curr Phar Des 2002;8(8):563–580.
83. D'Souza MP, Cairns JS, Plaeger SF. Current evidence and future direction for targeting HIV entry: therapeutic and prophylactic strategies. JAMA 2000;284(2):215–222.
84. Goss PE, Baker MA, Carver JP, Dennis JW. Inhibitors of carbohydrate processing: A new class of anticancer agents. Clin Cancer Res 1995;1(9):935–944.
85. Lehne G. P-glycoprotein as a drug target in the treatment of multidrug resistant cancer. Curr Drug Targets 2000;1(1):85–99.
86. Campbell KR, Ohman EM, Cantor W, Lincoff AM. The use of glycoprotein IIb/IIIa inhibitor therapy in acute ST-segment elevation myocardial infarction: current practice and future trends. Am J Cardiol 2000;85(8A):32C–38C.
87. Meier-Ewert HK, Nesto RW. Targeting the use of glycoprotein IIb/IIIa antagonists—the diabetic patient. Rev Cardiovasc Med 2002;3(Suppl 1):S20–S27.
88. Luscombe NM, Austin SE, Berman HM, Thornton JM. An overview of the structure of protein-DNA complexes. Genome Biol 2000;1(1):1–10.
89. Rhodes D, Schwabe JW, Chapman L, Fairall L. Towards an understanding of protein-DNA recognition. Philos Trans R Soc Lond Biol Sci 1996;351(1339):501–509.
90. Good L, Nielsen PE. Progress in developing PNA as a gene-targeted drug. Antisense Nucleic Acid Drug Dev 1997;7(4):431–437.
91. Welch JJ, Rauscher FJ, Beerman TA. Targeting DNA-binding drugs to sequence-specific transcription factor-DNA complexes. Differential effects of intercalating and minor groove binding drugs. J Biol Chem 1994;269(49):31051–31058.

92. Henderson D, Hurley H. Specific targeting of protein-DNA complexes by DNA-reactive drugs (+)-CC-1065 and pluramycin. J Mol Recog 1996;9(2):75–87.
93. Kondo S, Kondo Y, Li G, et al. Targeted therapy of human malignant glioma in a mouse model by 2-5A antisense directed against telomerase RNA. Oncogene 1998;16(25):3323–3330.
94. Stewart SA, Hahn WC. Prospects for anti-neoplastic therapies based on telomere biology. Curr Cancer Drug Targets 2001;2(1):1–17.
95. Blackburn EH. The telomere and telomerase: nucleic acid-protein complexes acting in a telomere homeostasis system: a review. Biochem (Mosc) 1997;62(11):1196–1201.
96. Neidle S, Kelland LR. Telomerase as an anti-cancer target: current status and future prospects. Anticancer Drug Des 1999;14(4):341–347.
97. Perry PJ, Arnold JR, Jenkins TC. Telomerase inhibitors for the treatment of cancer: the current perspective. Expert Opin Investig Drugs 2001;10(12):2141–2156.

III

Practice and Technology of Cellular Drug Delivery

12

Formulation Considerations for DNA-Based Therapeutics

Thomas J. Anchordoquy, Taylor K. Armstrong, Marion d.C. Molina, S. Dean Allison, Ye Zhang, Mayank M. Patel, Yvonne K. Lentz, and Gary S. Koe

1. Introduction

The ability to deliver genes to cells and tissues in vivo offers the potential to develop potent vaccines and treat many hereditary diseases that are currently considered incurable, e.g., cancer, cystic fibrosis (CF), severe combined immunodeficiency (SCID), and acquired immune deficiency syndrome (AIDS) *(1–6)*. Considering the tremendous promise of DNA-delivery technology, in addition to the extensive genetic information now available from the Human Genome Project, it is not surprising that gene therapy is being touted as the next revolution in medicine. Although a strict definition of "gene therapy" would be limited to therapeutic approaches that aim to use polynucleotides as a template for the in vivo production of proteins, the term is often used to refer to a wide variety of strategies that employ nucleotide-based molecules (e.g., vaccines, antisense, ribozymes, siRNA). To date, 70 clinical protocols have been approved for the delivery of naked DNA, comprising approx 11% of the total number of gene-therapy clinical protocols (http://www.wiley.co.uk/genetherapy/clinical). Slightly more studies (~12%) have employed nonviral, lipid-based vectors to facilitate DNA delivery. In comparison, the large majority of clinical gene therapy trials (>70%) utilizes viruses to deliver therapeutic genes because viruses are more efficient than contemporary synthetic gene-delivery systems. The higher efficiency of viruses should not be surprising if we recognize that these organisms have been evolving their gene-delivery machinery for billions of years. In contrast, the development of nonviral systems for therapeutic gene delivery can be traced back a mere 15 years *(7)*. Although more efficient nonviral gene-delivery systems continue to be developed, synthetic systems have yet to replicate the efficiency of viruses. One significant drawback of viral delivery is the immunogenicity of viruses, which causes significant inflammation in vivo *(8)*, and eliminates the potential for multiple dosing (anyone who has ever had a cold is familiar with the fever and inflammation associated with an immune response to viruses). In fact, the adverse reactions associated with viral delivery have been implicated as the cause of death in clinical trials *(9,10)*. Also, a clinical trial involving liver infusion of an adeno-associated virus (AAV) for the treatment of hemophilia B was halted because of the presence of the viral vector in patient semen, and the concern that the genetic alteration could be passed to offspring *(11)*. More recently, two patients treated with ex vivo gene

From: *Cellular Drug Delivery: Principles and Practice*
Edited by: D. R. Lu and S. Øie © Humana Press Inc., Totowa, NJ

therapy for the treatment of SCID have developed leukemia owing to insertional mutagenesis caused by the retroviral vector used in the study *(12,13)*. Considering the potential safety risk involved in employing viruses as a therapeutic moiety, there is renewed interest in developing safe, efficient, nonviral gene-delivery systems.

Recent studies have reported that synthetic vectors transfer genes at rates only three- to fourfold lower than adenoviruses, and that the overall efficacy of nonviral vectors is actually greater because of the high cytotoxicity and immunogenicity of the adenovirus *(14)*. In addition to the obvious safety advantages of avoiding viral vectors, the lack of a specific immune response to synthetic systems allows them to be readministered multiple times to increase transfection rates and prolong therapeutic gene expression *(15–18)*. Despite these advantages, nonviral gene-delivery systems have yet to demonstrate therapeutic levels of gene expression in human clinical trials. These disappointing clinical results have justified the predominant focus in nonviral gene-therapy research on improving transfection efficiency. It is intriguing that recent studies have suggested that delivery may not be as much of a barrier as is typically assumed, and that epigenetic events pertaining to the expression of the therapeutic gene deserve increased attention *(19–21)*. Furthermore, research on the plasmid DNA-mediated, specific, cytotoxic T-cell response to tumors *(22,23)* and small, interfering, double-stranded RNAs *(24,25)* have identified therapeutic approaches that may not require the high-transfection efficiencies necessary in traditional gene therapy. Although these developments represent promising avenues for nucleotide-based therapeutics, critical pharmaceutical problems (e.g., physicochemical instability of DNA-based therapeutics) will need to be resolved before these technologies are widely deployed. Unfortunately, the study of these less glamorous aspects of gene therapy has received little attention.

It is important to recognize that the widespread use of nucleotide-based therapeutics will require the bulk manufacturing of polynucleotides. For some purposes (e.g., antisense, siRNA, ribozymes), the relatively short strands (≤ 50 bases) of oligonucleotides can be chemically synthesized, and companies that specialize in the production of oligonucleotides are capable of producing tons of purified material. In contrast, larger sequences (typically > 3000 base pairs) encoding genes are synthesized in bacteria as double-stranded, supercoiled plasmids. Extensive purification procedures are subsequently employed to ensure that the isolated product contains the highest possible percentage of plasmids in the supercoiled form. A single cleavage event in the backbone of either polynucleotide chain causes the release of torsional energy in the supercoils, and the conversion to the open circle form *(26)*. Although backbone cleavage can result from any one of a variety of chemical mechanisms *(26–29)*, retention of the supercoiled form serves as an indicator of chemically stable conditions. In practice, industry strives to maintain the highest possible supercoil content (≥90%; *30*). In contrast to the enormous quantities previously mentioned for oligonucleotides, the production of *gram* quantities of purified supercoiled plasmid is usually sufficient for Phase I clinical trials for most applications.

As described earlier, the maintenance of supercoil content is used routinely as the primary measure of the chemical integrity/stability of naked DNA, and is routinely assessed with agarose gel electrophoresis. The stability of nonviral gene-delivery systems (i.e., DNA bound to agents that facilitate delivery) is more complicated, and involves aggregation of suspended particles and the retention of poorly characterized structural motifs, in addition to the maintenance of supercoil content. Considering the significantly different obstacles and corresponding strategies involved in the stabilization of a solution of naked DNA vs a suspension of DNA-containing particles, this chapter will address these challenges separately. In addition to discussing liquid formulations, we describe progress to date in the development of frozen and dehydrated preparations of both naked DNA and nonviral gene-delivery systems. Wherever possible, we attempt to provide some mechanistic insight into the findings described in the literature with the hope that future formulations may be rationally designed rather than empirically derived.

2. Naked DNA

Many cellular targets of gene therapy involve the use of delivery vehicles (e.g., viruses, cationic liposomes); however, Wolff et al. *(31)* have demonstrated that naked DNA is taken up readily by muscle cells in vivo. This observation has led to the development of DNA vaccines that are known to elicit both a humoral and cellular immune response *(4–6,32)*. These potent vaccines are currently being tested in Phase I/II clinical trials, and are being combined with traditional protein vaccines and/or adjuvants (e.g., interleukin-2 [IL-2]) to further boost immune responses to pathogenic parasites, viruses, bacteria, and cancer. With all of the promise surrounding nucleic acid-based therapeutics, surprisingly little attention has been paid to the chemical stability of DNA on a pharmaceutically relevant time-scale. Clearly, marketable nucleotide-based products (e.g., vaccines) will need to possess extended shelf-lives similar to other biopharmaceuticals (18 mo to 2 yr).

DNA, like other biopharmaceuticals (e.g., proteins, liposomes), is susceptible to many degradation reactions that are enhanced in aqueous formulations. However, unlike other biopharmaceuticals, the DNA used in gene therapy is utilized as a template for protein synthesis within the patient. It is imperative that we recognize this distinct difference between gene therapy and the administration of other biopharmaceuticals because chemical alterations of the nucleotide bases has been shown to result in mispairing that can ultimately lead to the intracellular production of mutated proteins *(33–35)*. Considering that mutations that result in altered protein folding can be recognized as foreign by the host's immune system, chemical alteration of DNA has the potential to trigger a life-threatening immune response to an endogeneously synthesized protein encoded by the "therapeutic" plasmid. Although the probability of this event is quite low, it behooves companies that are developing DNA-based therapeutics to utilize rigorous standards for maintaining chemical integrity.

To avoid chemical and enzymatic degradation in a laboratory setting, researchers commonly store DNA as a precipitate in ethanol at –80°C. Under these conditions, nucleic acids are stable for prolonged periods, but must be isolated from the ethanol, transferred to aqueous buffers, and quantified prior to use. These manipulations preclude ethanol precipitates from being used for clinical DNA preparations. Instead, it would be ideal if aqueous solutions could be formulated to resist depurination *(36)*, depyrimidination *(37)*, deamination *(38)*, and hydrolytic cleavage *(39,40)*. To this end, studies have attempted to formulate in alkaline conditions to inhibit these acid-catalyzed degradation mechanisms during prolonged storage *(26,27)*. Under such conditions (pH 8.5), the primary threat to DNA stability (excluding nuclease contamination) is oxidation *(26,27)*. (It is worth noting that lipids used in gene-delivery systems are rapidly degraded under these conditions *[41]*; a serious dilemma for companies developing lipoplexes as pharmaceutical products.) In general, oxidative damage is greatly enhanced by the presence of trace metals (e.g., Fe^{3+}, Cu^{2+}) that produce hydroxyl radicals via Fenton-type reactions *(26–29)*. In their classic study, Evans et al. demonstrated that transition metal impurities in various formulation components (including the purified DNA preparation) permitted significant levels of oxidation to occur during storage *(27)*. The rate of oxidation was attenuated when buffers were demetalated prior to storage. Although these findings demonstrate that trace-metal contaminants are responsible for the observed oxidation, the procedures used to remove trace metals for a small laboratory experiment are not practical for large-scale purification and production *(27,42)*. Therefore, chelators and antioxidants will need to be included in DNA formulations to reduce oxidation and maximize stability during storage. To this end, Evans et al. demonstrated that the combination of 200 μ*M* ethylenediaminetetraacetic acid (EDTA) and 1% ethanol is capable of preserving more than 90% of the initial supercoil content during 2 yr of storage at room temperature *(27)*. It is worth noting that EDTA does not bind metals in such a way that prevents hydroxyl radical formation and subsequent oxidation *(43)*, but the presence of ethanol as a radical scavenger is sufficient for prolonged storage in the liquid state. However, the use of ethanol is not compatible with some adjuvants and delivery systems that may be desirable in a DNA vaccine formulation *(27)*. For example, the presence of ethanol would solubilize lipids incorporated into the formulation. Therefore, it would be advantageous to develop formulation strategies that eliminate the need for ethanol. As shown by Evans et al., the use of 200 μ*M* diethylenetriaminepentaacetic acid (DTPA) enhances stability to an extent comparable to that observed with EDTA/ethanol *(27)*. These findings suggest that the employment of DTPA might circumvent the need for ethanol, thereby achieving stable liquid formulations that are more widely compatible with pharmaceutical applications.

To test the ability of DTPA to preserve DNA integrity during long-term storage, we monitored the room temperature stability of aqueous DNA formulations (50 μg/mL, 2.5 m*M* Tris-HCl, pH = 8.5) with and without 200 μ*M* DTPA (Fig. 1). The data indicate that the presence of DTPA significantly enhances the preservation of supercoil content during prolonged storage at room temperature. In addition, Fig. 2 shows that the preservation of biological activity of the plasmid correlates with the observed maintenance of supercoil content. However, the loss of supercoil content and transfection rates are much too rapid under these conditions, suggesting that aqueous formulations lacking ethanol do not provide sufficient stability during prolonged storage at room temperature. It is important to note that DNA stability can be a limiting factor even in instances where DNA is incorporated into a delivery system, and therefore the problems associated with chemical stability are not limited to naked DNA formulations. Although other chelators (e.g., desferal, inositol hexaphosphate) and scavengers (e.g., ascorbic acid, trolox, cysteine) might prove more effective and thereby eliminate the need for ethanol, sufficient stability on a pharmaceutical time scale may require the development of frozen or dehydrated formulations.

The effect of freeze-thawing on DNA was first reported by Shikama more than 35 yr ago *(44)*. This study demonstrated that the double-helix of calf thymus DNA remained intact during freezing to temperatures as low as –192°C *(44)*. Several years later, Lyscov and Moshkovsky described a mechanism of DNA degradation during freezing ("cryolysis"), that was dependent on the rate of cooling of the frozen sample *(45)*. These authors suggested that cryolysis resulted from the formation of cracks within the ice during freezing, and demonstrated that these cracks were more prevalent in rapidly cooled samples *(45)*. In contrast, Davis et al. showed that slow freezing is more damag-

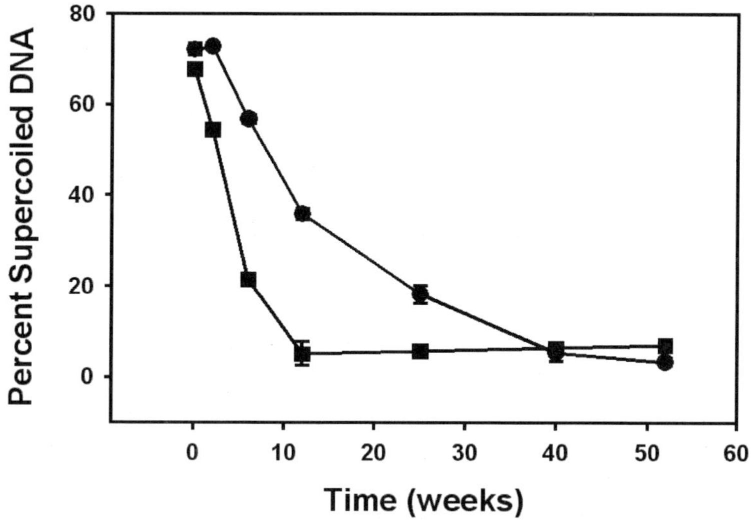

Fig. 1. Supercoil content of naked DNA after prolonged storage in an aqueous formulation. Plasmid DNA (50 μg/mL) was formulated in 2.5 m*M* Tris-HCl, pH 8.5, either with (circles) or without (squares) 200 μ*M* DTPA, and stored at room temperature for 1 yr. Supercoil content was assessed with agarose gels followed by ethidium bromide staining and quantification. Each symbol represents the mean ± 1 standard error of single measurements on triplicate samples.

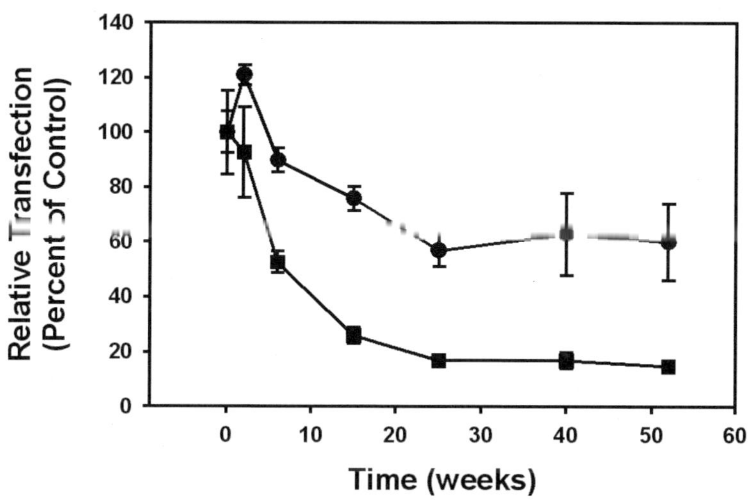

Fig. 2. Transfection rates of naked DNA after prolonged storage in an aqueous formulation. Plasmid DNA (50 μg/mL) encoding green fluorescent protein was formulated in 2.5 m*M* Tris-HCl, pH 8.5, either with (circles) or without (squares) 200 μ*M* DTPA, and stored at room temperature for 1 yr. At the indicated time-points, aliquots were removed and complexed with polyethylenimine (N/P = 7). The resulting complexes were used to transfect COS-7 cells in culture, and the percentage of cells expressing the encoded gene was quantified by FACS. Transfection rates are expressed as a percent of that for complexes prepared with fresh DNA (typically 40–60% of cells expressing). Each symbol represents the mean ± 1 standard error of single measurements on triplicate samples.

ing than rapid freezing, and that damage to oligonucleotides is dependent on base composition, structure, and length *(46)*. In a study by Ando et al., very rapid cooling by immersion of samples into liquid nitrogen was used to protect DNA from shear stress during homogenization *(47)*. However, these authors clearly demonstrated that both sugars and EDTA were required to preserve supercoil content during freezing *(47)*. Although it is not clear whether these excipients specifically protect against the cryolysis mechanism of degradation described by Lyscov and Moshkovsky, it seems possible that glass formation, caused by the presence of sugars *(48)*, would attenuate the cracking that is implicated in DNA degradation. As suggested by the authors, the presence of EDTA might inhibit DNase *(47)*. Additionally, EDTA could serve to chelate transition metals that are known to facilitate DNA degradation *(27)*. Surprisingly, we could not find a single report in the literature that investigated the storage stability of naked DNA in a frozen formulation. However, distribution of a frozen formulation is complicated by the potential for thawing, and the need for a dependable cold chain. It would therefore be advantageous to develop dehydrated preparations that could be shipped and stored at ambient temperatures.

With regards to the stability of DNA during dehydration, it is quite common for molecular biologists to dry down aqueous or ethanolic solutions of DNA for use in their experiments. Because dried samples are not stored for prolonged periods in the dried state, and because molecular biological techniques are based on gene amplification and precise selection of the desired products among large numbers of imperfect reaction products, DNA damage is not detected. Unfortunately, this situation leads to the common misconception that DNA is a stable molecule. This naïvetè on the part of molecular biologists permeates the field of gene therapy, and is largely responsible for the lack of studies investigating stability in a pharmaceutical context. Obviously, the sequence fidelity necessary to satisfy the experimental conditions in a typical molecular biology laboratory is drastically different than that required for a federally regulated, pharmaceutical production facility.

There are several methods of removing water from liquid preparations to produce dehydrated DNA formulations, e.g., spray-drying, spray freeze-drying, lyophilization. Although the small plasmids currently employed in gene therapy (~5 kb) are quite resistant to shear stress, spraying of DNA has largely been avoided owing to the common misperception that the shear-induced damage reported in early studies on genomic DNA is generally applicable to other forms, e.g., plasmids and oligonucleotides *(49–52)*. As a result, the majority of stabilization studies have employed freeze-drying (lyophilization) to generate dehydrated formulations. With the notable exception of lyophilization studies by Poxon and Hughes *(53)* and Pikal et al. *(54)*, we find it surprising that so little attention has been paid to the lyophilization of naked DNA *(53,54)*. Incidental reports available in the literature are contradictory; some suggest that dehydration might be stabilizing *(55,56)*, whereas others show that lyophilization induces damage *(47,57–59)*. Recent studies have demonstrated that sugars can protect DNA against damage (e.g., loss of supercoil) induced during lyophilization *(47,53)*. However, storage-stability studies on lyophilized DNA are lacking. In their study of lyophilized DNA, Poxon and Hughes *(53)* reported that lactose, glucose, and sucrose were able to preserve the biological activity of naked DNA stored at 75°C for 3 wk *(53)*. With the exception of that brief storage study, the only study to investigate the stability of naked DNA during prolonged storage in the dried state concluded that degradation occurs gradually over 14 mo at 4°C *(58)*. Considering that this study was conducted almost 30 yr ago, before the advent of modern molecular biology, there is an obvious need for extended storage studies on naked DNA dried in stabilizing sugar formulations.

It has been reported that lyophilization causes some denaturation of double-stranded DNA *(53)*. This finding is consistent with earlier work by Lindahl, who observed that DNA stored over phosphorous pentoxide does not retain its double-helical structure, which renders it more vulnerable to oxidative damage during storage *(60)*. It is interesting that spores that have evolved to survive prolonged periods in a quiescent state synthesize proteins that bind to the dehydrated A conformation, reducing rates of depurination by at least 20-fold *(60)*. Bacteria also respond to environmental stress by synthesizing proteins that co-crystallize with DNA, such that oxidation is greatly reduced *(61)*. A similar role has been suggested for chromatin in mammalian cells *(62)*. It follows that it may be possible to compact DNA such that bases become sequestered and less susceptible to reactive oxygen species (ROS) during prolonged storage. This suggestion is analogous to the "base-shielding hypothesis" that was first proposed by Ward and Kuo (1978) to explain the greater resistance of duplex DNA *(63)*, as compared with single-stranded polynucleotides, to chemical degradation *(37,64,65)*. Thus, the compaction of DNA may serve to enhance stability, but the use of cationic agents that bind to DNA is also used to facilitate delivery. Because DNA bound to compacting molecules technically is not "naked," it seems appropriate to consider the stability of DNA complexed with cationic agents in the following section on nonviral gene-delivery systems.

3. Nonviral Gene-Delivery Systems

Although improved nonviral vectors are continually being developed, it is generally recognized that the stability of these systems is insufficient for widespread application of this promising technology *(66–68)*. Therefore, it would be advantageous to develop methods of stabilizing nonviral vectors that allow preparations to be produced under optimal conditions, shipped, and stored until needed. Considering the massive research effort focused on improving the delivery efficiency of nonviral vectors, potent DNA-based therapeutics for the treatment of many diseases may soon become a reality. However, if synthetic delivery vehicles cannot be rendered sufficiently stable to be developed as pharmaceutical products, it is unlikely that these therapies will ever be employed to treat disease. Therefore, the realization of the enormous potential of nonviral gene therapy will ultimately require studies that investigate vector stability and characterization. Such studies will need to be conducted so that vectors with consistent physical and chemical properties can be manufactured on an industrial scale. Currently, nonviral vectors are made in relatively small batches by mixing a solution of DNA with a solution/suspension of an agent that facilitates delivery (e.g., lipid, polymer). Although there have been some interesting reports of neutral polymers enhancing transfection rates despite minimal interaction with the plasmid *(69,70)*, most delivery systems incorporate cationic agents that interact electrostatically with DNA to form complexes that are ultimately used for gene delivery *(71)*. This method of preparation typically results in a highly heterogenous suspension of particles possessing a wide range of +/– charge ratios and size *(72–76)*. Furthermore, the particle size, zeta potential, and other biophysical characteristics that are known to affect gene-delivery efficiency can be altered by many factors including mixing protocol, buffer components, and complexation time *(77–81)*. These issues complicate scale-up and manufacturing, and will clearly need to be addressed if nonviral gene-delivery proves clinically efficacious.

Assuming that adequate industrial-scale manufacturing procedures can be developed, it is imperative that the biophysical properties and gene-delivery efficiency of nonviral vectors be preserved until the product is administered in the clinic. It is important to emphasize that the biophysical properties (e.g., size) must be maintained in order to insure appropriate quality control of a pharmaceutical product, regardless of the ultimate effects on gene delivery. To this end, stability becomes an issue for the relatively small-scale preparations currently being used for Phase I trials, even though such studies are not designed to test therapeutic efficacy. Clearly, instability of the preparation after production will adversely affect quality control, safety, and transfection rates. Although other factors can dramatically alter the efficiency of gene transfer (e.g., route of administration, target tissue, rate of cell division, chemical integrity, polycation formulation), our discussion focuses on the physical stability of nonviral vectors, and the maintenance of particle characteristics during processing and storage.

It is well-known that aqueous suspensions of nonviral vectors have a tendency to aggregate over time *(67,73,82)*. The combination of multiply charged species in this manner results in incomplete charge neutralization, and a heterogenous suspension of vectors possessing a range of net charge and particle sizes *(73)*. It is this heterogeneity combined with unneutralized regions within complexes that fosters aggregation in aqueous suspensions *(83,84)*. The time-scale over which aggregation occurs is typically from minutes to hours, which is clearly insufficient for a pharmaceutical product *(68,85)*. Several reports claim to have developed "stable" nonviral vectors that do not aggregate for several hours *(17,86–88)*. Although this level of stability may be adequate for some experimental purposes, such short half-lives are clearly insufficient for widespread clinical use. Some clinical trials have attempted to circumvent this problem by preparing complexes at the bedside, immediately prior to injection *(15,89)*. This method of sample preparation and administration is not practical, and leads to significant variations in product quality, safety, and gene-delivery efficiency. This problem has stimulated an interest in developing stable, single-vial formulations that resist aggregation and possess consistent physical properties.

Studies have employed different approaches in an attempt to prevent aggregation of pre-assembled nonviral vectors. For example, Caplen et al. diluted complexes with an alkaline solution that was intended to enhance stability by weakening the interaction between the cationic lipid and DNA *(90,91)*. Using another approach, Hofland et al. and Gao and Huang employed sucrose density gradients to separate the heterogeneous population of particles, and isolate a stable fraction of lipid/DNA complexes *(92,93)*. These stable fractions could be stored under refrigerated conditions for 3 mo without detectable losses in transfection rates. It was subsequently demonstrated that these complexes effectively deliver genes in vivo *(94)*. Although this separation technique can be applied to research-scale projects, it is doubtful that sucrose density fractionation would be practical for bulk manufacturing.

In addition to isolating a stable fraction of complexes, the studies with sucrose gradients described earlier indicate that lipid/DNA complexes are more stable if they are separated from other components in the bulk suspension. Considering that preparations of lipid/DNA complexes typically contain free (i.e., uncomplexed) liposomes and/or naked DNA, these findings suggest that interactions between complexes and free components can result in decreased rates of gene delivery. Furthermore, it has been reported that aggregated nonviral vectors are significantly more toxic, demonstrating that instability ultimately can lead to safety concerns *(86)*. Such effects are not surprising if we consider the impact that incorporation of additional polynucleotides and/or fusion of excess cationic liposomes would have on a vector's charge, size, and lipid-DNA interactions. In order to enhance stability, it should be possible to develop methods of preventing interactions among components within a single suspension. In fact, Hong et al. and Eastman et al. utilized polyethylene glycol (PEG)-lipid conjugates to prevent aggregation of nonviral vectors and sterically stabilize particles in aqueous suspensions *(95,96)*. Studies have also utilized PEG derivatized to anionic peptides in order to coat cationic particles and reduce aggregation with serum components *(86,97,98)*. Although the use of PEGylated components appears to reduce aggregation of nonviral vectors under some conditions, PEGylated lipid/DNA complexes appear to be more sensitive to agitation stress than vectors lacking PEG *(99)*. It has been suggested that the PEG chains extending from vector surfaces become entangled during agitation, thereby facilitating aggregation and/or fusion *(99)*. In addition, steric stabilization by PEG is also known to curtail interactions with cells and interfere with cellular processing *(75,100)*. Therefore, nonviral vectors containing PEGylated components typically display lower transfection activities than their non-PEGylated counterparts, despite improved stability in physiological solutions *(99,100)*.

The use of condensing polyamines has also been shown to increase the stability of nonviral vectors, and prevent aggregation during storage *(95,101)*. Although it is doubtful that stabilization via condensation would be sufficient for prolonged storage as an aqueous preparation, studies by Cherng et al. report that suspensions of poly([2-dimethylamino]ethyl methacrylate)-based gene-delivery systems maintain particle size and in vitro transfection rates for 10 mo at 20°C *(102)*. This report of prolonged stability of an aqueous vector suspension is exceptional, and differs substantially from the short shelf lives (hours to days) that are typical of most synthetic gene-delivery systems. One factor that plays a significant role in determining stability in liquid suspensions is the concentration of both DNA and cationic agent(s) *(103)*. As the suspension gets more concentrated, there is a higher probability for aggregation. It follows that very dilute suspensions that are suitable for in vitro transfection studies exhibit markedly improved stability in liquid formulations as compared to when the identical vector is prepared under the concentrated conditions necessary for in vivo delivery *(103)*. In addition, it has been shown that prolonged storage causes lipid/DNA complexes to dissociate gradually over 70 d *(104)*. Although the increased stability of some of the aqueous formulations described previously is encouraging, it should be noted that storage-stability studies typically neglect stresses that are commonly encountered during processing and/or shipping, e.g., agitation and/or freeze-thawing *(99,105,106)*. Considering that both of these stresses can alter physical characteristics and significantly reduce transfection rates of nonviral vectors *(67,68,105,106)*, it is prudent to test preparations under these conditions and develop formulations that can resist shipping-induced damage.

One method to circumvent the sensitivity of liquid formulations to shipping stresses would be to develop frozen formulations that are resistant to agitation-induced aggregation (Table 1). If we consider the intricate associations between cationic agent and DNA within a complex that are required for optimal gene delivery (e.g., structures

Table 1
Freeze-Thawing Studies of DNA-Based Therapeutics

Cationic agent	References	Formulation	Freezing protocol	Storage
DOSPA/DOPE	*(92)*	5% dextrose	Frozen and stored at –20°C	3 mo, transfection retained
PDMAEMA	*(127,148)*	10% sucrose, trehalose, lactose	Frozen on precooled (–38°C) shelf	—
DOTAP/cholesterol/protamine	*(101)*	5% sucrose	Frozen at –20°C (30 min) then –80°C; Frozen directly at –80°C; liquid nitrogen	—
DMRIE/cholesterol DOTAP/DOPE DOSPA/DOPE	*(105)*	0.5 *M* sucrose (17%)	Frozen at –20°C; liquid nitrogen	—
DMRIE/DOPE	*(103)*	5% sorbitol + 20 m*M* sodium acetate	Frozen at –70°C, stored at –20°C	1 yr, transfection, lipid + DNA integrity retained
DMRIE/cholesterol	*(107)*	1% and 5% sucrose, trehalose, lactose, glucose, PEG, hydroxyethyl starch,	Frozen to –38°C at –1°C/min	—
DMRIE/cholesterol	*(108)*	Sucrose or glucose at 1000:1 sugar/DNA wt. ratio + 2.5 m*M* Tris-HCl	Frozen to –38°C at –1°C/min	—
DOTAP/cholesterol	*(109)*	No excipient	Frozen in liquid nitrogen, stored at –80°C	20 d, zeta potential, size, transfection retained

involved in compaction, membrane fusion, complex dissociation), it is not surprising that freezing alters vector characteristics and performance *(67,99,105)*. In addition to reductions in transfection activity, increases in particle size indicate that aggregation occurs during the freeze-thaw process *(99,105,107,108)*. Although this damage is extensive after a slow freeze-thaw cycle, rapid freezing and thawing results in minimal losses in transfection activity, and the maintenance of particle size *(103,105,109)*. These observations suggest that the increased time during which particles can diffuse during slow cooling promotes aggregation of nonviral vectors, which results in reduced transfection rates *(68,84,105,108)*. In fact, our experiments that assess particle size at different subzero temperatures after ice nucleation show that aggregation in the presence of excipients is attenuated at temperatures below –15°C (Fig. 3). Therefore, rapid freezing by immersion in liquid nitrogen presumably minimizes aggregation by reducing the time available for particles to collide in a liquid phase *(103,105,109)*. Furthermore, the fact that maintenance of particle size often correlates with the recovery of transfection rates indicates that aggregation during freezing represents a major mechanism of damage. It is worth noting that an increase in particle size can result from irreversible phenomena (e.g., lipid fusion) or simple "clumping" *(76)*. We have also observed that aggregation of some preparations is reversible (unpublished observations), but we have not determined whether dispersion of aggregates results in recovery of transfection activity. Because little attention has been paid to these different modes of "aggregation," further characterization of these different assemblies is clearly warranted.

The ability of "cryoprotectants" to prevent aggregation of proteins and liposomes during freezing is well-established *(110–113)*. The manifestation of employing stabilizing excipients (i.e., the prevention of aggregation) is the same for both proteins and liposomes, but the protective mechanism appears to be different for macromolecules dissolved in solution (e.g., proteins) as compared to suspended particles (e.g., liposomes). In the case of proteins, aggregation typically involves some "aggregation-competent" intermediate that is partially unfolded *(114)*. Therefore, aggregation can be prevented by employing excipients to inhibit the formation of partially unfolded species *(111,114)*. The mechanism by which stabilizing excipients, typically sugars, attenuate protein unfolding during freezing is the same mechanism that has been described by Timasheff and colleagues to explain solution stability, i.e., preferential exclusion *(115,116)*. This thermodynamic mechanism involves the exclusion of solutes from the hydration shell of the protein, and the subsequent stabilization of the protein's native structure. For the purposes of this discussion, it is important to recognize that the preferential exclusion mechanism derives from the inability of some solutes to access the surface of the protein (i.e., stabilizing solutes do not "bind" to the protein). In contrast, sugars are thought to stabilize liposomes by interacting directly with the membrane surface during freezing *(110,112,113,117)*. The role of binding in membrane protection was unequivocally demonstrated by Hincha et al. who used galactose-specific lectins to protect thylakoid membranes during freeze-thawing *(118)*. Although the implication of a direct interaction in stabilization does not preclude preferential exclusion of the excipient overall (i.e., a solute present at a high concentration on the membrane surface is still considered to be excluded if the bulk solute concentration is greater), it does suggest that the solute associated with the membrane surface is responsible for preserving vesicle integrity during freezing *(110,112,113,117)*. Such an interaction between sugars and various membrane systems has also been observed in the solution state *(119–122)*.

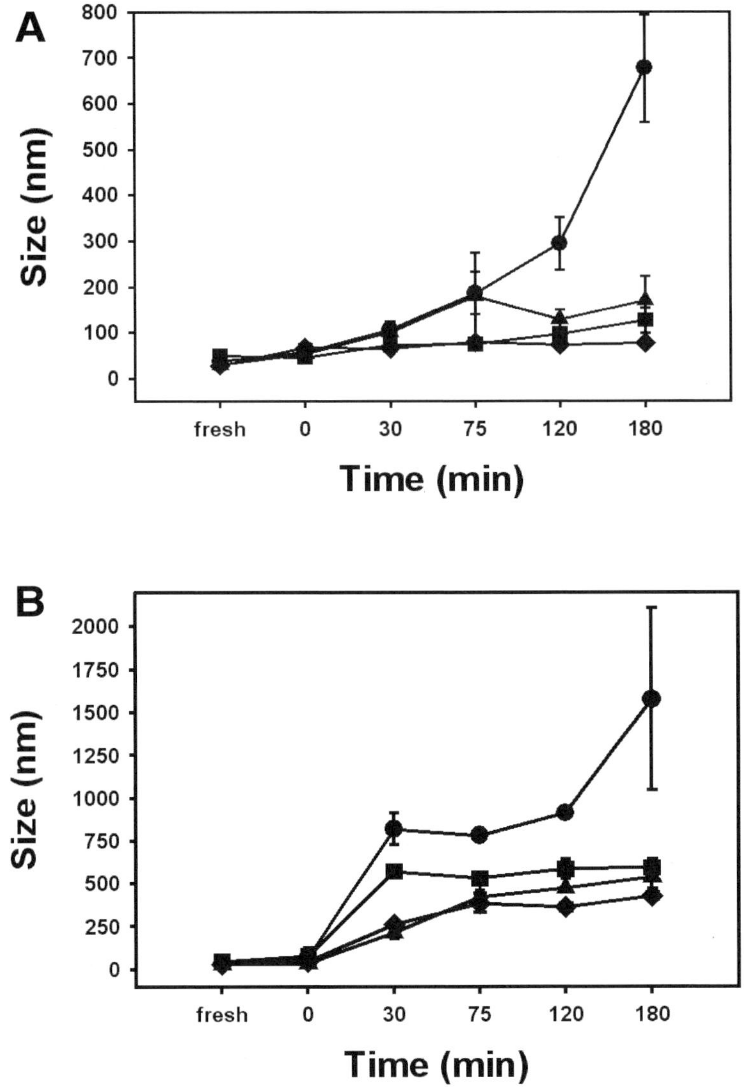

Fig. 3. Aggregation of polyethylenimine/DNA complexes after freezing. Vectors in either a sucrose (**A**; sugar/DNA = 200) or PEG (**B**; MW = 200 K, polymer/DNA = 125) formulation were cooled on a lyophilizer shelf that was ramped down to various temperatures at 2°C/min. When the sample temperatures reached approx –5°C, freezing was initiated by briefly dipping samples into liquid nitrogen such that ice was nucleated, but the sample was not extensively crystallized (time 0). Samples were then cooled on the lyophilizer shelf to the indicated sample temperatures, and held for 180 min. Under these conditions, sample temperatures reached –10°C (circles), –15°C (squares), –20°C (triangles), or –25°C (diamonds) by the 30-min time-point. Note that postnucleation sample temperatures of –10°C were insufficient to prevent aggregation. In contrast, vector size remained constant after achieving sample temperatures of ≤ –15°C over the time course of this experiment. Each symbol represents the mean ± 1 standard error of single measurements on triplicate samples.

In this way, solute-induced stabilization of suspended particles is analogous to the steric stabilization observed with PEGylated lipids *(123,124)*.

The application of the stabilization mechanisms described earlier to the preservation of nonviral vectors has not been extensively studied. It is clear that there are many types of damage that can be incurred by nonviral vectors, e.g., chemical damage, aggregation, and structural alterations. Although all of these types of damage are important for quality control and transfection activity, stability studies typically only monitor changes in particle size and/or transfection as indicators of "recovery." Therefore, the available data limits the discussion of preservation mechanism to a deliberation of how solutes prevent vector aggregation during freezing. Nonviral vector preparations are similar to liposomes in that they are a suspension of particles that can be stabilized sterically by the incorporation of PEGylated components *(86,95,97,98,124)*. Therefore, it seems plausible that the direct interaction mechanism described for liposomes (see earlier) might also apply to the stabilization of nonviral vectors. However, it should be pointed out that much greater amounts of sugar are required to prevent aggregation of nonviral vectors than that necessary to inhibit liposome fusion during freezing *(110,125)*. The relatively large quantities of excipients suggest that maintenance of particle size is owing to nonspecific, "bulk" characteristics of the formulation. We have suggested two mechanisms that could potentially contribute to the observed retention of nonviral vector size during freezing: (1) glass formation and (2) particle isolation.

3.1. Glass Formation

The formation of ice during the freezing process causes solutes and suspended particles to be concentrated in the unfrozen fraction *(126)*. At equilibrium, the composition of the unfrozen fraction must have a freezing point below that of the sample temperature. As cooling progresses to lower temperatures, the unfrozen solution will either crystallize or form an amorphous glass at a composition-dependent temperature (T_g'). Each of the sugars shown to be effective at preserving nonviral vectors readily form glasses during freezing *(48,68)*. In fact, Cherng et al. explicitly state that they were careful to freeze under conditions where the excipient forms a glassy state *(127)*. Thus, it could be argued that the observed maintenance of particle size derives from the immobilization of vectors within a glassy matrix that prevents aggregation.

Several studies have shown that maintenance of particle size is dependent on the concentration of stabilizing excipient *(105,108)*. Recent studies have further demonstrated that the excipient concentration that is sufficient to retain vector size during freezing is dependent on the particle concentration, i.e., more dilute suspensions require less excipient to prevent aggregation *(108)*. Because glass formation during freezing is independent of the initial excipient concentration, these findings suggest that mechanisms other than glass formation contribute to the maintenance of particle size.

Additional evidence against glass formation as the primary mechanism of stabilization comes from our work demonstrating that polymers that readily form glasses during freezing (e.g., hydroxyethyl starch) do not preserve particle size, whereas sugars (e.g., glucose) are able to maintain vector size at temperatures above T_g' *(108)*. Although immobilization of suspended particles within a glassy polymer matrix is sufficient to prevent aggregation, many hydrophilic polymers possess high surface-tension increments that may foster coalescence in the unfrozen solution prior to vitrification. This effect can be significant considering that excipients are dramatically concentrated upon ice formation, with concentrations in the unfrozen solution typically reaching more than 60% by weight *(108)*. The high concentration of excipients combined with

Table 2
Apparent Surface Tension of 5% PEG Solutions

Compound	MW	Surface tension	Standard error
Water	18	75.53	0.088
PEG	400	67.03	0.082
PEG	4600	64.03	0.216
PEG	20,000	63.93	0
PEG	200,000	60.36	0.525

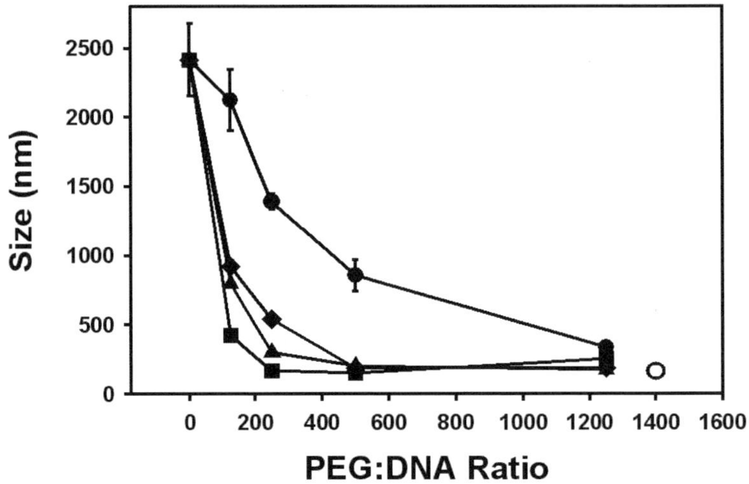

Fig. 4. Protection of DOTAP:DOPE/DNA complexes by PEG during freezing. Vectors were formulated in PEGs possessing molecular weights of 400 (circles), 4500 (triangles), 20 K (diamonds), and 200 K (squares). Samples were subjected to slow freezing (~2°C/min) to a shelf temperature of –40°C in a shelf lyophilizer. The shelf temperature was maintained overnight (~16 h), and vector size was monitored with dynamic light scattering after rapid thawing. The size of fresh vectors is indicated by the open circle at 1400 PEG/DNA ratio. Note that larger PEGs with lower surface tensions (Table 2) were progressively more effective at preserving particle size during freezing. Each symbol represents the mean ± 1 standard error of single measurements on triplicate samples.

subzero temperatures can result in surface tensions that are 30% higher than that observed in room-temperature solutions *(108)*. According to this argument, polymers that possess negative surface-tension increments (e.g., PEG) should not foster coalescence during freezing, and thus might serve as superior glass-forming cryoprotectants. Although we have yet to thoroughly investigate this hypothesis, our results with different molecular-weight PEGs demonstrate that cryoprotection correlates with surface tension (Table 2, Fig. 4). However, these results are complicated by the fact that T_g' *(48)* and the affinity for the vector surface *(128,129)* varies with polymer molecular weight. Regardless, these findings strongly suggest that stabilization during freezing cannot solely be the result of glass formation. In addition, the data in Fig. 3 show that aggregation is attenuated by sucrose at temperatures well above its T_g' (~–32°C). This suggests that the unfrozen solution at temperatures where particle size is maintained

(<–15°C) may be sufficiently viscous such that aggregation is inhibited on the experimental timescale (~16 h). Thus, glass formation, per se, may not be required under these conditions.

3.2. Particle Isolation

As described earlier, solutes and suspended particles are concentrated in the unfrozen fraction during the freezing process. The volume of the unfrozen fraction at any temperature is determined by the initial solute concentration *(126)*. Therefore, the observation that more dilute suspensions require lower initial excipient concentrations for cryoprotection suggests that "crowding" of particles facilitates aggregation *(108)*. In fact, Allison et al. demonstrated that there is a critical excipient:DNA ratio at which protection is observed *(108)*. Furthermore, the authors concluded that the volume of the unfrozen fraction at this ratio is sufficient to isolate lipid/DNA complexes in a viscous excipient matrix, thereby preventing aggregation. An excipient/DNA weight ratio of approx 1000 was sufficient to achieve complete protection by either sucrose or glucose, but only partial protection was observed in solutes that crystallize during freezing, e.g., mannitol *(108)*. These findings are consistent with reports from other laboratories, and suggest that results should be reported as a function of excipient/DNA ratio rather than the excipient concentration that is traditionally reported. Later studies reported that this excipient/DNA ratio can protect other lipid-DNA complexes and lipid-polymer-DNA complexes, but that polymer-DNA complexes require much greater ratios to prevent aggregation during freezing *(125)*. It was subsequently determined that the purity of the polymer and DNA can dramatically affect stability during physical stress, and that cleaner preparations of polymer/DNA complexes are also stabilized during freezing at an excipient/DNA weight ratio of 1000 (unpublished results).

As noted earlier, it is critical for quality control and safety that vector size be maintained during processing regardless of its effect on transfection rates *(99)*. As a result, size is typically monitored during stability studies. However, the ability of vectors to deliver genes (i.e., transfection rates) must also be preserved. Not surprisingly, we have demonstrated that simply preventing aggregation is not sufficient to maintain transfection rates of nonviral vectors during freezing *(125)*. These latter findings suggest that critical interactions between cationic agent and DNA within a complex can be perturbed by the freeze-thawing process. Clearly, more work is needed to characterize and understand these subtle interactions at the molecular level. Fortunately, recent work with a variety of biophysical techniques has begun to characterize interactions within nonviral vectors *(130–135)*, but it is still unclear which physical properties are critical for gene delivery.

Regardless of the mechanisms by which excipients might protect nonviral vectors during freezing, evidence to date suggests that frozen formulations may offer pharmaceutically relevant stability to some types of complexes *(103)*. However, the stability of frozen formulations can require strict maintenance of storage temperature to prevent thawing and/or crystallization of excipients *(68)*. It is important to note that prolonged stability in the frozen state can be extremely sensitive to very brief exposures to elevated subzero temperatures, e.g., during defrost cycles in a conventional freezer. Maintaining strict control of sample temperature during transportation is difficult, and can limit worldwide distribution, e.g., consider the shipment of a frozen formulation to the third world. Furthermore, the extra costs associated with the shipping and storage of frozen formulations can be substantial, even within the United States. These problems have generated interest in developing dehydrated formulations that are stable at ambient temperatures.

Dried preparations offer the potential for prolonged stability at room temperature, and can be ready to use after a simple rehydration step. In addition, rates of gene transfer to the lung have been shown to be increased by administration of a dried formulation *(136)*. Although there are several methods of removing water from liquid formulations, the spraying process involves high shear forces that can damage nonviral vectors *(52,137,138)*. However, several studies have demonstrated that the complexation of DNA with cationic agents greatly stabilizes polynucleotides during spraying *(139,140)*, but that cationic agents differ greatly in this regard *(137,138)*. These findings suggest that specific interactions with the polynucleotide chain may be necessary for optimal protection. The precise nature of these stabilizing interactions has yet to be investigated, but elucidating these protective mechanisms is of obvious importance to the development of gene-based therapeutics in general, and particularly relevant to pulmonary delivery.

Spray-drying of nonviral vectors is starting to receive more attention, and recent studies by Seville et al. suggest that spray-drying might offer significant advantages over traditional freeze-drying *(141)*. Our preliminary experiments have investigated spray freeze-drying and shown that, although vector size is maintained in the presence of sugars, transfection rates are greatly reduced (unpublished observations). The majority of stabilization studies have employed freeze-drying to generate dehydrated formulations (Table 3). Curiously, numerous studies have reported increases in transfection activity after freezing and/or dehydration *(84,101,107,141–143)*. This enigmatic effect typically is assumed to result from the cell-culture systems used for assaying transfection, i.e., expression levels are artificially elevated because aggregated complexes more readily sediment onto adherent cell lines. However, our previous work with suspended cells has demonstrated that aggregation cannot explain the observed increases in transfection *(84)*. In light of these findings, it is interesting that the lyophilization process has been shown to cause some denaturation of double-stranded DNA *(53)*. This observation is particularly intriguing considering that recent work has suggested that partial denaturation of the promoter region can actually stimulate transcription *(144)*. Although partial denaturation of naked DNA during dehydration would likely be reversed upon rehydration, strand separation could potentially be preserved in nonviral vectors if cationic agents rearrange to accommodate the altered structure. This notion is consistent with studies showing that lipids bind differently to DNA possessing different structures *(145)*. Alternatively, dehydration might cause structural changes in the complex (e.g., increased encapsulation) that facilitate transfection *(146)*. Although the increases in transfection that have been observed after freezing and drying are quite modest, such an effect complicates the assessment of recovery, but potentially could be used to bolster transgene expression in vivo.

The lyophilization process involves two stresses that are known to damage macromolecules, i.e., freezing and drying. Several studies have shown that slow cooling causes much more damage to vectors than rapid freezing *(103,105,109)*. As discussed earlier, this effect is owing to the greater amount of time that vectors spend in the concentrated unfrozen solution prior to solidification when cooling rates are slow *(68,84,105)*. Although the time available for coalescence in the unfrozen fraction can be altered significantly by the degree of supercooling, quick-freezing (e.g., plunging samples into liquid nitrogen) causes rapid heat removal and solidification that limits

Table 3
Lyophilization of Gene Delivery Systems

Delivery vehicle	References	Formulation	Freezing protocol	Storage
DOTAP/DOPE	*(143)*	0.5 *M* trehalose, sucrose, glucose	Liquid nitrogen	—
DMRIE/Cholesterol				
Transferrin/PEI	*(142)*			
Transferrin/polylysine/adenovirus adenovirus		10% sucrose in HBS	Precooled shelf to –40°C	—
PDMAEMA	*(127,148)*	10% sucrose, trehalose, maltose	Precooled shelf to –40°C	—
PDMAEMA	*(102)*	10% sucrose + 20 m*M* HEPES	Precooled shelf to –40°C; stored at 20°C	10 mo; size, supercoil content, transfection retained
		10% lactose, maltose,	–20°C (30 min) then –80°C (30 min);	
DOTAP/cholesterol/protamine retained	*(101)*	sucrose, trehalose	–80°C (30 min); liquid nitrogen; stored at 25°C	8 wk; size + transfection
DOTAP/protamine/DNA retained	*(141)*	3% lactose	Liquid nitrogen or spray-dried, stored under vacuum, 25°C	12 wk; transfection
		1%, 5% glucose, lactose, trehalose		
DMRIE/cholesterol	*(107)*	sucrose, mannitol, PEG, hydroxyethyl starch	–38°C at –1°C/min	—
Alkylated cys-Typ-Lys_{18}	*(151)*	5% sucrose, PEGylation	Dry ice	—
Polylysine/DNA	*(142)*	10% sucrose + HBS	Precooled shelf to –40°C, stored at –20°C	3 wk; transfection retained
		0.4% sucrose		
Adeno-associated virus	*(165)*	0.4% mannitol 1 mg/mL protamine	–40°C at –1°C/min, stored at 25°C	3 mo; titer maintained

the exposure of vectors to the highly concentrated unfrozen fraction. Although rapid cooling rates are achievable in commercial lyophilization, problems associated with liquid nitrogen sterilization and handling makes this approach less desirable *(147)*. Many studies have hastened freezing by loading filled vials on precooled shelves *(102,127,142,148)*, but cooling rates under these conditions are still slow when compared to that achieved with liquid nitrogen or by immersion in an ethanol/dry ice bath *(103,105,109,141)*. Accordingly, we must apply the knowledge gleaned from the cryoprotection experiments discussed previously to preserve vectors during the freezing step of the lyophilization process. However, it is critical that freezing studies mimic the cooling rates and incubation times encountered in a freeze-drying cycle if they are to be applicable to stability during lyophilization.

Although nature has perfected methods of protecting extremely complicated systems (i.e., whole organisms) during freezing and drying *(149)*, the prospect of stabilizing multicomponent assemblies (e.g., nonviral vectors) is a daunting challenge to a pharmaceutical scientist. Few things are more humbling than spending years trying to stabilize lipid-DNA complexes under carefully controlled conditions, while realizing that a lowly bacterium routinely preserves not only its membranes and DNA, but its entire host of metabolic machinery, during chaotic bouts with freezing and drying. A review of the literature reveals that lyophilization studies to date have predominantly focused on relatively simple systems, e.g., purified proteins, synthetic lipid vesicles. The fact that multicomponent systems are extremely difficult to stabilize should be recognized by scientists who advocate the use of targeting ligands, endosmolytic peptides, nuclear-localization peptides, and condensing polymers in the development of nonviral vectors *(150)*. However, it is encouraging that lyophilization studies to date have reported acute stabilization of polymer-based, lipid-based, and viral vectors (Table 3).

Similar to the stabilization achieved in freeze-thaw studies, protection of gene-delivery systems during lyophilization has been reported when sugars are included in the formulation *(101,102,107,127,141–143,148)*. It is clear that the choice of excipient is critical for protection during both the freezing and drying steps *(107,108)*. However, the dilution of vectors in concentrated excipient solutions is problematic because gene delivery is most efficacious at high vector concentration *(93,103)*. Although rehydration with reduced quantities of water can effectively re-concentrate the formulation, it is usually required that reconstituted preparations do not greatly exceed isotonicity (e.g., 5% glucose, 10% sucrose). High excipient concentrations can also make it necessary to extend the lyophilization cycle, which can significantly increase production costs *(84)*.

The mechanism by which excipients stabilize nonviral vectors during lyophilization has yet to be elucidated. In addition to stresses encountered during freezing (discussed previously), dehydration is known to perturb vector function *(107)*. In the case of aqueous formulations of nonviral vectors, the removal of water eliminates the suspending medium and facilitates contact between particles in the dried state. The incorporation of excipients prior to lyophilization would allow particles to be separated in a matrix that could potentially be maintained in the dried state *(108)*. This situation would minimize interactions between particles and inhibit aggregation. Apparently this separation can be maintained even if samples experience collapse, because we have repeatedly observed that collapse of the dried cake does not alter size or transfection rates upon rehydration (unpublished observations). This is consistent with a recent report by Kwok et al. showing that nonviral vectors can be stabilized without sugars during drying by incorporating high levels of PEGylated components into the particle *(151)*. However, similar experiments with PEGylated lipid-DNA complexes have shown clearly that formulation with stabilizing excipients is still necessary at lower PEGylation levels *(99)*.

Although the recovery of particle size often correlates with transfection rates, interactions between the cationic agent and DNA within a complex can be disrupted during lyophilization without altering particle dimensions *(125)*. Unfortunately, there is not more specific information about what structural characteristics are necessary for successful gene delivery, but it is clear that entrapment in a glassy matrix alone is unable to preserve macromolecular structure *(125)*. Studies on dried liposomes and proteins have demonstrated that sugars are able to preserve structural integrity by forming hydrogen bonds with the macromolecule, thereby mimicking the "hydrated" condition (152–154). This concept has been termed the "water replacement hypothesis" by Crowe and colleagues *(153,154)*, and has been implicated in the preservation of nonviral vectors in the dried state *(101,107,143)*. Recent studies have clearly documented that lyophilization in the presence of sugars does not completely preserve transfection activity during dehydration despite the maintenance of particle size *(99)*. The difference in the ability of sugars to protect complexes as compared with liposomes suggests that lipid-DNA complexes possess fundamentally different structures than traditional bilayers *(156)*. Although obvious differences include the DNA and the cationic headgroup (as compared with anionic and neutral headgroups in traditional liposomes), lipids within complexes are known to adopt a nonbilayer, inverted hexagonal phase that causes exposure of hydrophobic moieties to the aqueous solution *(106,131,155,156)*. Although the formation of such structures greatly facilitates membrane fusion and gene delivery, it is unclear whether water replacement via hydrogen bonding would be capable of stabilizing these hydrophobic assemblies. However, if vectors are designed such that the hexagonal phase is not induced until the vector fuses with a cellular membrane (as proposed in several studies *(157–159)*, particles would be inherently more stable, and the water-replacement mechanism should operate according to that proposed for lipid bilayers *(153,154)*. Clearly, more work is needed to understand the interactions between excipients and nonviral vectors so that lyophilized formulations can be rationally designed rather than empirically derived.

Considering the freezing and drying stresses discussed earlier, it is not surprising that preservation during lyophilization requires the use of stabilizing sugars (Table 3). Once in the dried state, it is typically assumed that vectors will be stable for indefinite periods of time. This naïve notion has caused leaders in the gene-therapy field to respond to questions about the shelf-life of their pet vector by stating: "It is lyophilized." Although stability for 10 mo at room temperature in the dried state has been reported, two published studies have also investigated accelerated stability at slightly higher storage temperatures *(101,102)*. In both of these studies, vectors stored at 37°–40°C exhibited decreased transfection rates as well as reduced supercoil content and increased particle size over the time course of the experiments. These results illustrate that degradation continues in the dried state, and that damage progressively accumulates even during storage at lower temperatures, albeit at slower rates. Our results from storage-stability experiments are consistent with this suggestion, and we observe a progressive degradation of lipid-DNA complexes in terms of transfection rates, supercoil content, particle size, and dye exclusion over 2 yr, even when samples are stored at –20°C (Molina et al., manuscript in preparation). Under conditions of prolonged storage, it is paramount to utilize nonreducing sugars (e.g., trehalose, sucrose), because reducing sugars (e.g., glucose, lactose, maltose) are known to form covalent adducts with nucleic acids *(160)*. In

Fig. 5. Lyophilized formulations of lipid-DNA complexes stored at 60°C for 10 mo. Note the obvious discoloration in the glucose formulation (left) and collapse in the sucrose formulation (center). In contrast, only minor shrinkage is evident in the lyophilized trehalose cake (right).

our experience, prolonged storage in a reducing sugar produces DNA products that do not migrate into a 0.8% agarose gel, and results in a brown cake indicative of nonenzymatic glycosylation via the Maillard reaction (Fig. 5). Although this deterioration is visible in lyophilized samples stored at 40°C and 60°C, the same chemistry occurs at lower temperatures even though cakes do not become visually discolored (at least not after 10 mo). However, even when nonreducing sugars are employed, progressive changes in transfection, supercoil content, particle size, and dye exclusion can be observed. Considering that vectors are immobilized in the glassy excipient phase, these findings suggest that degradation can occur at significant rates despite maintenance of the glassy phase. Although molecular relaxation rates are greatly reduced in the glassy state, the irregular arrangement of molecules in the amorphous phase causes an increase in free volume that can facilitate diffusion *(161)*. It follows that ROS generated during storage potentially could diffuse into and react with vectors in an amorphous matrix. Given that nucleic acids are known to be acutely sensitive to oxygen radicals *(33)*, formulations should be designed to minimize the potential for generating ROS during prolonged storage. As suggested by Middaugh et al. for the formulation of naked DNA in aqueous solutions, the use of chelators and antioxidants may prove to greatly enhance the shelf-life of nonviral vectors *(26)*. It should also be recognized that other vector components (e.g., lipids) are sensitive to oxidation, and that radicals generated from these components can subsequently react with the DNA *(162–164)*. Given the ability of oxidized bases to mispair *(33)*, deleterious reactions during storage can result in the in vivo production of mutated proteins with the potential for disastrous consequences; a situation that is unique to DNA-based therapeutics.

4. Conclusion

Aqueous formulations of nonviral vectors at the high concentrations necessary for clinical trials are very unstable, and it is doubtful that sufficient stability can be attained in liquid suspensions. Studies have shown that frozen formulations can offer prolonged stability in some cases, but freezing conditions and excipient mixtures that preserve one vector type do not necessarily stabilize other vectors *(109,125)*. In addition, there are concerns associated with the potential for thawing and the costs of maintaining the frozen state during shipping and storage. To date, lyophilization studies have reported that some nonviral vectors can be stabilized during acute freeze-drying stress, but prolonged storage studies are lacking. Unfortunately, most studies in this area have employed an empirical approach, and provide little insight into the mechanism(s) by which excipients stabilize DNA-based therapeutics. Future studies investigating the stability of nonviral vectors would be greatly enhanced if the interactions between cationic agents and DNA were better characterized such that stabilization strategies could focus on preserving specific nanoscale structures that are necessary for transfection. Considering the recent efforts toward understanding the subtle interactions within complexes *(130–135)*, we hope that stability studies will soon be able to focus on the preservation of specific biophysical characteristics.

Acknowledgments

The results described here were made possible owing to support from NIGMS (1 RO1 GM60587-01 to TJA), and Valentis Inc.

References

1. Cavazzana-Calvo M, Hacein-Bey S, Basile CD, et al. Gene therapy of human severe combined immunodeficiency (SCID)-X1 disease. Science 2000;288:669–672.
2. Blaese RM. Gene therapy for cancer. Sci Am 1997;276:111–115.
3. Anderson WF. Gene therapy. Sci Am 1995;273:124–128.
4. Boyer JD, Chattergoon MA, Ugen KE, et al. Enhancement of cellular immune response in HIV-1 seropositive individuals: a DNA-based trial. Clin Immunol 1999;90:100–107.
5. Tuteja R. DNA vaccines: a ray of hope. Crit Rev Biochem Mol Biol 1999;34:1–24.
6. Restifo NP, Ying H, Hwang L, Leitner WW. The promise of nucleic acid vaccines. Gene Ther 2000;7:89–92.
7. Felgner PL, Gadek TR, Holm MR, et al. Lipofection: a highly efficient, lipid-mediated DNA transfection procedure. Proc Natl Acad Sci USA 1987;87:7413–7417.
8. Dewey RA, Morrissey G, Cowsill CM, et al. Chronic brain inflammation and persistent herpes simplex virus 1 thymidine kinase expression in survivors of syngeneic glioma treated by adenovirus mediated gene therapy: implications for clinical trials. Nature Med 1999;5:1256–1263.
9. Fox JL. Gene-therapy death prompts broad civil lawsuit. Nat Biotech 2000;18:1136.
10. Nasto, B. Questions about systemic adenovirus delivery. Mol Ther 2002;5:652–653.
11. Biospace. Avigen reports status on liver infusion of gene therapy treatment for hemophilia B; Avigen's Coagulin-B clinical trial on hold. Biospace October 8 (2001) http://www.biospace.com/news.
12. Lemoine NR. Risks and benefits of gene therapy for immunodeficiency: a reality check. Gene Ther 2002;9:1561–1562.
13. Weiss R. Second boy receiving gene therapy develops cancer. Washington Post 1/15/03 p. A9.
14. Sen L, Hong Y-S, Luo H, et al. Efficiency, efficacy, and adverse effects of adenovirus- vs. liposome-mediated gene therapy in cardiac allografts. Am J Physiol Heart Circ Physiol 2001;281:H1433–H1441.
15. Nabel GJ, Nabel EG, Yang Z-Y, et al. Direct gene Transfer with DNA-liposome complexes in melanoma: expression, biologic activity, and lack of toxicity in humans. Proc Natl Acad Sci USA 1993;90:11307–11311.
16. Goddard CA, Ratcliff R, Anderson JR, et al. A second dose of a CFTR cDNA-liposome complex is as effective as the first dose in restoring cAMP-dependent chloride secretion to null CF mice trachea. Gene Ther 1997;4:1231–1236.
17. Meyer O, Schughart K, Pavirani A, Kolbe HVJ. Multiple systemic expression of human interferon-β in mice can be achieved upon repeated administration of optimized pcTG90-lipoplex. Gene Ther 2000;7:1606–1611.

18. Hyde SC, Southern KW, Gileadi U, et al. Repeat Administration of DNA/liposomes to the nasal epithelium of patients with cystic fibrosis. Gene Ther 2000;7:1156–1165.
19. Handumrongkul C, Zhong W, Debs R. Distinct sets of cellular gene control the expression of tranfected, nuclear-localized genes. Mol Ther 20002;5:186–194.
20. Chen Z-Y, Yant SR, He C-Y, et al. Linear DNAs concatemerize in vivo and result in sustained transgene expression in mouse liver. Mol Ther 2001;3:403–410.
21. Yew NS, Przybylska M, Ziegler RJ, et al. High and sustained transgene expression in vivo from plasmid vectors containing a hybrid ubiquitin promoter. Mol Ther 2001;4:75–82.
22. Dow SW, Fradkin LG, Liggitt, DH, et al. Lipid/DNA complexes induce a potent activation of innate immune responses and antitumor activity when administered intravenously. J Immunol 1999;163:1552–1561.
23. Siders WM, Vergillis K, Johnson C, et al. Tumor treatment with complexes of cationic lipid and noncoding plasmid DNA results in the induction of cytotoxic T cells and systemic tumor elimination. Mol Ther 2002;6:519–527.
24. Elbashir SM, Harborth J, Lendeckel W, et al. Duplexes of 21-nucleotide RNAs mediate RNA interference in cultured mammalian cells. Nature 2001;411:494–498.
25. Xia H, Mao Q, Paulson HL, Davidson BL. siRNA-mediated gene silencing in vitro and in vivo. Nature Biotech 2002;20:1006–1010.
26. Middaugh CR, Evans RK, Montgomery DL, Casimiro DR. Analysis of Plasmid DNA from a Pharmaceutical Perspective. J Pharm Sci 1998;87:130–146.
27. Evans RK, Xu Z, Bohannon KE, et al. Evaluation of degradation pathways for plasmid DNA in pharmaceutical formulations via accelerated stability studies. J Pharm Sci 2000;89:76–87.
28. Pogocki D, Schoneich C. Chemical stability of nucleic acid-derived drugs. J Pharm Sci 2000;89:443–456.
29. Hovorka SW, Schoneich C. Oxidative degradation of pharmaceuticals: theory, mechanisms and inhibition. J Pharm Sci 2001;90:253–269.
30. Durland RH, Eastman EM. Manufacturing and quality control of plasmid-based gene expression systems. Adv Drug Del Rev 1998;30:33–48.
31. Wolff JA, Malone RW, Williams P, et al. Direct gene transfer into mouse muscle in vivo. Science 1990;247:1465–1468.
32. Moelling K. Naked DNA: the poor man's gene therapy? Gene Ther 1998;5:573–574.
33. Breen AP, Murphy JA. Reactions of oxyl radicals with DNA. Free Rad Biol Med 1995;18:1033–1077.
34. Moriya M. Single-stranded shuttle phagemid for mutagenesis studies in mammalian cells: 8-Oxoguanine in DNA induces Targeted G·C → T·A transversions in simian kidney cells. Proc Natl Acad Sci USA 1993;90:1122–1126.
35. Cheng KC, Cahill DS, Kasai H, et al. 8-Hydroxyguanine, an Abundant Form of Oxidative DNA Damage, Causes G→T and A→C substitutions. J Biol Chem 1992;267:166–172.
36. Lindahl T, Nyberg B. Rate of depurination of native deoxyribonucleic acid. Biochemistry 1972;11:3610–3618.
37. Lindahl T, Karlstrom O. Heat-induced depyrimidination of deoxyribonucleic acid in neutral solution. Biochemistry 1973;12:5151–5154.
38. Shapiro R, Klein RS. The Deamination of cytidine and cytosine by acidic buffer solutions. Mutagenic implications. Biochemistry 1966;5:2358–2362.
39. Zoltewicz JA, Clark DF, Sharpless TW, Grahe G. Kinetics and mechanism of the acid-catalyzed hydrolysis of some purine nucleosides. J Am Chem Soc 1970;92:1741–1750.
40. Shapiro R, Danzig M. Acidic hydrolysis of deoxycytidine and deoxyuridine derivatives. The general mechanism of deoxyribonucleoside hydrolysis. Biochemistry 1972;11:23–29.
41. Vernooij EAAM, Kettenes-van den Bosch JJ, Underberg WJM, Crommelin DJA. Chemical hydrolysis of DOTAP and DOPE in a liposomal environment. J Cont Release 2002;79:299–303.
42. Marrot L, Giacomoni PU. Enhancement of oxidative DNA degradation by histidine: the role of stereochemical parameters. Mutation Res 1992;275:69–79.

43. Graf E, Mahoney JR, Bryant RG, Eaton JW. Iron-catalyzed hydroxyl radical formation. J Biol Chem 1984;259:3620–3624.
44. Shikama K. Effect of freezing and thawing on the stability of double helix of DNA. Nature 1965;207:529–530.
45. Lyscov VN, Moshkovsky YS. DNA cryolysis. Biochim Biophys Acta 1969;190:101–110.
46. Davis DL, O'Brien EP, Bentzley CM. Analysis of the degradation of oligonucleotide strands during the freezing/thawing processes using MALDI-MS. Anal Chem 2000;72:5092–5096.
47. Ando S, Putnam D, Pack DW, Langer R. PLGA microshperes containing plasmid DNA: preservation of Ssupercoiled DNA via cryopreparation and carbohydrate stabilization. J Pharm Sci 1999;88:126–130.
48. Levine H, Slade L. Principles of "cryostabilization" technology from structure/property relationships of carbohydrate/water systems: a review. Cryo-Lett 1988;9:21–63.
49. Davison BF. The effect of hydrodynamic shear on the deoxyribonucleic acid from T_2 and T_4 bacteriophages. Proc Natl Acad Sci USA 1959;45:1560–1568.
50. Levinthal C Davison PF. Degradation of deoxyribonucleic acid under hydrodynamic shearing forces. J Mol Biol 1961;3:674–683.
51. Bowman RD, Davidson N. Hydrodynamic shear breakage of DNA. Biopolymers 1972;11:2601–2624.
52. Lengsfeld CS, Anchordoquy TJ. Shear-induced degradation of plasmid DNA. J Pharm Sci 2002;91:1581–1589.
53. Poxon SW, Hughes JA. The effect of lyophilization on plasmid DNA activity. Pharm Dev Tech 2000;5:115–122.
54. Jameel F, Amsberry KL, Pikal MJ. Freeze drying properties of some oligonucleotides. Pharm Dev Tech 2001;6:151–157.
55. Tamm C, Shapiro HS, Chargaff E. Correlation between the action of pancreatic desoxyribonuclease and the nature of its substrates. J Biol Chem 1952;199:313–327.
56. Richards OC, Boyer D. Chemical mechanism of sonic, acid, alkaline and enzymatic degradation of DNA. J Mol Biol 1965;11:327–340.
57. Tamm C, Shapiro HS, Liphitz R, Chargaff E. Distribution density of nucleotides within a desoxyribonucleic acid chain. J Biol Chem 1953;203:673–688.
58. Kolobov AV, Vainberg YP. Stability of the physicochemical parameters of DNA during prolonged keeping. Bull Exp Biol Med 1976;81:360–363.
59. Gedik CM, Wood SG, Collins AR. Measuring oxidative damage to DNA; HPLC and the comet assay compared. Free Rad Res 1998;29:609–615.
60. Lindahl T. Instability and decay of the primary structure of DNA. Nature 1993;362:709–715.
61. Wolf SG, Frenkiel D, Arad T, et al. DNA protection by stress-induced biocrystalliation. Nature 1999;400:83–85.
62. Ljungman M, Hanawalt PC. Efficient protection against oxidative DNA damage in chromatin. Mol Carcinogen 1992;5:264–269.
63. Ward JF, Kuo I. Radiation damage to DNA in aqueous solution: a comparison of the response of the single-stranded form with that of double-stranded rorm. Rad Res 1978;75:278–285.
64. Fuciarelli AF, Wegher BJ, Blakely WF, Dizdaroglu M. Yields of radiation-induced base products in DNA: effects of DNA conformation and gassing conditions. Int J Radiat Biol 1990;58:397–415.
65. Ribeiro DT, Madzak C, Sarasin A, et al. Photochem Photobiol 1992;55:39–45.
66. Mahato RI, Rolland A, Tomlinson E. Cationic lipid-based gene delivery systems: pharmaceutical perspectives. Pharm Res 1997;14:853–859.
67. Mahato RI. Non-viral peptide-based approaches to gene therapy. J Drug Target 1999;7:249–268.
68. Anchordoquy TJ, Koe GS. Physical stability of nonviral plasmid-based therapeutics. J Pharm Sci 2000;89:289–296.
69. Anwer K, Earle KA, Shi M, et al. Synergistic effect of formulated plasmid and needle-free injection for genetic vaccines. Pharm Res 1999;16:889–895.
70. Lemieux P, Guerin N, Paradis G, et al. A combination of poloxamers increases gene expression of plasmid DNA in skeletal muscle. Gene Ther 2000;7:986–991.

71. Miedan VM, Cohen JS, Amariglio N, et al. Interaction of oligonucleotides with cationic lipids: the relationship between electrostatics, hydration and state of aggregation. Biochim Biophys Acta 2000;1464:251–261.
72. Zabner J, Fasbender AJ, Moninger T, et al. Cellular and molecular barriers to gene transfer by a cationic lipid. J Biol Chem 1995;270:18997–19007.
73. Gustafsson J, Arvidson G, Karlsson G, Almgren M. Complexes between cationic liposomes and DNA visualized by Cryo-TEM. Biochim Biophys Acta 1995;1235:305–312.
74. Smith JG, Wedeking T, Vernachio JH, et al. Characterization and in vivo testing of a hetergenous cationic lipid-DNA formulation. Pharm Res 1998;15:1356–1363.
75. Kwoh DY, Coffin CC, Lollo CP, et al. Stabilization of poly-L-lysine/DNA polyplexes for in vivo gene delivery to the liver. Biochim Biophys Acta 1999;1444:171–190.
76. Wasan EK, Harvie P, Edwards K, et al. A multi-step lipid mixing assay to model structural changes in cationic lipoplexes used for in vitro transfection. Biochim Biophys Acta 1999;1461:27–46.
77. Kichler A, Zauner W, Orgis M, Wagner E. Influence of the DNA complexation medium on the transfection efficiency of lipospermine/DNA particles. Gene Ther 1998;5:855–860.
78. Yang J-P, Huang L. Time-dependent maturation of cationic liposome-DNA complex for serum resistance. Gene Ther 1998;5:380–387.
79. Kennedy MT, Pozharski EV, Rakhmanova VA, MacDonald RC. Factors governing the assembly of cationic phospholipid-DNA complexes. Biophys J 2000;78:1620–1633.
80. Sakurai F, Inoue R, Nishino Y, et al. Effect of DNA/liposome mixing ratio on the physiochemical characteristics, cellular uptake and intracellular trafficking of plasmid DNA/cationic liposome complexes and subsequent gene expression. J Cont Rel 2000;66:255–269.
81. Sparrow JT, Edwards VV, Tung C, et al. Synthetic peptide-based DNA complexes for nonviral gene delivery. Adv Drug Delivery Rev 1998;30:115–131.
82. Tang MX, Szoka FC. The influence of polymer structure on the interactions of cationic polymers with DNA and morphology of the resulting complexes. Gene Ther 1997;4:823–832.
83. Kim J-S, Kim B-I, Maruyama A, et al. A new non-viral DNA delivery vector: the terplex system. J Cont Release 1998;53:175–182.
84. Allison SD, Anchordoquy TJ. Lyophilization of nonviral gene delivery systems. In Findeis M., ed. Nonviral Vectors for Gene Therapy. Methods in Molecular Medicine, Humana Press, Totowa, NJ, (2001), pp. 225–252.
85. Anchordoquy TJ. Physical stabilization of DNA-based therapeutics during freezing and drying. Am Pharm Rev 2001;4:34–41.
86. Finsinger D, Remy JS, Erbacher P, et al. Protective copolymers for nonviral gene vectors: synthesis, vector characterization and application in gene delivery. Gene Ther 2000;7:1183–1192.
87. Erbacher P, Zou S, Bettinger T, et al. Chitosan-based vector/DNA complexes for gene delivery: biophysical characteristics and transfection ability. Pharm Res 1998;15:1332–1339.
88. Yi SW, Yune TY, Kim TW, et al. A cationic lipid emulsion/DNA complex as a physically stable and serum-resistant gene delivery system. Pharm Res 2000;17:314–320.
89. Nabel GJ, Chang A, Nabel EG, et al. Immunotherapy of malignancy by in vivo gene transfer into tumors. Human Gene Ther 1992;3:399–410.
90. Caplen NJ, Kinrade E, Sorgi F, et al. In vitro liposome-mediated DNA transfection of epithelial cell lines using the cationic liposome DC-Chol/DOPE. Gene Ther 1995;2:603–613.
91. Caplen NJ, Alton EWFW, Middleton PG, et al. Liposome-mediated CFTR gene transfer to the nasal epithelium of patients with cystic fibrosis. Nature Med 1995;1:39–46.
92. Hofland HE, Shepard L, Sullivan SM. Formation of stable cationic lipid/DNA complexes for gene transfer. Proc Natl Acad Sci USA 1996;93:7305–7309.
93. Gao X, Huang L. Potentiation of cationic liposome-mediated gene delivery by polycations. Biochemistry 1996;35:1027–1036.
94. Hofland HE, Nagy D, Liu JJ, et al. In vivo gene transfer by intravenous administration of stable cationic lipid/DNA complex. Pharm Res 1997;14:742–749.
95. Hong K, Zheng W, Baker A, Papahadjopoulos D. Stabilization of cationic liposome-plasmid DNA complexes by polyamines and poly(ethylene glycol)-phospholipid conjugates for efficient in vivo gene delivery. FEBS Lett 1997;400:233–237.

96. Eastman SJ, Lukason MJ, Tousignant JD, et al. A concentrated and stable aerosol formulation of cationic lipid:DNA complexes giving high-level gene expression in mouse lung. Human Gene Ther 1997;8:765–773.
97. Ogris M, Brunner S, Schuller S, et al. PEGylated DNA/transferrin-PEI complexes: reduced interaction with blood components, extended circulation in blood and potential for systemic gene delivery. Gene Ther 1999;6:595–605.
98. Wheeler JJ, Palmer L, Ossanlou M, et al. Stabilized plasmid-lipid particles: construction and characterization. Gene Ther 1999;6:271–281.
99. Armstrong TC, Girouard LG, Anchordoquy TJ. PEGylation does not preserve cationic lipid/DNA complexes during freeze-thawing and lyophilization. J Pharm Sci 2002;91:2549–2558.
100. Harvie P, Wong FMP, Bally MB. Use of poly(ethylene glycol)-lipid conjugates to regulate the surface attributes and transfection activity of lipid-DNA particles. J Pharm Sci 2000;89:652–663.
101. Li B, Li S, Tan Y, et al. Lyophilization of cationic lipid-protamine-DNA (LPD) complexes. J Pharm Sci 2000;89:355–364.
102. Cherng J-Y, Talsma H, Crommelin DJA, Hennink WE. Long term stability of poly((2-dimethylamino)ethyl methacrylate)-based gene delivery systems. Pharm Res 1999;16:1417–1423.
103. Zelphati O, Nguyen C, Ferrari M, et al. Stable and monodisperse lipoplex formulations for gene delivery. Gene Ther 1998;5:1272–1282.
104. Lai E, van Zanten JH. Evidence of lipoplex dissociation in liquid formulations. J Pharm Sci 2002;91:1225–1232.
105. Anchordoquy TJ, Girouard LG, Carpenter JF, Kroll DK. Stability of lipid/DNA complexes during agitation and freeze-thawing. J Pharm Sci 1998;87:1046–1051.
106. Rosenzweig HS, Rakhmanova VA, McIntosh TJ, Macdonald RC. O-alkyl dioleoylphosphatidylcholinium compounds: the effect of varying alkyl chain length on their physical properties and in vitro DNA transfection activity. Bioconj Chem 2000;11:306–313.
107. Allison SD, Anchordoquy TJ, TJ. Mechanisms of protection of cationic lipid-DNA complexes during lyophilization. J Pharm Sci 2000;89:682–691.
108. Allison SD, Molina MdC, Anchordoquy TJ. Stabilization of lipid/DNA complexes during the freezing step of the lyophilization process: the particle isolation hypothesis. Biochim Biophys Acta 2000;1468:127–138.
109. Faneca H, Simoes S, Pedroso de Lima MC. Evaluation of lipid-based reagents to mediate intracellular gene delivery. Biochim Biophys Acta 2002;1567:23–33.
110. Anchordoguy TJ, Rudolph AS, Carpenter JF, Crowe JH. Modes of interaction of cryoprotectants with membrane phospholipids during freezing. Cryobiology 1987;24:324–331.
111. Carpenter JF, Crowe JH. The mechanism of cryoprotection of proteins by solutes. Cryobiology 1988;25:244–255.
112. Strauss G, Hauser H. Stabilization of lipid bilayer vesicles by sucrose during freezing. Proc Natl Acad Sci USA 198683:2422–2426.
113. Strauss G, Schurtenberger P, Hauser H. The interaction of saccharides with lipid bilayer vesicles: stabilization during freeze-thawing and freeze-drying. Biochim Biophys Acta 1986;858:169–180.
114. Prestrelski SJ, Arakawa T, Carpenter JF, JF. In: Cleland JL, Langer R, eds. Formulation and Delivery of Proteins and Peptides ACS Symposium Series 567, 1994, pp. 148–169.
115. Lee JC, Timasheff SN. The stabilization of proteins by sucrose. J Biol Chem 1981;256:7193–7201.
116. Arakawa T, Timasheff SN. Stabilization of protein structure by sugars. Biochemistry 1982;21:6536–6544.
117. Hincha DK. Low concentrations of trehalose protect isolated thylakoids against mechanical freeze-thaw damage. Biochim Biophys Acta 1989;987:231–234.
118. Hincha DK, Bakaltcheva I, Schmitt JM. Galactose-specific lectins protect isolated thylakoids against freeze-thaw damage. Plant Physiol 1993;103:59–65.
119. Sanatarius KA. The mechanism of cryoprotection of biomembrane systems by carbohydrates. In: Li PH, Sakai A, eds. Plant Cold Hardiness and Freezing Stress. New York, Academic Press, 1982, pp. 475–486.

120. Johnston DS, Coppard E, Parera GV, Chapman D. Langmuir film balance study of the interactions between carbohydrates and phospholipid monolayers. Biochemistry 1984;23:6912–6919.
121. Bummer P, Zografi G. The association of D-glucose with unilamellar phospholipid vesicles. Biophys Chem 1988;30:173–183.
122. Bechinger B, Macdonald PM, Seelig J. Deuterum NMR studies of the interactions of polyhydroxyl compounds and of glycolipids with lipid model membranes. Biochim Biophys Acta 1988;943:381–385.
123. PR Harrigan, Madden TD, Cullis PR. Protection of liposomes during dehydration or freezing. Chem Phys Lip 1990;52:139–149.
124. Papahadjopoulos D, Allen TM, Gabizon A, et al. Sterically Stabilized liposomes: improvements in pharmacokinetics and antitumor therapeutic efficacy. Proc Natl Acad Sci USA 1991;88(24):11460–11464.
125. Molina MdC, Allison SD, Anchordoquy TJ. Maintenance of particle size during the freezing step of the lyophilization process is insufficient for preservation of activity: insight from other structural indicators. J Pharm Sci 2001;90:1445–1455.
126. Luyet BJ. Physical changes occurring in frozen solutions during rewarming and melting. In: Wolstenholme GEW, O'Conner M, eds. The Frozen Cell. London, J. & A. Churchill, 1970, pp. 27–50.
127. Cherng JY, Wetering Pvd., Talsma H, et al. Freeze-drying of poly((2-dimethylamino) ethyl methacrylate)-based gene delivery systems. Pharm Res 1997;24:1838–1841.
128. Kuhl T, Guo Y, Alderfer JL, et al. Direct measurement of polyethylene glycol induced depletion attraction between lipid bilayers. Langmuir 1996;12:3003–3014.
129. Kawakami K, Nishihara Y, Hirano K. Effect of hydrophilic polymers on physical stability of liposome dispersions. J Phys Chem B 2001;105:2374–2385.
130. Hirsch-Lerner D, Barenholz Y. Probing DNA-cationic lipid interactions with the fluorophore trimethylammonium diphenyl-hexatriene (TMADPH). Biochim Biophys Acta 1998;1370:17–30.
131. Koltover I, Salditt T, Radler JO, Safinya CR. An inverted hexagonal phase of cationic liposome-DNA complexes related to DNA release and delivery. Science 1998;281:78–81.
132. Subramanian M, Holopainen JM, Paukku T, et al. Characterisation of three novel cationic lipids as liposomal complexes with DNA. Biochim Biophys Acta 2000;1466:289–305.
133. Choosakoonkriang S, Wiethoff CM, Anchordoquy TJ, et al. Infrared spectroscopic characterization of the interaction of cationic lipids with plasmid DNA. J Biol Chem 2001;276(11):8037–8043.
134. Wiethoff CM, Gill ML, Koe GS, et al. The structural organization of cationic lipid-DNA complexes. J Biol Chem 2002;277:44980–44987.
135. Braun CS, Jas GS, Choosakoonkriang S, et al. The structure of DNA within cationic lipid/DNA complexes. Biophys J 2003;84:1114–1123.
136. Sorgi F, Huang L. A dry powder formulation for gene therapy. Pharm Res 1995;12:S-266.
137. Densmore CL, Giddings TH, Waldrep JC, et al. Gene transfer by guanidinium-cholesterol: dioleoylphosphatidyl-ethanolamine liposome-DNA complexes in aerosol. J Gene Med 1999;1:251–264.
138. Densmore CL, Orson FM, Xu B, et al. Aerosol delivery of robust polyethyleneimine DNA complexes for gene therapy and genetic immunization. Mol Ther 2000;1:180–188.
139. Crook K, McLachlan G, Stevenson BJ, Porteous DJ. Plasmid DNA molecules complexed with cationic liposomes are protected from degradation by nucleases and shearing by aerosolisation. Gene Ther 1996;3:834–839.
140. Eastman SJ, Tousignant JD, Lukason MJ, et al. Optimization of formulations and conditions for the aerosol delivery of functional cationic lipid:DNA complexes. Human Gene Ther 1997;8:313–322.
141. Seville PC, Kellaway IW, Birchall JC. Preparation of dry powder dispersions for non-viral gene delivery by freeze-drying and spray-drying. J Gene Med 2002;4:428–437.
142. Talsma H, Cherng J-Y, Lehrmann H, et al. Stabilization of gene delivery systems by freeze-drying. Int J Pharm 1997;157:233–238.

143. Anchordoquy TJ, J Carpenter F, Kroll DJ. Maintenance of transfection rates and physical characterization of lipid/DNA complexes after freeze-drying and rehydration. Arch Biochem Biophys 1997;348:199–206.
144. Prasad TK, Gopal V, Rao NM. Structural changes in DNA mediated by cationic lipids alter in vitro transcriptional activity at low charge ratios. Biochim Biophys Acta 2003;1619:59–69.
145. Even-Chen S, Barenholz Y. DOTAP cationic liposomes prefer relaxed over Supercoiled plasmids. Biochim Biophys Acta 2000;1509:176–188.
146. Perrie Y, Gregoriadis G. Liposome-entrapped plasmid DNA: characterization studies. Biochim Biophys Acta 2000;1475:125–132.
147. McBurnie LD, Bardo B. Validation of sterile filtration of liquid nitrogen. Pharm Tech 2002;26:74–82.
148. Cherng J-Y, Wetering Pvd, Talsma H, et al. Stabilization of polymer-based gene delivery systems. Int J Pharm 1999;183:25–28.
149. Crowe JH, Clegg JS. Anhydrobiosis. Stroudsburg, PA, Dowden, Hutchinson and Ross, 1973.
150. Zuber G, Dauty E, Nothisen M, et al. Towards synthetic viruses. Adv Drug Del Rev 2001;52:245–253.
151. Kwok KY, Adami RC, Hester KC, et al. Strategies for maintaining the particle size of peptide DNA condensates following freeze-drying. Int J Pharm 2000;203:81–88.
152. Carpenter JF, Crowe JH. An infrared spectroscopic study of the interactions of carbohydrates with dried proteins. Biochemistry 1989;28:3916–3922.
153. Crowe JH, Crowe LM, Carpenter JF. Preserving dry biomaterials: the water replacement hypothesis, Part 1. BioPharm 1993;6(3):28–37.
154. Crowe JH, Crowe LM, Carpenter JF. Preserving dry biomaterials: the water replacement hypothesis, Part 2. BioPharm 19936(4):40–43.
155. Smisterova J, Wagenaar A, Stuart MCA, et al. Molecular shape of the cationic lipid controls the structure of cationic lipid/dioleylphosphatidylethanolamine-DNA complexes and the efficiency of gene delivery. J Biol Chem 2001;276:47615–47622.
156. Reimer DL, Zhang Y-P, Kong S, et al. Formation of novel hydrophobic complexes between cationic lipids and plasmid DNA. Biochemistry 1995;34:12877–12883.
157. Stamatatos L, Leventis R, Zuckerman MJ, Silvius JR. Interactions of cationic lipid vesicles with negatively charged phospholipid vesicles and biological membranes. Biochemistry 1988;27:3917–3925.
158. Bailey AL, Cullis PR. Membrane fusion with cationic liposomes: effects of target membrane lipid composition. Biochemistry 1997;36:1628–1634.
159. Hafez IM, Maurer N, Cullis PR. On the mechanism whereby cationic lipids promote intracellular delivery of polynucleic acids. Gene Ther 2001;8:1188–1196.
160. Bacala R, Model P, Cerami A. Modification of DNA by reducing sugars: a possible mechanism for nucleic acid aging and age-related dysfunction in gene expression. Proc Natl Acad Sci USA 1984;81:105–109.
161. Hancock BC, Zografi G. Characteristics and significance of the amorphous state in pharmaceutical systems. J Pharm Sci (1997)86:1–12.
162. Pietronigro DD, Jones WBG, Kalty K, Demopoulos HB. Interaction of DNA and liposomes as a model for membrane-mediated DNA damage. Nature 1977;267:78–79.
163. Li D, Wang M, Liehr JG, Randerath K. DNA adducts induced by lipids and lipid peroxidation products: possible relationships to I-compounds. Mut Res 1995;344:117–126.
164. Wheeler CJ, Sukhu L, Yang G, et al. Converting an alcohol to an amine in a cationic lipid dramatically alters the co-lipid requirement, cellular transfection activity and the ultrastructure of DNA-cytofectin complexes. Biochim Biophys Acta 1996;1280:1–11.
165. Croyle MA, Cheng X, Wilson JM. Development of formulations that enhance physical stability of viral vectors for gene therapy. Gene Ther 2001;8:1281–1290 (2001).

13

Cationic Microparticles and Emulsions As Effective Delivery Systems for Immune Stimulatory CpG DNA

Derek T. O'Hagan, Manmohan Singh, and Jeffrey B. Ulmer

1. Introduction: The Use of Microparticles and Emulsions for Vaccine Delivery

The adjuvant effect achieved as a consequence of the linking of antigens to synthetic particles has been known for many years and has been reviewed previously *(1)*. However, the particles used in early studies were nondegradable and therefore, were not appropriate for development as vaccine adjuvants for human use. In addition, antigens were often chemically conjugated to the nondegradable particles, adding an extra level of complexity and making commercial development even less likely *(2)*. Particulate delivery systems present multiple copies of antigens to the immune system and promote trapping and retention of antigens in local lymph nodes. The biodegradable and biocompatible polyesters, the polylactide-co-glycolides (PLG) are the primary candidates for the development of microparticles as adjuvants, because they have been used in humans for many years as resorbable suture material and as controlled release drug-delivery systems *(3,4)*. The adjuvant effect achieved through the encapsulation of antigens into PLG microparticles was first demonstrated by several groups in the early 1990s *(5–7)*. Microparticles were shown to be very potent for the induction of enhanced antibody responses and in contrast to alum, were also shown to be effective for the induction of cytotoxic T lymphocyte (CTL) responses in rodents *(8,9)*.

The mechanism of action for microparticles as adjuvants is predominantly through "antigen delivery" and is largely a consequence of the uptake of microparticles (<5 µm) by antigen-presenting cells (APC), which has been demonstrated on many occasions both in vitro and in vivo. In an early paper, Kanke et al. *(10)* described the uptake of microparticles (1–3 µm) by macrophages, but showed that microparticles of 12 µm were not taken up. Similarly, Tabata and Ikada showed that maximal uptake of microparticles into macrophages occurred with particles of <2 µm *(11,12)*. In addition, surface charge and hydrophobicity of the microparticles was shown to modify uptake *(12)*. It has also been reported that macrophages that carry microparticles to lymph nodes can mature into dendritic cells (DC) *(13)*. In addition, uptake of PLG microparticles into DC in vitro *(14)* and in vivo has been demonstrated *(15)*. It is assumed that the uptake of microparticles into APC underpins the ability of the particles to perform as vaccine-delivery systems/adjuvants.

From: *Cellular Drug Delivery: Principles and Practice*
Edited by: D. R. Lu and S. Øie © Humana Press Inc., Totowa, NJ

Emulsions have also been used as vaccine adjuvants for many years, but early studies were undertaken with nondegradable oils, which often tended to induce significant adverse effects, making these approaches unsuitable for commercial use *(16)*. Nevertheless, a potent oil-in-water (o/w) adjuvant, the syntex adjuvant formulation (SAF) *(17)* was developed using a biodegradable oil (squalane) in the 1980s, as a replacement for Freund's adjuvants. Freund's adjuvants are potent, but toxic water in mineral oil adjuvants, which may also contain killed mycobacteria *(16)*. However, SAF contained a bacterial cell wall-based synthetic adjuvant, threonyl muramyl dipeptide (MDP), and a nonionic surfactant, poloxamer L121, and proved too toxic for widespread use in humans *(18)*. Therefore, a squalene o/w emulsion was developed (MF59), without the presence of additional immunostimulatory adjuvants, which proved to be a potent adjuvant with an acceptable safety profile *(19,20)*. Uptake of the adjuvant into macrophages in vivo appears to be an important component of the mechanism of action of MF59 *(21)*. The safety and immunogenicity of MF59 adjuvanted influenza vaccine (FLUAD™) has been demonstrated in elderly subjects in clinical trials *(22)* and this data allowed the approval of this product for licensure in Europe in 1997. In summary, MF59 is a safe and well-tolerated vaccine adjuvant in humans and is effective for the induction of potent antibody responses. In many subsequent studies, emulsions similar to MF59 have been used as delivery systems for immunostimulatory adjuvants, including MPL and QS21.

2. Mechanistic Interpretation of the Activity of CpG DNA

2.1. Stimulatory CpG Oligonucleotides

The immunostimulatory effects of bacterial DNA have been known since the mid-1980s and, in retrospect, possibly since the time of William Coley, who used bacterial extracts to treat cancer patients more than 100 years ago. Tokunaga and colleagues first demonstrated that DNA, isolated from mycobacteria, could activate natural killer (NK) cells and elicit antitumor effects in animal models *(23)*. They further showed the importance of a CpG element within active DNA sequences *(24)*. Pisetsky et al. then reported that bacterial DNA and synthetic poly C-G oligonucleotides had strong B-cell mitogenic properties, inducing them to proliferate and produce immunoglobulins *(25)*. They also demonstrated the need for unmethylated cytosines in active oligonucleotides *(26)*. Through a series of studies to define the basic CpG motif, Krieg and colleagues identified purine-purine-C-G-pyrimidine-pyrimidine as generally the most active *(27)*. The potency and nature of immune stimulation by CpG oligonucleotides was subsequently shown to depend on the size, number, spacing, and base context of the CpG motifs, the nature of the oligonucleotide backbone, and the species from which the activated cells were derived (for review *see* ref. *[28]*). There appears to be two basic classes of active CpG oligonucleotides based on their physical nature and the type of activity induced *(29,30)*. CpG-A oligonucleotides activate NK and plasmacytoid dendritic cells (pDC) to secrete interferon-α (IFN-α) and typically have phosphodiester backbones. CpG-B oligonucleotides have phosphorothioate backbones and strong activity on B cells. A third type of CpG oligonucleotide has counteracting effects on active sequences and has been termed CpG-N *(31)*.

2.2. Mode of Action of CpG Oligonucleotides

Active CpG oligonucleotides can exert their effects, either directly or indirectly, on many cell types. CpG-B act directly on B cells to induce proliferation, secrete cytokines and immunoglobulins, and upregulate the cell-surface expression of Fc receptors, major histocompatibility complex (MHC) class II, and costimulatory molecules *(27,32)*. These effects can be potentiated by an anti-apoptotic effect of CpG on B cells mediated by downregulation of Fas *(33)*. pDC are also affected directly by CpG, resulting in cell-surface expression of activation markers, maturation, and secretion of cytokines (such as interleukin-6 [IL-6], tumor necrosis factor-α [TNF-α], IFN-α, IFN-γ, IL-12) and chemokines *(34)*. CpG have little or no effect on myeloid DC (mDC). CpG has indirect activating effects on monocytes, macrophages, NK cells and T cells in a paracrine fashion mediated by cytokines secreted by other cells. It has recently been shown that CpG acts specifically through Toll-like receptor 9 (TLR9) *(35)*. The distribution of TLR9 on various cells is consistent with direct activation of B cells and pDC, but not mDC, NK cells, or T cells.

The mode of action of CpG oligonucleotides involves a rapid (within minutes) signal-transduction cascade leading to upregulation of mitogen-actived protein kinases (MAPKs), NF-κB, and transcription factors controlling the expression of cytokines, chemokines and costimulatory molecules. TLR9 is both necessary and sufficient for these effects *(36)*. Thus, CpG oligonucleotides are recognized by the innate immune system as pathogen-associated molecular patterns (PAMPs), in a manner similar to other PAMPs such as lipopolysaccharide (LPS), dsRNA, and other microbial signatures.

2.3. Utility of CpG Oligonucleotides

CpG oligonucleotides have been shown to enhance or modulate the innate and adaptive immune systems. Innate immune protection has been demonstrated by pretreatment with CpG in animal models of *Listeria moncytogenes (37)*, *Francisella* tularensis *(38)*, *Leishmania major (39)*, and Ebola virus *(40)*. In some cases, amelioration of disease was noted with therapeutic application after onset of disease. CpG has been widely used as a vaccine adjuvant to increase adaptive immune responses. CpG as an adjuvant has a strong tendency to induce a Th-1 type of helper T-cell response, as seen by IFN-γ production by CD4 and CD8 T cells and CTL activity. Antibody responses are also increased, owing to enhanced T-cell help and/or direct activation of B cells. These effects have been reported for many antigens in animal models of infectious disease (for review, *see* ref. *41*), and in human clinical trials. CpG has also been used extensively in cancer immunotherapy. It has been effective as a stand-alone treatment to suppress tumor growth via cytokines and activated lymphocytes *(42)*, as an adjunct to monoclonal antibody (MAb) therapy via antibody-dependent cellular toxicity *(43)*, and as a cancer vaccine adjuvant to enhance immune responses against tumor antigens *(44)*. Finally, the strong Th1-type of responses induced by CpG has been used effectively to modulate the negative effects of Th2-type responses in allergy and asthma *(45)*.

3. Barriers to Cellular Uptake of CpG

Toll-like receptor 9 (TLR9) is a transmembrane receptor that, unlike certain other TLRs, appears to reside in endosomes, rather than on the plasma membrane *(35)*. Studies on the uptake of CpG by cells have shown it to similarly localize in endosomes *(46)*. There does not appear to be a specific "DNA receptor" on the cell surface, hence this uptake is probably mediated by fluid-phase endoyctosis. Activity of CpG requires cellular uptake *(27)* and endosomal acidification *(47)*. Taken together, these data suggest that the site of CpG-TLR9 interaction and initiation of the signal-transduction cascade occur in endosomes. However, means to disrupt endosomal membranes or directly transfect CpG oligonucleotides into cells enhances activity and results in an accumulation in the nucleus *(48,49)*, suggesting an additional effect of CpG downstream of TLR9. With respect to efficient use of CpG as a vaccine adjuvant, there are several possible important considerations, including delivery of CpG to the appropriate cells (e.g., B cells and pDC), achieving a sufficient critical mass of CpG molecules locally (i.e., in endosomes), and temporal and spatial association with the antigen for optimal induction of antigen-specific immune responses.

4. Delivery Systems for CpG DNA

Recently, we described a novel approach designed to improve the potency of CpG as an adjuvant, through the use of cationic microparticles, which were able to adsorb the CpG onto the surface *(50)*. This approach had previously been shown to be successful as a delivery systems for DNA vaccines, which were also adsorbed onto the cationic surface *(51,52)*. The objective of this work was to ensure that both the CpG adjuvant and the antigen of interest would be taken up into the same population of APC. To achieve this, both the CpG and the antigen were adsorbed onto charged PLG microparticles, either cationic for the CpG, or anionic for the antigen *(53)*. We had previously shown that the cationic microparticles with adsorbed DNA were able to transfect DCs in culture, whereas naked DNA was unable to transfect DC *(54)*. Moreover, the cationic microparticles were able to efficiently adsorb DNA at high loading levels and were able to offer protection against degradation in vitro and in vivo *(55)*. The ability of cationic microparticles to efficiently adsorb and deliver DNA and CpG is a consequence of the use of the cationic surfactant cetyl trimethylammonium bromide in the microparticle preparation process *(51,55,56)*. A similar approach was also described recently in which we used cationic emulsions to adsorb the CpG and ensure effective delivery into APC *(57)*. The potency of cationic microparticles and emulsions for delivery of adjuvant active CpG will be described in more detail in Subheadings 5 and 6.

Several groups have recently described the use of liposomes as delivery systems for CpG *(58–61)*. The approaches used varied from co-encapsulating the antigen and CpG in cationic liposomes *(58)*, the encapsulation of CpG alone in novel lipid vesicles *(59)*, the encapsulation of CpG in large multilamellar vesicles *(61)*, and encapsulation in haptenized liposomes *(60)*. However, the principal observation was consistent and showed that the potency of CpG as an adjuvant was enhanced by its encapsulation in liposomes. In addition to liposomes, nanoparticles prepared from PLG have also been used as a delivery system for CpG, in which the antigen was also entrapped *(62)*. This work followed on from earlier work, which had showed that the delivery of antisense oligonucleotides to macrophages in vitro could be improved by microencapsulation in PLG *(63)*. Similarly, the work on liposomal delivery of CpG was preceded by the use of liposomes as delivery systems for antisense oligonucleotides *(64)*.

In addition to the use of synthetic particulate formulations as delivery systems for CpG, several groups have also described the use of "natural" particulates as delivery systems. The potency of hepatitis B core antigen (HBcAg) was significantly improved when it was expressed with a C-terminal cationic peptide region, which was able to bind small amounts of CpG *(65)*. Similarly, Storni et al. *(66)* showed that the potency of HBcAg was improved by its administration in conjunction with CpG. In a recent study undertaken to evaluate the use of CpG in combination with conventional adju-

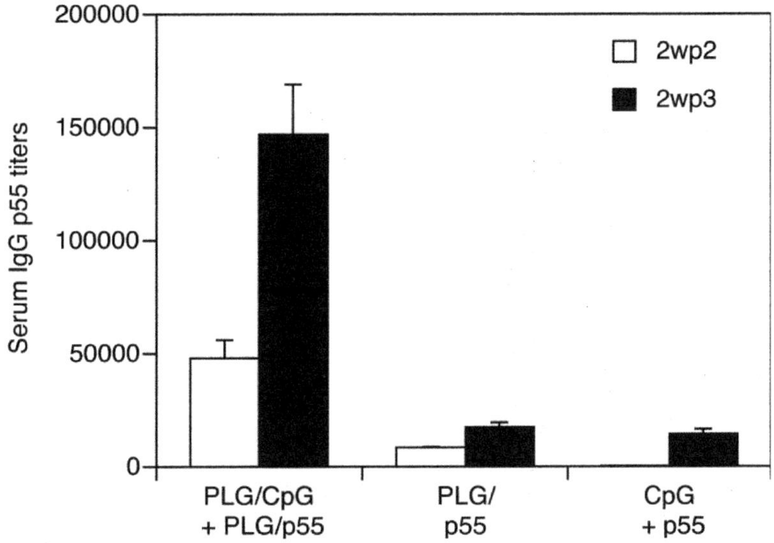

Fig. 1. The adjuvant effect of CpG for the induction of antibody responses against p55 gag. Both CpG and antigen were evaluated as soluble components, and adsorbed to PLG microparticles.

vants, it was clear that the use of CpG in conjunction with a variety of emulsion formulations was capable of inducing potent immune responses *(67)*.

5. Cationic PLG Microparticles for CpG Delivery

As described earlier, we have used novel cationic PLG microparticles as delivery systems for CpG, which have been used in combination with antigens adsorbed onto similar anionic PLG microparticles *(50)*. CpG alone did not appear to be a potent adjuvant for a recombinant HIV-1 p55 gag protein antigen, because the responses induced were only comparable to those induced by the protein adsorbed to PLG (Fig. 1). The CpG + p55 gag protein combination did induce an enhanced response in comparison to immunization with a control inactive oligonucleotide. However, when the CpG was adsorbed to the cationic PLG microparticles, there was a clear and significant adjuvant effect at 2 wk post second immunization (2wp2) and at 2 wk post third immunization (2wp3), in comparison to the alternative groups. In addition, the adsorption of CpG to PLG microparticles induced a dramatic switch in the ratio of IgG2a/IGg1, which was not achieved when CpG was simply added to the p55 gag antigen (Fig. 2). Hence, the optimal mouse CpG sequence *(68)* was only potent as an adjuvant for p55 gag, when it was used in combination with PLG microparticles.

Subsequent studies evaluating the antibody responses with CpG and a recombinant env gp120 from HIV-1 showed similar responses to those described above for p55 gag (Fig. 3). CpG alone was an ineffective adjuvant for gp120, but was very effective when adsorbed to cationic PLG microparticles. Notably, PLG/gp120 in combination with PLG/CpG induced significantly enhanced antibody responses in comparison to MF59 + gp120, a well-established potent adjuvant formulation. Hence for the induction of enhanced responses to env gp120, it appeared that both the antigen and the CpG adjuvant needed to be presented to the immune system on the surface of PLG microparticles.

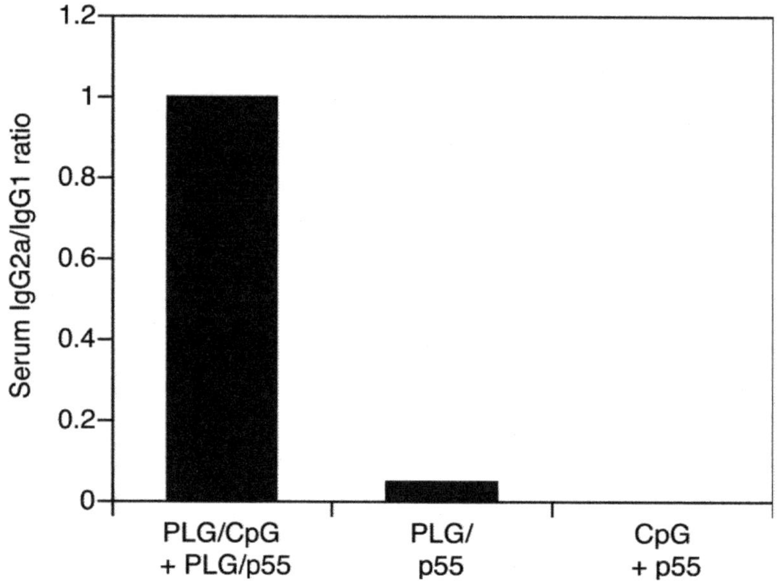

Fig. 2. Induction of higher IgG2a titers with the PLG microparticle formulation with adsorbed p55 gag antigen and CpG adjuvant. Both CpG and antigen were also evaluated as soluble components.

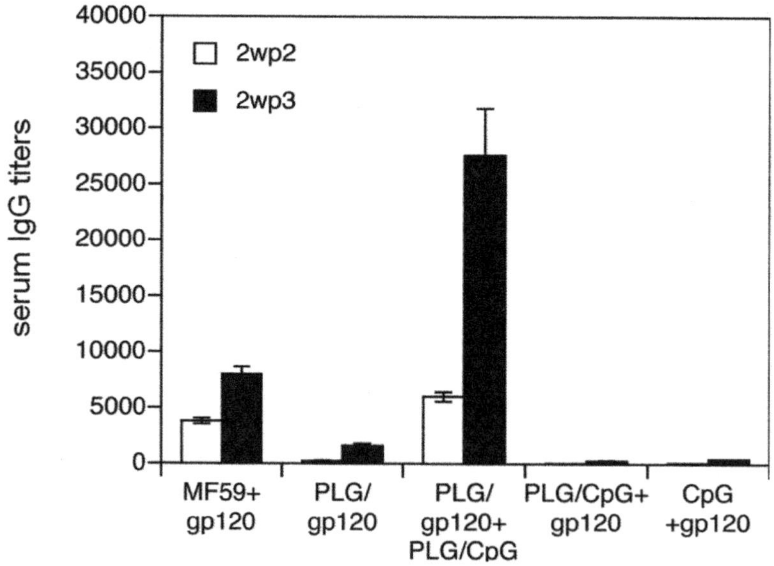

Fig. 3. The adjuvant effect of CpG for the induction of antibody responses against gp120. Both CpG and gp120 were evaluated as soluble components, and adsorbed to PLG microparticles.

The observations on antibody induction with CpG were reproduced in studies to evaluate the induction of CTL responses to an epitope in p55 gag. CpG was effective for CTL induction against p55 gag only when adsorbed onto cationic PLG microparticles (Fig. 4). In contrast, p55 + CpG was not effective for CTL induction. Remarkably, the

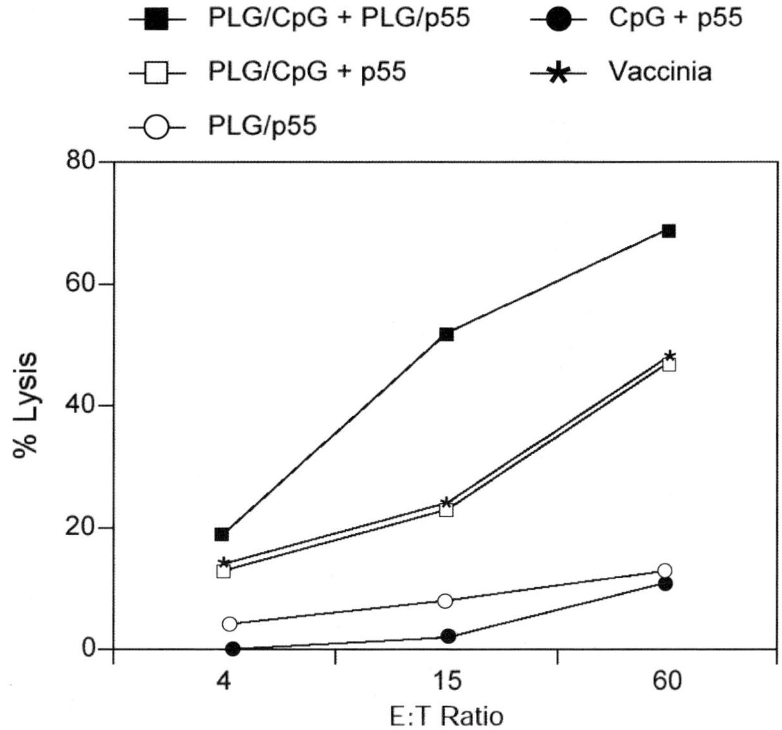

Fig. 4. The adjuvant effect of CpG for the induction of CTL responses against p55 gag. Both CpG and antigen were evaluated as soluble components, and adsorbed to PLG microparticles. Infection with vaccinia gag-pol was used as a positive control.

PLG/p55 + PLG/CpG group, which had both antigen and adjuvant adsorbed to separate PLG microparticles, was more potent than the positive control, involving intraperitoneal (ip) infection with vaccinia virus.

Hence, although it was clear that CpG DNA was capable of performing as a vaccine adjuvant for the HIV-1 proteins evaluated, it was ineffective when used as an adjuvant in the absence of microparticles. The cationic microparticles described in the current studies may offer an attractive, practical, and potent approach for the delivery of adjuvant active CpG DNA sequences. However, the range of antigens for which microparticle formulations are necessary to ensure the potency of CpG adjuvants remains to be determined.

6. Cationic Emulsion for CpG Delivery

In addition to charged microparticles, we also developed cationic submicron emulsions for CpG delivery *(57)*, which were prepared in the presence of DOTAP (1,2-Dioleoyl-3-Trimethylammonium-propane) or DDA (dimethyl dioctadecyl ammonium) as cationic agents and used squalene as the oil phase, as used previously for MF59 *(20)*. Similar to the work described earlier on microparticles, CpG was adsorbed onto the cationic emulsion droplets to ensure optimal delivery into APC and the formulations were used in conjunction with a recombinant p55 gag antigen from HIV-1 (Fig. 5). The simple combination of MF59 and CpG resulted in the induction of a significantly greater antibody response than either component alone. However, adsorbing the CpG to the cationic emulsion did not result in an increase in antibody titer over what was

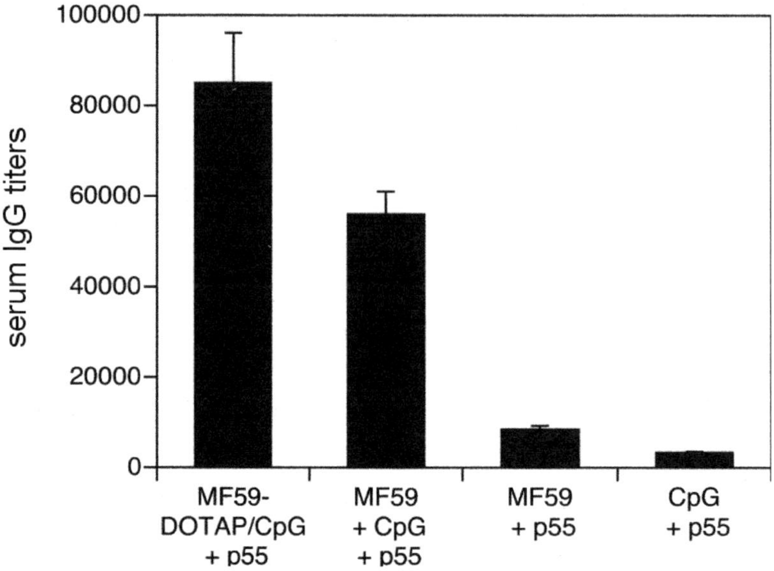

Fig. 5. The adjuvant effect of CpG for the induction of antibody responses against p55 gag. Both CpG and antigen were evaluated as soluble components and adsorbed to MF59/DOTAP emulsion.

achieved with the simple combination. Nevertheless, the cationic emulsions with adsorbed CpG were able to induce CTL responses against p55 gag, while simple combinations of MF59 and CpG were not *(69)*. Hence, the studies described here demonstrated that the potency of CpG as a vaccine adjuvant for HIV-1 antigens was enhanced by the presentation of the CpG adsorbed onto the surface of cationic emulsions.

The hypothesis that led to development of cationic emulsions was that direct delivery of CpG to the endosomal compartment of DC would enhance activity, because MF59 emulsion has previously been shown to be taken up into endosomes of phagocytic cells after intramuscular injection. *(70)*. In addition, cationic PLG particles with similar zeta potential to the emulsions, and ability to adsorb CpG and DNA have been shown to be able to deliver adsorbed DNA into DCs *(54)*. Recently, we have also shown that cationic emulsions are an effective delivery systems for DNA vaccines *(71)*.

7. Conclusions and Future Directions

In summary, the studies we have undertaken and described here have clearly demonstrated that the potency of CpG as a vaccine adjuvant for HIV-1 antigens can be significantly enhanced by its presentation on the surface of cationic microparticles or emulsions. These results are entirely consistent with similar work done elsewhere with alternative antigens, which have described the use of various particulate-delivery systems, including liposomes, emulsions, and nanoparticles, to improve the potency of CpG as an adjuvant. It is postulated that the CpG adjuvant is more potent when used in conjunction with delivery systems, largely as a consequence of its improved delivery into the endosomes of APC, where it can more easily interact with TLR9. However, other factors may also contribute to the enhanced potency of CpG when used in conjunction with delivery systems, including the ability of the various particulate carriers to protect CpG against degradation in vivo. In addition, the detergents used in the preparation of the cationic microparticles and emulsions described here may help to disrupt endosomes, allowing release of CpG into the cytoplasm and enhancing nuclear delivery. However, this represents a mere speculation, which needs to be proved experimentally in dedicated studies.

References

1. O'Hagan DT. Microparticles as oral vaccines. In: O'Hagan DT, ed. Novel Delivery Systems for Oral Vaccines. Boca Raton: CRC Press, Inc., 1994, pp. 175–205.
2. Kovacsovics-Bankowski M, Clark K, Benacerraf B, Rock KL. Efficient major histocompatibility complex class I presentation of exogenous antigen upon phagocytosis by macrophages. Proc Natl Acad Sci USA 1993;90:4942–4946.
3. Okada H, Toguchi H. Biodegradable microspheres in drug delivery. 1995;12:1–99.
4. Putney SD, Burke PA. Improving protein therapeutics with sustained-release formulations [published erratum appears in Nat Biotechnol 1998 May;16(5):478]. Nat Biotechnol 1998;16:153–157.
5. Eldridge JH, Staas JK, Meulbroek JA, Tice TR, Gilley RM. Biodegradable and biocompatible poly(DL-lactide-co-glycolide) microspheres as an adjuvant for staphylococcal enterotoxin B toxoid which enhances the level of toxin-neutralizing antibodies. Infect Immun 1991;59:2978–2986.
6. O'Hagan DT, Jeffery H, Roberts MJ, et al. Controlled release microparticles for vaccine development. Vaccine 1991;9:768–771.
7. Singh M, Singh A, Talwar GP. Controlled delivery of diphtheria toxoid using biodegradable poly(D,L-lactide) microcapsules. Pharm Res 1991;8:958–961.
8. Maloy KJ, Donachie AM, O'Hagan DT, Mowat AM. Induction of mucosal and systemic immune responses by immunization with ovalbumin entrapped in poly(lactide-co-glycolide) microparticles. Immunology 1994;81:661–667.
9. Moore A, McGuirk P, Adams S, et al. Immunization with a soluble recombinant HIV protein entrapped in biodegradable microparticles induces HIV-specific CD8+ cytotoxic T lymphocytes and CD4+ Th1 cells. Vaccine 1995;13:1741–1749.
10. Kanke M, Sniecinski I, DeLuca PP. Interaction of microspheres with blood constituents: I. Uptake of polystyrene spheres by monocytes and granulocytes and effect on immune responsiveness of lymphocytes. J Parenter Sci Technol 1983;37:210–217.
11. Tabata Y, Ikada Y. Macrophage phagocytosis of biodegradable microspheres composed of L-lactic acid/glycolic acid homo- and copolymers. J Biomed Mater Res 1988;22:837–858.
12. Tabata Y, Ikada Y. Phagocytosis of polymer microspheres by macrophages. Adv Polymer Sci 1990;94:107–141.
13. Randolph GJ, Inaba K, Robbiani DF, et al. Differentiation of phagocytic monocytes into lymph node dendritic cells in vivo. Immunity 1999;11:753–761.
14. Lutsiak ME, Robinson DR, Coester C, et al. Analysis of poly(D,L-lactic-co-glycolic acid) nanosphere uptake by human dendritic cells and macrophages in vitro. Pharm Res 2002;19:1480–1487.
15. Newman KD, Elamanchili P, Kwon GS, Samuel J. Uptake of poly(D,L-lactic-co-glycolic acid) microspheres by antigen-presenting cells in vivo. J Biomed Mater Res 2002;60:480–486.
16. Lindblad EB. Freund's Adjuvants. In: O'Hagan D, ed. Vaccine Adjuvants: Preparation Methods and Research Protocols, vol. 42. Totowa, NJ, Humana Press, 2000, pp. 49–64.
17. Allison AC, Byars NE. An adjuvant formulation that selectively elicits the formation of antibodies of protective isotypes and of cell-mediated immunity. J Immunol Methods 1986;95:157–168.
18. Edelman R. Adjuvants for the future. In: Levine MM, Woodrow GC, Kaper JB, Cobon GS, eds. New Generation Vaccines. 2nd vol. New York, Marcel Dekker, Inc., 1997, pp. 173–192.
19. Ott G, Barchfeld GL, Chernoff D, Radhakrishnan R, et al. MF59: Design and evaluation of a safe and potent adjuvant for human vaccines. In: Powell MF, Newman MJ, eds. Vaccine Design: The Subunit and Adjuvant Approach. New York, Plenum Press, 1995, pp. 277–296.

20. Ott G, Radhakrishnan R, Fang J-H, Hora M. The adjuvant MF59: a 10-year perspective. In: O'Hagan DT, ed. Vaccine Adjuvants: Preparation Methods and Research Protocols. Totowa, NJ, Humana Press Inc., 2000, pp. 211–228.
21. Dupuis M, Denis-Mize K, LaBarbara A, et al. Immunization with the adjuvant MF59 induces macrophage trafficking and apoptosis. Eur J Immunol 2001;31:2910–2918.
22. Podda A. The adjuvanted influenza vaccines with novel adjuvants: experience with the MF59-adjuvanted vaccine. Vaccine 2001;19:2673–2680.
23. Tokunaga T, Yamamoto H, Shimada S, et al. Antitumor activity of deoxyribonucleic acid fraction from Mycobacterium bovis BCG. I. Isolation, physicochemical characterization, and antitumor activity. 1984;72:955–962.
24. Kuramoto E, Yano O, Kimura Y, et al. Oligonucleotide sequences required for natural killer cell activation. Jpn J Cancer Res 1992;83:1128–1131.
25. Messina JP, Gilkeson GS, Pisetsky DS. Stimulation of in vitro murine lymphocyte proliferation by bacterial DNA. J Immunol 1991;147:1759–1764.
26. Messina JP, Gilkeson GS, Pisetsky DS. The influence of DNA structure on the in vitro stimulation of murine lymphocytes by natural and synthetic polynucleotide antigens. Cell Immunol 1993;147:148–157.
27. Krieg AM, Yi AK, Matson S, et al. CpG motifs in bacterial DNA trigger direct B-cell activation. Nature 1995;374:6546–6549.
28. Krieg AM. CpG motifs in bacterial DNA and their immune effects. Annu Rev Immunol 2002;20:709–760.
29. Verthelyi D, Ishii K, Gursel M, et al. Human peripheral blood cells differentially recognize and respond to two distinct CPG motifs. J Immunol 2001;166:2372–2377.
30. Krug A, Rothenfusser S, Hornung V, et al. Identification of CpG oligonucleotide sequences with high induction of IFN-alpha/beta in plasmacytoid dendritic cells. Eur J Immunol 2001;31:2154–2163.
31. Krieg AM, Wu T, Weeratna R, et al. Sequence motifs in adenoviral DNA block immune activation by stimulatory CpG motifs. Proc Natl Acad Sci USA 1998;95:12631–12636.
32. Decker T, Schneller F, Sparwasser T, et al. Immunostimulatory CpG-oligonucleotides cause proliferation, cytokine production, and an immunogenic phenotype in chronic lymphocytic leukemia B cells. Blood 2000;95:999–1006.
33. Wang Z, Karras JG, Colarusso TP, et al. Unmethylated CpG motifs protect murine B lymphocytes against Fas-mediated apoptosis. Cell Immunol 1997;180:162–167.
34. Hartmann G, Weiner GJ, Krieg AM. CpG DNA: a potent signal for growth, activation, and maturation of human dendritic cells. Proc Natl Acad Sci USA 1999;96:9305–9310.
35. Hemmi H, Takeuchi O, Kawai T, et al. A Toll-like receptor recognizes bacterial DNA. Nature 2000;408:740–745.
36. Bauer S, Kirschning CJ, Hacker H, et al. Human TLR9 confers responsiveness to bacterial DNA via species-specific CpG motif recognition. Proc Natl Acad Sci USA 2001;98:9237–9242.
37. Krieg AM, Love-Homan L, Yi AK, Harty JT. CpG DNA induces sustained IL-12 expression in vivo and resistance to Listeria monocytogenes challenge. J Immunol 1998;161:2428–2434.
38. Elkins KL, Rhinehart-Jones TR, Stibitz S, et al. Bacterial DNA containing CpG motifs stimulates lymphocyte-dependent protection of mice against lethal infection with intracellular bacteria. J Immunol 1999;162:2291–2298.
39. Walker PS, Scharton-Kersten T, Krieg AM, et al. Immunostimulatory oligodeoxynucleotides promote protective immunity and provide systemic therapy for leishmaniasis via IL-12- and IFN-gamma-dependent mechanisms. Proc Natl Acad Sci USA 1999;96:6970–6975.
40. Klinman DM, Verthelyi D, Takeshita F, Ishii KJ. Immune recognition of foreign DNA: a cure for bioterrorism? Immunity 1999;11:123–129.
41. Krieg AM, Davis HL. Enhancing vaccines with immune stimulatory CpG DNA. Curr Opin Mol Ther 2001;3:15–24.
42. Wickstrom E. Antisense c-myc inhibition of lymphoma growth. Antisense Nucleic Acid Drug Dev 1997;7:225–228.

43. Wooldridge JE, Ballas Z, Krieg AM, Weiner GJ. Immunostimulatory oligodeoxynucleotides containing CpG motifs enhance the efficacy of monoclonal antibody therapy of lymphoma. Blood 1997;89:2994–2998.
44. Kim SK, Ragupathi G, Musselli C, et al. Comparison of the effect of different immunological adjuvants on the antibody and T-cell response to immunization with MUC1-KLH and GD3-KLH conjugate cancer vaccines. Vaccine 1999;18:597–603.
45. Serebrisky D, Teper AA, Huang CK, et al. CpG oligodeoxynucleotides can reverse Th2-associated allergic airway responses and alter the B7.1/B7.2 expression in a murine model of asthma. J Immunol 2000;165:5906–5912.
46. Pichon C, Monsigny M, Roche AC. Intracellular localization of oligonucleotides: influence of fixative protocols. Antisense Nucleic Acid Drug Dev 1999;9:89–93.
47. Yi AK, Tuetken R, Redford T, et al. CpG motifs in bacterial DNA activate leukocytes through the pH-dependent generation of reactive oxygen species. J Immunol 1998;160:4755–4761.
48. Krieg AM, Tonkinson J, Matson S, et al. Modification of antisense phosphodiester oligodeoxynucleotides by a 5' cholesteryl moiety increases cellular association and improves efficacy. Proc Natl Acad Sci USA 1993;90:1048–1052.
49. Yamamoto T, Yamamoto S, Kataoka T, Tokunaga T. Lipofection of synthetic oligodeoxyribonucleotide having a palindromic sequence of AACGTT to murine splenocytes enhances interferon production and natural killer activity. Microbiol Immunol 1994;38:831–836.
50. Singh M, Ott G, Kazzaz J, et al. Cationic microparticles are an effective delivery system for immune stimulatory CpG DNA. Pharm Res 2001;18:1476–1479.
51. Singh M, Briones M, Ott G, O'Hagan D. Cationic microparticles: a potent delivery system for DNA vaccines. Proc Natl Acad Sci USA 2000;97:811–816.
52. O'Hagan D, Singh M, Ugozzoli M, et al. Induction of potent immune responses by cationic microparticles with adsorbed HIV DNA vaccines. J Virol 2001;75:9037–9043.
53. Kazzaz J, Neidleman J, Singh M, et al. Novel anionic microparticles are a potent adjuvant for the induction of cytotoxic T lymphocytes against recombinant p55 gag from HIV-1. J Control Release 2000;67:347–356.
54. Denis-Mize KS, Dupuis M, MacKichan ML, et al. Plasmid DNA adsorbed onto cationic microparticles mediates target gene expression and antigen presentation by dendritic cells. Gene Ther 2000;7:2105–2112.
55. Briones M, Singh M, Ugozzoli M, et al. The preparation, characterization, and evaluation of cationic microparticles for DNA vaccine delivery. Pharm Res 2001;18:709–711.
56. Singh M, Ugozzoli M, Briones M, et al. The effect of CTAB concentration in cationic PLG microparticles on DNA adsorption and in vivo performance. Pharmaceut Res 2003;20:244–248.
57. O'Hagan DT, Singh M, Kazzaz J, et al. Synergistic adjuvant activity of immunostimulatory DNA and oil/water emulsions for immunization with HIV p55 gag antigen. Vaccine 2002;20:3389–3398.
58. Gursel I, Gursel M, Ishii KJ, Klinman DM. Sterically stabilized cationic liposomes improve the uptake and immunostimulatory activity of CpG oligonucleotides. J Immunol 2001;167:3324–3328.
59. Mui B, Raney SG, Semple SC, Hope MJ. Immune stimulation by a CpG-containing oligodeoxynucleotide is enhanced when encapsulated and delivered in lipid particles. J Pharmacol Exp Ther 2001;298:1185–1192.
60. Li WM, Bally M, Schutze-Redelmeier M-P. Enhanced immune response to t-independent antigen by using CpG oligodeoxynucleotides encapsulated in liposomes. Vaccine 2002;20:148–157.
61. Joseph A, Louria-Hayon I, Plis-Finarov A, et al. Liposomal immunostimulatory DNA sequence (ISS-ODN): an efficient parenteral and mucosal adjuvant for influenza and hepatitis B vaccines. Vaccine 2002;20:3342–3354.
62. Diwan M, Tafaghodi M, Samuel J. Enhancement of immune responses by co-delivery of a CpG oligodeoxynucleotide and tetanus toxoid in biodegradable nanospheres. J Control Release 2002;85:247–262.

63. Freytag T, Dashevsky A, Tillman L, et al. Improvement of the encapsulation efficiency of oligonucleotide-containing biodegradable microspheres. J Control Release 2000;69:197–207.
64. Zelphati O, Szoka FC, Jr. Intracellular distribution and mechanism of delivery of oligonucleotides mediated by cationic lipids. Pharm Res 1996;13:1367–1372.
65. Riedl P, Buschle M, Reimann J, Schirmbeck R. Binding immune-stimulating oligonucleotides to cationic peptides from viral core antigen enhances their potency as adjuvants. Eur J Immunol 2002;32:1709–1716.
66. Storni T, Lechner F, Erdmann I, et al. Critical role for activation of antigen-presenting cells in priming of cytotoxic T cell responses after vaccination with virus-like particles. J Immunol 2002;168:2880–2886.
67. Ioannou XP, Gomis SM, Karvonen B, et al. CpG-containing oligodeoxynucleotides, in combination with conventional adjuvants, enhance the magnitude and change the bias of the immune responses to a herpesvirus glycoprotein. Vaccine 2002;21:127–137.
68. Davis HL, Weeranta R, Waldschmidt TJ, et al. CpG DNA is a potent enhancer of specific immunity in mice immunized with recombinant hepatitis B surface antigen. J Immunol 1998;160:870–876.
69. O'Hagan D, Valiante NM. Recent advances in vaccine adjuvants and delivery systems. Pharmaceut News 2002;9:397–405.
70. Dupuis M, Denis-Mize K, Woo C, et al. Distribution of DNA vaccines determines their immunogenicity after intramuscular injection in mice. J Immunol 2000;165:2850–2858.
71. Ott G, Singh M, Kazzaz J, et al. A cationic sub-micron emulsion (MF59/DOTAP) is an effective delivery system for DNA vaccines. J Control Release 2002;79:1–5.

14

Transdermal Delivery of Antisense Oligonucleotides

WeiQi Lin, Michel Cormier, and Peter E. Daddona

1. Introduction

In recent years, antisense oligodeoxynucleotides (ODNs) have emerged as promising biopharmaceutical therapeutics. These agents specifically target genes or gene transcripts involved in pathogenesis. Several clinical trials have demonstrated the therapeutic value and low toxicity of ODNs *(1–5)*. Because ODNs, like most biopharmaceuticals, are highly susceptible to degradation in the gastrointestinal environment and only poorly absorbed through biomembranes, they have been delivered primarily by parenteral infusion. This route of administration, however, is invasive and inconvenient, and repeated injections or lengthy infusions can be painful and disruptive to normal life. Therefore, transdermal delivery is being explored as a more patient-friendly alternative *(6)*.

Transdermal drug delivery has a long history dating back to the use of plasters and poultices to introduce medicinal compounds into local skin sites or directly underlying tissue. More recently, the potential of the transdermal route for systemic drug delivery has been recognized and exploited. Transdermal delivery offers many of the same advantages as drug delivery by parenteral infusion: both routes of administration can be used in nauseated patients, are minimally affected by intake of food, can be easily interrupted, and allow compounds to escape first-pass metabolism in the liver. Added advantages of transdermal delivery systems over parenteral infusion include noninvasiveness and ease of use.

Transdermal delivery of ODNs is complicated by the presence of nucleases in the skin, which are known to degrade topically delivered ODNs *(7)*. Modification with phosphorothioate (PS) can render an ODN more nuclease-resistant, protecting it from any significant degradation in the skin within a 24-h period *(8)* and increasing its metabolic half-life in the liver to greater than 24 h *(9,10)*. Structural modification of an ODN at the 2'-position of the carbohydrate moieties or the C(5) position of any pyrimide moieties can also increase its nuclease resistance *(11–14)*.

The major obstacles to transdermal delivery of ODNs, however, are their large size and their hydrophilicity. Because of its specialized structure, skin is generally only permeable to small, lipophilic drugs *(15)*. This chapter will discuss the various strategies that have been employed to increase permeability of the skin to ODNs without causing undue irritation.

From: *Cellular Drug Delivery: Principles and Practice*
Edited by: D. R. Lu and S. Øie © Humana Press Inc., Totowa, NJ

2. Structure and Function of Skin

Skin is composed of two main layers, the inner dermis and the outer epidermis. To become systemically available, a transdermally delivered compound has to traverse the avascular epidermis and reach the dermis, where abundant capillaries provide a vascular surface of approx 1–2 cm^2 per cm^2 of skin surface for exchange of substances between blood and skin. In contrast to the largely acellular dermis, the epidermis is composed entirely of cells. Of those cells, 80% are derived from keratinocytes, which owe their name to an intracellular abundance of keratin, a water-insoluble, filamentous protein. Within the epidermis, the keratinocytes are organized into defined strata. The innermost layer of the epidermis, the stratum germinativum, continuously generates new keratinocytes, leading to a renewal of the entire epidermis on average every 14 d. As the keratinocytes move outward from the stratum germinativum, they undergo a series of differentiations. In the stratum spinosum, the flattening cells become filled with cellular organelles (lamellar granules), which extrude their lipid contents into the intercellular space in the next layer, the stratum granulosum. Keratinocytes reach their terminally differentiated state as corneocytes in the stratum corneum, the next and outermost epidermal layer, which represents the main permeation barrier of the skin.

Corneocytes are devoid of all organelles and filled instead with keratin filaments enmeshed in a protein matrix. The normal plasma membrane is replaced by the cornified cell envelope, a unique structure consisting of a highly insoluble, cross-linked protein inner surface and a ω-hydroxyceramide monolayer outer surface that anchors the corneocytes in a lipid matrix of unusual composition. The corneocyte "bricks" are arranged in a regular, interdigitated pattern in this lipid "mortar," forming a "wall" about 15–20 cell layers deep. This formidable barrier is only disrupted by skin appendages such as hair follicles and sebaceous and sweat glands, which cover less than 1% of the surface area.

The primary function of the skin, the body's largest organ, is protection from the environment. The physical barrier created by the stratum corneum prevents loss of water and other vital substances from the body and entry of foreign substances or microorganisms into the body. A second line of defense is created by antigen-presenting Langerhans cells in the epidermis and dendritic cells (DCs) in the dermis, which can activate the innate and/or acquired immune system in response to any invading microorganisms or foreign substances *(16)*.

3. Enhancement of Transdermal Transport

In principle, transdermal transport can proceed along three different pathways: the transcellular, intercellular, and appendageal pathways *(17)*. The specialized structure of the stratum corneum, however, limits transcellular and intercellular transport to small and hydrophobic molecules. Hydrophilic macromolecules such as ODNs can only find entry through skin appendages disrupting the stratum corneum.

Chemical additives or mechanical abrasion can be used to disrupt an area of the stratum corneum, allowing ODNs to penetrate into the dermis through the trans- or intercellular routes; this strategy, however, inherently involves the risk of skin irritation. More promising strategies focus on enhancing ODN transport through skin pores, either by increasing the rate of transport through existing skin appendages or by amplifying the number of available skin pores through creation of transient artificial micropores. Because ODNs are highly water-soluble and charged, their flux through skin pores can be increased by application of a low-level electrical field, a process known as iontophoresis. New conduits for ODN transport can be opened by electroporation and low-frequency ultrasound, which cause a localized and temporary breakdown of the structure of the stratum corneum. Alternatively, artificial aqueous pathways through the superficial layers of the skin can be generated thermally or mechanically using miniaturized wires or precision-fabricated microprojections, respectively.

3.1. Chemical Enhancement

Use of a mixture of polyethylene glycol (PEG) and linoleic acid to reduce the barrier function of the stratum corneum resulted in a marked increase in the passive flux of an uncharged antisense molecule through mouse skin *(6)*. Chemical enhancement of ODN transport could also be achieved through addition of ethanol or 1-menthol or through delipidization of the stratum corneum by solvent treatment *(18)*. However, these approaches may be limited by the irritancy potential of the additives *(19)*.

Mehta et al. *(20)* formulated a cream containing anti-ICAM-1 phosphorothioate ODNs. At 2% concentration, this cream suppressed about 66% of ICAM-1 expression in the treated skin in a mouse model and lead to a noticeable reduction of ICAM-1 expression in human skin grafted onto severe combined immunodeficient (SCID) mice.

3.2. Iontophoretic Delivery

In iontophoresis, charged molecules are driven into the skin by application of a low level electrical field over long periods of time. Transport of chemical agents across a membrane under the influence of an electrical field is governed by three forces *(21)*: passive diffusion, electromigration, and electro-osmosis. For charged molecules, electromigration, which is believed to proceed predominantly through appendageal pathways *(22)*, represents the major driving force. Electro-osmosis, the convective flow of solvent and species dissolved in it, may become important for the transport of large molecules including ODNs *(23)*. In contrast, the contribution of passive diffusion to the transdermal transport of large, charged molecules such as ODNs is negligible owing to the lipophilic nature of the stratum corneum.

The high water-solubility and charged nature of ODNs characterize them as good candidates for iontophoresis. In vitro iontophoretic ODN delivery through hairless-mouse skin, hairless rat skin, or human skin has been used to determine the influence of pH, salt content, current density, and ODN structure on transport. Oldenberg and colleagues *(7)* measured the effect of nucleotide composition on the flux of ODNs with a length of 15 nucleotides (15mers). Comparable fluxes were found for oligomers containing only deoxyadenine $[(dA)_{15}]$, only thymidine $[(dT)_{15}]$, or all four nucleotides in a random sequence; in contrast, lower fluxes were recorded for the corresponding $(dG)_{15}$ or $(dC)_{15}$ homo-oligomers or for aptamers. This result suggests that the three-dimensional (3D) structure of an ODN plays an important role in iontophoresis. The significance of structure formation for the iontophoretic mobility of ODNs is supported further by the findings that reversing the sequence of bases in an ODN affected the flux *(14)* and that single nucleotides crossed the dermo-epidermal junction less readily than ODNs did *(8)*.

Brand and Iversen demonstrated that PS modification of ODNs had little or no effect on iontophoretic transport through hairless mouse skin *(24)*. In contrast, 2'-*O*-methoxyethylation of the carbohydrate moieties in a 20-mer decreased the flux rate to approximately one-fourth of that observed for a PS ODN with the same sequence (unpublished results from authors' lab). In terms of therapeutic value, however, an up to 10-fold increase in potency for the 2'-*O*-methoxyethyl ODN more than compensated for its reduced iontophoretic mobility. A C(5)-propyne modified PS ODN not only showed more than 67-fold higher potency than the unmodified PS ODN, but also exhibited a fivefold higher iontophoretic flux rate *(25)*.

In previous in vivo studies in mice *(7,8,14,22,25)*, iontophoresis enhanced the transdermal delivery of ODNs, but did not result in delivery of therapeutically relevant doses. However, the emergence of highly potent modified ODNs with increased iontophoretic flux rates has generated new interest in this mode of transdermal ODN delivery *(25)*.

3.3. Electroporation and Sonophoresis

Electroporation employs brief high-voltage pulses to create highly transient, aqueous pores in the lipid network of the stratum corneum. Upon electroporation, a FITC-labeled anti-c-myb ODN applied to heat-stripped human skin in vitro localized in regions approx 30 μm in diameter, demarcating the transient skin pores *(26)*. Regnier and Preat and coworkers examined the parameters controlling transdermal delivery of ODNs through hairless rat skin by electroporation *(27)* and found an increase in ODN delivery of two orders of magnitude over passive diffusion *(28)*. In addition to its effect on the stratum corneum, electroporation also increased the permeability of cell membranes, resulting in enhanced uptake of topically applied ODN into keratinocytes in the deeper layers of the epidermis *(8,29–31)*. Nuclear localization of ODNs in skin cells after electroporation indicated the feasibility of this method for local skin treatment with ODNs *(22)*. Use of electroporation in the clinic, however, may be limited by patient compliance and device requirement issues.

In sonophoresis, pulses of low-frequency ultrasound are used to increase skin permeability to macromolecules through thermal, mechanical, and cavitational effects on the stratum corneum *(32)*. Experimental results indicate that cavitation of keratinocytes might be the key mechanism in enhancing transdermal transport by low-frequency ultrasound. Use of sonophoresis for transdermal delivery of macromolecules, in particular protein hormones, is under active evaluation (Sontra Medical Corporation, Franklin, MA). Parameters such as frequency, intensity, medium, and time of application affect the efficiency and safety of this method.

3.4. Mechanical Enhancement

Removal of the stratum corneum by either mechanical abrasion or tape stripping significantly enhanced permeation through the skin for a wide range of pharmaceuticals, including PS ODNs *(18,21,22)*. However, practical applications of this method of enhancement have been limited owing to lack of control and reproducibility. In contrast, artificial microchannels or micropores through the stratum corneum can be generated in a highly defined and reproducible manner. These artificial channels form aqueous transport pathways for the passive diffusion or iontophoretic delivery of ODNs. Because the micropores are confined to the epidermis, which does not contain nerve endings or blood vessels, their formation causes little or no pain or bleeding. Depending on their size and whether or not the site is under occlusion, microchannels will self-seal after 1–24 h *(33,34)*.

Two types of devices are currently used to generate micropores through the skin. In the thermal MicroPor™ system (Altea Therapeutics, Tucker, GA), a pulse of thermal

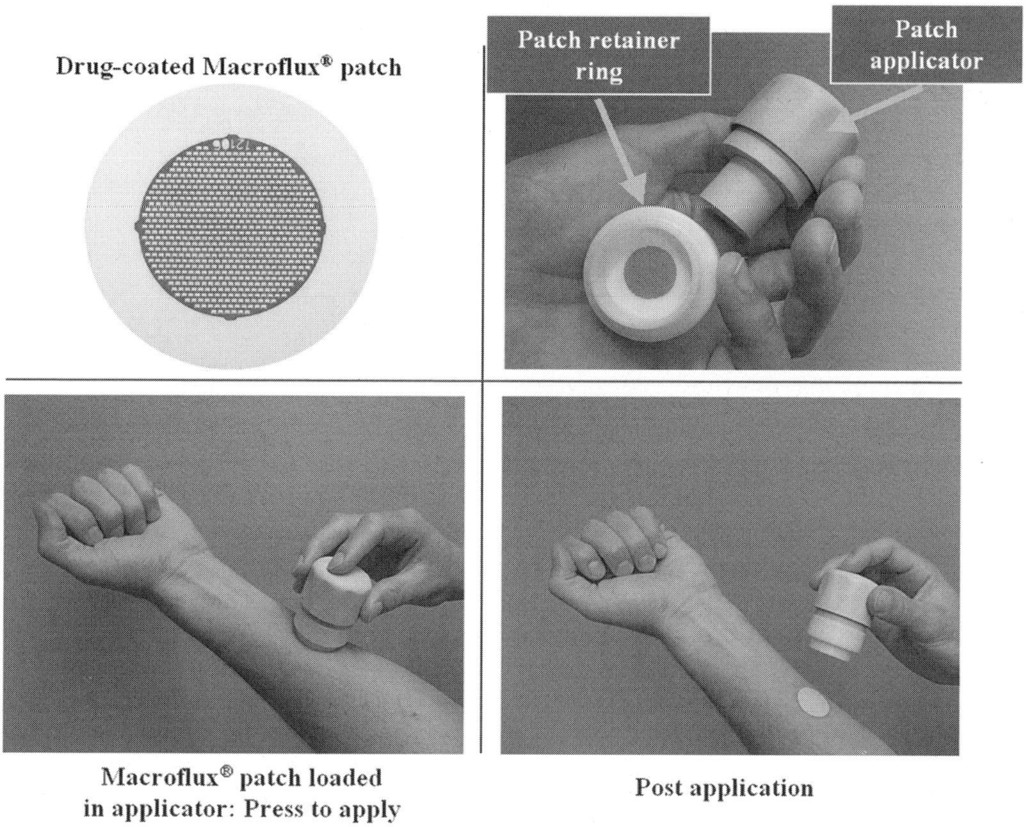

Fig. 1. A prototype Macroflux® transdermal system.

energy delivered from an array of metallic heating filaments placed directly against the skin causes the formation of micropores through the stratum corneum.

Mechanical creation of artificial transport pathways is achieved with microfabricated microprojection arrays of up to 970 projections/cm^2. The shape of the titanium or silicone microprojections and their angle relative to the skin surface varies according to the application, but is uniform within the array. Microprojections are generally less than 225 µm long to avoid penetration into the dermis. For bolus delivery, drug can be coated directly onto the microprojections *(35)*. Alternatively, the arrays can be integrated with a drug reservoir for passive or iontophoretic delivery *(36)*.

Macroflux® systems (ALZA Corporation, Mountain View, CA) use a spring-loaded, self-actuated device to apply the microprojection array to the skin, resulting in uniform and reproducible penetration (Fig. 1) *(34)*. When the application site of a microprojection array was stained with India ink, the skin surface showed a uniform penetration pattern. Depth analysis revealed that fewer than 30% of the microprojections penetrated deeper than 100 µm, and none went beyond 350 µm. The average depth of penetration was less than 100 µm and thus lay within the 50–150 µm thickness range of both human and hairless guinea pig epidermis *(35)*. As expected, this very shallow microprojection penetration did not result in any topical adverse effects.

We have successfully used the Macroflux® system for in vivo transdermal ODN delivery in hairless guinea pigs *(36)*. Iontophoretic flux of an anti-ICAM-1

A One hour passive delivery with Macroflux®

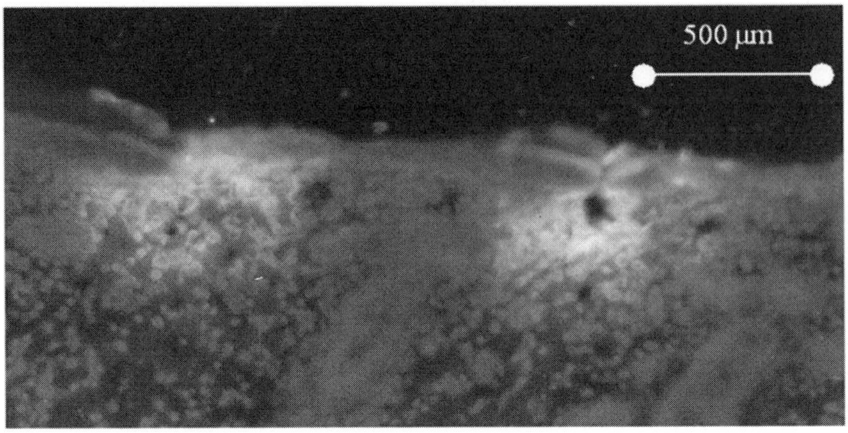

B One hour iontophoretic delivery with Macroflux®

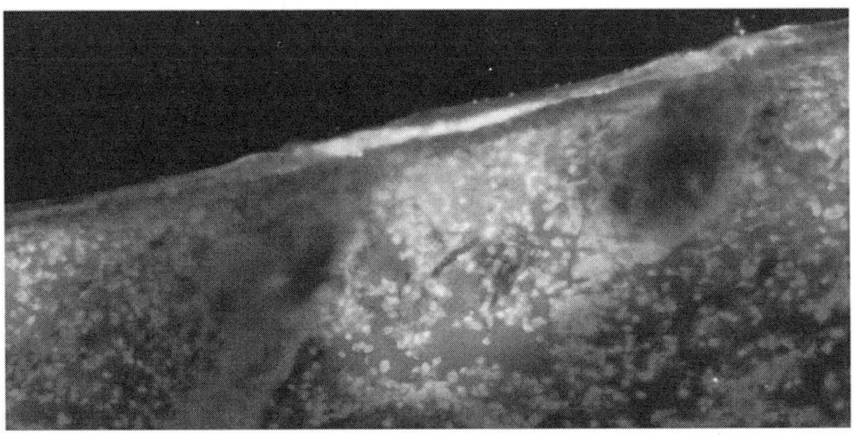

Fig. 2. Distribution of a randomized FITC-labeled phosphorothioate ODN in the skin following transdermal delivery by Macroflux®. Skin samples from hairless guinea pigs were taken after 1 h of Macroflux®-aided delivery with either a passive system **(A)** or an iontophoretic system **(B)**. Skin samples were also counterstained with propidium iodine (PI), a red fluorescent and DNA-binding dye, to highlight cells in the skin.

phosphorothioate ODN (ISIS 2302) was increased more than 100-fold by pretreatment of the skin with a Macroflux® microprojection array. After Macroflux®-assisted delivery of a randomized FITC-labeled phosphorothioate ODN, fluorescence was found superimposed on the pathways created by the microprojections (Fig. 2); enhanced lateral diffusion in the epidermal layers underneath the stratum corneum was also apparent. In contrast, fluorescence was not observed in areas of intact skin. These results indicate that ODN transport primarily proceeds through the artificial microchannels/micropores created by the microprojection array, whereas skin appendages play only a minimal role in ODN delivery under these conditions. Surprisingly, Macroflux®-assisted iontophoretic delivery of ODNs appeared to be independent of nucleotide sequence or com-

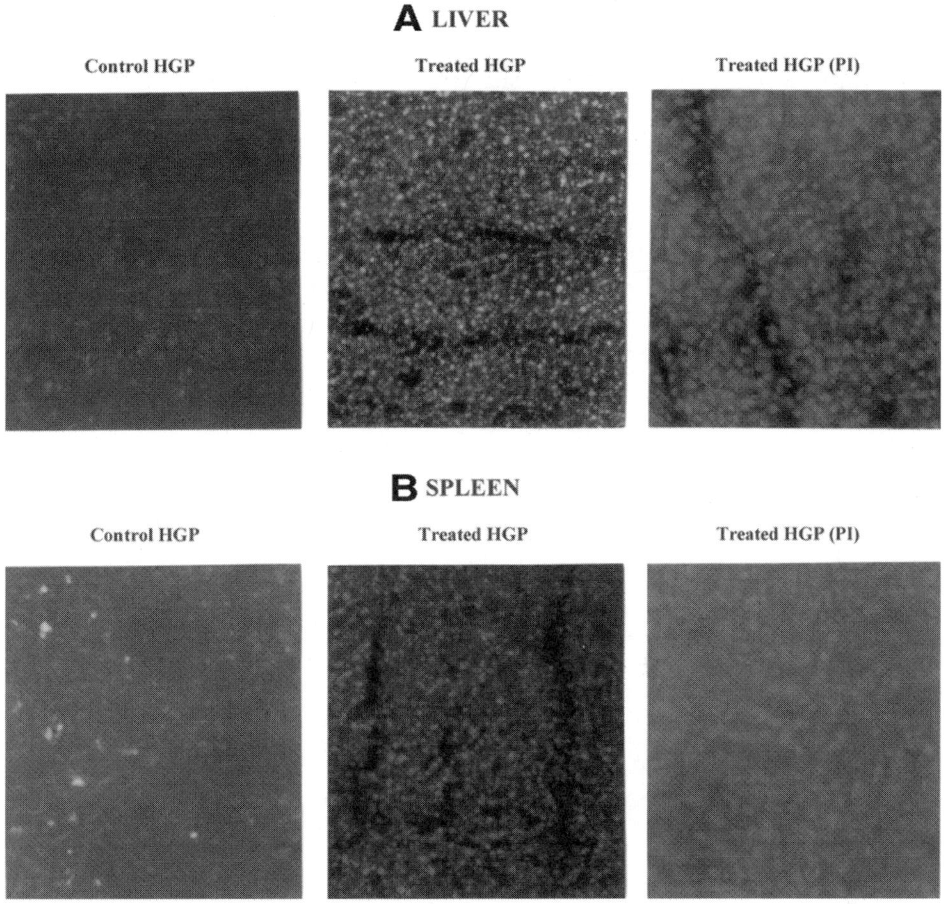

Fig. 3. Systemic localization of a randomized FITC-labeled phosphorothioate ODN following transdermal delivery. Organ samples were taken from hairless guinea pigs 2 h after receiving no ODN (control) or ODN delivered by a Macroflux® system. Tissues were immediately frozen, then cryotomed and viewed under a fluorescent microscopy. Some tissue slides were counterstained with PI to reveal cellular nuclei. Green FITC fluorescence appeared to be localized intracellularly in the liver **(A)**, as well as in the spleen **(B)**.

position, which were shown to affect the iontophoretic mobility of ODNs in the absence of microchannels *(14)*. This finding suggests that micropores created by the Macroflux® system are less selective than skin appendages, which appear to favor small and positively charged species *(23,24)*.

In skin biopsies from Macroflux® application sites, ODN was detected at concentrations of up to 1 mg/cm^3 skin (full thickness) *(36)*, a value that is pharmacologically relevant for local therapy *(18)*. Following Macroflux® delivery, fluorescence from the randomized FITC-labeled phosphorothioate ODN could also be detected in samples taken from the liver, the spleen, and the duodenum, where it was located mostly intracellularly (Fig. 3). The intensity of the green FITC was rather strong in the liver (Fig. 3A), slightly weaker in the spleen (Fig. 3B), and even weaker in the duodenum (not shown). The quantity of intact phosphorothioate ODN detected by capillary electrophoresis in liver samples after Macroflux®-assisted iontophoretic delivery was similar to that seen after intravenous injections *(36)*. The amount of the ODN in livers of hairless guinea pigs could reach more than 4 mg following transdermal delivery, which is approx 0.2 mg/g tissue. These results suggest that the Macroflux® system is effective in both local and systemic delivery of antisense ODNs.

4. Conclusion

Therapeutic use of ODNs would greatly benefit from the availability of a patient-friendly and noninvasive delivery method. Recent advances in increasing the potency of ODNs through structural modifications have made these macromolecules suitable for transdermal delivery. The most promising transdermal delivery method to date is based on the controlled generation of aqueous pathways through the stratum corneum, either mechanically or thermally, to enhance passive or iontophoretic delivery. Using the Macroflux® system, a mechanical device for the controlled creation of micropores through the superficial layers of the skin, we demonstrated successful transdermal delivery of antisense ODNs in vivo in hairless guinea pigs. Following Macroflux®-assisted transdermal delivery, ODNs were detected both in the skin and systemically distributed to various other organs. At the application site, a pharmacologically relevant concentration of ODN could be achieved. Given the ongoing improvements in ODN chemistry, transdermal or topical delivery of ODNs for therapeutic use is now not only attractive but also feasible.

Acknowledgment

The authors are grateful to Dr. Ute Geigenmuller, who has made useful suggestions and has done a wonderful job in improving the manuscript.

References

1. Agrawal S, Iyer RP. Perspectives in antisense therapeutics. Pharmacol Ther 1997;76:151–160.
2. Agrawal S, Kandimalla ER. Antisense and/or immunostimulatory oligonucleotide therapeutics. Curr Cancer Drug Targets 2001;1:197–209.
3. Gewirtz AM. Oligonucleotide therapeutics: clothing the emperor. Curr Opin Mol Ther 1999;1:297–306.
4. Holmlund JT, Monia BP, Kwoh TJ, Dorr FA. Toward antisense oligonucleotide therapy for cancer: ISIS compounds in clinical development. Curr Opin Mol Ther 1999;1:372–385.
5. Jansen B, Zangemeister-Wittke U. Antisense therapy for cancer: the time of truth. Lancet Oncol 2002;3:672–683.
6. Brand RM, Iversen PL. Transdermal delivery of antisense compounds. Adv Drug Delivery Rev 2000;44:51–57.
7. Oldenberg KR, Vo KT, Smith GA, Selick HE. Iontophoretic delivery of oligonucleotides across full thickness hairless mouse skin. J Pharm Sci 1995;84:915–921.
8. White PJ, Fogarty RD, Liepe IJ, et al. Live confocal microscopy of oligonucleotide uptake by keratinocytes in human skin grafts on nude mice. J Invest Dermatol 1999;112:887–892.
9. Crooke ST. Delivery of oligonucleotides and polynucleotides. J Drug Target 1995;3:185–190.
10. Crooke RM, Graham MJ, Crooke ME, Crooke ST. In vitro pharmacokinetics of phosphorothioate antisense oligonucleotides. J Pharmacol Exp Ther 1995;275:462–473.
11. Cook PD. Medicinal chemistry of antisense oligonucleotides: future opportunities. Anticancer Drug Res 1991;6:585–607.
12. Micklefield J. Backbone modification of nucleic acids: synthesis, structure and therapeutic applications. Curr Med Chem 2001;8:1157–1179.
13. Urban E, Noe CR. Structural modification of antisense oligonucleotides. Farmaco 2003;58:243–258.

14. Brand RM, Wahl A, Iversen PL. Effects of size and sequence on the iontophoretic delivery of oligonucleotides. J Pharm Sci 1998;87:49–52.
15. Mitragotri S, Johnson ME, Blankschtein D, Langer R. An analysis of the size selectivity of solute partitioning, diffusion, and permeation across lipid bilayers. Biophys J 1999;77:1268-1283.
16. Chuong CM, Nickoloff BJ, Elias PM, et al. What is the 'true' function of skin? Exp Dermatol 2002;11:159–187.
17. Asbill CS, El-Kattan AF, Michniak B. Enhancement of transdermal drug delivery: chemical and physical approaches. Crit Rev Ther Drug Carrier Syst 2000;17:621–658.
18. Kitagawa S, Li H. Effects of removal of stratum corneum, delipidization and addition of enhancers, ethanol and l-menthol, on skin permeation of benzoic acid and its 4-n-alkyl substituents in excised guinea pig dorsal skin. Chem Pharm Bull (Tokyo) 1999;47:44–47.
19. Hadgraft J. Passive enhancement strategies in topical and transdermal drug delivery. Int J Pharm 1999;184:1–6.
20. Mehta RC, Stecker KK, Cooper SR, et al. Intercellular adhesion molecule-1 suppression in skin by topical delivery of antisense oligonucleotides. J Invest Dermatol 2000;115:805–812.
21. Bronaugh RL, Stewart RF. Methods for in vitro percutaneous absorption studies V: permeation through damaged skin. J Pharm Sci 1985;74:1062–1066.
22. Regnier V, Preat V. Localization of a FITC-labeled phosphorothioate oligodeoxynucleotide in the skin after topical delivery by iontophoresis and electroporation. Pharm Res 1998;15:1596–1602.
23. Regnier V, De Morre N, Jadoul A, Preat V. Mechanisms of a phosphorothioate oligonucleotide delivery by skin electroporation. Int J Pharm 1999;184:147–156.
24. Brand RM, Iversen PL. Iontophoretic delivery of a telomeric oligonucleotide. Pharm Res 1996;13:851–854.
25. Brand RM, Hannah TL, Norris J, Iverson PL. Transdermal delivery of antisense oligonucleotides can induce changes in gene expression in vivo. Antisense Nucleic Acid Drug Dev 2001;11:1–6.
26. Zewert TE, Pliquett UF, Langer R, Weaver JC. Transdermal transport of DNA antisense oligonucleotides by electroporation. Biochem Biophys Res Comm 1995;212:286–292.
27. Regnier V, Le Doan T, Preat V. Parameters controlling topical delivery of oligonucleotides by electroporation. Drug Target 1998;5:275–289.
28. Regnier V, Tahiri A, Andre N, et al. Electroporation-mediated delivery of 3'-protected phosphodiester oligodeoxynucleotides to the skin. J Control Release 2000;67:337–346.
[illegible] An experimental model for interpreting percutaneous penetration of oligonucleotides that incorporates the role of keratinocytes. J Invest Dermatol 1998;111:1166–1171.
30. Noonberg SB, Garovoy MR, Hunt CA. Characteristics of oligonucleotide uptake in human keratinocyte cultures. J Invest Dermatol 1993;101:727–731.
31. Nestle FO, Mitra RS, Bennett CF, et al. Cationic lipid is not required for uptake and selective inhibitory activity of ICAM-1 phosphorothioate antisense oligonucleotides in keratinocytes. J Invest Dermatol 1994;103: 569–575.
32. Langer R, Blankschtein D, Mitragotri S, Tang H. Theoretical description of transdermal transport of hydrophilic permeants: application to low-frequency sonophoresis. J Pharm Sci 2001;90:545–568.
33. Henry S, McAllister DV, Allen MG, Prausnitz MR. Microfabricated microneedles: a novel approach to transdermal drug delivery. J Pharm Sci 1998;87:922–925.
34. Cormier M, Daddona PE. Macroflux technology for transdermal delivery of therapeutic proteins and vaccines. In: Roberts M, Hadgraft J, Rathbone MJ, eds. Modified-Release Drug Delivery Technology. New York, Marcel Dekker, 2002, pp. 589–598.
35. Matriano JA, Cormier M, Johnson J, et al. Macroflux® microprojection array patch technology: a new and efficient approach for intracutaneous immunization. Pharm Res 2002;19:63–70.
36. Lin WQ, Cormier M, Samiee A, et al. Transdermal delivery of antisense oligonucleotides with microprojection patch (Macroflux®) technology. Pharm Res 2001;18:1789–1793.

15

Fc Receptor Targeting With Recombinant Immunoglobulins and Immunoglobulin Formulations

Adrian Bot and Dan Smith

1. Introduction

Manipulation of immunity for the purpose of controlling pathogenic responses or prevent diseases such as infections can be achieved by two different approaches: (1) an antigen- or epitope-based strategy, designed to influence the immune system by interaction with lymphocytes that express only receptors specific for such antigens (e.g., vaccines); or (2) the use of agents that decrease or increase immunity by acting via nonantigen specific receptors on lymphocytes (e.g., immune-suppressive drugs for the treatment of autoimmune diseases or immune stimulatory cytokines in cancer). Advances during the last two decades in particular have resulted in the understanding that noninfectious diseases such as autoimmunity and certain cancers can be influenced in a manner depending on antigen-specific recognition. This finding, combined with advances in target discovery and validation catalyzed by the elucidation of the human genetic map, has resulted in an unprecedented world-wide effort in designing safer drugs or therapeutic strategies that give the concept of vaccination a new meaning.

1.1. Dual Role of Antigen-Presenting Cells (APC) in the Control of Immunity

APC are critical in the regulation of immune responses from at least two points of view. First, they present the major histocompatibility complex (MHC) class I- and II-restricted epitopes to specific T-cells. This event in itself, although necessary, is not sufficient in triggering a full-blown immune response. An additional prerequisite for induction of strong immunity is co-stimulation via nonantigen-specific receptors. Thus, the immune responsiveness is regulated by a set of complementary cognate and noncognate factors under APC control.

The immune response is essentially controlled by the T-cells that recognize epitopes presented by APC in the context of the appropriate MHC molecules. T helper (Th) cells recognizing MHC class II (MHC-II)-restricted epitopes fall into different functional categories depending on the cytokine profile they acquire during the differentiation process. More than a decade ago, when the seminal observation of functional dichotomy of Th cells was made *(1)*, it was proposed that the Th cells fall into two categories:

From: *Cellular Drug Delivery: Principles and Practice*
Edited by: D. R. Lu and S. Øie © Humana Press Inc., Totowa, NJ

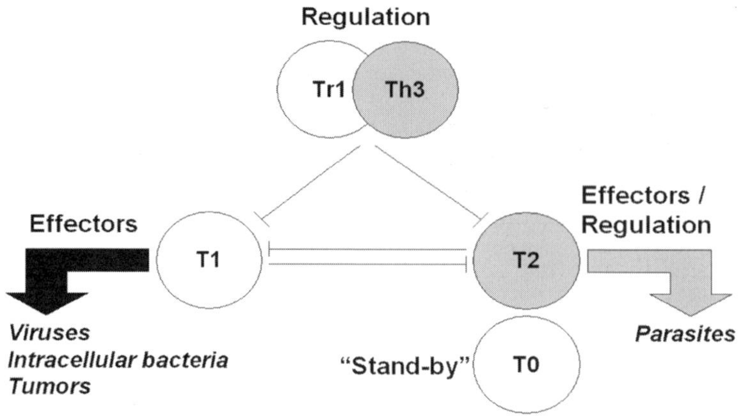

Fig. 1. Schematic representation of the functional organization of the T-cell compartment, as currently understood. Effector and regulatory T cells differentiate from naïve, competent precursors in the peripheral lymphoid organs depending on the presence of antigens, costimulation, and stochastic factors. They may transiently assume an ambivalent T_0 phenotype, consisting in production of limited levels, broad range cytokines ("stand-by" stage). From this stage, they could rapidly differentiate based on micro-environmental clues, to T2 cells (producing mostly IL-4, IL-5, and/or IL-13), T1 cells (producing IL-2, IFN-γ, TNF-β), Th3 (TGF-β), Tr1 (IL-10), or minute subsets such as $CD4^+CD25^+$ regulatory cells. T1 and T2 cells execute dual roles: T1 cells fight viral infections, certain bacterial infections, and tumors. In addition, they keep in check the differentiation and expansion of T2 cells. In contrast, T2 cells fight parasitic infections and they keep in check the differentiation and expansion of the T1 pool. Specialized, regulatory T cells such as Th3 and Tr1 are believed to be essential in maintaining under control the expansion of effector cells (T1 and T2) in general, facilitating a non-inflammatory status in basal conditions (chronic exposure to innocuous antigens) and preventing immunopathology during the response to pathogenic agents. Of note, the pleiotropism and versatility of this organizational structure, that is designed to mobilize as rapidly as possible an antimicrobial response but limit unnecessary immunopathology.

1. Th1 cells, which produce mainly interferon-γ (IFN-γ) and interleukin-2 (IL-2), and enhance the anti-viral response
2. Th2 cells, which produce IL-4 and are critical mediators of anti-parasitic immunity

Following more than one decade of intensive research in this area, it appears that the functional structure of the T-cell compartment is at least tripolar, rather than bipolar, as proposed initially (Fig. 1) and that it extends to MHC-I-restricted T cells (Tc) *(2)*. A feature of Th1 and Th2 cells is that they are involved in both immune regulation and the effector arm of immunity. In addition to Th1 and Th2 cells, there are T regulatory cells, devoid of overt effector capability but endowed with potent immune modulatory functions via production of IL-10 (Tr1 cells) *(3)* or transforming growth factor-β (TGF-β) (Th3 cells) *(4)*. T regulatory cells are mostly involved in maintaining the homeostasis of the immune response at mucosal sites, where a tight control of tolerance vs immunity must be maintained in order to allow effective immune defense without organ-associated immune pathology *(5)*. It is currently believed that a failure of effective reciprocal regulation between different functional subsets of T cells is responsible for T1-mediated pathology (autoimmune diseases such as multiple sclerosis [MS] and type 1 diabetes) or T2-mediated diseases (asthma, allergies).

The induction of T-cell responses followed by differentiation is governed by specific rules that are believed to be essential to the process of self, nonself-discrimi-nation. The system of MHC is a molecular-sorting mechanism that selects proteasome-processed peptides of potential significance for immunosurveillance. Mere exposure of T cells to peptides on MHC molecules expressed by APC in the presence of appropriate co-stimulation does not necessarily result in generation of effector cells. There are at least two additional homeostatic mechanisms involved in preventing immunopathology, independent of MHC-restricted recognition. Thymic negative selection ensures that the T-cell precursors colonizing the peripheral lymphoid organs do not recognize, with high affinity, most of the self epitopes expressed on MHC molecules by default *(6)*. However, there is a set of T-cell epitopes that is not expressed in the thymus, and thus do not participate in central negative selection. Such epitopes may be tissue-restricted, cryptic, or developmental. In addition, the immune system is faced continuously with substantial exposure to innocuous nonself antigens associated, for example, with gut-flora or saprophytes colonizing the other mucosal areas. Thus, an additional mechanism ensures that full-blown autoimmune or inflammatory responses, respectively, are elicited only when supplemental conditions associated with pathogenic infection *(7)* and/or lesions *(8)* are present. Such a checkpoint is based on MHC and T-cell receptor (TCR)-independent recognition by APC of microbial-derived pattern motifs (e.g., carbohydrates, lipids, lipopeptides, or nucleic acids) *(7)*, or triggered by endogenous mediators ("danger motifs") liberated following cellular injury (e.g., tumor necrosis factor-α [TNF-α], IL-1β or heat-shock proteins [hsps]) *(8)*. In addition, penetration of antigen beyond the mucosal-epithelial barrier that may be caused by, for example, virulent microbes contributes to enhanced immunity by mere "geographic" co-localization with cells capable of co-stimulation *(9)*. It is believed that only recognition of MHC-peptide complexes, in context of simultaneous co-stimulation triggered by microbial or endogenous danger signals, results in full-blown differentiation of effector T cells. For example, IL-12, a critical factor for differentiation of IFN-γ-producing T cells via STAT-4-mediated activation *(10)*, is produced by innate immune cells only in conditions of immunological or cellular stress. In stark contrast, innate immune cells associated with mucosal areas produce by default immune regulatory cytokines such as IL-10 and TGF-β that increase the threshold associated with induction of inflammation and assist in generation of regulatory T cells. Thus, continuous exposure to innocuous antigens in the absence of strong co-stimulation triggers nonpathogenic regulatory T cells rather than T-cell effectors, further increasing the stimulation threshold required for inflammation *(5)*.

During the last few years, the dichotomy between the "danger-model" and microbial motif recognition model has been overemphasized *(11,12)* and questioned, fueled by the different nature (endogenous vs exogenous) of the pro-inflammatory mediators required for differentiation of T-cell effectors. For example, recent data brought into focus a new category of motifs—noncoding RNA-associated motifs—that are produced within cells infected with negative-stranded RNA viruses or certain DNA viruses *(13)*. In contrast to previously defined microbial-associated motifs, such as CpG nucleotides, lipopeptides, and microbial carbohydrates, naturally occurring dsRNA motifs from infected cells are not borne by microbes and thus cannot be categorized strictly as microbial motifs. Instead, they are produced by the transcriptional machinery of the infected cell using nucleotides of endogenous origin, with critical participation from viral proteins such as RNA-dependent polymerases. However, such motifs cannot be categorized as strictly endogenous either, because their effect is dependent on the nucleotide composition *(13)*, which is, in turn, determined by the genetic structure of the virus.

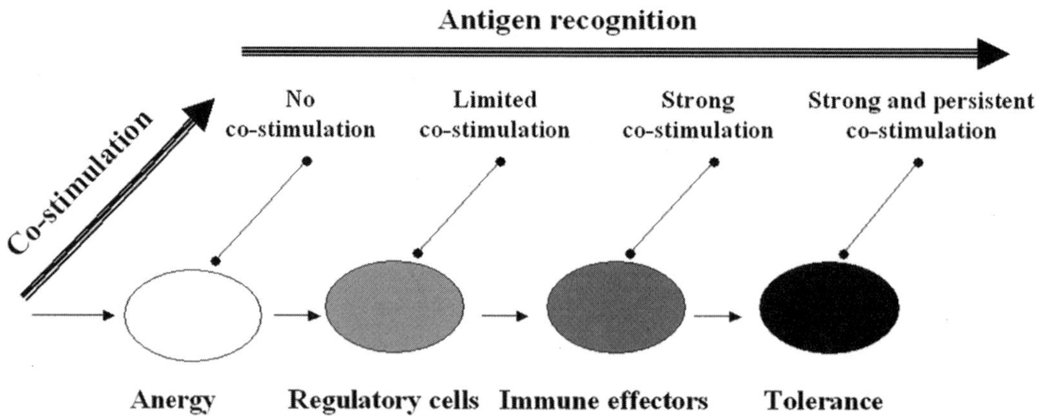

Fig. 2. Two categories of factors influence, essentially, the differentiation of T cells to functional/regulatory subsets. Exposure to antigen (peptides located on specialized MHC) or "cognate" interaction. The strength and persistence of cognate interaction has a decisive influence on whether the antigen-bearing material is largely neglected or effected upon. In addition, the co-stimulatory environment signaling a potential or real pathogenic process, dictates the functional profile of the differentiating T cells. In the presence of reduced "threat"-associated signals reminiscent of saprophytic or "normal" flora, the assumed T-cell profile will be largely regulatory. Upon exposure to pathogens, the T cells assume an effector phenotype. To limit uncontrollable expansion of effectors and thus fulminant immunopathology, excessive stimulation of T cells results in their death ("antigen-induced cell death"). Thus, the ability of T cells to decode different categories of signals is coupled to a complex but versatile functional organization.

Thus, neither the danger model nor the microbial motif-recognition model can accommodate this class of immune regulatory motifs. A unified perspective, as the one depicted in the Fig. 2, could integrate all these aspects if "endogenous," "exogenous," and "hybrid" motifs, such as dsRNAs, are all considered in fact "immune stress molecules" (ISM) with the ability to anticipate, prime, or signal pathologic processes associated with infection or other factors.

1.2. Targeting the APC: The Need

A prerequisite for the manipulation of the immune response in an antigen-specific manner is the use of molecules encompassing epitopes that are to be recognized by TCR. A method to circumvent the requirement for antigen processing within APC is the use of peptide epitopes. In addition, compared with recombinant proteins and live vectors, the use of peptides is currently more cost-effective from a production point of view. There is a substantial drawback, however, related to the use of peptides that consists of their unfavorable pharmacokinetics (PK) profile stemming from rapid elimination via the kidneys and/or peptidase-mediated degradation.

The PK profile of peptides, as limiting factor for the loading of APC in vivo, is dramatically revealed by comparing the immune responses triggered by injection of a T-cell epitope vs APC loaded ex vivo with dose-matched peptide. Whereas the adoptive transfer of mouse I-E^d+ dendritic cells (DC) pulsed in vitro with a Th peptide (homagglutinin [HA] 110-120 of PR8 strain of influenza virus) triggered detectable peptide-specific IL-4 and IL-2-producing T cells, direct injection of dose-matched HA

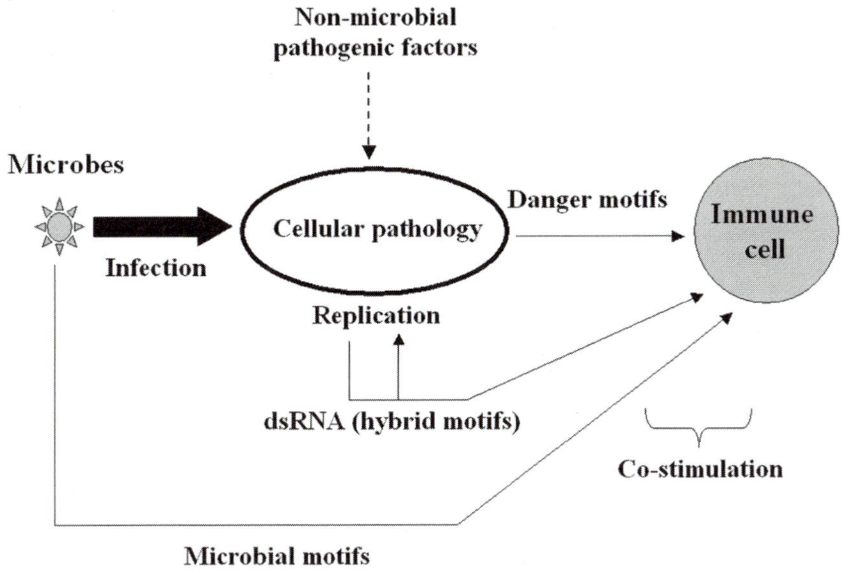

Fig. 3. The immune system decodes different categories of factors associated with potential or ongoing pathogenic processes. This results in the capability to pre-empt or modulate appropriate effector mechanisms. There are essentially three such categories of factors: (1) microbial-associated molecular motifs, present in the structure of the pathogen (such as peptidoglycans, lipopeptides, polysaccharides, endotoxin, certain species of DNAs); (2) mediators of strict endogenous origin (encoded in the host genome), produced subsequent to cellular injury (infectious or noninfectious); "danger motifs" such as TNF-α, and heat-shock proteins; and (3) "hybrid" factors such as dsRNA that are produced by infected cells but encoded in the microbial genome.

peptide failed to prime immunity (Fig. 3). Failure to retrieve substantial numbers of APC loaded in vivo with peptide (*see* Section 2) strongly supports this interpretation.

In general, the literature supports the conclusion that whereas peptides are being considered effective agents to manipulate in vitro T-cell response, the in vivo efficiency in most of the systems is suboptimal, unless methods such as depot formulations or targeted delivery are used. Various targeted delivery strategies have been tested in the past: from microparticles that are able to deliver immunodominant epitopes to APC *(14)*, to targeted administration of peptides into areas rich in APC *(15)*, culminating with the use of recombinant vectors that deliver immunmodulatory peptides to APC, in a manner depending on ligand-receptor interaction *(16)*. In the present chapter, we summarize our experience in preclinical models, with two novel categories of specialized targeting vectors: recombinant immunoglobulins that deliver immune modulatory peptides via FcγRs on APC and secondly, spray-dried lipid-based microparticles (SDM) bearing ligands for cellular receptors on APC.

2. Targeted Delivery of Epitopes to APC Via Recombinant Immunoglobulins

The main drawback of peptides as potential drugs derives from their poor PK profile. The ability of peptides representing T-cell epitopes to load endogenous MHC molecules depends on whether APC are exposed *in situ* to persisting, high concentrations of peptide. On the other hand, mere prolongation of the half-life of peptides in vivo by chemical modification, is likely to interfere with the on/off rate of the peptide-MHC complex and does not necessarily target the epitope to APC. Thus, a molecular-targeting procedure that troubleshoots the drawbacks described earlier, may transform peptides into more effective drugs. Such a strategy should be based on vectors with the following properties:

1. The vector molecule should be biocompatible, in particular devoid of intrinsic immunogenic effects. Thus, it would allow repeated administration of the epitope without build-up in deleterious antivector antibodies. In addition, use of such a vector will be compatible with treatment of inflammatory disorders or, with some modifications in the vaccination strategy (co-administration of adjuvant), could be applied in antimicrobial or tumoral vaccination.
2. The vector should be able to correct the poor PK of peptide epitopes. This depends largely on the molecular weight of the carrier, as well as on the relative resistance to proteolytic degradation.
3. The vector should be able to localize nearby or target preferentially the APC and display reduced affinity for non-APC "somatic" cells. This would optimize the efficiency, limit requirement for large doses, and minimize safety concerns.
4. Finally, such a vector should be able to promote, if possible, more effective loading of MHC on APC, ideally by targeting the epitope in cellular compartments where this process occurs with maximal efficiency.

Based on such considerations, various groups started to consider immunoglobulins as a choice as epitope-delivery vectors *(17–19)*.

2.1. Control of APC Loading With Peptide Epitopes

Use of immunoglobulins to deliver T-cell epitopes to APC requires preservation of the conformation of the carrier, such that neither the half-life nor the FcγR-targeting ability of the molecule is altered. To achieve this aim, one could use the complementarity determining regions (CDRs) located within the variable regions of the heavy and light chains for targeted binding to specific antigens. The immunoglobulin molecule is able to accommodate extensive amino acid residue variability in CDRs, without deleterious impact on the functionality of the molecule. The length of CDRs varies, from less than 9 to approx 15 amino acids, with few exceptions. This allows replacement of at least one CDR/Ig chain *(20)* and in certain cases, of two of them *(21)* with epitopes, without substantial interference in the assembling of the molecule and its ability to bind to FcγRs, via the constant domains C_H2 and 3 of the heavy chain. In addition, the original binding specificity of the Fab segment is abrogated, which is not, in itself, a drawback.

An alternative engineering strategy is the insertion of epitopes into a C_H (e.g., CH2) *(22)*. This strategy interferes with the functional conformation of the Fc portion of immunoglobulin and may result in the inability to bind FcγRs on APC. This can be compensated by using, as a scaffold, immunoglobulins that bind to receptors on APC, such as MHC-II or FcγRs. The N-terminal end of the heavy chain is another area where T-cell epitopes were introduced *(23)*. Finally, peptides could be chemically coupled to carbohydrate moieties on immunoglobulin, which allows processing-free presentation of peptides by MHC-II molecules on the surface of APC *(24)*.

Several studies evaluated the efficiency of delivering MHC-II-restricted epitopes by IgG, to APC in vitro. It was clearly demonstrated, using transformed and primary APC cell types, that FcγR-mediated internalization of a MHC-II-restricted epitope results in effective processing and presentation to specific T cells. In fact, on a molar basis, IgG-mediated internalization of peptide was a few orders of magnitude more effective in formation of MHC-class II-epitope complexes, as compared to nonengineered peptide *(25,26)*. This results from the more optimal loading of nascent MHC-class II molecules in endolysosomes, where processing of the recombinant immunoglobulin occurs subsequent to FcγR-mediated internalization. However, the enhancement in the generation of MHC-peptide complexes by professional (differentiated or immortalized) APC was 100–1000-fold more substantial—as compared with nonactivated APC harvested from central or peripheral lymphoid organs.

In addition to differential cellular handling, other factors may influence the immunomodulatory potency of peptide epitopes in or out of context of IgG vectors, subsequent to in vivo administration. More recently, when T-cell lines or hybridomas specific for defined MHC-II-restricted epitopes were generated and characterized, it became possible to evaluate the in vivo APC loading and subsequent generation of MHC-II-peptide complexes.

This allowed us to dissect a complex of factors responsible for the efficiency of epitope presentation subsequent to in vivo delivery by using IgG backbone or more conventional means. It was clear that on molar basis, delivery of peptides within IgG vector was roughly 2 orders of magnitude more potent than peptide alone, in forming MHC-peptide complexes in vivo when measured indirectly, by the ability of retrieved APC to stimulate, ex vivo, a co-stimulation-independent, antigen-specific T-cell clone. One of the major questions was, however, whether elevating the injection dose of nonengineered peptide could restore the potency of IgG-peptide molecule. Even 1000-fold molar excess of peptide in saline could not reach the APC-targeting efficiency of the IgG vector (*see* Table 1), suggesting that multiple factors limit the bioactivity of immune modulatory peptide epitopes in vivo, in an absolute fashion. A similar result was recently reported in a slightly different model *(27)*. Because the PK of peptides is negatively influenced by the short half-life (owing to protease-dependent degradation and glomerular filtration) reducing the exposure time of APC, depot formulations may ameliorate the in vivo properties of peptide epitopes. As shown in the Table 1, use of oil-in-water depot-type emulsions was not able to parallel the APC-loading efficiency noted with Ig-based vector, although an improvement over peptide in saline was evident in regard to loading of APC in the draining lymph nodes. This modest effect could be explained by the fact that despite prolonging the exposure time of local APC to peptide, the cellular handling itself still remains a limiting factor. Strikingly, parenteral administration (subcutaneous [sc] or intraperitoneal [ip]) of Ig-peptide resulted in effective formation of MHC-peptide complexes on both draining lymph node and splenic APCs, suggesting excellent systemic exposure. The amount of MHC-peptide complexes on thymus-derived APC (expressed as equivalent percent loaded APC) was rather limited in all cases. Ex vivo experiments suggested that this may reflect an inherent deficiency of thymic APC in processing and presenting the epitope from this backbone rather than a more limited exposure of central lymphoid organs to antigen. Finally, this type of assay allows the estimation of the in vivo persistence of immunogenic/immune modulatory MHC-peptide complexes on APC, which is on the order of 1–2 wk on professional APC such as DC. This obviously has important consequences on the design of product profiles.

Table 1
IgG-Mediated Delivery of MHC-II-Restricted Epitopes: Published Studies

Type of construct	Reference	Nature of data
Peptide epitope inserted within CDR of IgG_H	Zanetti, 1992 *(19)*	Description of the general concept
	Zaghouani et al., 1993 *(17)*	More effective handling by APC ex vivo
	Brumeanu et al., 1996 *(21)*	Immunity to double epitope (T, B)-engineered IgG
Peptide epitope inserted within C_H of IgG	Lunde et al., 1997 *(18)*	In vitro APC processing
	Lunde et al., 2001 *(30)*	Targeting via MHC-II and sIgD
Peptide inserted at the N-termini of H chain	Zambidis and Scott, 1996 *(23)*	Induction of tolerance to foreign T and B epitopes
Peptide epitopes chemically coupled to IgG	Brumeanu et al., 1997 *(32)*; Bona et al., 1998 *(16)*	APC handling and immune response

More recently, use of antibodies specific for sIgD on B cells or MHC-II on APC, further improved the ability of IgG vectors to deliver MHC-II-restricted epitopes inserted within the C_H region, to APC in vivo (*see* Table 2) *(29,30)*. Such antibodies were orders of magnitude more potent than Fc-FcγR targeted constructs, although additional cumbersome steps of humanizing the construct are required in this case. In addition, it is important to note that in such cases, cross-linking of membrane-bound receptors may lead to effects beyond just epitope-loading of APC. Finally, insertion of peptides at the N terminus of the heavy chain *(31)* or just chemical coupling onto Ig carbohydrates *(32)* may allow accommodation of multiple epitopes or larger peptides, as opposed to manipulating just CDR.

In summary, use of advanced immunological techniques allowed the delineation of a matrix of factors responsible for the substantial increase in the efficiency of IgG vectors in targeting APC and resulting in formation of potentially immunogenic MHC-peptide complexes: (1) uptake by $Fc\gamma R^+$ cells, which are essentially expressed by APC; (2) improved cellular handling owing to receptor-mediated internalization into endolysosomes and avoidance of rapid clearance via the glomerular filter and finally, (3) improved PK owing to protection from protease-mediated degradation. Thus, IgG vectors ameliorate important limiting factors associated with the PK of small peptides that represent T-cell epitopes.

2.2. Manipulation of T-Cell Responses Against MHC-II-Restricted Epitopes

A major question relates to the outcome of epitope expression on APC subsequent to use of various delivery vectors that correct the PK profile of peptides. The usefulness of a vector category is directly proportional to the ability to generate a broad, controllable spectrum of immune responses that could be applied as desired in various clinical circumstances. For example, use of replicating agents as epitope-delivery vectors which unavoidably encompasses or results in generation of "danger motifs," limits their usefulness to vaccination, immunotherapy of chronic infections/diseases, or cancer pur-

Table 2
IgG-Mediated Delivery of MHC-I-Restricted Epitopes: Published Studies

Type of construct	Reference	Nature of data
Peptide epitope inserted within CDR of IgG_H	Zaghouani, 1992 *(40)*	Ex vivo lysis by CTL of transfected target cells
	Kuzu et al., 1993 *(41)*	Immunity triggered by transfected cells
	Zaghouani et al., 1993 *(39)*	Lack of ex vivo presentation by APC to MHC-I restricted T cells
	Billetta et al., 1995 *(42)*	Immunogenicity of transfected B cells but not Ig-peptide
Polypeptide replacing CH2 and CH3	Wallace et al., 2001 *(46)*	Targeting via anti-FcγRI and ex vivo processing by APC

poses rather than to autoimmune or inflammatory diseases. In contrast, vectors that allow generation of antiinflammatory effectors and eventually tolerance, could be used in autoimmune diseases. In addition, with minimal modifications, such as the delivery in conjunction with T1-biasing adjuvants, the vector's use could be extended to antimicrobial or tumoral vaccination. In summary, epitope-delivery vectors that allow for the instruction of the resulting immune response are more promising from a commercial perspective.

From this point of view, the concept of IgG as an epitope-delivery vector has raised considerable interest. Such interest is amplified by the recent observations that in certain conditions, engagement of FcγR results in induction of IL-10 production by APC *(33)*. Nevertheless, the in vivo Th-cell profile subsequent to use of Ig as delivery vectors was not explored in detail until recently. Our studies showed that injection of BALB/c mice with a recombinant IgG bearing Th and B cell influenza virus epitopes within CDR3 and CDR2 of V_H segment, resulted in induction of T cell epitope-specific proliferative and antibody responses *(28)*. Use of similar constructs to deliver self or self-modified epitopes such as multiple sclerosis (MS)-associated PLP-derived epitopes resulted in suppression of autoimmunity in the experimental allergic encephalomyelitis mouse model *(34,35)*, in a manner dependent upon endogenous production of IL-10 *(36)*. This immediately raised the possibility that formation of MHC-peptide complexes on APC subsequent to IgG-mediated delivery of epitopes, triggers an anti-inflammatory or regulatory immune response that can overcome T1-mediated pathogenic responses. The immune profile of the T-cell response to a foreign MHC-II-restricted epitope (HA 110-120 of PR8 influenza virus) was characterized by ELISPOT analysis and found to comprise IL-4 producing (T2), IL-10 (Tr1), and TGF-β-producing T cells. Subsequent to adoptive transfer of APC subsets in vivo loaded with Ig-peptide, it became evident that different types of cells are responsible for different aspects of the T cell profile (Fig. 4). Thus, monocytes were mostly responsible for the induction of IL-4 Th2 and IL-10 Tr1 cells, whereas DC were mostly capable of inducing Th3 cells. The Th1 response, however, was not significant. This observation reinforces that a pro-inflammatory immune response does not necessarily result when the antigen-loading of APC is increased and the nature of the epitope is foreign. Additional factors are

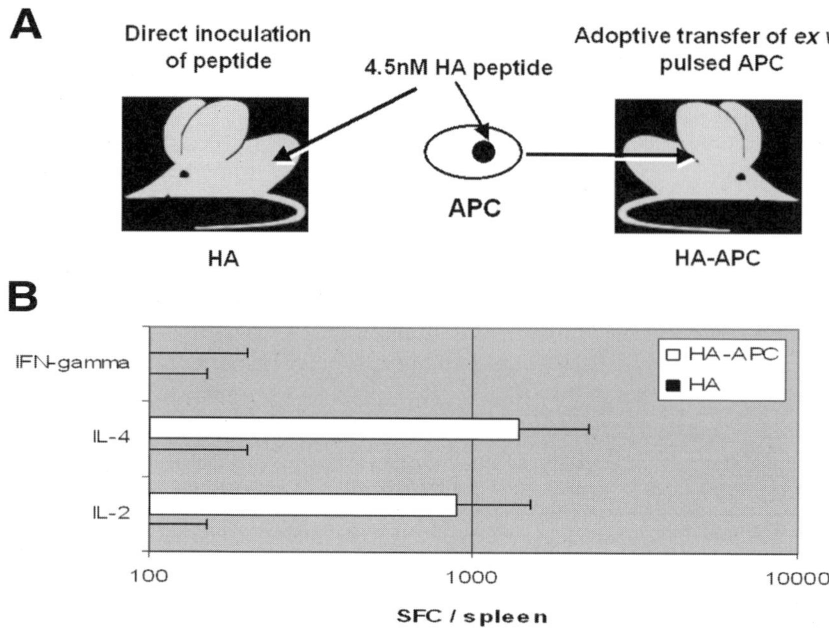

Fig. 4. The PK profile is a limiting factor for immune-modulatory peptides. The diagram describes a simple experiment that supports this conclusion: mice were injected with either 4.5 n*M* of an immune modulating peptide (HA 110-120, restricted to a mouse MHC-II allele), or alternatively, with APCs pulsed ex vivo with the same amount of peptide (**A**). The immune response developed by syngeneic mice was measured by ELISPOT analysis and represented in (**B**) as the average number/spleen of HA peptide-specific, cytokine-producing T cells (spot-forming cells [SFC]). In contrast to the mice inoculated with ex vivo pulsed APC (HA-APC), the mice injected with dose-matched peptide (HA) failed to develop a significant immune response.

needed, such as factors that stimulate the activity of APC, including IL-12 production. Finally, when the number of peripheral T cells is reduced by, for example, depletion procedures, effective Ig-mediated loading of APC with a Th epitope results in anergy rather than immune deviation *(37)*. This observation raises the possibility that the ratio between the epitope-presenting APC and the specific T cells is critical to the outcome of immune modulation, an observation of potential practicality. Based on these observations, it was not surprising that Ig-mediated delivery of epitopes resulted in induction of immune unresponsiveness to Th- and B-cell epitopes in a distinct model, comprising bacteriophage-derived epitopes *(23)*. Subsequent work from the same laboratory implicated the B cells as likely mediators of this effect *(31)*. Our results demonstrate that obliteration of autoreactive T cells by anergy or peripheral deletion is not a prerequisite for suppression of autoimmunity such as EAE. Thus, multiple administrations of a mixture of IgG vectors bearing two dominant epitopes (MBP- and PLP-derived) into SJL mice with chronic progressive disease resulted in suppression of disease advancement; however, in mice that did not progress to a paralytic stage, the frequency of cytokine-producing autoreactive T cells was actually increased (Fig. 5). A closer look at the T-cell profile revealed T2 immune deviation because lymphocytes

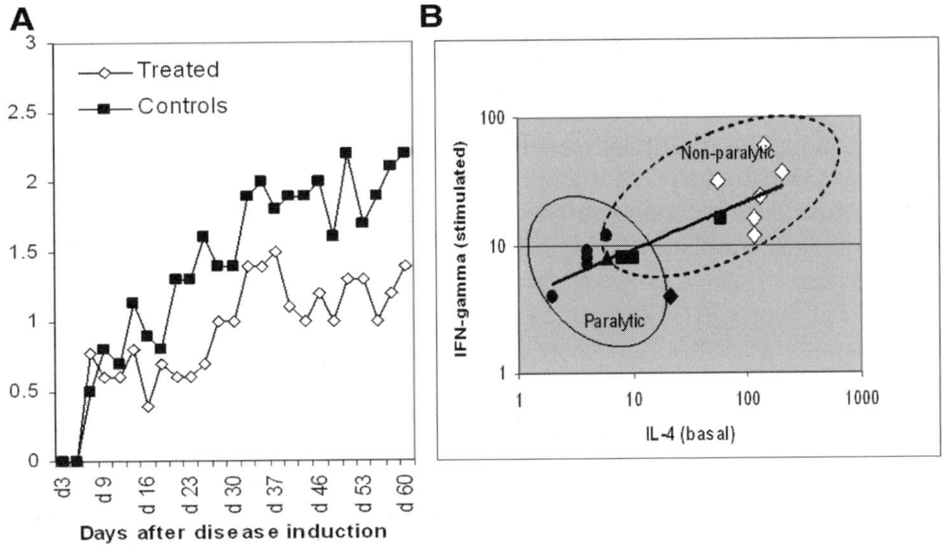

Fig. 5. In vivo epitope targeting of disease-associated epitopes to FcγR+ APC results in suppression of autoimmune disease associated with expansion rather than diminution of autoreactive T-cell pools. Chronic-progressive encephalomyelitis was induced in SJL mice by injection of brain-homogenate in complete Freund's adjuvant. At various intervals after the induction of disease, a mixture of immunoglobulins carrying dominant epitopes (PLP- and MBP-derived) were administered to the mice (four times, at d 9, 14, 19, and 23, by sc injection; 150 μg/construct/dose). **(A)** The mean clinical score ($n = 8$/group) representative for the degree of paralysis, was represented against time. A clinical score above 1.5 is associated with partial or complete paralysis of the hind-limbs. The T-cell reactivity against a major PLP-derived epitope included in the immunotherapeutic regimen was measured by ELISPOT analysis at d 60 after induction of disease **(B)**. On the *y* axis, we represented the frequency (per million splenocytes) of IFN-γ producing T cells, upon ex vivo peptide stimulation (the default production was nil). On the *x* axis, we represented the frequency of IL-4-producing T cell assessed in "default" conditions (incubation with media, without peptide). With diamonds, treated mice; other symbols, control mice. There was a correlation between expansion of cytokine-producing cells and protection from paralysis.

isolated from treated mice produced large amounts of IL-4 in a basal state but little IFN-γ, unless ex vivo stimulated with large concentrations of antigen.

In conclusion, Ig-mediated delivery of Th epitopes in vivo does not result in induction of Th1 immunity unless specific biasing adjuvants are used. Instead, depending on the ratio between the number of loaded APC to specific T cells, it results in anergy/tolerance in case of a high APC to T-cell ratio, or induction of Th2, Tr1, and Th3 responses when this particular ratio is more optimal. This creates the possibility of using Ig-mediated delivery of MHC-II-restricted epitopes for the immunotherapy of T1-controlled autoimmune diseases strongly associated with MHC alleles and specific antigens, such as type 1 diabetes and multiple sclerosis.

2.3. Induction of MHC Class I-Restricted Immunity by Targeted Delivery of Peptides

MHC-I-restricted T cells are instrumental in mediating antiviral and antitumoral immunity. In question is whether the incorporation of peptide epitopes within IgG vectors permits induction of MHC-I-restricted T cells. It was previously thought that for effective processing and presentation via MHC-I-restricted pathway, the epitope(s) should be produced inside the cell (e.g., subsequent to infection). During the last few years, however, accumulating evidence pointed to the possibility that in certain differentiation stages, APC can process and present exogenous nonreplicating antigens via the MHC-I-pathway *(38)*. From an evolutionary standpoint, this alternative pathway may have stemmed from the need to complement intracellular immunity leading to rapid apoptosis of infected cells, and/or counteract immune evasion mechanisms deployed by microbes that avoid infection of APC.

Initial animal testing of an immunoglobulin carrying a nucleoprotein (NP)-derived MHC-I-restricted peptide within CDR3 (IgNP) showed that mouse APCs were unable to process the construct and present the peptide ex vivo, in contrast to Ig bearing an MHC-II-restricted peptide *(39)*. Transfectomas expressing IgNP were able to process and present the NP-derived epitope *(40)*, which suggested that when fed as protein, IgNP does not access the same cellular compartments accessible in the case of intracellular production and required for MHC-I processing and presentation. In addition, adoptive transfer of IgNP-expressing transfectoma to naïve mice resulted in generation of MHC-I restricted cytotoxic cells *(41)* with similar results obtained recently in a different model *(42)*. The eventual interference of flanking residues with processing of the MHC-I-epitope was thus ruled out. Overall, the initial assessment was that Ig-mediated delivery of MHC-I-peptide does not result in immune response unless the protein is expressed within the APC, with implications toward gene-expression strategies.

There are two substantial drawbacks of these early experiments. First, the read-out used was conventional cytotoxicity. As more recent studies showed, there is a functional diversity at the level of Tc cells as well, with some of them capable of producing cytokines but largely devoid of conventional cytotoxicity *(43)*. Also, it may be that the ability of APC to process and present MHC-I-restricted peptides depends largely on the type and degree of activation of APC. Indeed, more recently, it was shown that in certain differentiation stages, professional APC acquire the ability to cross-prime $CD8^+$ T cells subsequent to effective processing of exogenous, nonreplicating antigen *(44)*. A re-evaluation of the possibility to induce processing and presentation of MHC-I-restricted T cells by targeting nonreplicating agents via FcγR resulted in a few interesting recent studies. One of them *(45)* demonstrated that immune complexes are able to induce effective cross-priming, in a manner dependent on the engagement of FcγRs bearing activatory $ITAM^+$ motifs. In addition, use of bispecific antibodies against an APC receptor and a tumor-associated antigen (Her-2) facilitated ex vivo cross-processing and activation of $CD8^+$ T cells *(46)*. Finally, in a more related system, engineered anti-FcγRI antibodies bearing MHC-I restricted tumor-associated epitopes within C_H segment were able to induce activation of specific T cells subsequent to ex vivo pulsing of APC (*see* Table 3) *(47)*.

One possibility is that cross-linking of FcγR or, in general, activating receptors on APC, triggers activation of APC to a stage compatible with cross-processing. Using Fc-directed targeting of an MHC-I-restricted, viral-derived epitope via a recombinant

Table 3
Particle Engineering for Passive and Active Immunization

Excipients	In vitro dissolution[a]	Passive immunization		Active immunization	
		Peak lung concentration	Anti-viral activity	BA cell targeting	Immune response
DPPC, Ca^{2+}, Lactose, Tyloxapol	Rapid (<5 min)	81 ± 22[b]	95×[c]	0.9 ± 0.8[d]	10–100×[e]
DPPC, Ca^{2+}, HES	Slow (>15 min)	38 ± 6	2.5×	5.1 ± 0.7	100×

[a]Stability of the particles in saline, reflected by matrix dissolution and release of actives.
[b]The concentration of active (IgG) in pulmonary interstitial tissue of rodents, subsequent to pulmonary delivery (μg/mL, mean ± SEM).
[c]Fold decrease of the pulmonary influenza virus titer subsequent to respiratory delivery of specific IgG, compared with control mock-treated animals.
[d]Percentage of cells from bronchoalveolar lavage (BA) that internalized IgG-loaded microparticles, subsequent to pulmonary aerosolization (mean ± SEM).
[e]Magnitude of anti-IgG immune response subsequent to aerosolization (fold increase, over nonformulated IgG).

IgG bearing the nonself peptide within CDR3, we showed that mere loading of APC with antigen via FcγR results in presentation of the peptide in context of MHC-I. The APCs, thus loaded with antigen ex vivo, were able to induce an IL-4 Tc2 response upon adoptive transfer, but not a full-blown Tc1 response usually associated with cytotoxicity (Fig. 6). This may explain the earlier, negative assessment of the immunogenicity of such constructs. Interestingly, additional activation of APC by using a TLR-ligand polyA:polyU (pA:pU) resulted in acquisition of ability of APC to prime IFN-γ Tc1 cells.

Together, these results suggest a previous underestimation of the ability of APC to present MHC-I-restricted peptides from the context of FcγR-targeted IgG. Thus, rather than controlling the presentation, by modulating the degree of activation of APC, one could influence the nature of the effector cells triggered subsequent to epitope recognition (Fig. 7). This has obvious practical implications for the use of IgG as vectors in vaccination and immunotherapy.

3. Targeted Delivery of Immunoglobulin-Engineered Microparticles

One of the most spectacular advances in the area of biotechnology during the last couple of decades was associated with microparticle engineering and delivery of small drugs and macromolecules. If molecular vectors such as IgG can shuttle peptides to APC, microparticles could be used to improve bioavailability, optimize the therapeutic index, or target certain drugs to specific cells. Among those drugs, peptide epitopes and, in general, immune modulators are of considerable interest for vaccination or immunotherapy of inflammatory diseases and cancer. From this perspective, microparticle formulation of peptide epitopes can be viewed as an alternative to the use of molecular vectors, circumventing the need to design, scale-up, and purify expensive recombinant

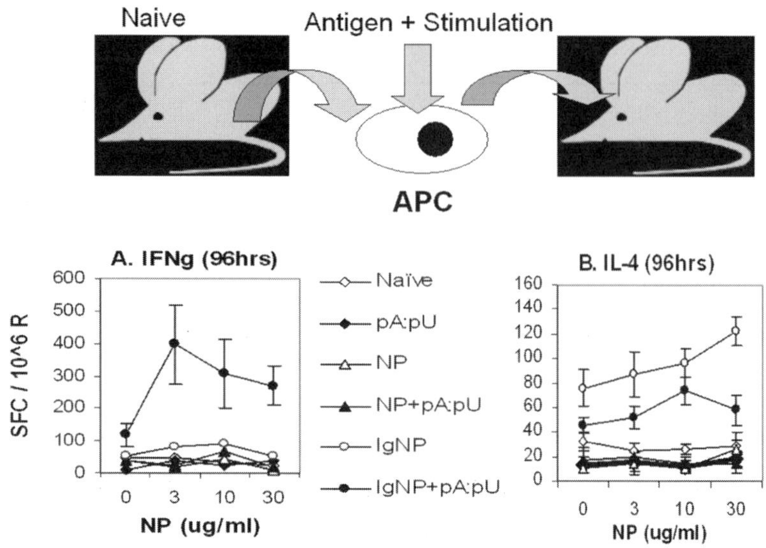

Fig. 6. Differential induction of Tc1 and Tc2 immunity to an MHC-I-restricted epitope delivered via recombinant IgG. The experimental protocol is represented in the top panel. APC were harvested from naïve mice and pulsed ex vivo with antigens (recombinant Ig carrying an influenza NP-derived epitope [IgNP]; or molar dose-matched peptide [NP]) plus or minus immune-stimulating motifs (dsRNA, pA:pU). Subsequently, the APC were adoptively transferred to naïve syngeneic mice and the T-cell response measured by ELISPOT analysis. The results were expressed as average frequency of SFC in spleen, per million responder cells: **(A)** IFN-γ^+ SFC and **(B)** IL-4^+ SFC for various concentrations of NP peptide.

molecules. However, there are limitations as well because most microparticle formulations are usually compatible only with mucosal and/or a limited number of parenteral routes (e.g., sc) owing to safety concerns or nonspecific scavenging by the reticuloendothelial system.

Microparticle formulations may optimize the immune modulatory properties of peptides or antigens by one or more of following mechanisms:

1. Creating a depot effect (slow-release) and by this, increasing the time of exposure of local APC to peptides;
2. Resulting supramolecular complexes (peptide-peptide or peptide-matrix-peptide aggregates) that are more effectively dealt with by APC;
3. Facilitating receptor-mediated internalization of ligand-engineered microparticles with their content, into specific cells; and
4. With addition of immune stimulating co-excipients, instruction of magnitude and profile of resulting immune response to the peptide or antigen co-formulated.

A substantial amount of work has been focused on designing biocompatible microformulations, such as liposomes, able to dock onto specific cells in a manner dependent on receptor–ligand interaction. Such vesicles, denominated immunoliposomes *(48)*, were originally designed as a lipid-based structure (mostly phosphatidylethanolamine) co-formulated with IgG monoclonal antibodies (MAbs) coupled to palmitic acid. Studies carried out two decades ago showed that such liposomes were able to dock onto antigen$^+$ but not antigen$^-$ target cells, opening an avenue in microparticle engi-

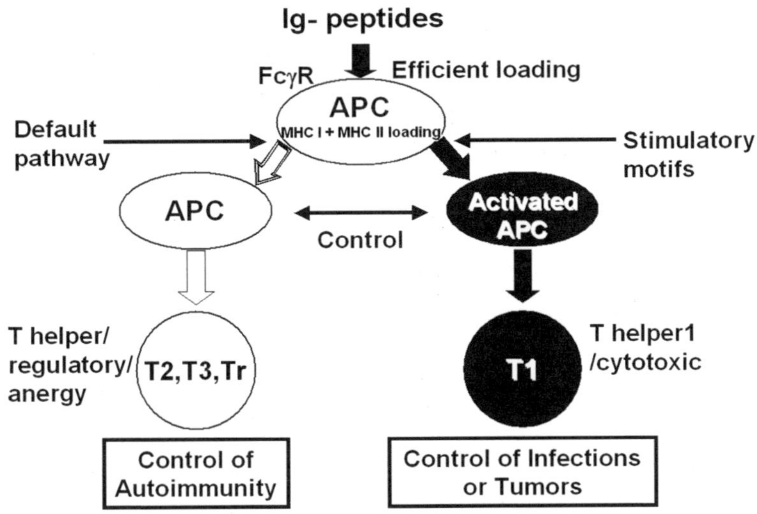

Fig. 7. In vivo manipulation of magnitude and quality of the T-cell response. Effective in vivo loading of APC by FcγR-targeted delivery of epitopes is compatible with both MHC-I and -II-restricted presentation. In the absence of immune-stimulating factors, however, the default pathway consists in generation of T2, T3, and Tr responses with potential application in the immunotherapy of autoimmune diseases. In contrast, when the effective loading of APC is paralleled by strong activation resulting in enhanced co-stimulation, the result is differentiation of T1 cells with potential applicability in viral infections and cancer.

neering and drug delivery. Subsequently, it has been shown that antibody-coupled immunoliposomes were effective in mediating specific intracellular delivery of cytotoxic drugs, with a major challenge: the potential inactivation of the drugs in the liposomes *(49)*. Nevertheless, these early observations stimulated more work in the area of immunoliposome-mediated delivery of cytotoxic drugs, with the overall aim of ameliorating the efficacy and safety profile *(50)*.

Interestingly, increasing the number of antibodies per liposome, although resulting in enhanced cellular internalization, diminished the specificity of the immunoliposome *(51)*. Efforts were made, in parallel, to optimize immunoliposome formulations by exploring alternative lipid excipients, carbohydrates as minor excipients and various antibody-anchoring strategies *(52)*. At the level of targeting molecules (the antibody), it was demonstrated that immunoglobulins recognizing foreign, virus-derived antigens on infected cells could be used to specifically target drugs to infected cells (e.g., the herpes simplex virus [HSV] system) *(53)*. More recently, with the understanding that a potential complicating interaction limiting the targeting efficiency of immunoliposomes occurs between the Fc portion of formulated antibodies and FcγR on reticuloendothelial cells, effort has been put into designing such particles with single-chain or antibody fragments devoid of Fc *(54,55)*. In contrast to the effort in designing immunoliposomes for delivery of therapeutic or diagnostic compounds to tumor cells expressing a wide variety of receptors (from folate receptors to Her-2/Neu), considerably less work was done in the field of using immunoliposomes for specific targeting of antigens or peptides. A notable exception is represented by studies involving targeting of M cells, critical APC involved in antigen translocation via mucosa *(56,57)*, although in such cases microbial derived compounds such as cholera toxin subunit B were used as targeting agents. One potential drawback of conventional liposomal technologies is the limited compatibility with mucosal and, in particular, respiratory delivery owing to their nature (suspensions in saline or liquid vehicle).

With the emergence of alternative technologies to create targeted particles, the opportunity to create targeted structures that could be delivered to or via the mucosal area increased considerably. Some efforts in this direction are summarized next.

3.1. The Challenge of Delivering Immunoglobulins by Solid Aerosols

There are two potential methods to target epitopes to mucosal-associated lymphoid cells by the airway route:

1. Delivery of molecules formulated within aerosols, capable of rapidly releasing the payload and allowing specific internalization into specialized cells; or
2. Delivery of metastable aerosols that interact in a receptor-ligand specific manner with target cells and allow the internalization of the payload.

Immunoglobulin delivery to or via the respiratory tract is appealing from the point of view of active or passive local or systemic immunotherapy. Recombinant immunoglobulins could carry disease-associated epitopes for active immuneprophylaxis or therapy, as described earlier. In contrast, passive immunization comprises delivery of specific antibodies, able to recognize microbial-associated antigens or mediators involved in disease processes.

If the aim is to deliver immunoglobulins in a fashion that maximizes the exposure of mucosal-associated lymphoid tissue or, in general, the pulmonary interstitial tissue, then airway delivery may be more effective than systemic administration (e.g., iv injection). In order for this to be accomplished, a formulation technology that preserves the functional structure of antibodies and results in optimal aerodynamic characteristics is warranted. Spray-dried lipid-based microparticles (SDM; major excipient dipalmitoylphosphatidylcholine, DPPC) containing human IgG could be formulated accordingly *(58)*. Bioactive IgG formulated in such particles (SDM), retain their functionality upon in vitro or in vivo release. There are multiple formulation challenges associated with delivery of large molecules such as immunoglobulins: (1) The stability of microparticles: The design needs to be more refractory to water. Inclusion of high molecular-weight excipients such as hydroxyethylstarch or of cationic divalent ions (Ca^{2+} or Mg^{2+}), facilitates this task, (2) The aerodynamic profile: the density of microparticles should be ideally limited. (3) The kinetics of immunoglobulin release: this factor is negatively impacted by making the particles more stable. A slower in vivo release is associated with more substantial expelling of loaded microparticles by mucociliary escalator, early after airway delivery. In addition, internalization by airway phagocytes is increased, owing to interaction of IgG associated with the matrix with the FcγR on cells. This limits the amount of immunoglobulin released within airways and able to penetrate into the pulmonary interstitial tissue. Thus, as shown in Fig. 8, there is a trade-off among the structural characteristics of microparticles, important for the delivery of immunoglobulins to or via the respiratory tract. Increasing the kinetics of release to minimize clearance of loaded particles implies making them less stable/aerodynamic and vise versa. Potential strategies to ameliorate this problem are: (1) use of a major excipient that is less prone to scavenging by bronchoalveolar macrophages (such as DPPC); or (2) use of strong surfactant detergents, approved for human use, such as tyloxapol, that promote particle dissolution but do not interfere

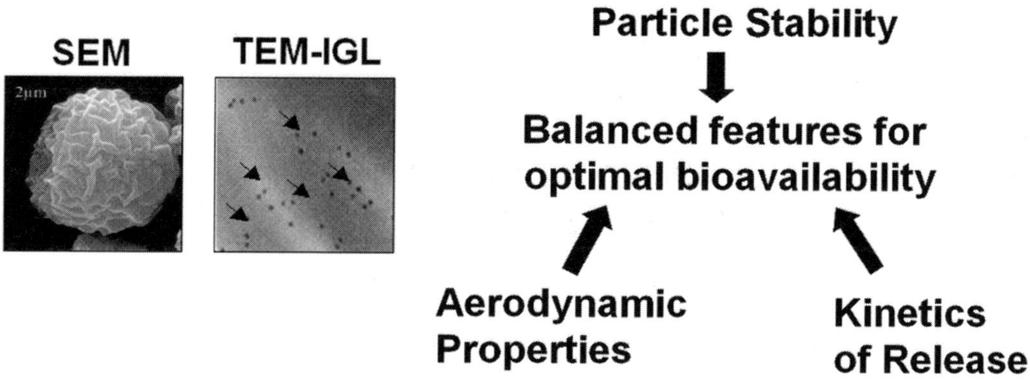

Fig. 8. Engineering microparticles for optimal delivery of macromolecules via the respiratory tract requires achieving a balance among different parameters. The electron micrographs depict spray-dried lipid microparticles (scanning and transmission electron micrographs [SEM, TEM]), loaded with IgG. The samples for TEM were prepared by immune gold-labeling with tagged anti-IgG reagents. In order to maximize the bioavailability of macromolecules delivered by microparticles, the following conditions must be achieved: (1) increased kinetics of release (which limits shelf stability owing to the association with increased hydrophilic nature); and (2) proper aerodynamic properties, allowing delivery to the bronchoalveolar region. Thus, a compromise must be achieved between stability of the particles and the other features required for optimal bioavailability, thus limiting the choice of excipients and engineering technologies that are in fact compatible with the task of microparticle-mediated delivery of macromolecules.

negatively with the stability and aerodynamic parameters. This last possibility may reduce the formation of protein-lipid-protein aggregates that, owing to their inherent immunogenicity, negatively interfere with the efficacy and safety profiles.

A parameter of interest in case of immunoglobulin delivery is the local vs systemic biodistribution. Results obtained in rodent models of aerosol delivery using particles optimized for aerosol delivery of immunoglobulins show that airway delivery achieves peak concentrations of Ig in the pulmonary interstitial tissue that are 50 times higher that those achieved by iv injection. In fact, less that 1% of the immunoglobulin delivered by injection localized in the lung tissue, compared with approx 40% of the immunoglobulin delivered via airways. Interestingly, although intra-airway depot effect is impossible to achieve with this generation of microparticles, the lung interstitial tissue itself facilitates slow-release into the systemic circulation. The peak concentration of Ig is achieved at 2–3 d after airway delivery, with approx 25% of the antibody delivered via airways being translocated slowly into the blood. Thus, the peak concentration of IgG achieved by airway administration is only 1% of that achieved by iv injection, but the slow release from the pulmonary tissue results in prolonged detectable levels. By subtraction, it results that approx 15% of the total amount of immunoglobulin delivered to the airways were cleared within the pulmonary interstitial tissue, where the mucosal lymphoid tissue is located. In addition, pulmonary interstitial tissue is drained via lymph mainly into mediastinal lymphoid nodes. Thus, in conclusion, airway delivery of immunoglobulin targets at least 10 times more efficiently the mucosal-associated lymphoid tissue of the respiratory tract, as compared with conventional intravascular administration (Fig. 9).

These results have implications for active and passive immunotherapy with immunoglobulins aimed for lung-associated molecular targets. There are complex factors influencing the bioavailability of such macromolecules that must be taken into account (some of them specific for Ig) beyond the delivery component. Optimized formulations, compatible with solid aerosol delivery, could achieve significant loading of pulmonary interstitial and lymphoid tissue in a manner that is practical in conditions of permissive airways. For example, if only 50% of a 10-mg dose of a dry-powder formulation containing 70% IgG reaches the bronchoalveolar area, this would result in approx one-half mg of immunoglobulin actually targeted to pulmonary-associated lymphoid tissue.

3.2. Receptor-Mediated Targeting of Spray-Dried Microparticles to APC

An alternative to using recombinant molecules to deliver disease-associated epitopes to APC is to employ microparticles loaded with antigens that are able to deliver the immune modulatory payload to specific cells. The difference is that the targeting agent and the active are brought together in a noncovalent fashion, via a biocompatible matrix. As described in detail earlier, the interest in immunoliposomes was channeled specifically in the area of targeted delivery of cytostatic or cytotoxic compounds to, with some exceptions, M cells, specialized APCs in the mucosal epithelial layer.

A significant practical improvement in microparticle targeting to the respiratory tract is represented by use of solid aerosols in the form of dry powder, obtained by spray drying. If formulated with certain lipids, such as DPPC or Di-Stearoyl Phosphatial Choline (DSPC), the dry particles quickly hydrate and transition into structures that may be reminiscent of liposomes in vivo *(59)*. Thus, APCs associated with the mucosal respiratory tract could be accessed in a more practical and effective fashion, as opposed to iv delivery or airway administration of liposomes in saline. There are three critical parameters that should be taken into account in this context: (1) the intrinsic immunologic properties of excipients that may impact the activity of innate and immune response, as well as the overall outcome; (2) the specific design of particles, to ensure receptor–ligand-mediated targeting; and (3) the function of the targeted APC in the induction/regulation of the immune response.

Microparticle excipients such as DPPC, which are poorly phagocytosed by macrophages and are part of the normal composition of the lung surfactant, are desirable because they allow for enhanced targeting specificity upon additional engineering of the vector, as well as a higher capability of manipulating the immune response via supplementation with immune modulating co-excipients. Even in conditions when the excipients are separately immunologically inert, formulation of antigens in solid aerosols modifies their immune properties. This is owing to the fact that during the formulation process (e.g., spray drying, emulsification, micronization, or lyophilization), there is a certain degree of conformational denaturation, combined with protein-matrix interaction. The immune system is handling microaggregates or nanoaggregates in a different fashion, as opposed to monomeric, dissolved antigen in saline. These aggregates are handled more effectively by specialized APC, with substantial impact on the T-cell response. Our studies using formulations of spray-dried antigens in an animal model demonstrated that even in the case of immune-inert excipients, the response to formulated antigens was elevated as represented by an MHC-II-restricted, T2-controlled increase inantibody titers *(60)*. When T cells were disabled (e.g., in MHC-II or CD3ε deficient animals), the increase in IgG1 response was obliterated. However, the APC were more able to present specific epitopes subsequent to in vivo administration

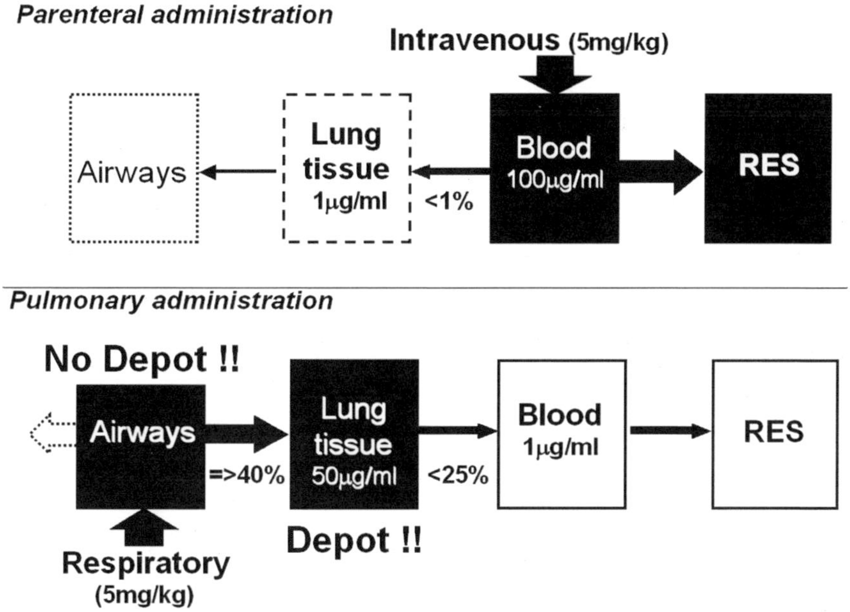

Fig. 9. Optimized technology to deliver IgG to the respiratory tract vs conventional parenteral delivery: differential biodistribution. Upper panel, iv administration of IgG in bolus; lower panel, IgG administration as dry powder (dose-matched). The biodistribution was assessed in rodents. Less than 1% of iv-injected IgG localized in the pulmonary interstitial tissue; most of it is internalized by the reticuloendothelial system (RES) including APC. In contrast, airway delivery of IgG by spray-dried microparticles results in approx 40% of immunoglobulin penetrating into the pulmonary interstitial tissue. There is no depot effect within the airway owing to active macrophage scavenging and mucociliary escalator expelling loaded microparticles. In contrast, there is a pronounced depot effect within the interstitial tissue. Less than 25% of IgG is slowly released into the systemic circulation (peak blood concentration achieved at 2–3 d). The peak IgG concentration in the lungs achieved by airway administration was roughly two orders of magnitude higher than the one resulting from iv injection. The opposite was true for the peak concentration in the blood.

of formulated antigen, supporting the concept that antigen handling is more effective. In addition, the engagement of B cells by formulated antigen was more effective, suggesting exposure of additional epitopes and/or multivalent interaction of protein aggregates with immunoglobulin receptors on lymphocytes (Fig. 10). Overall, in the absence of specific immune-stimulating adjuvants, the default immunity to formulated antigen consists of T2-dependent immunity while co-formulation with T1-driving agents, such as IL-12, reprograms the differentiation of lymphocytes in vivo *(60)*. Thus, mere formulation of antigen in spray-dried lipid microparticles, without ligand engineering, results in amplification of immunity to a certain extent and in a given direction.

Specific cell-targeting capabilities of spray-dried microparticles may further amplify the immune properties of antigens, in vivo. One limitation in this concern is the rapid reorganization of the dried-lipid excipient subsequent to interaction with the respiratory mucosa. During this process, ligand, antigens and soluble co-excipient could be lost and the targeting ability of the complex diminished. Consequently, it is clear that microparticles designed to achieve higher bioavailability by having more rapid release-kinet-

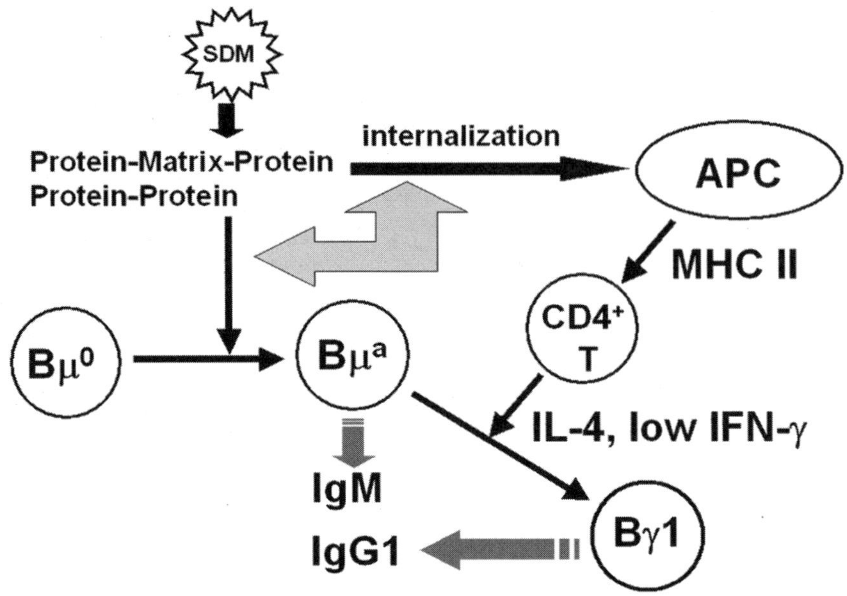

Fig. 10. Mechanisms responsible for the immunogenicity of spray-dried lipid microparticles loaded with antigenic material. Upon interaction with the mucosal environment, protein-protein, and protein-matrix-protein micro and nano-aggregates are released. Such aggregates have dual effects. First, they are more effectively internalized by APC, resulting in more effective processing and presentation of resulting epitopes in context of MHC-II molecules. Secondly, such aggregates interact with B-cells localized in the mucosal-associated respiratory tract, resulting in more effective surface receptor cross-linking and activation of IgM-producing B-cells (from resting *B*μ^0 to activated Bμ^a stage). The increased activation of CD4$^+$ T2 cells in the absence of immune-stimulating factors (T1-biasing adjuvants) promotes B-cell expansion and differentiation to IgG1-producing cells. This model based on studies carried out in wild-type and genetically manipulated mice, explains the increased immunogenicity of antigens formulated in spray-dried lipid microparticles.

ics profile, are not optimal for cellular targeting (*see* Table 3). In contrast, additional stabilization of the antigen-matrix-ligand structures facilitates a receptor-mediated delivery of the epitopes. The SDM-mediated technology is compatible with the design of microparticles fit for either one or the other of these tasks. As summarized in the Table 3, and previously reported *(58–61)*, more stable, compact SDM structures, incorporating chelating ions and high-molecular-weight carbohydrates, as opposed to low-molecular-weight carbohydrates and surfactant detergents, can facilitate receptor–ligand mediated co-internalization of the ligand, the active, and the matrix into phagocytic cells. Importantly, immunoglobulin-ligand-engineered particles mediated a more effective handling of the antigen (influenza virus-derived) by APC in vivo, with a beneficial impact on the overall immune response *(61)*. Strikingly, the APC isolated from the pulmonary interstitial tissue or by bronchoalveolar lavage, were able to stimulate peptide-specific T cells ex vivo *(61)*. Finally, targeted delivery of influenza virus antigens remained compatible with induction of antibody responses, suggesting that the increased Th response compensated for the more limited exposure of B cells to antigen.

The ability of the mucosal-associated APC, targeted with antigen, to prime T cells in vivo, is a prerequisite for such a delivery strategy to be successful. In contrast to separate molecules, microparticles cannot usually penetrate the airways into the submucosal space, thus at least two contrasting delivery pathways could be exploited: first, as described earlier, rapid-release particles facilitating penetration of individual molecules into the interstitial tissue; or alternatively, ligand-mediated targeting of the microparticles to airway-associated APC such as M cells or bronchoalveolar APC (macrophages or DC). We previously showed that airway APC isolated by lavage can be targeted by IgG-engineered spray-dried microparticles and subsequently display an ex vivo priming capability *(61)*. Combined with the increased immunity via such APC-targeted microparticles, our data suggest that airway APC may be involved in the priming of T-cell responses. Supported by other studies *(62,63)*, this may occur owing to retrograde trafficking of microparticle-loaded APC or exocytosis of processed antigens followed by more effective handling by APC associated with the lymphoid structures in the lung. It is difficult to reconcile, however, the previously described immunosuppressive role of bronchoalveolar macrophages *(64)* with the putative involvement in priming a T-cell response subsequent to microparticle-mediated targeting. Based on these published data, we propose the following, testable model with practical implications in the area of microparticle-mediated delivery of antigens, or immune-modulating peptides. In this model, the default function of airway APC is to maintain immune homeostasis by producing regulatory cytokines such as IL-10, TGF-β, or prostaglandins as well as scavenge immunologically inert particles or saprophytes. The fate of these APC (macrophages mostly) would be clearance via the mucociliary escalator. In contrast, exposure of such airway APC to infection-related agents, or "danger-motifs," abrogates their immune-suppressor capabilities, which may result in acquisition of the ability to traffic back into the interstitial tissue and further into T-cell dependent areas in the draining lymphoid organs, followed by priming of specific T cells. Thus, in this model, the airway APC are an intricate part of the active mucosal immunity, executing a dual role: maintenance of immune homeostasis and active surveillance/sampling of the extracorporeal environment. If this model proves to be valid, it will extend the known range of cellular immunity from intracorporeal to extracorporeal domain, with potential practical implications.

4. Conclusions

The area of APC targeting is rapidly evolving owing to the need to circumvent the poor PK profile of low molecular-weight peptide epitopes that are otherwise active in vitro. In addition, the intrinsic efficiency of immune-modulating peptides and antigens in general, may be enhanced by specific targeting to APC and particularly, subcellular compartments involved in antigen processing. Significant discoveries around the role of epitope-specific T cells in the pathogenesis of major inflammatory disorders, such as diabetes, MS, allergies, and asthma, further increased the interest in novel APC-targeting strategies. The initial interest in using APC targeting as a method to increase immunity to microbial and tumor-associated antigens expanded with the discovery that the resulting immune response could comprise MHC-I in addition to MHC-II-restricted T cells. Finally, a deep understanding of the functional organization of the APCs allows in vivo manipulation of the resulting T-cell profile irrespective of the targeting strategy (e.g., recombinant molecules or microparticles). Thus, APC targeting of peptides or antigens creates the potential for an enhanced impact on immune response with co-administered modulating agents directing the immune profile when necessary.

References

1. Mosmann TR, Cherwinski H, Bond MW, et al. Two types of murine helper T cell clone. I. Definition according to profiles of lymphokine activities ans secreted proteins. J Immunol 1986;136(7):2348–2357.
2. Sad S, Marcotte R, Mosmann TR. Cytokine-induced differentiation of precursor mouse CD8+ T cells into cytotoxic CD8+ T cells secreting Th1 or Th2 cytokines. Immunity 1995;2(3):271–279.
3. Asseman C, Powrie F. Interleukin 10 is a growth factor for a population of regulatory T cells. Gut 1998;42(2):157–158.
4. Gorelik L, Flavell RA. Transforming growth factor-beta in T-cell biology. Nat Rev Immunol 2002;2(1):46–53.
5. Bachmann MF, Kopf M. Balancing protective immunity and immunopathology. Curr Opin Immunol 2002;14(4):413–419.
6. Kisielow P, von Boehmer H. Negative and positive selection of immature thymocytes: timing and the role of the ligand for alpha beta T cell receptor. Semin Immunol 1990;2(1):35–44.
7. Janeway CA Jr. The immune system evolved to discriminate infectious nonself from noninfectious self. Immunol Today 1992;13(1):11–16.
8. Matzinger P. Tolerance, danger, and the extended family. Annu Rev Immunol 1994;12:991–1045.
9. Zinkernagel RM. Localization dose and time of antigens determine immune reactivity. Semin Immunol 2000;12(3):163–171; discussion 257–344.
10. Wurster AL, Tanaka T, Grusby MJ. The biology of Stat4 and Stat6. Oncogene 2000;19(21):2577–2784.
11. Medzhitov R, Janeway CA Jr. Decoding the patterns of self and nonself by the innate immune system. Science 2002;296(5566):298–300.
12. Matzinger P. The danger model: a renewed sense of self. Science 2002;296(5566):301–305.
13. Wang L, Smith D, Bot S, et al. Noncoding RNA danger motifs bridge innate and adaptive immunity and are potent adjuvants for vaccination. J Clin Invest 2002;110(8):1175–1184.
14. Mora AL, Tam JP. Controlled lipidation and encapsulation of peptides as a useful approach to mucosal immunizations. J Immunol 1998;161(7):3616–3623.
15. Finerty S, Stokes CR, Gruffydd-Jones TJ, et al. Targeted lymph node immunization can protect cats from a mucosal challenge with feline immunodeficiency virus. Vaccine 2001;20(1–2):49–58.
16. Bona CA, Casares S, Brumeanu TD. Towards development of T-cell vaccines. Immunol Today 1998;19(3):126–133.
17. Zaghouani H, Steinman R, Nonacs R, et al. Presentation of a viral T cell epitope expressed in the CDR3 region of a self immunoglobulin molecule. Science 1993;259(5092):224–227.
18. Lunde E, Bogen B, Sandlie I. Immunoglobulin as a vehicle for foreign antigenic peptides immunogenic to T cells. Mol Immunol 1997;34(16–17):1167–1176.
19. Zanetti M. Antigenized antibodies. Nature 1992;355(6359):476–477.
20. Zaghouani H, Kuzo Y, Kuzo H, et al. Engineered immunoglobulin molecules as vehicles for T cell epitopes. Int Rev Immunol 1993;10(2–3):265–278.
21. Brumeanu TD, Bot A, Bona CA, et al. Engineering of doubly antigenized immunoglobulins expressing T and B viral epitopes. Immunotechnology 1996;2(2):85–95.
22. Eidem JK, Rasmussen IB, Lunde E, et al. Recombinant antibodies as carrier proteins for subunit vaccines: influence of mode of fusion on protein production and T-cell activation. J Immunol Methods 2000;245(1–2):119–131.
23. Zambidis ET, Scott DW. Epitope-specific tolerance induction with an engineered immunoglobulin. Proc Natl Acad Sci USA 1996;93(10):5019–5024.
24. Brumeanu TD, Dehazya P, Wolf I, Bona CA. Enzymatically mediated, glycosidic conjugation of immunoglobulins with viral epitopes. J Immunol Methods 1995;183(2):185–197.

25. Brumeanu TD, Swiggard WJ, Steinman RM, et al. Efficient loading of identical viral peptide onto class II molecules by antigenized immunoglobulin and influenza virus. J Exp Med 1993;178(5):1795–1799.
26. Legge KL, Min B, Potter NT, Zaghouani H. Presentation of a T cell receptor antagonist peptide by immunoglobulins ablates activation of T cells by a synthetic peptide or proteins requiring endocytic processing. J Exp Med 1997;185(6):1043–1053.
27. Rasmussen IB, Lunde E, Michaelsen TE, et al. The principle of delivery of T cell epitopes to antigen-presenting cells applied to peptides from influenza virus, ovalbumin, and hen egg lysozyme: implications for peptide vaccination. Proc Natl Acad Sci USA 2001 Aug 28;98(18):10296–10301.
28. Bona CA, Bot A, Brumeanu TD. Immunogenicity of viral epitopes expressed on genetically and enzymatically engineered immunoglobulins. Chem Immunol 1997;65:179–206.
29. Lunde E, Munthe LA, Vabo A, et al. Antibodies engineered with IgD specificity efficiently deliver integrated T-cell epitopes for antigen presentation by B cells. Nat Biotechnol 1999;17(7):670–675.
30. Lunde E, Rasmussen IB, Eidem JK, et al. "Troy-bodies": antibodies as vector proteins for T cell epitopes. Biomol Eng 2001;18(3):109–116.
31. Zambidis ET, Barth RK, Scott DW. Both resting and activated B lymphocytes expressing engineered peptide-Ig molecules serve as highly efficient tolerogenic vehicles in immunocompetent adult recipients. J Immunol 1997;158(5):2174–2182.
32. Brumeanu TD, Casares S, Dehazya P, et al. Presentation of a viral peptide assembled on the carbohydrate moieties of immunoglobulin does not require processing. Eur J Immunol 1997;27(9):2408–2416.
33. Legge KL, Gregg RK, Maldonado-Lopez R, et al. On the role of dendritic cells in peripheral T cell tolerance and modulation of autoimmunity. J Exp Med 2002;196(2):217–227.
34. Min B, Legge KL, Pack C, Zaghouani H. Neonatal exposure to a self-peptide-immunoglobulin chimera circumvents the use of adjuvant and confers resistance to autoimmune disease by a novel mechanism involving interleukin 4 lymph node deviation and interferon gamma-mediated splenic anergy. J Exp Med 1998;188(11):2007–2017.
35. Legge KL, Min B, Cestra AE, et al. TCR agonist and antagonist exert in vivo cross-regulation when presented on Igs. J Immunol 1998;161(1):106–111.
36. Legge KL, Min B, Bell JJ, et al. Coupling of peripheral tolerance to endogenous interleukin 10 promotes effective modulation of myelin-activated T cells and ameliorates experimental allergic encephalomyelitis. J Exp Med 2000;191(12):2039–2052.
37. Lanoue A, Bona C, von Boehmer H, Sarukhan A. Conditions that induce tolerance in mature CD4+ T cells. J Exp Med 1997;185(3):405–414.
38. den Haan JM, Bevan MJ. Antigen presentation to CD8+ T cells: cross-priming in infectious diseases. Curr Opin Immunol 2001;13(4):437–441.
39. Zaghouani H, Kuzu Y, Kuzu H, et al. Contrasting efficacy of presentation by major histocompatibility complex class I and class II products when peptides are administered within a common protein carrier, self immunoglobulin. Eur J Immunol 1993;23(11):2746–2750.
40. Zaghouani H, Krystal M, Kuzu H, et al. Cells expressing an H chain Ig gene carrying a viral T cell epitope are lysed by specific cytolytic T cells. J Immunol 1992;148(11):3604–3609.
41. Kuzu Y, Kuzu H, Zaghouani H, Bona C. Priming of cytotoxic T lymphocytes at various stages of ontogeny with transfectoma cells expressing a chimeric Ig heavy chain gene bearing an influenza virus nucleoprotein peptide. Int Immunol 1993;5(10):1301–1307.
42. Billetta R, Filaci G, Zanetti M. Major histocompatibility complex class I-restricted presentation of influenza virus nucleoprotein peptide by B lymphoma cells harboring an antibody gene antigenized with the virus peptide. Eur J Immunol 1995;25(3):776–783.
43. Mosmann TR, Li L, Sad S. Functions of CD8 T-cell subsets secreting different cytokine patterns. Semin Immunol 1997;9(2):87–92.
44. Heath WR, Carbone FR. Cross-presentation, dendritic cells, tolerance and immunity. Annu Rev Immunol 2001;19:47–64.

45. Clynes R, Maizes JS, Guinamard R, et al. Modulation of immune complex-induced inflammation in vivo by the coordinate expression of activation and inhibitory Fc receptors. J Exp Med 1999;189(1):179–185.
46. Wallace PK, Kaufman PA, Lewis LD, et al. Bispecific antibody-targeted phagocytosis of HER-2/new expressing tumor cells by myeloid cells activated in vivo. J Immunol Methods 2001;248(1–2):167–182.
47. Guyre CA, Barreda ME, Swink SL, Fanger MW. Colocalization of Fc gamma RI-targeted antigen with class I MHC: implications for antigen processing. J Immunol 2001;166(4):2469–2478.
48. Huang A, Kennel SJ, Huang L. Immunoliposome labeling: a sensitive and specific method for cell surface labeling. J Immunol Methods 1981;46(2):141–151.
49. Huang A, Kennel SJ, Huang L. Interactions of immunoliposomes with target cells. J Biol Chem 1983;258(22):14034–14040.
50. Allen TM, Ahmad I, Lopes de Menezes DE, Moase EH. Immunoliposome-mediated targeting of anti-cancer drugs in vivo. Biochem Soc Trans 199523(4):1073–1079.
51. Houck KS, Huang L. The role of multivalency in antibody mediated liposome targeting. Biochem Biophys Res Commun 1987;145(3):1205–1210.
52. Sunamoto J, Sato T, Hirota M, et al. A newly developed immunoliposome—an egg phosphatidylcholine liposome coated with pullulan bearing both a cholesterol moiety and an IgMs fragment. Biochim Biophys Acta 1987;898(3):323–330.
53. Norley SG, Huang L, Rouse BT. Targeting of drug loaded immunoliposomes to herpes simplex virus infected corneal cells: an effective means of inhibiting virus replication in vitro. J Immunol 1986136(2):681–685.
54. Laukkanen ML, Alfthan K, Keinanen K. Functional immunoliposomes harboring a biosynthetically lipid-tagged single-chain antibody. Biochemistry 1994;33(38):11664–11670.
55. Nielsen UB, Kirpotin DB, Pickering EM, et al. Therapeutic efficacy of anti-ErbB2 immunoliposomes targeted by a phage antibody selected for cellular endocytosis. Biochim Biophys Acta 2002;1591(1–3):109–118.
56. Frey A, Giannasca KT, Weltzin R, et al. Role of the glycocalyx in regulating access of microparticles to apical plasma membranes of intestinal epithelial cells: implications for microbial attachment and oral vaccine targeting. J Exp Med 1996;184(3):1045–1059.
57. Lian T, Ho RJ. Cholera toxin B-mediated targeting of lipid vesicles containing ganglioside GM1 to mucosal epithelial cells. Pharm Res 1997;14(10):1309–1315.
58. Bot AI, Tarara TE, Smith DJ, et al. Novel lipid-based hollow-porous microparticles as a platform for immunoglobulin delivery to the respiratory tract. Pharm Res 2000;17(3):275–283.
59. Dellamary LA, Tarara TE, Smith DJ, et al. Hollow porous particles in metered dose inhalers. Pharm Res 2000;17(2):168–174.
60. Wang L, Bot S, Smith D, et al. Immunological properties of spray-dried lipid microparticles developed for nasal and systemic administration of vaccines. Pharma Res 2002;12(1):53–61.
61. Bot AI, Smith DJ, Bot S, et al. Receptor-mediated targeting of spray-dried lipid particles coformulated with immunoglobulin and loaded with a prototype vaccine. Pharm Res 2001;18(7):971–979.
62. Corry D, Kulkarni P, Lipscomb MF. The migration of bronchoalveolar macrophages into hilar lymph nodes. Am J Pathol 1984;115(3):321–328.
63. Havenith CE, van Miert PP, Breedijk AJ, et al. Migration of dendritic cells into the draining lymph nodes of the lung after intratracheal instillation. Am J Respir Cell Mol Biol 1993;9(5):484–488.
64. Lyons CR, Ball EJ, Toews GB, et al. Inability of human alveolar macrophages to stimulate resting T cells correlates with decreased antigen specific T cell-macrophage binding. J Immunol 1986;137(4):1173–1180.

16

Antibody-Mediated Drug Delivery in Cancer Therapy

Zhenping Zhu and Daniel J. Hicklin

1. Introduction

The lack of specificity of currently available chemo- and radiotherapeutic agents constitutes a major obstacle to the treatment of cancer. The difference between malignant and normal cells in regards to their sensitivity to cytotoxic therapies is, in most cases, not sufficient to allow potentially curative doses of drugs or radiation to be administered without unacceptable toxicity to normal cells. Thus, the administration of therapeutic doses of these agents during the treatment of cancer patients also damages rapidly proliferating cells of host tissues such as hematopoietic cells, hair follicles, and the epithelial lining of the gastrointestinal tract. It is generally accepted that the usefulness of many cytotoxic therapies would be enhanced if these agents were rendered more tumor-selective.

Targeted therapy of cancer is based on the use of specific carriers to deliver cytotoxic agents, including chemotherapeutic drugs, radioisotopes, or toxins, to their preferred site of action (i.e., tumors). For targeted delivery of cytotoxic agents, it is important to select a carrier that can be delivered selectively to tumor cells. The drug or radioisotope can be attached to this carrier by a number of synthetic or biochemical means to form a tumor-selective drug conjugate. Administration of such conjugate should lead to the accumulation of drug or radioisotope preferentially in the tumor without significant distribution to normal tissues, followed by selective damage to the tumor cells. Alternatively, the drug conjugate may be selectively "activated" in tumor tissue. Several classes of specific carriers have been evaluated for the selective delivery of drugs or radioisotopes to tumors. Such specific carriers include antibodies *(1–4)*, cytokines such as interleukin-2 (IL-2) and granulocyte-macrophage colony-stimulating factor (GM-CSF) *(5,6)*, growth factors such as epidermal growth factor (EGF) and vascular endothelial growth factor (VEGF) *(7–9)*, and hormones such as gonadotropin-releasing hormone *(10,11)*. The use of antibodies as carriers of cytotoxic drugs is particularly attractive because of their unique specificity and high affinity for tumor antigens. In this chapter we will review the various classes of immunoconjugates developed for delivery of drugs to tumor cells.

From: *Cellular Drug Delivery: Principles and Practice*
Edited by: D. R. Lu and S. Øie © Humana Press Inc., Totowa, NJ

2. Antibodies As Tumor-Targeting Agents

2.1. History of Antibody-Based Tumor Targeting

At the beginning of the 20th century, Paul Ehrlich proposed the use of polyclonal antitumor antibodies bound to diphtheria toxin as a "magic bullet" against malignant cells *(12)*. However, the use of polyclonal antibodies (PAbs) for targeted delivery at that time was limited by their heterogeneity with respect to size, charge, antigen specificity, affinity, low immunoreactive fraction, and the contamination of unwanted immunoglobulins (Ig). Furthermore, the production of highly specific polyclonal antitumor antisera/antibodies is difficult and unreliable; different lots of antisera may have various specificity and affinity, and quantity is usually limited. Despite these limitations, several investigators pursued the use of polyclonal antitumor antibodies as specific carriers for drug delivery in cancer therapy. Early studies in experimental animal models involved the use of a few available anticancer drugs available at the time that lend themselves to simple covalent coupling to antibodies. The first report on the use of an antibody-drug conjugate was published in 1958, when Mathe et al. linked methotrexate by diazotization to antibodies raised against murine L1210 leukemia cells, and used the conjugate in the successful treatment of L1210-bearing DBA/2 mice *(13)*. Although the result appeared to be promising, no serious efforts to extend this therapeutic modality to patients were made until 1967, when Ghose and colleagues demonstrated that therapeutic amounts of radioactive ^{131}I could be linked to a PAb, with retention of antibody activity, and that the radiolabeled antibody could eradicate experimental tumors in vivo *(14,15)*. In 1972, the same group coupled chlorambucil to polyclonal antitumor antibodies and used the preparation to treat melanoma patients in the first human study with antitumor antibodies *(16–20)*. Thirteen patients with inoperable recurrent malignant melanoma were treated with chlorambucil bound to polyclonal antitumor antibodies raised in rabbits and goats, and compared to a control arm of 11 patients treated with conventional chemotherapy using dimethyltriazenoimidazole carboxamide. Objective responses were achieved in 2 patients and 5 other patients showed disease stabilization in the antibody/chlorambucil-treated group, whereas all 11 patients in the control arm showed disease progression. The median survival of the responders and the stabilizers were 20 mo, compared to that of 3 mo for nonresponders in the antibody/chlorambucil group and patients in the control group. During the same period, Ghose's group also demonstrated that radiolabeled antitumor antibodies could localize preferentially not only in tumors in mouse models (both syngeneic and xenografted human tumors), but also in tumors in patients *(21–23)*. More impressively, administration of a radiolabeled antibody led to cure of lymphoma-inoculated mice *(24)*. For excellent reviews on the early history of antibody-based targeted therapy of cancer, *see* refs. *1* and *23*.

In 1976, Kohler and Milstein, by employing a method of somatic hybridization, successfully generated "hybridoma" cell lines producing monoclonal antibodies (MAbs) of defined specificities *(25)*. The principle advantages of mAb over the conventional PAbs are obvious, including the defined specificity, homogeneity, and availability of MAbs in practically unlimited quantities. These properties of MAbs render them as the most attractive carriers for the selective delivery of therapeutic agents to malignant tumors. To date, numerous MAbs have been produced against virtually every malignant tumor of human tissues. Many of these MAbs have been used as tumor-specific carriers of cytotoxic agents and evaluated either in animal models and/or patients.

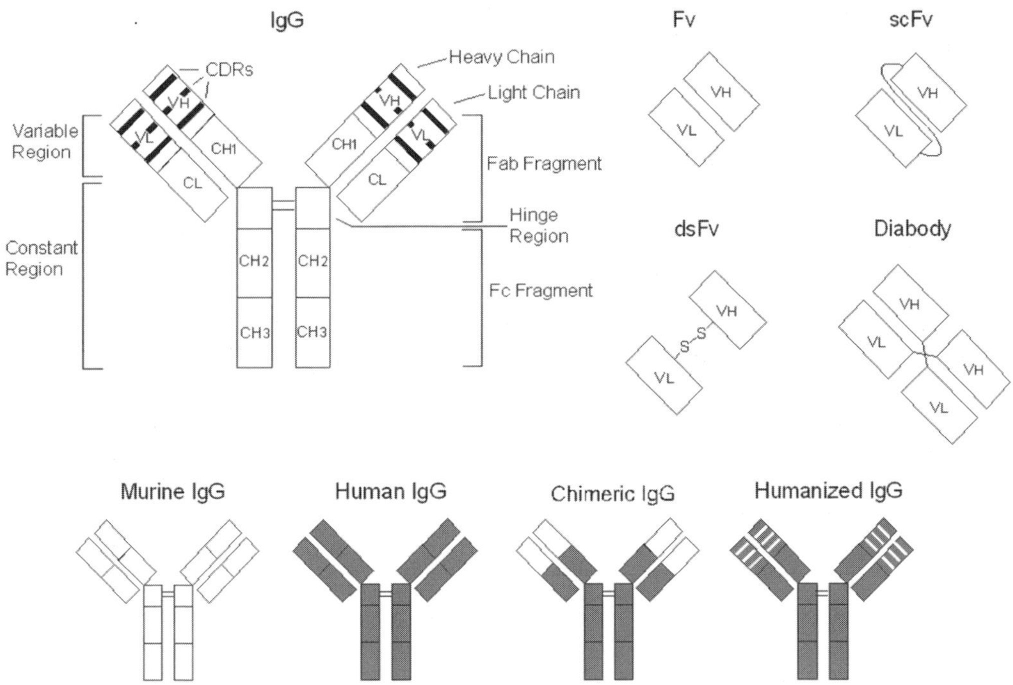

Fig. 1 Schematic representation of an IgG antibody and its fragments. Top panel, Each IgG molecules consists of four polypeptide chains: two identical light and two identical heavy chains paired together by interchain disulfide bonds. The Y-shape IgG contains two Fab (antigen-binding) fragments and a Fc (crystalline) fragment linked together via the hinge region. The smallest module of an antibody required for specific binding is Fv fragment comprising only the variable domains of light (V_L) and heavy (V_H) chains. Each V_L and V_H domains contain three hypervariable regions called complementarity-determining regions (CDRs) that form the antigen-binding surface and are responsible for antibody specificity and diversity. Introduction of a polypeptide linker between the V_L and the V_H domains (in either orientation) results in the formation of a scFv or diabody (or triabody or tetrabody) depending on the length of the linker. Alternatively, a disulfide bond can be introduced into the interface between the V_L and the V_H domain to form a disulfide bond-stabilized Fv (dsFv). Bottom panel, MAbs produced from the traditional hybridoma technique are generally of murine origin. Chimeric antibody is generated by joining the V_L and the V_H domains of a murine MAb to human constant domains: mouse V_L to human C_L and mouse V_H to human C_H1-hinge-C_H2-C_H3, respectively. In antibody humanization, only the CDRs of the murine MAb, along with one to several other mouse residues determined to be critical in maintaining the antibody affinity, are grafted into a human framework. Fully human antibodies can be routinely obtained nowadays with the availability of human antibody phage-display library and human Ig transgenic mouse. (Note: all drawings are not to scale.)

2.2. Antibody Structure and Engineering

The majority of antibodies that are used for targeted therapy fall into the IgG class of immunoglobulins. IgG is a tetrameric glycoprotein, consisting of four polypeptide chains: two identical light chains and two identical heavy chains. Each IgG contains two antigen-binding fragments (Fab) and an Fc domain joined together by a hinge region (Fig. 1). The Fab fragment is responsible for specific antigen binding, whereas the Fc domain binds to the Fc receptors on effector cells, fixes complement, and elicits other in vivo biological responses. The variable light (VL) region and the variable heavy (VH) region within the Fab fragment directly contact antigen and are responsible for the unique specificity and diversity of antibodies. Within each VL and VH, there exist three hypervariable regions, also called complementarity determining regions (CDRs), that form the binding surface that contacts the antigen. The smallest functional module of an antibody required for specific antigen binding is the Fv fragment. The Fv fragment is a heterodimer of VL and VH domains; the fragment is usually unstable in solution, because the two domains are noncovalently linked. The instability of Fv fragments was overcome by the invention of single-chain Fv (scFv) in which the VL and the VH domains are connected via a peptide linker *(26)*. The scFv fragment in solution exists mostly as monomer, when the linker is 12–15 or more amino acids, or as a dimer (as so-called "diabody") when the linker is between 5 and 12 amino acids *(27,28)*. Interestingly, several groups reported recently that the scFv mostly form trimers ("triabody") and/or tetramers ("tetrabody") when the VL and VH domains are fused together with a linker of 0–2 amino acids (for review, *see* ref. *29*). An alternative approach to increasing the stability of an Fv fragment is to introduce an interchain disulfide bond between the interface of the VL and the VH domains (for review, *see* ref. *30*). The residues in the interface to be mutated into cysteine, ideally locate outside the CDR regions, were identified with the aid of computer molecular modeling *(31,32)*. Several disulfide bond-stabilized Fv fragments have been produced with increased stability and/or antigen-binding affinity *(33,34)*.

To date, the majority of hybridoma-derived antitumor MAbs are of murine origin. These antibodies are immunogenic in humans and elicit a human anti-murine antibody (HAMA) response. The HAMA can form immuncomplexes with subsequent administrated therapeutic antibody, leading to increased clearance accompanied by decreased tumor localization of the antibody, and in some case, serious side effects such as an allergic anaphylactic reaction. Smaller antibody fragments, such as Fab, Fv, and scFv, are usually less immunogenic than the intact IgG in human owing to the lack of the Fc domain. Other approaches attempting to reduce the immunogenicity of rodent-derived antibodies include chemical modification of the antibodies such as conjugation of the antibodies to polyethylene glycol (PEG) or oxidized dextran, and co-administration to patients of immunosuppressive agents such as cyclosporin A, cyclophosphamide, and steroids. Recent advancement in antibody-engineering technologies has not only enabled the ability to tailor-make antibody molecules with predefined characteristics such as size, valency, and multispecificities to suit the intended applications, but also led to the production of chimeric and humanized antibodies with greatly reduced immunogenicity. Chimeric antibodies were the first generation of this genetic-engineering approach where the variable domains (both VL and VH) of a murine MAb were cloned and fused to the constant domains of a human IgG to create a new hybrid IgG molecule that retains the original antigen binding affinity and specificity, but only contains, in theory, one-third of murine amino acid sequences *(35)* (Fig. 1). Humanization of antibodies takes this approach one step further by genetically grafting only the CDRs of the murine antibody, along with a few murine residues outside the CDRs believed to be important for antigen-binding affinity, into a human IgG framework (Fig. 1). These residues are identified with the help of computer-based modeling and site-directed mutagenesis because of their critical roles in maintaining the correct conformation of the antigen-binding surface and/or direct contacting the antigen *(36)*. Approximately

90–95% of the amino acid sequence of a humanized antibody is of human origin. Since the mid-1990s, with the advancement in human antibody phage-display libraries *(37,38)* and human Ig transgenic mice *(39,40)*, fully human antibodies with desired specificities can now be readily isolated. This new generation of antibody therapeutics, including chimeric, humanized, and fully human antibodies, has proved to be much less immunogenic in multiple clinical studies (for review, *see* ref. *41–45*).

2.3. Classes of MAb Therapy and Mechanisms of Action

Three major classes of MAbs have been developed as cancer therapeutics: (1) antibodies that act as molecular antagonists that modulate the function of key regulatory molecules on tumor cells, such as blocking growth factor/receptor interaction and/or downregulating expression of oncogenic proteins (or receptors) on the cell surface; (2) antibodies that recruit effector mechanisms of the immune system, such as the antibody-dependent cellular cytotoxicity (ADCC) and complement-mediated cytotoxicity (CMC), and (3) antibodies used as targeting devices (immunoconjugates) to specifically deliver cytotoxic agents to tumor sites. Functional blockade is thought be one of the main antitumor mechanisms for several antibodies, including those directed against EGF receptor (also called HER1) and HER2 (*erb*B2/neu) on tumor-cell surface *(46,47)*, and receptors for VEGF on endothelial-cell surface *(48)*. By interfering with important growth factor/receptor pathways, these antibodies can influence the growth and survival of tumor cells. In addition, antibodies that inhibit function of regulatory pathways may potentiate the cytotoxic effects of chemotherapeutic drugs and radiation. For example, a number of studies have demonstrated that both Rituxan™ and Herceptin™ could significantly enhance the therapeutic efficacy of various cytotoxic agents in both laboratory and clinical settings *(49,50)*. There is evidence to suggest that immune-effector mechanisms may also play an important role in the clinical antitumor efficacy of the antilymphoma antibody Rituxan™ and the antibreast-cancer antibody Herceptin™. For example, both Rituxan™ and Herceptin™ have been shown to mediate significant levels of ADCC and CMC effects on a number of malignant cell lines *(51,52)*. Further, the antitumor effect of these two MAbs in vivo was severely diminished in mice with deficiency of Fc receptor on its effector-cell surface or when mutants of the MAb with reduced Fc-binding efficiency were used *(53)*.

A large number of MAbs specific for a diverse set of tumor targets have been utiltized for the development of immunoconjugates. Several types of cytotoxic agents, including toxins, chemotherapeutic agents and radioisotopes, have been conjugated to antitumor antibodies and tested preclinically and in various clinical studies. For example, clinical trials have been performed using immunotoxins (ITs) in patients with carcinomas of breast, ovary, and colon, as well as several lymphomas and leukemia (for reviews, *see* refs. *58–60*). An anti-CD33-calicheamicin conjugate, Mylotarg™, has gained FDA approval for the treatment of acute myeloid leukemia (AML) in patients over 60 yr of age *(61)*. In addition, a ^{90}Y-labeled anti-CD20 antibody (Zevalin™) was approved by the US Food and Drug Administration (FDA) for non-Hodgkin's lymphoma (NHL) *(62)*, and a ^{131}I-labeled anti-CD20 antibody (Bexxar™) has been tested in multiple Phase III clinical trials *(63)* and is waiting for FDA approval for the treatment of B-cell malignancies.

Other approaches that utilize MAbs in cancer therapy but do not fit into the classes outlined earlier include anti-idiotypic MAbs that mimic tumor antigens to stimulate the anti-idiotype network to generate antitumor anti-anti-idiotypic antibody response *(54)*;

catalytic antibodies that function as catalysts to induce hydrolysis of cell membrane or proteins, or to convert molecular oxygen into hydrogen peroxide to achieve cell killing *(55)*, antibodies that enhance patient's immune response to tumors by stimulating cytotoxic T lymphocytes via CD40 *(56)* or by antagonizing endogenous immune inhibitory factors such as CTLA-4 *(57)*.

2.4. Selection Criteria for Antibodies As Immunoconjugates

One of the most important aspects of MAb-based targeting is the selection of an appropriate antibody to be used as the carrier of the cytotoxic agent. The choice of an ideal MAb depends on characteristics of the target antigen, various requirements for the conjugation method, and the mechanism of action for a cytotoxic agent. The choice of MAb also may be dependent on the intended use of the immunoconjugate. For example, for radioimmunoimaging the radiolabeled antibody should specifically and rapidly localize in tumor tissue to create a high tumor/nontumor ratio that is required for ideal imaging soon after the administration of the radioimmunoconjugate. It is also desirable that the radiolabeled antibody be cleared rapidly from the body once the tumor images have been obtained. On the other hand, for targeted antitumor therapy, in addition to its specific localization, the immunoconjugate should remain in the tumor for an adequate period of time to allow the MAb-linked cytotoxic agent to exert its effect. The nature of the cytotoxic moiety used in the conjugate and its mode of action also has significant impact on the selection of antibody carrier. For example, a radioisotope decaying by beta emission can kill cells within an area of several diameters of the targeted tumor cells, whereas most chemotherapeutic agents and toxins require internalization to exert their cytotoxic effect. Thus, a MAb that is ideal for use as a carrier for delivering radioisotopes may not necessarily be a good candidate for the delivery of chemotherapeutic agents or toxins. Several other factors that deserve careful consideration in selecting MAbs as effective carriers for tumor targeting are briefly discussed next.

2.4.1. Intact Antibody vs Antibody Fragments

The choice of intact antibody or its fragments mostly depends on the intended use of the conjugate. An intact antibody possesses a longer circulation half-life (up to 1–3 wk in humans as demonstrated by several humanized or human antibodies), and is able to elicit effector mechanisms, including ADCC and CMC, that may contribute to the overall antitumor activity of the conjugate. For the purpose of delivering cytotoxic agents, most investigators prefer the intact MAb because, compared to its fragments, intact MAb produces a higher percentage of injected conjugate that localizes in the tumor *(64–67)*. Further, a longer residence time of the intact MAb in tumor may be beneficial for the targeted drug or radioisotope to exert its cytotoxic effects *(64–67)*. In contrast, smaller antibody fragments such as scFv and Fab have a much shorter half-life and these fragments are excreted quickly through the kidney. It is believed that these antibody fragments are, in general, better candidates than intact MAbs for tumor imaging because they penetrate tumor faster and more homogeneously, clear faster from the circulation, and accumulate less in organs of the reticular endothelial systems (such as liver and spleen) *(64–67)*. Another advantage of using antibody fragments is that they are usually less immunogenic than the intact MAb owing to the lack of the Fc domain.

2.4.2. Antibody Specificity

The importance of MAb specificity is obvious because it allows the antibody or its immunoconjugate to bind the target antigen selectively. Although high specificity of

MAbs is universally recognized as an advantageous attribute for MAb-based drug targeting, there are only a handful of antigens that are truly considered tumor-specific, including clone-specific idiotypic Ig on the surface of malignant B cells and T-cell antigen-receptor protein on malignant T cells. Most antitumor antibodies studied to date are, in fact, antibodies directed against tumor-associated antigens (TAA). These antigens are not strictly tumor-specific but they have restricted distribution in normal tissues. Some examples of TAA include carcinoembryonic antigen (CEA), mucin-associated antigen (MUC1), prostate specific membrane antigen (PSMA), HER2/neu, tyrosinase, gp75, and gp100. In most cases, the differential expression of TAA between tumors and normal tissues allows sufficient selectivity for cytotoxic targeting. Sometimes normal organ-specific antigens, like prostate-specific antigen (PSA), or lineage-specific antigen like CD20 and CD19 on B-cells, can also be used for tumor targeting, because damage to these cells or tissues is not life-threatening.

2.4.3. Antibody-Binding Affinity

It is generally accepted that to achieve effective targeting, the affinity of a MAb needs to be of 10^8 M^{-1} or better *(68)*. Although some investigators presume that high-affinity antibody is desirable *(69)*, there is considerable debate whether improving MAb affinity beyond 10^9 M^{-1} would result in a further increase in its tumor localization. Recently, Adams et al. employed several anti-HER2 antibodies of varying affinities derived from the same parent antibody by side-directed mutations, and showed a positive correlation between the amount of antibody localized in tumor and antibody affinity up to 10^9 M^{-1} *(70)*. Further increase in affinity beyond 10^9 M^{-1} did not, however, result in higher tumor localization of the antibody *(71)*. Histological study of the tumor specimens demonstrated that antibodies with moderate affinity (between 10^8 and 10^9 M^{-1}) penetrated deeper in to tumors and percolated more homogeneously into tumor mass than antibodies with high affinity ($>10^{10}$ M^{-1}), which were trapped around the perivascular tumor cells with much less localization in the distant tumor tissues *(71)*.

2.4.4. Antigen Expression

For antibody-based drug targeting, the antigen should be expressed on the surface of target cells and must be accessible to the administered antibody or its conjugate. Ideally, the target antigen should be present only on tumor cells, and on all tumor cells. Unfortunately, most TAA are expressed in a heterogeneous manner and only a proportion of tumor cells express the antigen. This is not a critical limitation for radioimmunotherapy, especially when beta emitters were used; the radiation can effectively reach and therefore damage cells within an area of several diameters of the targeted tumor cells ("bystander effect"). It is also important that the tumor antigen be expressed on the target cells at high density. The usefulness of targeting would be greatly compromised if the antigen were expressed on the cell surface at a very low level that would not permit the binding of adequate amounts of conjugate needed to achieve a therapeutic response.

An emerging new concept to overcome the heterogeneity of tumor antigen expression is to use antibodies directed against angiogenic markers that are selectively upregulated in tumor vasculature, such as receptors for VEGF *(72–74)*, integrins *(75)*, VE-cadherin *(76)*, and fibronectin isoforms *(77,78)*. Because angiogenesis is required for both tumor growth beyond certain size and metastasis, local damage to tumor vasculature may detrimentally affect all tumor cells that are dependent on the targeted vasculature for nutrients, irrespective of the antigen-expression profile of individual tumor cells. Anti-angiogenic antibody-based therapies should, therefore, be applicable to all types of cancers (for reviews, *see* refs. *48,73,74,77,78*). However, it is yet to be established whether true "tumor vasculature-specific" targets can be identified and whether these targets can be exploited for immunoconjugates approaches to cancer therapy.

2.4.5. Antigen Modulation

Ideally, the target antigen should not be secreted or shed from the tumor cell surface into tumor stroma and/or the circulation in amounts that could affect binding of the carrier antibody to tumor cells. Soluble antigen may form immune complexes with administered antibody or its conjugate, leading to rapid clearance from the circulation and thus impeding delivery of the antibody or its conjugate to tumor. The rate of antigen internalization after antibody binding is another critical factor that may influence the effectiveness of antibody-based drug targeting. Because most currently used chemotherapeutic agents and toxins exert their cytotoxic effect on intracellular targets, the ideal antibody conjugate should be rapidly internalized (endocytosed) by the tumor cells and release active form of these cytotoxic agents inside the cells. Thus, antibodies specific for antigens that trigger efficient internalization upon binding may represent more effective carriers for intracellular drug delivery. On the other hand, internalization is not required for the cytotoxic effects of (β and γ) radiation and agents that act through cell membrane such as cytolysis enzymes. In fact, antibodies with high internalization rate may be poorly retained by the tumor cells after initial binding because of the intracellular degradation and the efflux of lower molecular-weight metabolites following the rapid endocytosis process. This has been shown to be disadvantageous for radioimmunotherapy with ^{131}I-labeled antibodies; the degradation of endocytosed radioiodinated antibody, including dehalogenation and efflux of the catabolic small fragments, resulted in rapid loss of radioactivity from the tumors *(79)*. Novel protein radioiodination techniques have been developed to alleviate the loss of radioactivity owing to dehalogenation *(80,81)*. Alternatively, radiometal isotopes, such as ^{90}Y and ^{111}In, may prove beneficial for targeting rapidly modulating antigens because these isotopes appear to be selectively retained intracellularly after degradation of the carrier antibody *(80,81)*.

2.5. Factors That Influence the Distribution of Immunoconjugates

A key obstacle in MAb-based drug targeting is the inability of the antibody or its conjugate to reach all regions within a tumor in adequate quantity. Despite the fact that accumulation up to 20–30% of injected dose per gram (% ID/g) tissue has been achieved in human tumor xenografts in mice, only very small fractions of the administrated MAb or its conjugate accumulated in tumors, usually in the range of 0.001–0.01% ID/g tissues, in most clinical studies *(82,83)*. Several important factors that affect the localization of MAb and its conjugate in tumors are discussed briefly next.

2.5.1. Physiological Factors

Several tumor-related physiological factors can adversely affect MAb localization in tumors (for reviews, *see* refs. *84,85*). These factors include the heterogeneity of blood supply, elevated intratumoral interstitial pressure, and the large transport distance, i.e., from the site of antibody transvasation to cells in the peripheral region of the tumor.

The tumor vasculature is highly heterogeneous and may be completely different from host's normal vasculature depending on the tumor type, growth rate, and location

(84,85). It consists of vessels recruited from the preexisting network and vessels resulted from the angiogenic response of the host vasculature to growth factors released by cancer cells. One of the key differences between normal and tumor vessels is that the latter are dilated, saccular, and tortuous and may contain tumor cells within the endothelial lining of the vessel wall. The heterogeneity of tumor vasculature would reduce the chance of delivery of MAb or its conjugate to some area of tumor owing to poor blood supply. In addition, an increase in the intercapillary distance would require the MAb or its conjugate to traverse a longer distance in the interstitium to reach tumor cells in the peripheral regions. Tumor necrosis has a similar adverse effect on antibody localization in the tumor because it destroys or narrows intratumoral vessels by thrombosis.

Once the MAb or its conjugate have reached an exchange vessel in tumor, its extravasation occurs by diffusion (solute movement resulting from concentration gradient) and convection (solute movement associated with bulk solvent movement) and to some extent, transcytosis. The rate of extravasation of MAb or its conjugate is dependent on: (1) surface area of exchange; (2) the transvascular concentration and pressure gradient; and (3) other transport parameters, such as vascular permeability and hydraulic conductivity (a constant related to fluid leakage and pressure gradient). Despite higher permeability of vessels in tumors than that in normal tissues, a number of factors can adversely affect extravasation of the circulating immunoconjugate, resulting in poor tumor localization by the antibody *(86,87)*. First, tumors contain regions of high interstitial pressure owing to rapid tumor growth, necrosis, and/or infiltration, and increased concentration of plasma proteins in tumor interstitium resulted from vascular leakage and poor of lymphatic drainage. Because the transvascular passage of macromolecules under normal conditions occurs primarily by convection, a high intratumoral interstitial pressure would impede the fluid extravasation, leading to a decrease in extravasation of MAb or its conjugate. Second, the average area of vascular space per gram of tumor tissues decreases with tumor growth and/or tumor necrosis, therefore resulting in a reduction in the transvascular exchange.

Extravasated immunoconjugate could move through the interstitial space by diffusion and convection to reach target tumor cells *(87,88)*. The interstitial space in tumors usually is larger than that in normal tissues. This means immunoconjugates have to transverse longer distance in the tumor interstitium to reach tumor cells in the periphery regions of vascular supply. This may increase the probability of catabolism of MAb or its conjugate in tumor interstitium before it could bind to tumor cells. Tumor metabolism of immunoconjugate may adversely affect the amount of accumulation as well as the residence time of the immunoconjugate within the tumor. Further, in addition to specific binding to tumor cells, the immunoconjugate may also bind nonspecifically to proteins or other tissue components during its transportation in tumor interstitium. These binding reactions may not only slower the diffusion rate of immunoconjugate, but also retard its movement towards tumor periphery regions, especially when a high-affinity MAb is used.

2.5.2. Pharmacological Factors

A number of pharmacological factors influence the accumulation of immunoconjugate in target tumors, including the antibody classes (e.g., IgG vs IgM) and the forms (e.g., intact MAb vs fragments) of the antibodies (see discussion in Subheading 2.4.), the dosage and the route of administration, and the metabolism and clearance of immunoconjugate from the circulation.

The effect of the administered dose of an immunoconjugate on the % ID/g tumor localization is difficult to predict. Increase of the dose of radiolabeled MAbs given to xenograft-bearing mice has produced conflicting results, with increased, unchanged, or decreased tumor uptake reported (as regard to % ID/g tumor tissues) *(89–91)*. Studies conducted to date in patients have used a wide range of antibody doses varying from less than 1.0 mg to more than 1.0 g *(92)*. The optimal antibody dose may vary depending on many factors, such as the quality of the MAb (e.g., specificity, affinity, and immunoreactivity), the quantity of antigen (e.g., the level of antigen expression and the tumor burden), the rate of antigen/antibody modulation (e.g., shedding and internalization), and different tumor types as well as hosts *(92)*. It should be noted that for a tumor of defined type and size, the number of antigenic sites is limited and may be saturated with appropriate amount of MAb. Once saturated, increasing the dose of MAb given to the tumor-bearing host will not increase the absolute amount of MAb that localized in the tumor. Under this situation, the value of % ID/g tumor tissue decreases, although the absolute amount of tumor-localized MAb remains the same.

The most common route of antibody administration is by intravenous (iv) bolus injection or infusion. Uptake and catabolism by organs of the reticular endothelial system (e.g., liver and spleen) and clearance of small antibody fragments through kidney appear to be the major routes of elimination of circulating immunoconjugate *(93)*. For intracavitary tumors, injection of immunoconjugate via the relevant intracavitary route may provide several theoretical advantages, including increased concentration of MAb at the tumor site and potentially decreased toxicity to those normal organs that are involved in the metabolism of the MAb. For example, studies in both animal models and patients have shown that the intraperitoneal injection was superior to the intravenous route for targeting peritoneal tumors *(94)*.

2.5.3. Immunological Factors

In addition to the properties of MAb such as the quality, the forms, and the species of the antibodies, which were discussed earlier, several important characteristics of antigens may also significantly affect MAb localization in both tumor and normal tissues. For example, high-level circulating antigen, resulted from antigen secretion and/or shedding, may form complexes with administered MAb and prevent the MAb from reaching the tumor sites. Furthermore, for antigen with a rapid shedding or internalization rate, if the shed or internalized antigen is not re-expressed, subsequent administration of immunoconjugate would not produce significant tumor localization. Finally, antigen heterogeneity may also have a significant impact on tumor localization of MAb or its conjugate. Tumor-antigen heterogeneity includes the difference in: (1) the proportion of cells within the tumor expressing the antigen; (2) the levels of antigen expression in different tumor cells within the tumor; and (3) the exact location of the antigen (surface vs intracellular), its accessibility, and its availability in the environment of the cells.

2.5.4. Physical and Technical Factors

In the preparation of an immunoconjugate, the biological properties of the MAb must correlate with the physical properties of the cytotoxic agents chosen for conjugation. For example, a MAb with a high rate of internalization upon binding to target antigen on tumor-cell surface is a desirable carrier for toxins and most chemotherapeutic drugs, because internalization is required for the cytotoxic effects of these agents. In the cases of radioimaging and radioimmunotherapy, it is important to carefully match the biological half-life of the carrier MAb and the physical half-life of the radioisotope. For example, a long half-life radioisotope should be used when a long time is required for the carrier MAb to accumulate in the tumor, or vice versa, to ensure the best therapeutic effect or quality of imaging. Other limitations that deserve consideration in the construction of radioimmunoconjugates include the availability of radioisotopes, the choice of appropriate methods for radiolabeling, and the availability of instrumentation and facility for the handling of radioisotopes.

3. Antibody-Drug Conjugates (Chemoimmunoconjugates)

Antibody-based targeting of conventional chemotherapeutic agents has been an especially attractive concept. There are several advantages of using chemotherapeutic agents for the construction of immunoconjugates, including: (1) these clinically used agents have established profile of antitumor activity and mode of action; (2) the mechanisms underlying the dose-limiting toxicity and the protocol to control and/or reverse the toxicity have been well-established; and (3) the pharmacokinetics of the agents are known, and the methods to monitor the drug level in circulation and tumors are readily available. Conjugation of chemotherapeutic agents to antitumor antibodies may improve their therapeutic index by increasing both the specific tissue distribution and the retention time of these agents within tumors, and accordingly reducing normal tissue toxicity. A variety of chemotherapeutic agents with different mechanisms of action, including alkylating agents (e.g., chlorambucil and melphalan), DNA intercalators (e.g., doxorubicin and daunorubicin), mitotic inhibitors (e.g., vinblastine and vindesin), and antimetabolites (e.g., methotrexate and 5-fluorouracil [5-FU]), have been conjugated to antitumor MAb and the resulting immunoconjugates tested for antitumor activity both in vitro and in vivo (for review, *see* refs. *1–4*).

Methods for coupling chemotherapeutic agents to MAbs must ensure the retention of activity of both the MAb and the cytotoxic agents. Optimal coupling method should allow controlled drug incorporation, and avoid formation of homopolymers of the MAb or drugs and aggregates of the conjugate. Practically, the methodology for conjugation should be technically feasible and reproducible. Further, the drug must be delivered to the target site in a form that is active or can be activated *in situ*. Because both antibodies and most antitumor drugs have certain functional groups that are essential for binding to target molecule, chemical groups chosen for linkage should not be required for antibody and drug action, or alternatively, these should become available following endocytosis and intracellular catabolism of the immunoconjugate. Drugs have been linked to MAbs using a number of chemical groups including amino, carboxyl, hydroxyl, and sulphydryl residues (*see* reviews in refs. *3,95–97*). In some instance, these chemical groups exist in drugs or MAb molecule, whereas in other cases, they are introduced in the drug and/or MAb molecules by chemical modification. The chemical nature of the linkage commonly used includes peptide bond; aldehyde/Schiff's base (followed by reduction reagents such as sodium borohydride or sodium cyanoborohydride to stabilize the linkage); disulfide bond; and so-called acid-labile linkages, including hydrazone bond and *cis*-aconityl or related linkages (*see* reviews in refs. *3,95–97*). It should be considered that, in some cases, a direct coupling between MAbs and drug molecules may create a steric hindrance that can negatively affect the rate of drug release from the immunoconjugate by blocking enzyme access. In this regard, the introduction of a spacer between the MAb and the drug molecule can overcome this steric hindrance and allow the drug to be released in a given milieu. For example, both lysosomotropic spacer and acid-sensitive spacers have been used as means to release active drug molecules inside lysosomal compartment *(98,99)*.

The site of attachment of the drug molecules to MAbs is also an important point to consider in the preparation of chemoimmunoconjugates. The most commonly used site in an antibody is the ε-amino group of the lysine residues, because they are convenient to use for linking to drug molecules via either amide or thiolester bonds. Different MAbs have a variable number of lysine residues, which are spread over the whole antibody molecule. It is difficult to control both the site and the amount of drug attachment to the antibody molecule, thus leading to the potential for production of a heterogeneous population of immunoconjugates. Furthermore, random conjugation of drug molecule to lysine residues within the antibody-binding site may result in the loss of antibody binding to target. In this regard, a site-specific linkage has been exploited by attaching the drug molecules to the sugar residues of the antibody *(100,101)*. Oligosaccharide residues are infrequent in the antigen-binding site, but are universally found in the Fc domain of the antibodies. Thus, site-specific linkage via the sugar residues may avoid drug loading in the antigen binding sites of the antibody. Most chemical reactions for site-specific linkage are based on oxidation of the oligosaccharide residues followed by formation of Schiff's base or hydrazone bond with the drug molecules *(100,101)*.

To achieve therapeutic benefit in chemoimmunoconjugates, it is essential that they deliver sufficient cytotoxic load to the target tumor cells. The number of molecules required for each cytotoxic agent to kill a tumor cell varies depending on its potency. For example, although a single toxin molecule is enough to kill a cell upon entering the cytosol, approx 10^6 and 10^7 molecules of methotrexate and doxorubicin or daunorubicin, respectively, are needed to accomplish the same task *(3,102)*. Because most of the chemoimmunoconjugates studied to date have a low drug to MAb ratio (in the range of 3–10 drug molecules per antibody molecule), combined with the fact that the number of antigen sites on each tumor cell is usually in the range of 10^5–10^7, it is likely that most antitumor MAb cannot deliver adequate cytotoxic molecules necessary for cell killing under steady-state conditions. Several strategies have been exploited to enhance the effectiveness of MAb-drug conjugates. One of the strategies is to construct chemoimmunoconjugates with highly potent cytotoxic agents so that even a small number of targeted molecules can achieve cell killing. Examples of these highly toxic agents include maytansinoids (microtubule-targeting agent) *(103)* and the enediyene family of antibiotics (DNA-targeting agents) *(104)*; these compounds are approx 100–1000-fold more toxic than conventional chemotherapeutic agents. Several MAb-maytansinoids immunoconjugates have been successfully constructed and have demonstrated potent antitumor activity in eradicating established xenografts in animal models *(105)*. Two compounds of the enediyene family of antibiotics, neocarzinostatin and calicheamicin, have been used in MAb-based targeting *(106,107)*. In fact, an anti-CD33 MAb-calicheamicin conjugate, Mylotarg, has demonstrated significant activity in clinical trials and was approved by the FDA in year 2000 for the treatment of AML *(61)*.

Another strategy to enhance the antitumor activity of chemoimmunoconjugates is to increase drug loading, or incorporation, on each antibody molecule. The antibodies tend, however, to lose their immunoreactivity and solubility after incorporation of about 10 moles of drug per mole of IgG under direct conjugation *(108,109)*. The most commonly used method to address this limitation is the use of an intermediate carrier; the drug molecules are first loaded at high levels of drug substitution onto a multivalent intermediate carrier, such as oxidized dextran and human serum albumin, followed by conjugation of one or two of the preloaded intermediate carrier to an antibody molecule. For example, doxorubicin was conjugated to an antihuman chronic lymphocytic leukemia (CLL) MAb via oxidized dextran at a drug/MAb ratio of approx 25–40 without affecting antibody activity, compared with a drug/MAb ratio 4–6 when using a direct linkage. The MAb-dextran-doxorubicin conjugate was much more potent than the direct conjugate in inhibiting the growth of human leukemia cells both in vitro and in vivo in an animal model *(110)*. Incorporating a spacer between the antibody and the intermediate carrier may prove to be further beneficial; the spacer can be specifically designed to not only alleviate the potential for steric hindrance caused by the MAb and the intermediate carrier, but also to provide a preferable cleavage site, e.g., lysomotropic and pH-sensitive spacers, to control the release of the conjugated drug in a given milieu, such as in the lysosomal compartment *(110)*. The criteria for the selection of an intermediate carrier include the molecular size and shape (e.g., globular or linear), homogeneity, charge, solubility, ease of handling, number of available functional groups, coupling stability, toxicity, and biodegradability.

Recently, liposomes have attracted considerable interest in serving as the intermediate drug carrier, mainly owing to their ability to load a large amount of drug molecules without the requirement for chemical modification, i.e., the drug molecules packed inside the liposome remain in their native form. In the preparation of immunoliposomes, the cytotoxic agents are first loaded into the liposome followed by chemical conjugation of the liposome to MAb molecules via the functional groups that are pre-incorporated in the liposome surface (for reviews, *see* refs. *111–113*). A promising method developed recently involves a "postinsertion" technique by which the clinically approved liposomal drug, such as Doxil, can be easily converted into a tumor-targeting preparation by a simple step of incubation with the antitumor antibodies or its fragments *(114,115)*. The advantage of this approach is that by using clinically approved liposomal drugs, it not only confers tumor specificity to the nontargeted liposomal drug, but also theoretically can be applied to a variety of tumor targeting antibodies and also avoids issues associated with the manufacturing of multiple chemoimmunoconjugates. However, there are also several limitations associated with liposome therapy, including slow tumor penetration owing to the large size of liposome and premature removal of liposome from circulation by macrophages in the liver and spleen *(111–113)*. Increased tumor localization of immunoliposome, as well as other MAb-based conjugates, can be achieved by enhancing tumor vascular permeability by methods such as hyperthermia *(116)*, external radiation *(117)*, and administration of pharmacological agents or cytokines including IL-2 and tumor necrosis factor-α (TNF-α) *(118,119)*. Furthermore, the invention of sterically stabilized liposomes, or "stealth liposomes," has greatly improved the pharmacokinetics and tumor localization of immunoliposomes by reducing their uptake and degradation by macrophages *(120,121)*. Several promising immunoliposome preparations, including those targeting HER2 and CD19, have been studied in animal models and shown potent antitumor activity *(122–124)*.

Mylotarg (Gemtuzumab ozogamicin, also called CMA-676) is the first, and to date, the only chemoimmunoconjugate to receive FDA approval for clinical application in patients of over 60 yr of age with AML *(61,125)*. Mylotarg consists of a humanized anti-CD33 MAb (HP67.6) coupled to calicheamicin. In a Phase I clinical trial, 40 patients with refractory or relapsed CD33+ AML were treated with escalating doses, from 0.25 mg/m^2 up to 9.0 mg/m^2, of Mylotarg as 2-h iv infusions at 2-wk intervals for one to three cycles. Three patients achieved blood count normalization (complete response [CR]) and leukemia cells were eliminated from the blood and bone marrow of another five patients (partial response [PR]) *(126)*. All the doses were well-tolerated, with side effects including reversible fever, chills, hypotension, and hepatic transaminasemia and hyperbilirubinemia. No antibody responses to the carrier humanized anti-CD33 antibody were observed. In several Phase II studies, 142 patients with relapsed CD33+ AML were given 9.0 mg/m^2 on d 1 and 15. CRs were seen in 23 patients, and PRs were achieved in another 19 patients, for an overall response rate (ORR) of 30% (42/142) *(127)*. The progression-free survival at 30 mo was 35%, and overall 1-yr survival was 31%; both rates compare favorably to rates of other salvage therapy for relapsed/refractory AML. Again, no antibody responses to the carrier humanized anti-CD33 antibody were observed in these trials. Apart from those toxicities seen in the Phase I trial, grade 3–4 level of myelosuppression was observed in 98% of patients *(127)*.

Another chemoimmunoconjugate that has been studied extensively is BR96-doxorubicin conjugate (BR96-DOX) (for review, *see* ref. *128*). BR96 is an anti-Le^y MAb that internalizes following binding to cell-surface antigen. In the preparation of BR96-DOX conjugate, a chimeric version of BR96 was used coupled with doxorubicin-6-maleimido-caproylhydrazone derivative via reduced thiol groups on the BR96 antibody molecule. Approximately eight molecules of doxorubicin were linked to each MAb molecule. Preclinical evaluation of BR96-DOX was carried out in human tumor xenograft-bearing nude mice and rats *(129,130)*. At doses of 200 mg/kg or more, the conjugates led to regression and cures of mice or rats bearing various human xenografts including lung, colon, and breast carcinomas, without any signs of toxicity. On the other hand, doxorubicin either linked to a control antibody, or used by itself at the maximum tolerated dose (10 mg/kg), had little antitumor activity. Unmodified BR96, alone or together with doxorubicin, was not curative *(129,130)*. Toxicology studies performed on dogs that expressed Le^y antigen in the epithelial cells from the gastrointestinal tract that was cross-reactive to BR96, showed that BR96-DOX and unmodified BR96 had the same dose-limiting toxicity, i.e., severe vomiting and bloody diarrhea, when given at doses greater than 400 mg/kg *(128)*. In viewing of the promising animal studies, two Phase I clinical trials and a Phase II clinical trial were performed in patients with advanced solid tumors. In the Phase I trials, the patients were given BR96-DOX either by iv infusion over 24 h every 3 wk or by bolus injection weekly. The maximum tolerated doses were established as approx 700 mg/m^2 (~18 mg/kg) and 200 mg/m^2 (~5 mg/kg) for the infusion and the bolus injection protocols, respectively *(131,132)*. In the Phase II trial, 14 patients with metastatic breast cancer were given BR96-DOX at 700 mg/m^2 every 3 wk. Only one patient showed PR after treatment with the conjugate, compared with 4 PRs in 9 patients treated with free doxorubicin *(133)*. The dose-limiting toxicity was, as observed in the dogs, gastrointestinal, particularly vomiting and bloody diarrhea. Biopsy studies demonstrated significant BR96-DOX binding to epithelial cells in the gastrointestinal tract, with the damage most likely a result of complement activation *(133)*. Further dose escalation was limited because of the gastrointestinal toxicity. Several approaches are being investigated, aiming to enhance the therapeutic index by increasing the potency of BR96-DOX conjugate and/or reducing its toxicity. For example, new coupling chemistries have been explored to increase drug incorporation and conjugate stability. In addition, genetic modifications of BR96 have been made to abolish its capacity to fix complement *(128)*. Another approach is to combine the BR96-DOX with other conventional chemotherapeutic agents, such as Taxotere and Gemzar. It has been observed that BR96-DOX may synergize with these chemotherapeutic agents, so that a lower dose of the conjugate can be used to achieve the same curative effect observed in animal models *(134)*. In this regard, several Phase I and II studies have been completed in patients with breast and colon cancers using BR96-DOX in combination with Taxotere. Other Phase II trials are being conducted in patients of prostate cancer and nonsmall cell lung cancer (NSCLC) using BR96-DOX in combination with Taxotere or Gemzar.

4. Antibody-Radioisotope Conjugates (Radioimmunoconjugates)

The use of radiolabeled MAb for cancer treatment offers several theoretical advantages over that of chemoimmunoconjugates and the immunotoxins to be discussed later. Once bound to tumor cells, the radioimmunoconjugates can kill not only the MAb-coated cells, but potentially also those surrounding tumor cells ("bystander effect") via radioisotope emission within an area defined by the path length of the isotope and the composition of the tumor tissues. This allows the killing of antigen-negative tumor-cell variants and tumor cells that are inaccessible to the radioimmunoconjugate. Furthermore, internalization of the MAb, which is necessary for the cytotoxicity of most chemoimmunoconjugates and immunotoxins, is generally not required for that of the radioimmunoconjugates. One exception to this is radioimmunoconjugates containing alpha emitters that have a very short path length (approximately the distance of one cell).

Several important issues to consider in choosing an appropriate radioisotope for preparation of radioimmunoconjugate include the properties of the carrier MAb and the intended use of the conjugate. It is important to carefully match the physical properties of the radioisotope to the biological properties of the carrier MAb. For example, for radioimmunoimaging purpose, an antibody fragment with short circulation half-life coupled with a gamma emitter of matched physic half-life would be the ideal choice. On the other hand, for radioimmunotherapy using an intact MAb, the half-life of the radioisotope should be long enough to allow the MAb to localize and accumulate in the tumor to exert its cytotoxic effect. There is still considerable debate regarding the best class of radioisotopes suitable for MAb-targeted cancer therapy. Beta emitters such as ^{131}I, ^{90}Y, and ^{188}Re release radiation that penetrates several millimeters and therefore could be beneficial in killing tumor cells that are either antigen-negative or inaccessible to the radiolabeled MAb *(135–137)*. However, use of beta emitters somewhat compromises the selectivity of the targeted therapy by damaging nearby normal cells. Alpha emitters, such as ^{211}At and ^{212}Bt, are more effective in inducing cell damage once delivered to the vicinity of the tumor cells, but because their energy dissipates over a very short range, the damage is likely to be limited to a single cell *(138,139)*. Some isotopes decay by electron capture and subsequent Auger cascade, which is believed to be highly efficient in producing lethal effect (by damaging DNA molecules) if initiated near the cell nucleus. The antitumor activity of Auger emitters, such as ^{125}I, therefore could be enhanced considerably when linked to a carrier MAb that undergoes active internalization once binding to tumor-surface antigen *(140,141)*. Some of the radioisotopes, along with their relevant properties, that are being used or that are believed to have potential use in radioimmunotherapy are listed in Table 1.

Among these radioisotopes, ^{131}I has been used most often because of its availability, ease of manipulation for MAb labeling, and favorable emission characteristics. Most

Table 1
Radioisotopes Used for Radioimmunotherapy of Cancers

Isotope	Mode of decay	Half-life (h)	maximum tissue path length (mm)	Particle energy (MeV)
^{131}I	β,γ	193	2.4	0.61
^{90}Y	β	64.2	11.9	2.28
^{186}Re	β,γ	91.2	5.0	1.07
^{188}Re	β,γ	16.7	11.0	2.12
^{67}Cu	β,γ	58.6	2.2	0.577
^{177}Lu	β,γ	160.8	2.2	0.5
^{212}Bi	α	1.01	0.04–0.08	6.09
^{211}At	α	7.2	0.04–0.08	5.8
^{125}I	Auger	60.2 d	0.01–0.02	0.35

of the currently used iodination methods, including chloramine T and the iodogen methods, involve covalent attachment of radioiodine to the tyrosine residues in the antibody *(142,143)*. These methods oxidize iodine to a cationic species capable of replacing a hydrogen atom in the phenolic ring of tyrosine in the ortho position. They are relatively easy to perform and can yield high iodine incorporation, but may damage MAb activity by iodinating the tyrosine residues in the CDRs that are important for antigen binding *(144)*. One nonoxidative method for protein iodination is the Bolton-Hunter method, which attaches the radioiodine to the lysine residues in the antibody. A common major drawback of the aforementioned methods is the relative instability of the radiolabel in vivo owing to a process called dehalogenation *(142,143)*. This rapid degradation of iodinated MAb can lead not only to decreased localization of the MAb in tumor, but also to unwanted accumulation of radioactivity in the thyroid and stomach. It has been postulated that the dehalogenation of radiolabeled MAb is mediated via enzymes involved in thyroid hormone metabolism, such as deiodinases, that cannot distinguish iodine-labeled tyrosine from thyroxine *(145)*. Therefore, it is plausible that dehalogenation could be minimized if the iodine is linked to proteins in a way that is not susceptible to deiodinases *(146–148)*. In this regard, several methods have been developed for protein iodination via nonphenolic aromatic rings. In these methods, a radiolabeled intermediate, such as *N*-succinimidyl (tri-*n*-butylstannyl) benzoate, 3-tri-*n*-butylstannylphenylisothiocyanate, or *N*-succinimidyl 3-iodobenzoate, is first prepared, followed by covalent conjugation to MAb, usually via the lysine residues *(149–152)*. Antibodies iodinated via these methods were shown to be superior to those labeled by chloramines T method in regard to immunoreactivity, stability both in vitro and in vivo, and localization in tumor xenografts, along with reduced iodine accumulation in the thyroid and stomach *(149–152)*.

An undesired issue in using ^{131}I-labeled MAb for cancer therapy is that the radioisotope, in addition to its therapeutic beta radiation, also emits long-range gamma radiation that may pose an unwanted risk to patient's family members and health care workers. In this regard, a pure beta emitter, such as ^{90}Y, has gained considerable interest in recent years *(135–137)*. In contrast to ^{131}I, ^{90}Y is a metallic radioisotope that is usually linked to MAb via the use of bifunctional or heterobifunctional chelating agents. A chelator, such as DTPA and EDTA, is first labeled with the metallic radioisotopes followed by its attachment to the MAb by a reactive functional group (for reviews, *see* refs. *153,154*). Although bifunctional chelators can be designed for linkage to different functional groups in the antibody molecule, e.g., amino group, sulfhydryl group, or tyrosines, the most commonly used bifunctional chelators are those designed specifically for coupling through lysine residues. A major obstacle to the extensive use of metallic radioisotopes has been the difficulty in the development of high-affinity chelators that can form stable bonds to the carrier MAb without affecting its antigen-binding activity *(153–155)*. The main drawback associated with the use of metallic isotope-labeled MAb is their tendency for accumulation of radioactivity in the liver and bone *(156)*.

Zevalin™, a ^{90}Y-labeled murine anti-CD20 antibody (Ibritumomab Tiuxetan, or IDEC-Y2B8), is the first radiolabeled MAb to receive approval for therapeutic applications from the FDA *(62)*. Zevalin™ is composed of a murine anti-CD20 MAb (IgG1, κ), Ibritumomab, covalently bound to the metal chelator tiuxetan, which links ^{90}Y. The use of a murine MAb rather than a human or humanized version in the conjugate proved to be advantageous; the circulation half-life of Ibritumomab in human beings is approx 48–72 h (compared with that of 1–3 wk for most humanized and human MAb) *(157,158)*, which matches very well with the physical half-life of ^{90}Y (~64 h). Thus, the use of a murine MAb minimizes nonspecific radiation to nontarget cells/organs. Extensive clinical trials have been performed using Zevalin™ in patients with relapsed/refractory low- and intermediate-grade NHL. Several Phase I/II studies demonstrated that the optimal regimen for Zevalin therapy consists of two pretreatments with Rituxan, at 250 mg/m^2 7 d prior to and immediately prior to the administration of Zevalin. The maximum tolerated doses of Zevalin are 0.4 mCi/kg for patients with platelet count higher than 150,000 and less than 25% bone-marrow involvement with NHL, and 0.3 mCi/kg for patients with platelet count between 100,000 and 149,000 *(159,160)*. In a randomized Phase III trial, 143 patients were treated with Zevalin (73 patients, 0.4 mCi/kg) or Rituxam alone (70 patients) in which the patients were given 4 weekly doses of 375 mg/m^2. The ORRs was 80% in the Zevalin treatment group, compared with 56% in the Rituxan group, with CRs of 30% vs 16%, respectively *(161)*. In a Phase II trial in patients with mild thrombocytopenia (platelet count from 100,000 to 149,000), Zevalin, given at 0.3 mCi/kg, yielded an ORR of 82% with 37% CRs *(162)*. Furthermore, in an open-label, nonrandomized trial, 57 patients who had failed prior Rituxan therapy, Zevalin regimen (0.4 mCi/kg) demonstrated an ORR of 74%, which was significantly higher than the ORR obtained from the last cycle of Rituxan therapy (32%) *(163)*. More importantly, 51% (19 out of 37) patients who did not respond to their last Rituxan therapy showed a response to the Zevalin regimen. The primary toxicities associated with Zevalin treatment are hematological, including neutropenia, thrombocytopenia, and depletion of peripheral B lymphocytes, all of which are transient and reversible. Nonhematologic toxicities are similar between Zevalin and Rituxan, which include asthenia, nausea, chills, fever, throat irritation, headache, and vomiting. It is noteworthy that plasma immunoglobulin levels remain in the normal range during Zevalin therapy, and the overall incidence of serious clinical infections was low. Further, HAMA was only observed in three out of 211 patients tested (1.4%) *(62)*.

Another radioimmunoconjugate that has been tested extensively in clinical trials is Bexxar™, a ^{131}I-labeled murine anti-CD20 MAb (also known as anti-B1 antibody, or Tositumomab) *(63)*. Bexxar, like that of Zevalin, is given to patients in three stages over 1–2 wk, but with various doses calculated for each individual patient according to the pharmacokinetic (PK) parameters measured prior to the therapeutic intervention. The patient is administered 450 mg of unlabeled MAb on d 0, along with 5 mCi of Bexxar (35 mg MAb) to determine the PK of the antibody. Whole-body gamma counting is performed on d 0, 2, 3, or 4, and d 6 or 7, to calculate the radiation dose delivered to whole body and bone marrow. On d 7–14, the therapeutic dose, comprising 450 mg unlabeled MAb with sufficient dose of Bexxar (35 mg of labeled MAb), is given to deliver the predetermined maximum-tolerated radiation dose to the whole body and bone marrow, which is 75 cGy in patients with a normal platelet count and 65 cGy in patients with platelet count between 100,000 and 149,000 *(63,164)*. Accordingly, the actual ^{131}I radioactivity administered to each patient varies from 45–177 mCi. In two Phase II trials in patients with relapsed/refractory low-grade or transformed NHL, Bexxar regimen yielded ORRs of 71% and 60%, with CRs of 34% and 32%, respectively *(165,166)*. When given as the initial therapy to 76 patients with follicular NHL, Bexxar treatment resulted in an ORR of 97% with 63% CRs *(167)*. In a multicenter pivotal Phase III trial in patients with low-grade or transformed NHL, and who had failed at least two prior regimens and had relapsed within 6 mo of completing those therapies, ORRs to the Bexxar regimen was reported in 65% of the patients with 20% CRs *(168)*. In a Phase II trial in patients who had failed Rituxan therapy, Bexxar treatment yielded 57% ORRs with 14% CRs *(169)*. The major toxicities of Bexxar are mild to moderate transient neutropenia, thrombocytopenia, and anemia. Approximately 6% of patients developed AML and myelodysplasia syndrome, but this may be related to prior chemotherapy *(63)*. Other common side effects of Bexxar are similar to those of Zevalin, including asthenia, fever, nausea, and headache. Bexxar gained approval recommendation from The Oncologic Drugs Advisory Committee in December 2002 and is awaiting marketing approval from the FDA.

5. Antibody-Toxin Conjugates (Immunotoxins)

One of the advantages of using toxins as the cytotoxic moiety in MAb-based tumor targeting is their extreme potency. In many cases, a single toxin molecule is able to kill a cell once entering the cytosol *(170)*. Three major types of toxins are used in the preparation of immunotoxins: bacteria toxins, plant toxins, and fungal toxins. All three types of toxins catalytically inhibit protein synthesis in eukaryotes, but each at a distinct step during translation. Based on their biochemical characteristics, plant toxins can be grouped as: type I, single-chain ribosomal inhibitory polypeptides such as pokeweed antiviral protein and saporin; and type II, such as ricin and abrin, which are heterodimers. The A chain of type II toxins are the toxic moiety, whereas the B chain contains the binding site for carbohydrates on cell surface through which A chain gains access to cell cytosol. Type I plant toxins only contain the catalytic domains without the binding domains. Bacteria toxins such as pseudomonas exotoxin A (PE) produced by *Pseudmonas aeruginosa* and diphtheria toxin (DT) produced by *Corynebacteria diphtheria*, are produced as single-chain polypeptides, each containing a binding and a catalytic domains separated by a translocation domain. Fungal toxins such as alpha sarcin are also single-chain proteins, but are functionally different from the type I plant toxins. For example, α sarcin is a phosphodiesterase, whereas type I plant toxins are *N*-glycosidases.

Originally ITs were constructed via chemical coupling methods using the intact toxin molecules. These ITs were extremely cytotoxic to target cells, but the specificity was suboptimal, because the binding moiety of the toxins, e.g., the B chain of ricin or the Ia (binding) domain of PE, may bind to the surface of normal cells via carbohydrate residues *(171,172)*. Several approaches have been developed to circumvent this problem: (1) Construction of ITs that contain only the toxic moiety, such as the A chain of ricin or the catalytic domain of PE. While enhancing the specificity, the potency of these ITs was significantly reduced owing to lack of the binding and translocation unit *(173,174)*. Furthermore, the glycosylated side residues of toxin A chain may still cause nonspecific uptake by hepatic nonparenchymal cells via the mannose receptor and by macrophages *(175,176)*. Chemical deglycosylation has been used to modify the A chain in order to reduce the nonspecific uptake *(177,178)*. (2) Construction of ITs with the intact toxins that have been chemically modified to block the oligosaccharide-binding sites on the B chains, for example, blocked ricin *(179,180)*. (3) Delivery of the B chain together with the A chain-containing ITs either directly or indirectly via a second MAb homing to the same target cells *(181,182)*. (4) Construction of ITs using single-chain ribosomal inhibitory proteins, such as pokeweed antiviral protein and saporin, that do not contain any cell-binding components *(183,184)*. Several deglycosylated ricin A chain (dgA)-based ITs, for example, the anti-CD22 ITs, RFB4-dgA, and RFB4-Fab'-dgA; the anti-CD25 IT, RFT5-dgA; and the anti-CD19 IT, HD37-dgA, have been produced via chemical coupling and tested in both animal models and limited Phase I trials. Furthermore, an anti-CD19 IT, comprising of the anti-B4 antibody coupled to an intact ricin with blocked B chain, was also constructed and tested in at least two Phase I trials. (For detailed reviews on the preclinical development and clinical investigation on these ITs, *see* refs. *58–60*.)

Several major drawbacks in preparation of ITs via chemical coupling include low efficiency of the coupling processes, loss of antibody and/or toxin activity owing to chemical modification, generation of heterogeneous products, and difficulties in large-scale production. Recent developments in general molecular-biology techniques and antibody engineering, along with the structural knowledge of several toxin molecules including PE, DT, and ricin, has made it possible to produce second-generations ITs via recombinant methods. In this approach, the MAb or its fragment is fused genetically to the catalytic domain of a toxin, followed by expression of the fusion protein in *Escherichia coli* or other hosts. For example, in the preparation of PE-based ITs, the recombinant antibody fragment, either as a scFv or a disulfide bond-stabilized Fv fragment, is fused to the N-terminus of the truncated derivatives of PE, e.g., PE40 (amino acid 253–613, with the binding domain Ia [amino acid 1–252] deleted) or PE38 (a PE40 derivative with additional deletion of portion of domain Ib [amino acid 365–379]) *(185–187)*. Interestingly, ITs produced by fusion of the antibody fragment to the C-terminus of PE40 or PE38 are not active *(188)*. For DT-based ITs, the inverse arrangement is required for activity. In this case, the antibody fragment must be fused to the C-terminus of DT to free the N-terminus for translocation and catalytic activity *(189)*.

DAB389IL2 (Ontak™), a recombinant fusion protein comprising IL-2 fused to the first 389 amino acid of DT, is the first and the only IT approved so far by the FDA for clinical application in the treatment of cutaneous T-cell lymphoma (CTCL) *(190,191)*. In a Phase I trial, Ontak produced 5 CRs and 8 PRs in 35 CTCL patients and 1 CR and 2 PRs in 17 patients with NHL *(192,193)*. In a Phase III trial of 71 CTCL patients, 7 CRs and 14 PRs were achieved *(194)*. Several PE40 or PE38-based recombinant ITs have also been successfully produced and tested in a number of clinical studies, including the anti-Le^y antibody B3(dsFv)-PE38 (also known as LMB-9) and BR96(sFv)-

PE40, the anti-^{erb}B2 (HER2/neu) antibody e23(dsFv)-PE38, the anti-CD25 antibody anti-Tac(Fv)-PE38 (also known as LMB2), and the anti-CD22 antibody RFB4(dsFv)-PE38 (also known as BL22) (for reviews, *see* refs. *58–60*). LMB2 is very potent against human leukemia cells both in vitro and in animal models. In Phase I clinical trials, LMB2 was given to 35 patients for a total of 59 treatments, and resulted in CR of one patient with hairy-cell leukemia (HCL) and PRs in seven other patients *(195)*. It is noteworthy that all four HCL patients responded to LMB2 (one CR and three PRs) therapy. A Phase II trial of LMB2 is planned in patients with CD25+ hematologic malignancies. BL22 is a disulfide bond-stabilized IT, and is currently in a dose-escalating phase I trial *(196)*. Of 16 HCL patients who were resistant to cladribine, 11 had CRs and 2 had PRs; the remaining three non-responders either received low doses of BL22 or had neutralizing antibodies to the toxin *(197)*.

The dose-limiting toxicity of many IT therapies is vascular leak syndrome (VLS), with symptoms ranging from weight gain and mild peripheral edema to hypertension and pulmonary edema *(198)*. Several mechanisms have been proposed to be responsible for VLS, including cytokine-induced activation of macrophages and endothelial cells, inflammation, and direct damage to the endothelial cells by the ITs *(199–201)*. Several recent studies suggested that a three amino acid motif present in both protein toxins and IL-2 might be responsible for VLS *(202–204)*. Mutation of this sequence in toxin molecules, or use of anti-inflammatory agents and/or cytokine antagonists, therefore, may prove to be beneficial in future clinical trials *(201,204)*. Other side effects associated with IT therapies include: (1) nonspecific toxicity resulted from binding of either the MAb or the toxin molecule, and (2) immunogenicity of the bacteria or plant-derived toxin molecules. Molecular modeling and engineering of the MAb and/or the toxins may help to construct ITs with reduced nonspecific binding to normal cells. Concurrent administration of immunosuppressive agents such as CTLA-4, anti-CD20 MAb, or steroids may alleviate the immune response to the ITs *(201,205)*. Further, the use of human-derived toxin-like molecules, such as RNase, along with humanized or human antibodies, may prove to be nonimmunogenic in human therapy *(206,207)*.

6. Antibody-Directed Enzyme-Prodrug Therapy (ADEPT)

ADEPT, the use of antibodies to carry enzymes to tumors for *in situ* generation of cytotoxic molecules from relatively nontoxic precursors, was first proposed by Philpott et al. in 1973 *(208,209)*, and further developed by Bagshawe and Senter and their colleagues *(210,211)*. In theory, an enzyme molecule can activate an unlimited number of precursors leading to the local amplification and accumulation of cytotoxic molecules within a tumor. Thus, ADEPT would potentially overcome the problem associated with low uptake of the chemoimmunoconjugates by tumor cells. Furthermore, because the cytotoxic molecules are generated within the tumors, nonspecific toxicities generally associated with conventional chemotherapies are minimized owing to low drug distribution to normal tissues such as bone marrow and intestinal epithelium. ADEPT usually involves two steps: first, the antitumor antibody-enzyme conjugate (or fusion protein) is given to patients. After sufficient amount of time to allow adequate tumor localization and clearance of the conjugate from the circulation (for an optimal tumor/nontumor localization ratio), a prodrug (or precursor) is administered for its conversion into an active cytotoxic molecule by enzyme pre-localized in the tumors. Activation of prodrug at tumor sites results in a high local concentration of the cytotoxic molecules, which may not only kill the antigen-positive tumor cells, but also reach and kill nearby antigen-negative tumor cells (the "bystander effect") (for reviews, *see* refs. *212–215*). An additional step can be added between the administration of antibody-enzyme conjugate and the prodrug; that is, to give the patients a clearing agent that will remove the circulating conjugate at a faster rate, thereby further minimizing prodrug activation outside of tumors.

Unlike for chemoimmunoconjugates and immunotoxins, the ideal carrier MAb for ADEPT approach should be a noninternalizing antibody because antigen modulation upon antibody binding will decrease the enzyme concentrations on tumor-cell surface, and as a result, reduce the rate of prodrug activation. The MAb should also possess high specificity to tumor antigens (ideally surface antigens) with high binding affinity. To minimize nonspecific toxicity, it is essential that the prodrug-activating enzyme is not present in any normal tissues. Other criteria for enzyme selection include high substrate specificity, optimal enzyme activity at or near physiological pH and temperature, nonrequirement of cofactors, absence of enzyme inhibitors in tumors, and stability during conjugation and in circulation. So far, a number of enzymes have been employed in the ADEPT approach, including carboxypeptidase, alkaline phosphatase, aminopeptidase, cytosine deaminase, α-galactosidase, β-glucuronidase, β-lactamase, nitroredutase, and penicillin amidase (for reviews, *see* refs. *212–215*). Most of these enzymes are of plant or bacteria origin, thus are immunogenic in humans. The use of PEG-modified enzyme, or administration of immunosuppressive agents such as cyclosporin and FK506, may reduce the immunogenicity of the MAb-enzyme conjugate. The ultimate resolution to the immunogenicity issue is to use humanized or human antibodies *(216)* and human enzymes, e.g., human glucuronidase and carboxypeptidase *(217,218)*. Finally, catalytic MAb, or abzyme, can be used to replace the enzyme in the conjugate *(219,220)*. Catalytic antibodies are antibodies against antigens that mimic the transition-state of a chemical reaction and thus they catalyze a given reaction by preferentially binding to and stabilizing the transition state configuration *(221)*. In ADEPT, the catalytic MAb is coupled to an antitumor MAb to form a bispecific antibody molecule. One arm of the molecule binds to tumor cells while the other arm catalyzes prodrug into active agent *(222,223)*. One of the advantages of this approach is its high specificity because an abzyme can be generated to catalyze a chemical reaction that is not mediated by a naturally occurring enzyme *(223)*. Further, the immunogenicity of the MAb-enzyme fusion protein can be greatly reduced by using a humanized or a human abzyme as the catalytic unit.

A number of MAb-enzyme conjugates have been produced, either by chemical coupling or by recombinant methods, and tested both in vitro and in vivo in animal models (for reviews, *see* refs. *213–215*). A limited number of clinical trials have been performed to date, mostly with conjugates containing the enzymes of bacteria origin. For example, in a Phase I trial in 10 patients with advanced colorectal cancer, the patients were treated with an anti-CEA antibody (A5B7) F(ab')2 fragment conjugated to carboxypeptidase G2. A benzoic acid mustard-glutamate prodrug was given and converted by CPG2 in the tumor into a cytotoxic form. PR was seen in one patient, with stable disease seen in six other patients *(224)*. In another study, where 27 cancer patients were given A5CP antibody-carboxypeptidase G2 conjugate and ZD2676P prodrug (bis-iodo phenol mustard), disease stabilization was seen in three patients *(225)*. In both studies, significant HAMA and antibody to carboxypeptidase G2 were developed in all patients. Current studies in ADEPT focus on the use recombinant fusion protein comprising of small antibody fragments, e.g., scFv, Fv, or Fab, fused to relatively nonimmunogenic enzymes *(213–215)*. The clinical utility of these new generation MAb-enzyme fusion proteins remains to be demonstrated in future trials.

7. Bispecific Antibody-Mediated Drug Delivery

Bispecific antibodies (BsAb) are immunoglobulin-based molecules that bind to two different epitopes on either the same or distinct antigens *(226,227)*. The majority of studies with BsAb have been focused on their use to redirect cytotoxic immune effector cells to kill tumor cells. In this context, one arm of the BsAb binds to antigen expressed on the tumor-cell surface, whereas the other arm binds to a determinant expressed on effector cells. Cross-linking between the tumor and the effector cells triggers the activation of the effector cells, resulting in killing of the tumor cell. A number of cell-surface molecules expressed by effector cells have been explored for efficient cell activation in order to achieve potent tumor-killing. For example, CD3, CD16, and CD64 expressed on T lymphocytes and natural killer (NK) cells/macrophages and other mononuclear cells (MNCs), respectively *(228–230)*. Preliminary results from a number of early-stage clinical trials using these antitumor/anti-effector cell BsAb appeared to be very promising. Interested readers are referred to several excellent recent reviews on this topic (for reviews, *see* refs. *231–234*). In addition to effector cell-targeting, BsAb has also been used to target cytotoxic agents, such as radionuclides, drugs, and toxins, to tumors. In this approach, the effector-cell specificity of an antitumor BsAb is replaced by an antibody directed against either the cytotoxic agent directly or the carrier of the agent, such as a hapten or liposome *(235–240)*. Like the ADEPT, the BsAb-based targeting approach usually involves two steps: the BsAb is first administered to the patients, and after an adequate interval for the systemic clearance of the antibody to allow an optimal tumor/normal tissue localization ratio, the cytotoxic modalities is given, which leads to rapid accumulation of the cytotoxic agent within the tumor owing to its binding to the tumor-bound BsAb. A variety of cytotoxic modality have been tested in BsAb-based targeting, including chemotherapeutic agents such as vinca alkaloids *(235)* and anthracycline antibiotics *(236)*; toxins such as ricin A chain, saporin, and gelonin *(237)*; chelated radioisotopes *(238)*; prodrug-activating enzymes *(239)*; and drug-loaded liposomes *(240)*. A similar approach to the BsAb-based targeting is the use of a biotinylated antitumor MAb, followed by a streptavidin-labeled carrier, such as a heptan or a liposome preloaded with cytotoxic agents *(241,242)*.

A major obstacle in the development of BsAb has been the difficulty in producing the antibody in sufficient quantity and quality for clinical studies via traditional hybrid hybridoma methodology or chemical conjugation *(243)*. Coexpression of two different sets of IgG light and heavy chains in hybridomas may produce up to 10 different light- and heavy-chain pairs, with only one of these pairs the functional bispecific heterodimer *(244)*. On the other hand, chemical crosslinking of two IgGs or their fragments is often inefficient and can lead to the loss of antibody activity *(245)*. In both methods, purification of the BsAb from the nonfunctional species, such as homodimers and mispaired heterodimers of noncognate Ig light and heavy chains produced by the hybridoma, and multimeric aggregates resulting from chemical conjugation, is often difficult and the yield is usually low *(246)*. In recent years, a variety of recombinant methods have been developed for efficient production of BsAb, both as antibody fragments (for reviews, *see* refs. *29,243,247*) and full-length IgG formats *(248)*. In addition, production up to one gram of antibody per liter of culture via *E. coli* fermentation has been reported for a number of BsAb fragments *(249–251)*. Taken together, these novel technologies in BsAb production will greatly facilitate its application in tumor-targeting drug delivery.

References

1. Ghose T, Blair AH. Antibody-linked cytotoxic agents in the treatment of cancer: current status and future prospects. J Natl Cancer Inst 1978;61:657–676.
2. Ghose T, Blair AH, Uadia P, et al. Antibodies as carriers of cancer chemotherapeutic agents. Ann NY Acad Sci 1985;446:213–227.
3. Garnett MC. Targeted drug conjugates: principles and progress. Adv Drug Deliv Rev 2001;53:171–216.
4. Ghose T. The current status of tumor targeting. A review. In: Page M, ed. Cancer Drug Discovery and Development: Tumor Targeting in Cancer Therapy. Totowa, Humana Press Inc., 2002, pp. 3–78.
5. Foss FM. DAB(389)IL-2 (ONTAK): a novel fusion toxin therapy for lymphoma. Clin Lymphoma 2000;1:110–116.
6. Frankel AE, Powell BL, Hall PD, et al. Phase I trial of a novel diphtheria toxin/granulocyte macrophage colony-stimulating factor fusion protein (DT388GMCSF) for refractory or relapsed acute myeloid leukemia. Clin Cancer Res 2002;8:1004–1013.
7. Lappi DA, Baird A. Mitotoxins: growth factor-targeted cytotoxic molecules. Prog Growth Factor Res 1990;2:223–236.
8. Siegall CB, FitzGerald DJ, Pastan I. Selective killing of tumor cells using EGF or TGF alpha-Pseudomonas exotoxin chimeric molecules. Semin Cancer Biol 1990;1:345–350.
9. Ramakrishnan S, Olson TA, Bautch VL, Mohanraj D. Vascular endothelial growth factor-toxin conjugate specifically inhibits KDR/flk-1-positive endothelial cell proliferation in vitro and angiogenesis in vivo. Cancer Res 1996;56:1324–1330.
10. Nechushtan A, Yarkoni S, Marianovsky I, Lorberboum-Galski H. Adenocarcinoma cells are targeted by the new GnRH-PE66 chimeric toxin through specific gonadotropin-releasing hormone binding sites. J Biol Chem 1997;272:11597–11603.
11. Schlick J, Dulieu P, Desvoyes B, et al. Cytotoxic activity of a recombinant GnRH-PAP fusion toxin on human tumor cell lines. FEBS Lett 2000;472:241–246.
12. Ehrlich P. On immunity with specific reference to cell life. Proc R Soc London 1900;66:429.
13. Mathe G, Log T, Bernard J. Effet sur la leucemie 1210 de la souris d'un combinaison par diazotation d'A-methopterine et de γ-globulines de hamsters porteurs de cette leucemie par heterogreffe. C R Acad Sci 1958;246:1626–1628.
14. Ghose T, Cerini M, Carter M, Nairn RC. Immunoradioactive agent against cancer. Br Med J 1967;1:90–93.
15. Ghose T, Cerini M. Radiosensitization of Ehrlich ascites tumour cells by a specific antibody. Nature 1969;222:993–994.
16. Ghose T, Nigam SP. Antibody as carrier of chlorambucil. Cancer 1972;29:1398–1400.
17. Ghose T, Norvell ST, Guclu A, et al. Immunochemotherapy of cancer with chlorambucil-carrying antibody. Br Med J 1972;3:495–499.
18. Ghose T, Norvell ST, Guclu A, Macdonald AS. Immunochemotherapy of human malignant melanoma with chlorambucil-carrying antibody. Eur J Cancer 1975;11:321–326.
19. Ghose T, Norvell ST, Guclu A, et al. Immunochemotherapy of malignant melanoma with chlorambucil-bound antimelanoma globulins: preliminary results in patients with disseminated disease. J Natl Cancer Inst 1977;58:845–852.
20. Tai J, Blair AH, Ghose T. Tumor inhibition by chlorambucil covalently linked to antitumor globulin. Eur J Cancer 1979;15:1357–1363.
21. Ghose T, Guclu A, Tai J, et al. Antibody as carrier of ^{131}I in cancer diagnosis and treatment. Cancer 1975;36:1646–1657.
22. Ghose T, Tai J, Aquino J, et al. Tumor localization of 131-I-labeled antibodies by radionuclide imaging. Radiology 1975;116:445–448.

23. Ghose T, Tai J, Guclu A, et al. Antibodies as carriers of radionuclides and cytotoxic drugs in the treatment and diagnosis of cancer. Ann NY Acad Sci 1976;277:671–689.
24. Ghose T, Guclu A. Cure of a mouse lymphoma with radio-iodinated antibody. Eur J Cancer 1974;10:787–792.
25. Kohler G, Milstein C. Continuous cultures of fused cells secreting antibody of predefined specificity. Nature 1975;256:495–497.
26. Huston JS, Levinson D, Mudgett-Hunter M, et al. Protein engineering of antibody binding sites: recovery of specific activity in an anti-digoxin single-chain Fv analogue produced in *Escherichia coli*. Proc Natl Acad Sci USA 1988;85:5879–5883.
27. Holliger P, Prospero T, Winter G. "Diabodies": small bivalent and bispecific antibody fragments. Proc Nat Acad Sci USA 1993;90:6444–6448.
28. Zhu Z, Zapata G, Shalaby MR, et al. High level secretion of a humanized bispecific diabody from *Escherichia coli*. Bio/Technology (NY) 1996;14:192–196.
29. Todorovska A, Roovers RC, Dolezal O, et al. Design and application of diabodies, triabodies, and tetrabodies for cancer targeting. J Immunol Methods 2001;248:47–66.
30. Reiter Y, Brinkmann U, Lee B, Pastan I. Engineering antibody Fv fragments for cancer detection and therapy: disulfide-stabilized Fv fragments. Nat Biotechnol 1996;14:1239–1245.
31. Reiter Y, Brinkmann U, Webber KO, et al. Engineering interchain disulfide bonds into conserved framework regions of Fv fragments: improved biochemical characteristics of recombinant immunotoxins containing disulfide-stabilized Fv. Protein Eng 1994;7:697–704.
32. Reiter Y, Brinkmann U, Jung SH, et al. Disulfide stabilization of antibody Fv: computer predictions and experimental evaluation. Protein Eng 1995;8:1323–1331.
33. Reiter Y, Brinkmann U, Jung SH, et al. Improved binding and antitumor activity of a recombinant anti-erbB2 immunotoxin by disulfide stabilization of the Fv fragment. J Biol Chem 1994;269:18327–18331.
34. Reiter Y, Pastan I. Antibody engineering of recombinant Fv immunotoxins for improved targeting of cancer: disulfide-stabilized Fv immunotoxins. Clin Cancer Res 1996;2:245–252.
35. Morrison SL, Johnson MJ, Herzenberg LA, Oi VT. Chimeric human antibody molecules: mouse antigen-binding domains with human constant region domains. Proc Natl Acad Sci USA 1984;81:6851–6855.
36. Jones PT, Dear PH, Foote J, et al. Replacing the complementarity-determining regions in a human antibody with those from a mouse. Nature 1986;321:522–555.
37. McCafferty J, Griffiths AD, Winter G, Chiswell DJ. Phage antibodies: filamentous phage displaying antibody variable domains. Nature 1990;348:552–554.
38. Clackson T, Griffiths AD, Hoogenboom HR, Winter G. Making antibody fragments using phage display libraries. Nature 1991;352:624–628.
39. Green LL, Hardy MC, Maynard-Currie CE, et al. Antigen-specific human monoclonal antibodies from mice engineered with human Ig heavy and light chain YACs. Nat Genet 1994;7:13–21.
40. Lonberg N, Taylor LD, Harding FA, et al. Antigen-specific human antibodies from mice comprising four distinct genetic modifications. Nature 1994;368:856–859.
41. Shin SU, Morrison SL. Production and properties of chimeric antibody molecules. Methods Enzymol 1989;178:459–76.
42. Winter G, Harris WJ. Humanized antibodies. Immunol Today 1993;14:243–246.
43. Winter G, Griffiths AD, Hawkins RE, Hoogenboom HR. Making antibodies by phage display technology. Annu Rev Immunol 1994;12:433–455.
44. Lonberg N, Huszar D. Human antibodies from transgenic mice. Int Rev Immunol 1995;13:65–93.
45. Jakobovits A. Production and selection of antigen-specific fully human monoclonal antibodies from mice engineered with human Ig loci. Adv Drug Deliv Rev 1998;31:33–42.
46. Waksal H. Role of an anti-epidermal growth factor receptor in treating cancer. Cancer Metastasis Rev 1999;18:427–436.
47. Pegram MD, Konecny G, Slamon DJ. The molecular and cellular biology of HER2/neu gene amplification/overexpression and the clinical development of herceptin (trastuzumab) therapy for breast cancer. Cancer Treat Res 2000;103:57–75.

48. Zhu Z, Bohlen P, Witte L. Clinical development of angiogenesis inhibitors to vascular endothelial growth factor and its receptors as cancer therapeutics. Curr Cancer Drug Targets 2002;2:135–156.
49. Coiffier B. Rituximab in combination with CHOP improves survival in elderly patients with aggressive non-Hodgkin's lymphoma. Semin Oncol 2002;29(Suppl 6):18–22.
50. Ligibel JA, Winer EP. Trastuzumab/chemotherapy combinations in metastatic breast cancer. Semin Oncol 2002;29(Suppl 11):38–43.
51. Lewis GD, Figari I, Fendly B, et al. Differential responses of human tumor cell lines to anti-p185HER2 monoclonal antibodies. Cancer Immunol Immunother 1993;37:255–263.
52. Reff ME, Carner K, Chambers KS, et al. Depletion of B cells in vivo by a chimeric mouse human monoclonal antibody to CD20. Blood 1994;83:435–445.
53. Clynes RA, Towers TL, Presta LG, Ravetch JV. Inhibitory Fc receptors modulate in vivo cytoxicity against tumor targets. Nat Med 2000;6:443–446.
54. DeFreitas E, Suzuki H, Herlyn D, et al. Human antibody induction to the idiotypic and anti-idiotypic determinants of a monoclonal antibody against a gastrointestinal carcinoma antigen. Curr Top Microbiol Immunol 1985;119:75–89.
55. Wentworth AD, Jones LH, Wentworth P Jr, et al. Antibodies have the intrinsic capacity to destroy antigens. Proc Natl Acad Sci USA 2000;97:10930–10935.
56. French RR, Chan HT, Tutt AL, Glennie MJ. CD40 antibody evokes a cytotoxic T-cell response that eradicates lymphoma and bypasses T-cell help. Nat Med 1999;5:548–553.
57. Hurwitz AA, Foster BA, Kwon ED, et al. Combination immunotherapy of primary prostate cancer in a transgenic mouse model using CTLA-4 blockade. Cancer Res 2000;60:2444–2448.
58. Brinkmann U. Recombinant antibody fragments and immunotoxin fusions for cancer therapy. In Vivo 2000;14:21–27.
59. Kreitman RJ. Toxin-labeled monoclonal antibodies. Curr Pharmacies Biotech 2001;2:313–325.
60. Reiter Y. Recombinant immunotoxins in targeted cancer cell therapy. Adv Cancer Res 2001;81:93–124.
61. Sievers EL, Linenberger M. Mylotarg: antibody-targeted chemotherapy comes of age. Curr Opin Oncol 2001;13:522–527.
62. Krasner C, Joyce RM. Zevalin: 90yttrium labeled anti-CD20 (ibritumomab tiuxetan), a new treatment for non-Hodgkin's lymphoma. Curr Pharm Biotechnol 2001;2:341–349.
63. Cheson B. Bexxar (Corixa/GlaxoSmithKline). Curr Opin Invest Drugs 2002;3:165–170.
64. Holton OD 3rd, Black CD, Parker RJ, et al. Biodistribution of monoclonal IgG1, F(ab')2, and Fab' in mice after intravenous injection. Comparison between anti-B cell (anti-Lyb8.2) and irrelevant (MOPC-21) antibodies. J Immunol 1987;139:3041–3049.
65. Zhu Z, Ghose T, Iles S, et al. Pharmacokinetics, biodistribution and tumor localization of two anti-human B-cell chronic lymphocytic leukemia monoclonal antibodies and their F(ab)'2 fragments in a xenograft model. Cancer Lett 1994;76:31–44.
66. Dykes PW, Bradwell AR, Chapman CE, Vaughan AT. Radioimmunotherapy of cancer: clinical studies and limiting factors. Cancer Treat Rev 1987;14:87–106.
67. Thomas GD, Chappell MJ, Dykes PW, et al. Effect of dose, molecular size, affinity, and protein binding on tumor uptake of antibody or ligand: a biomathematical model. Cancer Res 1989;49:3290–3296.
68. Cannon JB, Hui HW. Immunoconjugates in drug delivery systems. Targeted Diagn Ther 1990;3:121–139.
69. Schlom J, Eggensperger D, Colcher D, et al. Therapeutic advantage of high-affinity anticarcinoma radioimmunoconjugates. Cancer Res 1992;52:1067–1072.
70. Adams GP, Schier R, Marshall K, et al. Increased affinity leads to improved selective tumor delivery of single-chain Fv antibodies. Cancer Res 1998;58:485–490.
71. Adams GP, Schier R, McCall AM, et al. High affinity restricts the localization and tumor penetration of single-chain fv antibody molecules. Cancer Res 2001;61:4750–4755.
72. Prewett M, Huber J, Li Y, et al. Antivascular endothelial growth factor receptor (fetal liver kinase 1) monoclonal antibody inhibits tumor angiogenesis and growth of several mouse and human tumors. Cancer Res 1999;59:5209–5218.

73. Hicklin DJ, Witte L, Zhu Z, et al. Monoclonal antibody strategies to block angiogenesis. Drug Discov Today 2001;6:517–528.
74. Brekken RA, Thorpe PE. Vascular endothelial growth factor and vascular targeting of solid tumors. Anticancer Res 2001;21:4221–4229.
75. Varner JA, Cheresh DA. Integrins and cancer. Curr Opin Cell Biol 1996;8:724–730.
76. Liao F, Li Y, O'Connor W, et al. Monoclonal antibody to vascular endothelial-cadherin is a potent inhibitor of angiogenesis, tumor growth, and metastasis. Cancer Res 2000;60:6805–6810.
77. Viti F, Giovannoni L, Neri D. Recombinant antibodies for the selective targeting of tumor neovasculature. Curr Opin Drug Discov Devel 2002;5:204–213.
78. Halin C, Neri D. Antibody-based targeting of angiogenesis. Crit Rev Ther Drug Carrier Syst 2001;18:299–339.
79. Press OW, Hansen JA, Farr A, Martin PJ. Endocytosis and degradation of murine anti-human CD3 monoclonal antibodies by normal and malignant T-lymphocytes. Cancer Res 1988;48:2249–2257.
80. Hiltunen JV. Search for new and improved radiolabeling methods for monoclonal antibodies. A review of different methods. Acta Oncol 1993;32:831–839.
81. Garg PK, Slade SK, Harrison CL, Zalutsky MR. Labeling proteins using aryl iodide acylation agents: influence of meta vs para substitution on in vivo stability. Int J Rad Appl Instrum [B] 1989;16:669–673.
82. Epenetos AA, Snook D, Durbin H, Johnson PM, Taylor-Papadimitriou J. Limitations of radiolabeled monoclonal antibodies for localization of human neoplasms. Cancer Res 1986;46:3183–3191.
83. Sands H, Jones PL. Physiology of monoclonal antibody accretion by tumors. Cancer Treat Res 1990;51:97–122.
84. Jain RK. Transport of molecules, particles, and cells in solid tumors. Annu Rev Biomed Eng 1999;1:241–263.
85. Jain RK. Delivery of molecular and cellular medicine to solid tumors. Adv Drug Deliv Rev 2001;46:149–168.
86. Jain RK. Transport of molecules across tumor vasculature. Cancer Metastasis Rev 1987;6:559–593.
87. Jain RK. Vascular and interstitial barriers to delivery of therapeutic agents in tumors. Cancer Metastasis Rev 1990;9:253–266.
88. Jain RK. Transport of molecules in the tumor interstitium: a review. Cancer Res 1987;47:3039–3051.
89. Carrasquillo JA, Abrams PG, Schroff RW, et al. Effect of antibody dose on the imaging and biodistribution of indium-111 9.2.27 anti-melanoma monoclonal antibody. J Nucl Med 1988;29:39–47.
90. Fenwick JR, Philpott GW, Connett JM. Biodistribution and histological localization of anti-human colon cancer monoclonal antibody (MAb) 1A3: the influence of administered MAb dose on tumor uptake. Int J Cancer 1989;44:1017–1027.
91. Boerman OC, Sharkey RM, Wong GY, et al. Influence of antibody protein dose on therapeutic efficacy of radioiodinated antibodies in nude mice bearing GW-39 human tumor. Cancer Immunol Immunother 1992;35:127–134.
92. Grossbard ML, Press OW, Appelbaum FR, et al. Monoclonal antibody-based therapies of leukemia and lymphoma. Blood 1992;80:863–878.
93. Jones PL, Brown BA, Sands H. Uptake and metabolism of 111In-labeled monoclonal antibody B6.2 by the rat liver. Cancer Res 1990;50(Suppl):852s–856s.
94. Rowlinson G, Snook D, Busza A, Epenetos AA. Antibody-guided localization of intraperitoneal tumors following intraperitoneal or intravenous antibody administration. Cancer Res 1987;47:6528–6531.
95. Ghose T, Blair AH. The design of cytotoxic-agent-antibody conjugates. Crit Rev Ther Drug Carrier Syst 1987;3:263–359.
96. Blair AH, Ghose T. Linkage of cytotoxic agents to immunoglobulins. J Immunol Methods 1983;59:129–143.

97. Ghose T, Blair AH, Kulkarni PN. Preparation of antibody-linked cytotoxic agents. Methods Enzymol 1983;93:280–333.
98. Fitzpatrick JJ, Garnett MC. Design, synthesis and in vitro testing of methotrexate carrier conjugates linked via oligopeptide spacers. Anticancer Drug Des 1995;10:1–9.
99. Dillman RO, Johnson DE, Shawler DL, Koziol JA. Superiority of an acid-labile daunorubicin-monoclonal antibody immunoconjugate compared to free drug. Cancer Res 1988;48:6097–6102.
100. Rodwell JD, Alvarez VL, Lee C, et al. Site-specific covalent modification of monoclonal antibodies: in vitro and in vivo evaluations. Proc Natl Acad Sci USA 1986;83:2632–2636.
101. Kralovec J, Singh M, Mammen M, et al. Synthesis of site-specific methotrexate-IgG conjugates. Comparison of stability and antitumor activity with active-ester-based conjugates. Cancer Immunol Immunother 1989;29:293–302.
102. Aboud-Pirak E, Lesur B, Rao KS, et al. Cytotoxic activity of daunorubicin or vindesin conjugated to a monoclonal antibody on cultured MCF-7 breast carcinoma cells. Biochem Pharmacol 1989;38:641–648.
103. Chari RV, Martell BA, Gross JL, et al. Immunoconjugates containing novel maytansinoids: promising anticancer drugs. Cancer Res 1992;52:127–131.
104. Smith AL, Nicolaou KC. The enediyne antibiotics. J Med Chem 1996;39:2103–2117.
105. Liu C, Tadayoni BM, Bourret LA, et al. Eradication of large colon tumor xenografts by targeted delivery of maytansinoids. Proc Natl Acad Sci USA 1996;93:8618–8623.
106. Kondo S, Nakatsu S, Sakahara H, et al. Antitumour activity of an immunoconjugate composed of anti-human astrocytoma monoclonal antibody and neocarzinostatin. Eur J Cancer 1993;29A:420–423.
107. Lode HN, Reisfeld RA, Handgretinger R, et al. Targeted therapy with a novel enediyene antibiotic calicheamicin q(I)1 effectively suppresses growth and dissemination of liver metastases in a syngeneic model of murine neuroblastoma. Cancer Res 1998;58:2925–2928.
108. Kulkarni PN, Blair AH, Ghose T. Covalent binding of methotrexate to immunoglobulins and the effect of antibody-linked drug on tumor growth in vivo. Cancer Res 1981;41:2700–2706.
109. Trail PA, Bianchi AB. Monoclonal antibody drug conjugates in the treatment of cancer. Curr Opin Immunol 1999;11:584–588.
110. Zhu Z, Kralovec J, Ghose T, Mammen M. Inhibition of Epstein-Barr-virus-transformed human chronic lymphocytic leukaemic B cells with monoclonal-antibody-adriamycin (doxorubicin) conjugates. Cancer Immunol Immunother 1995;40:257–267.
111. Ghose T, Singh M, Faulkner G, et al. Antibody-aided liposomal drug delivery. In: Gregoriadis G, ed. Liposomes as Drug Carriers. New York: John Wiley & Sons, 1988, pp. 697–708.
112. Mastrobattista E, Koning GA, Storm G. Immunoliposomes for the targeted delivery of antitumor drugs. Adv Drug Deliv Rev 1999;40:103–127.
113. Bendas G. Immunoliposomes: a promising approach to targeting cancer therapy. BioDrugs 2001;15:215–224.
114. Ishida T, Iden DL, Allen TM. A combinatorial approach to producing sterically stabilized (Stealth) immunoliposomal drugs. FEBS Lett 1999;460:129–133.
115. Iden DL, Allen TM. In vitro and in vivo comparison of immunoliposomes made by conventional coupling techniques with those made by a new post-insertion approach. Biochim Biophys Acta 2001;1513:207–216.
116. Kong G, Anyarambhatla G, Petros WP, Braun RD, Colvin OM, Needham D, Dewhirst MW. Efficacy of liposomes and hyperthermia in a human tumor xenograft model: importance of triggered drug release. Cancer Res 2000;60:6950–6957.
117. Stickney DR, Gridley DS, Kirk GA, Slater JM. Enhancement of monoclonal antibody binding to melanoma with single dose radiation or hyperthermia. NCI Monogr 1987;3:47–52.
118. Hennigan TW, Begent RH, Allen-Mersh TG. Histamine, leukotriene C4 and interleukin-2 increase antibody uptake into a human carcinoma xenograft model. Br J Cancer 1991;64:872–874.
119. Khawli LA, Miller GK, Epstein AL. Effect of seven new vasoactive immunoconjugates on the enhancement of monoclonal antibody uptake in tumors. Cancer 1994;73(Suppl):824–831.
120. Papahadjopoulos D, Allen TM, Gabizon A, Mayhew E, Matthay K, Huang SK, Lee KD, Woodle MC, Lasic DD, Redemann C, et al. Sterically stabilized liposomes: improvements in

pharmacokinetics and antitumor therapeutic efficacy. Proc Natl Acad Sci USA 1991;88:11460–11464.
121. Tseng YL, Hong RL, Tao MH, Chang FH. Sterically stabilized anti-idiotype immunoliposomes improve the therapeutic efficacy of doxorubicin in a murine B-cell lymphoma model. Int J Cancer 1999;80:723–730.
122. Park JW, Hong K, Kirpotin DB, et al. Anti-HER2 immunoliposomes: enhanced efficacy attributable to targeted delivery. Clin Cancer Res 2002;8:1172–1181.
123. Park JW, Kirpotin DB, Hong K, et al. Tumor targeting using anti-her2 immunoliposomes. J Control Release 2001;74:95–113.
124. Lopes de Menezes DE, Pilarski LM, Allen TM. In vitro and in vivo targeting of immunoliposomal doxorubicin to human B-cell lymphoma. Cancer Res 1998;58:3320–3330.
125. Voutsadakis IA. Gemtuzumab Ozogamicin (CMA-676, Mylotarg) for the treatment of CD33+ acute myeloid leukemia. Anticancer Drugs 2002;13:685–892.
126. Sievers EL, Appelbaum FR, Spielberger RT, et al. Selective ablation of acute myeloid leukemia using antibody-targeted chemotherapy: a phase I study of an anti-CD33 calicheamicin immunoconjugate. Blood 1999;93:3678–3684.
127. Sievers EL, Larson RA, Stadtmauer EA, et al. Efficacy and safety of gemtuzumab ozogamicin in patients with CD33-positive acute myeloid leukemia in first relapse. J Clin Oncol 2001;19:3244–3254.
128. Hellstrom I, Hellstrom KE, Senter PD. Development and activities of the BR96-doxorubicin immunoconjugate. Methods Mol Biol 2001;166:3–16.
129. Trail PA, Willner D, Lasch SJ, et al. Cure of xenografted human carcinomas by BR96-doxorubicin immunoconjugates. Science 1993;261:212–215.
130. Sjogren HO, Isaksson M, Willner D, et al. Antitumor activity of carcinoma-reactive BR96-doxorubicin conjugate against human carcinomas in athymic mice and rats and syngeneic rat carcinomas in immunocompetent rats. Cancer Res 1997;57:4530–4536.
131. Slichenmyer WJ, Saleh MN, Bookman MA, et al. Phase I studies of BR96 doxorubicin in patients with advanced solid tumors that express the Lewis Y antigen. Anti-Cancer Treatment, Sixth International Congress (Abstract), 1996;95:6–9.
132. Saleh MN, LoBuglio AF, Trail PA. Immunoconjugate therapy of solid tumors: studies with BR96-doxorubicin. In: Grossbard M, ed. Basic and Clinical Oncology: Monoclonal Antibody-Based Therapy of Cancer, vol. 15. New York, Marcel Dekker, 1998, pp. 397–416.
133. Tolcher AW, Sugarman S, Gelmon KA, et al. Randomized phase II study of BR96-doxorubicin conjugate in patients with metastatic breast cancer. J Clin Oncol 1999;17:478–484.
134. Trail PA, Willner D, Bianchi AB, et al. Enhanced antitumor activity of paclitaxel in combination with the anticarcinoma immunoconjugate BR96-doxorubicin. Clin Cancer Res 1999;5:3632–3638.
135. Buchsbaum DJ, Lawrence TS, Roberson PL, et al. Comparison of 131I- and 90Y-labeled monoclonal antibody 17-1A for treatment of human colon cancer xenografts. Int J Radiat Oncol Biol Phys 1993;25:629–638.
136. Stein R, Chen S, Haim S, Goldenberg DM. Advantage of yttrium-90-labeled over iodine-131-labeled monoclonal antibodies in the treatment of a human lung carcinoma xenograft. Cancer 1997;80(Suppl):2636–2641.
137. Stein R, Govindan SV, Chen S, et al. Radioimmunotherapy of a human lung cancer xenograft with monoclonal antibody RS7: evaluation of (177)Lu and comparison of its efficacy with that of (90)Y and residualizing (131)I. J Nucl Med 2001;42:967–974.
138. Zalutsky MR, Bigner DD. Radioimmunotherapy with alpha-particle emitting radioimmunoconjugates. Acta Oncol 1996;35:373–379.
139. Imam SK. Advancements in cancer therapy with alpha-emitters: a review. Int J Radiat Oncol Biol Phys 2001;51:271–278.
140. Mattes MJ. Radionuclide-antibody conjugates for single-cell cytotoxicity. Cancer 2002;94(Suppl):1215–1223.
141. Welt S, Scott AM, Divgi CR, et al. Phase I/II study of iodine 125-labeled monoclonal antibody A33 in patients with advanced colon cancer. J Clin Oncol 1996;14:1787–1797.

142. Wilbur DS. Radiohalogenation of proteins: an overview of radionuclides, labeling methods, and reagents for conjugate labeling. Bioconjug Chem 1992;3:433–470.
143. Bhargava KK, Acharya SA. Labeling of monoclonal antibodies with radionuclides. Semin Nucl Med 1989;19:187–201.
144. Nikula TK, Bocchia M, Curcio MJ, et al. Impact of the high tyrosine fraction in complementarity determining regions: measured and predicted effects of radioiodination on IgG immunoreactivity. Mol Immunol 1995;32:865–872.
145. Zalutsky MR, Narula AS. A method for the radiohalogenation of proteins resulting in decreased thyroid uptake of radioiodine. Int J Rad Appl Instrum [A] 1987;38:1051–1055.
146. Zalutsky MR, Garg PK, Narula AS. Labeling monoclonal antibodies with halogen nuclides. Acta Radiol Suppl 1990;374:141–145.
147. Wilbur DS, Hadley SW, Hylarides MD, et al. Development of a stable radioiodinating reagent to label monoclonal antibodies for radiotherapy of cancer. J Nucl Med 1989;30:216–226.
148. Stein R, Govindan SV, Mattes MJ, et al. Improved iodine radiolabels for monoclonal antibody therapy. Cancer Res 2003;63:111–118.
149. Zalutsky MR, Narula AS. Radiohalogenation of a monoclonal antibody using an N-succinimidyl 3-(tri-n-butylstannyl)benzoate intermediate. Cancer Res 1988;48:1446–1450.
150. Ram S, Buchsbaum DJ. Radioiodination of monoclonal antibodies D612 and 17-1A with 3-iodophenylisothiocyanate and their biodistribution in tumor-bearing nude mice. Cancer 1994;73(Suppl):808–815.
151. Garg PK, Alston KL, Welsh PC, Zalutsky MR. Enhanced binding and inertness to dehalogenation of alpha-melanotropic peptides labeled using N-succinimidyl 3-iodobenzoate. Bioconjug Chem 1996;7:233–239.
152. Zhu Z, Ghose T, Kralovec Y, Yang C. Immunoreactivity, stability, pharmacokinetics and biodistribution of a monoclonal antibody to human leukemic B cells after three different methods of radioiodination. Nucl Med Biol 1994;21:873–882.
153. Subramanian R, Meares CF. Bifunctional chelating agents for radiometal-labeled monoclonal antibodies. Cancer Treat Res 1990;51:183–199.
154. Gansow OA. Newer approaches to the radiolabeling of monoclonal antibodies by use of metal chelates. Int J Rad Appl Instrum [B] 1991;18:369–381.
155. Wu RS. Novel bifunctional linkers for antibody chelation with radiometals. Cancer Treat Res 1990;51:215–232.
156. Deshpande SV, Denardo SJ, Meares CF, et al. Effect of different linkages between chelates and monoclonal antibodies on levels of radioactivity in the liver. Int J Rad Appl Instrum [B] 1989;16:587–597.
157. Wiseman GA, White CA, Stabin M, et al. Phase I/II 90Y-Zevalin (yttrium-90 ibritumomab tiuxetan, IDEC-Y2B8) radioimmunotherapy dosimetry results in relapsed or refractory non-Hodgkin's lymphoma. Eur J Nucl Med 2000;27:766–777.
158. Tobinai K, Kobayashi Y, Narabayashi M, et al. Feasibility and pharmacokinetic study of a chimeric anti-CD20 monoclonal antibody (IDEC-C2B8, rituximab) in relapsed B-cell lymphoma. The IDEC-C2B8 Study Group. Ann Oncol 1998;9:527–534.
159. Knox SJ, Goris ML, Trisler K, et al. Yttrium-90-labeled anti-CD20 monoclonal antibody therapy of recurrent B-cell lymphoma. Clin Cancer Res 1996;2:457–470.
160. Witzig TE, White CA, Wiseman GA, et al. Phase I/II trial of IDEC-Y2B8 radioimmunotherapy for treatment of relapsed or refractory CD20(+) B-cell non-Hodgkin's lymphoma. J Clin Oncol 1999;17:3793–3803.
161. Witzig TE, Gordon LI, Cabanillas F, et al. Randomized controlled trial of yttrium-90-labeled ibritumomab tiuxetan radioimmunotherapy versus rituximab immunotherapy for patients with relapsed or refractory low-grade, follicular, or transformed B-cell non-Hodgkin's lymphoma. J Clin Oncol 2002;20:2453–2463.
162. Wiseman GA, Gordon LI, Multani PS, et al. Ibritumomab tiuxetan radioimmunotherapy for patients with relapsed or refractory non-Hodgkin lymphoma and mild thrombocytopenia: a phase II multicenter trial. Blood 2002;99:4336–4342.

163. Witzig TE, Flinn IW, Gordon LI, et al. Treatment with ibritumomab tiuxetan radioimmunotherapy in patients with rituximab-refractory follicular non-Hodgkin's lymphoma. J Clin Oncol 2002;20:3262–3269.
164. Kaminski MS, Zasadny KR, Francis IR, et al. Iodine-131-anti-B1 radioimmunotherapy for B-cell lymphoma. J Clin Oncol 1996;14:1974–1981.
165. Kaminski MS, Estes J, Zasadny KR, et al. Radioimmunotherapy with iodine (131)I tositumomab for relapsed or refractory B-cell non-Hodgkin lymphoma: updated results and long-term follow-up of the University of Michigan experience. Blood 2000;96:1259–1266.
166. Vose JM, Wahl RL, Saleh M, et al. Multicenter phase II study of iodine-131 tositumomab for chemotherapy-relapsed/refractory low-grade and transformed low-grade B-cell non-Hodgkin's lymphomas. J Clin Oncol 2000;18:1316–1323.
167. Kaminski MS, Estes J, Tuck M, et al. Iodine 131 tositumomab therapy for previously untreated follicular lymphoma (FL). Proc Am Soc Clin Oncol 2000;19:abstract 11.
168. Kaminski MS, Zelenetz AD, Press OW, et al. Pivotal study of iodine I 131 tositumomab for chemotherapy-refractory low-grade or transformed low-grade B-cell non-Hodgkin's lymphomas. J Clin Oncol 2001;19:3918–3928.
169. Horning SJ, Lucas JB, Younes A, et al. Iodine-131 tositumomab for non-Hodgkin's lymphoma (NHL) patients who progressed after treatment with rituximab: result of multicenter phase II study. Blood 2000;96(suppl 1):abstract 2184.
170. Yamaizumi M, Mekada E, Uchida T, Okada Y. One molecule of diphtheria toxin fragment A introduced into a cell can kill the cell. Cell 1978;15:245–250.
171. Pai LH, Batra JK, FitzGerald DJ, et al. Anti-tumor activities of immunotoxins made of monoclonal antibody B3 and various forms of Pseudomonas exotoxin. Proc Natl Acad Sci USA 1991;88:3358–3362.
172. Steeves RM, Denton ME, Barnard FC, et al. Identification of three oligosaccharide binding sites in ricin. Biochemistry 1999;38:11677–11685.
173. Vallera DA, Quinones RR, Azemove SM, Soderling CC. Monoclonal antibody-toxin conjugates reactive against human T lymphocytes. A comparison of antibody linked to intact ricin toxin with antibody linked to ricin A chain. Transplantation 1984;37:387–392.
174. Eccles SA, McIntosh DP, Purvies HP, et al. An ineffective monoclonal antibody-ricin A chain conjugate is converted to a tumouricidal agent in vivo by subsequent systemic administration of ricin B chain. Cancer Immunol Immunother 1987;24:37–41.
175. Worrell NR, Skilleter DN, Cumber AJ, Price RJ. Mannose receptor dependent uptake of a ricin A chain: antibody conjugate by rat liver non-parenchymal cells. Biochem Biophys Res Commun 1986;137:892–896.
176. Bourrie BJ, Casellas P, Blythman HE, Jansen FK. Study of the plasma clearance of antibody: ricin-A-chain immunotoxins. Evidence for specific recognition sites on the A chain that mediate rapid clearance of the immunotoxin. Eur J Biochem 1986;155:1–10.
177. Blakey DC, Watson GJ, Knowles PP, Thorpe PE. Effect of chemical deglycosylation of ricin A chain on the in vivo fate and cytotoxic activity of an immunotoxin composed of ricin A chain and anti-Thy 1.1 antibody. Cancer Res 1987;47:947–952.
178. Fulton RJ, Tucker TF, Vitetta ES, Uhr JW. Pharmacokinetics of tumor-reactive immunotoxins in tumor-bearing mice: effect of antibody valency and deglycosylation of the ricin A chain on clearance and tumor localization. Cancer Res 1988;48:2618–2625.
179. Lambert JM, McIntyre G, Gauthier MN, et al. The galactose-binding sites of the cytotoxic lectin ricin can be chemically blocked in high yield with reactive ligands prepared by chemical modification of glycopeptides containing triantennary N-linked oligosaccharides. Biochemistry 1991;30:3234–3247.
180. Grossbard ML, Lambert JM, Goldmacher VS, et al. Correlation between in vivo toxicity and preclinical in vitro parameters for the immunotoxin anti-B4-blocked ricin. Cancer Res 1992;52:4200–4207.
181. Wawrzynczak EJ, Watson GJ, Cumber AJ, et al. Blocked and non-blocked ricin immunotoxins against the CD4 antigen exhibit higher cytotoxic potency than a ricin A chain immunotoxin potentiated with ricin B chain or with a ricin B chain immunotoxin. Cancer Immunol Immunother 1991;32:289–295.

182. McIntosh DP, Edwards DC, Cumber AJ, et al. Ricin B chain converts a non-cytotoxic antibody-ricin A chain conjugate into a potent and specific cytotoxic agent. FEBS Lett 1983;164:17–20.
183. Bolognesi A, Tazzari PL, Olivieri F, et al. Evaluation of immunotoxins containing single-chain ribosome-inactivating proteins and an anti-CD22 monoclonal antibody (OM124): in vitro and in vivo studies. Br J Haematol 1998;101:179–188.
184. Barbieri L, Bolognesi A, Valbonesi P, et al. Polynucleotide: adenosine glycosidase activity of immunotoxins containing ribosome-inactivating proteins. J Drug Target 2000;8:281–288.
185. Chaudhary VK, Queen C, Junghans RP, et al. A recombinant immunotoxin consisting of two antibody variable domains fused to Pseudomonas exotoxin. Nature 1989;339:394–397.
186. Brinkmann U, Pai LH, FitzGerald DJ, et al. B3(Fv)-PE38KDEL, a single-chain immunotoxin that causes complete regression of a human carcinoma in mice. Proc Natl Acad Sci USA 1991;88:8616–8620.
187. Brinkmann U, Reiter Y, Jung SH, et al. A recombinant immunotoxin containing a disulfide-stabilized Fv fragment. Proc Natl Acad Sci USA 1993;90:7538–7542.
188. Pastan I, Chaudhary VK, FitzGerald DJ. Recombinant toxins as novel therapeutic agents. Annu Rev Biochem 1992;61:331–354.
189. Chaudhary VK, FitzGerald DJ, Pastan I. A proper amino terminus of diphtheria toxin is important for cytotoxicity. Biochem Biophys Res Commun 1991;180:545–551.
190. Figgitt DP, Lamb HM, Goa KL. Denileukin diftitox. Am J Clin Dermatol 2000;1:67–72.
191. Foss FM. Interleukin-2 fusion toxin: targeted therapy for cutaneous T cell lymphoma. Ann NY Acad Sci 2001;941:166–176.
192. LeMaistre CF, Saleh MN, Kuzel TM, et al. Phase I trial of a ligand fusion-protein (DAB389IL-2) in lymphomas expressing the receptor for interleukin-2. Blood 1998;91:399–405.
193. Saleh MN, LeMaistre CF, Kuzel TM, et al. Antitumor activity of DAB389IL-2 fusion toxin in mycosis fungoides. J Am Acad Dermatol 1998;39:63–73.
194. Olsen E, Duvic M, Frankel A, et al. Pivotal phase III trial of two dose levels of denileukin diftitox for the treatment of cutaneous T-cell lymphoma. J Clin Oncol 2001;19:376–388.
195. Kreitman RJ, Wilson WH, White JD, et al. Phase I trial of recombinant immunotoxin anti-Tac(Fv)-PE38 (LMB-2) in patients with hematologic malignancies. J Clin Oncol 2000;18:1622–1636.
196. Barth S. Technology evaluation: BL22, NCI. Curr Opin Mol Ther 2002;4:72–75.
197. Kreitman RJ, Wilson WH, Bergeron K, et al. Efficacy of the anti-CD22 recombinant immunotoxin BL22 in chemotherapy-resistant hairy-cell leukemia. N Engl J Med 2001;345:241–247.
198. Baluna R, Vitetta ES. Vascular leak syndrome: a side effect of immunotherapy. Immunopharmacology 1997;37:117–132.
199. Kuan CT, Pai LH, Pastan I. Immunotoxins containing Pseudomonas exotoxin that target LeY damage human endothelial cells in an antibody-specific mode: relevance to vascular leak syndrome. Clin Cancer Res 1995;1:1589–1594.
200. Lindstrom AL, Erlandsen SL, Kersey JH, Pennell CA. An in vitro model for toxin-mediated vascular leak syndrome: ricin toxin A chain increases the permeability of human endothelial cell monolayers. Blood 1997;90:2323–2334.
201. Siegall CB, Liggitt D, Chace D, et al. Characterization of vascular leak syndrome induced by the toxin component of Pseudomonas exotoxin-based immunotoxins and its potential inhibition with nonsteroidal anti-inflammatory drugs. Clin Cancer Res 1997;3:339–345.
202. Baluna R, Rizo J, Gordon BE, et al. Evidence for a structural motif in toxins and interleukin-2 that may be responsible for binding to endothelial cells and initiating vascular leak syndrome. Proc Natl Acad Sci USA 1999;96:3957–3962.
203. Baluna R, Coleman E, Jones C, et al. The effect of a monoclonal antibody coupled to ricin A chain-derived peptides on endothelial cells in vitro: insights into toxin-mediated vascular damage. Exp Cell Res 2000;258:417–424.
204. Vitetta ES. Immunotoxins and vascular leak syndrome. Cancer J 2000;6(Suppl 3):S218–S224.
205. Siegall CB, Haggerty HG, Warner GL, et al. Prevention of immunotoxin-induced immunogenicity by coadministration with CTLA4Ig enhances antitumor efficacy. J Immunol 1997;159:5168–5173.

206. Rybak SM, Hoogenboom HR, Meade HM, et al. Humanization of immunotoxins. Proc Natl Acad Sci USA 1992;89:3165–3169.
207. Psarras K, Ueda M, Yamamura T, et al. Human pancreatic RNase1-human epidermal growth factor fusion: an entirely human 'immunotoxin analog' with cytotoxic properties against squamous cell carcinomas. Protein Eng 1998;11:1285–1292.
208. Philpott GW, Bower RJ, Parker CW. Selective iodination and cytotoxicity of tumor cells with an antibody-enzyme conjugate. Surgery 1973;74:51–58.
209. Philpott GW, Shearer WT, Bower RJ, Parker CW. Selective cytotoxicity of hapten-substituted cells with an antibody-enzyme conjugate. J Immunol 1973;111:921–929.
210. Bragshawe KD. Antibody directed enzymes revive anti-cancer prodrugs concept. Br J Cancer 1987;56:531–532.
211. Senter PD, Saulnier MG, Schreiber GJ, et al. Anti-tumor effects of antibody-alkaline phosphatase conjugates in combination with etoposide phosphate. Proc Natl Acad Sci USA 1988;85:4842–4846.
212. Bragshawe KD. The First Bagshawe lecture. Towards generating cytotoxic agents at cancer sites. Br J Cancer 1989;60:275–281.
213. Senter PD. Activation of prodrugs by antibody-enzyme conjugates: a new approach to cancer therapy. FASEB J 1990;4:188–193.
214. Bagshawe KD, Sharma SK, Springer CJ, Rogers GT. Antibody directed enzyme prodrug therapy (ADEPT). A review of some theoretical, experimental and clinical aspects. Ann Oncol 1994;5:879–891.
215. Senter PD, Springer CJ. Selective activation of anticancer prodrugs by monoclonal antibody-enzyme conjugates. Adv Drug Deliv Rev 2001;53:247–264.
216. Rodrigues ML, Presta LG, Kotts CE, et al. Development of a humanized disulfide-stabilized anti-p185HER2 Fv-beta-lactamase fusion protein for activation of a cephalosporin doxorubicin prodrug. Cancer Res 1995;55:63–70.
217. Wolfe LA, Mullin RJ, Laethem R, et al. Antibody-directed enzyme prodrug therapy with the T268G mutant of human carboxypeptidase A1: in vitro and in vivo studies with prodrugs of methotrexate and the thymidylate synthase inhibitors GW1031 and GW1843. Bioconjug Chem 1999;10:38–48.
218. Haisma HJ, Sernee MF, Hooijberg E, et al. Construction and characterization of a fusion protein of single-chain anti-CD20 antibody and human beta-glucuronidase for antibody-directed enzyme prodrug therapy. Blood 1998;92:184–190.
219. Rader C, List B. Catalytic antibodies as magic bullets. Chemistry 2000;6:2091–2095.
220. Driggers EM, Schultz PG. Catalytic antibodies. Adv Protein Chem 1996;49:261–287.
221. Wade H, Scanlan TS. The structural and functional basis of antibody catalysis. Annu Rev Biophys Biomol Struct 1997;26:461–493.
222. Wentworth P, Datta A, Blakey D, et al. Toward antibody-directed "abzyme" prodrug therapy, ADAPT: carbamate prodrug activation by a catalytic antibody and its in vitro application to human tumor cell killing. Proc Natl Acad Sci USA 1996;93:799–803.
223. Kakinuma H, Fujii I, Nishi Y. Selective chemotherapeutic strategies using catalytic antibodies: a common pro-moiety for antibody-directed abzyme prodrug therapy. J Immunol Methods 2002;269:269–281.
224. Napier MP, Sharma SK, Springer CJ, et al. Antibody-directed enzyme prodrug therapy: efficacy and mechanism of action in colorectal carcinoma. Clin Cancer Res 2000;6:765–772.
225. Francis RJ, Sharma SK, Springer C, et al. A phase I trial of antibody directed enzyme prodrug therapy (ADEPT) in patients with advanced colorectal carcinoma or other CEA producing tumours. Br J Cancer 2002;87:600–607.
226. Cao Y, Lam L. Bispecific antibody conjugates in therapeutics. Adv Drug Deliv Rev 2003;55:171–197.
227. Peipp M, Valerius T. Bispecific antibodies targeting cancer cells. Biochem Soc Trans 2002;30:507–511.
228. Xiong D, Xu Y, Liu H, et al. Efficient inhibition of human B-cell lymphoma xenografts with an anti-CD20 x anti-CD3 bispecific diabody. Cancer Lett 2002;177:29–39.

229. Kipriyanov SM, Cochlovius B, Schafer HJ, et al. Synergistic antitumor effect of bispecific CD19 x CD3 and CD19 x CD16 diabodies in a preclinical model of non-Hodgkin's lymphoma. J Immunol 2002;169:137–144.
230. James ND, Atherton PJ, Jones J, et al. A phase II study of the bispecific antibody MDX-H210 (anti-HER2 x CD64) with GM-CSF in HER2+ advanced prostate cancer. Br J Cancer 2001;85:152–156.
231. van Spriel AB, van Ojik HH, van De Winkel JG. Immunotherapeutic perspective for bispecific antibodies. Immunol Today 2000;21:391–397.
232. Weiner LM, Alpaugh RK, von Mehren M. Redirected cellular cytotoxicity employing bispecific antibodies and other multifunctional binding proteins. Cancer Immunol Immunother 1997;45:190–192.
233. Curnow RT. Clinical experience with CD64-directed immunotherapy. An overview. Cancer Immunol Immunother 1997;45:210–215.
234. Segal DM, Weiner GJ, Weiner LM. Bispecific antibodies in cancer therapy. Curr Opin Immunol 1999;11:558–562.
235. Corvalan JR, Smith W, Gore VA. Tumour therapy with Vinca alkaloids targeted by a hybrid-hybrid monoclonal antibody recognising both CEA and Vinca alkaloids. Int J Cancer Suppl 1988;2:22–25.
236. Ford CH, Osborne PA, Rego BG, Mathew A. Bispecific antibody targeting of doxorubicin to carcinoembryonic antigen-expressing colon cancer cell lines in vitro and in vivo. Int J Cancer 2001;92:851–855.
237. Ferrini S, Sforzini S, Canevari S. Bispecific monoclonal antibodies for the targeting of type I ribosome-inactivating proteins against hematological malignancies. Methods Mol Biol 2001;166:177–192.
238. Chang CH, Sharkey RM, Rossi EA, et al. Molecular advances in pretargeting radio-imunotherapy with bispecific antibodies. Mol Cancer Ther 2002;1:553–563.
239. Sahin U, Hartmann F, Senter P, et al. Specific activation of the prodrug mitomycin phosphate by a bispecific anti-CD30/anti-alkaline phosphatase monoclonal antibody. Cancer Res 1990;50:6944–6948.
240. Cao Y, Suresh MR. Bispecific MAb aided liposomal drug delivery. J Drug Target 2000;8:257–266.
241. Rosebrough SF. Two-step immunological approaches for imaging and therapy. Q J Nucl Med 1996;40:234–251.
242. Yao Z, Zhang M, Kobayashi H, et al. Improved targeting of radiolabeled streptavidin in tumors pretargeted with biotinylated monoclonal antibodies through an avidin chase. J Nucl Med 1995;36:837–841.
243. Carter P, Ridgway J, Zhu Z. Toward the production of bispecific antibody fragments for clinical applications. J Hematother 1995;4:463–470.
244. Suresh MR, Cuello AC, Milstein C. Bispecific monoclonal antibodies from hybrid hybridomas. Methods Enzymol 1986;121:210–228.
245. Zhu Z, Lewis GD, Carter P. Engineering high affinity humanized anti-p185HER2/anti-CD3 bispecific F(ab')2 for efficient lysis of p185HER2 overexpressing tumor cells. Int J Cancer 1995;62:319–324.
246. Cao Y, Suresh MR. Bispecific antibodies as novel bioconjugates. Bioconjug Chem 1998;9:635–644.
247. Pluckthun A, Pack P. New protein engineering approaches to multivalent and bispecific antibody fragments. Immunotechnology 1997;3:83–105.
248. Carter P. Bispecific human IgG by design. J Immunol Methods 2001;248:7–15.
249. Horn U, Strittmatter W, Krebber A, et al. High volumetric yields of functional dimeric miniantibodies in Escherichia coli, using an optimized expression vector and high-cell-density fermentation under non-limited growth conditions. Appl Microbiol Biotechnol 1996;46;524–532.
250. Pack P, Kujau M, Schroeckh V, et al. Improved bivalent minantibodies, with identical avidity as whole antibodies, produced by high cell density fermentation of *Escherichia coli*. Bio/Technology 1993;11:1271–1277.

251. Zhu Z, Zapata G, Shalaby R, et al. High level secretion of a humanized bispecific diabody from Escherichia coli. Biotechnology (NY) 1996;14:192–196.

17

Automated Property Profiling for Cellular Drug Delivery

Edward H. Kerns and Li Di

1. Introduction

Drug discovery research has been primarily focused on structure-activity relationships (SAR). Compounds from libraries are screened for activity and the "hits" are used as the starting point for the synthesis of analogs that explore the chemical space around the core structure in order to increase activity. Thus, the relationship of structure to function is mapped using various bio-assays of increasing sophistication. Molecular modeling assists in the planning of structural modifications by visualizing the active site of the target protein and modeling small-molecule structures that would have higher binding affinity for the target.

An increasing interest has arisen in an alternate means of increasing activity: the improved delivery of drug to the therapeutic target *(1–12)*. Many barriers oppose the access of drug molecules to the target. In living systems, access to the target can be estimated via pharmacokinetics (PK). Thus, PK is used as a tool during drug discovery to monitor drug exposure after dosing. When PK characteristics are acceptable, then delivery is not considered a hurdle. However, in a large number of cases, PK is a hurdle and improvement must be undertaken for the compound series to be improved to reach the goal of discovering an effective development candidate. In order to diagnose the cause(s) of inadequate PK *(1,3–5,13,14)*, discovery researchers recognized the role of physicochemical and structural ("drug-like") properties. Some of the physicochemical properties most responsible for systemic exposure are solubility, permeability, lipophilicity, pKa, and stability. Screening of these properties is often termed "pharmaceutical profiling." Measurement of properties in vitro was initiated in order to diagnose or predict PK properties.

The availability of data from the measurement from pharmaceutical profiling has opened new opportunities in other areas of pharmaceutical research beyond in vivo animal experiments. Properties also affect in vitro experiments using enzymes/receptors and cell-based assays *(9,15)*. In the past, the lack of compound activity in an in vitro assay led to its removal from further consideration. However, binding to the active site of an enzyme or receptor target is not the only variable in an in vitro assay. Activity is also affected by barriers that attenuate compound delivery to the protein target. As with in vivo systems, physicochemical properties affect how the compound performs

From: *Cellular Drug Delivery: Principles and Practice*
Edited by: D. R. Lu and S. Øie © Humana Press Inc., Totowa, NJ

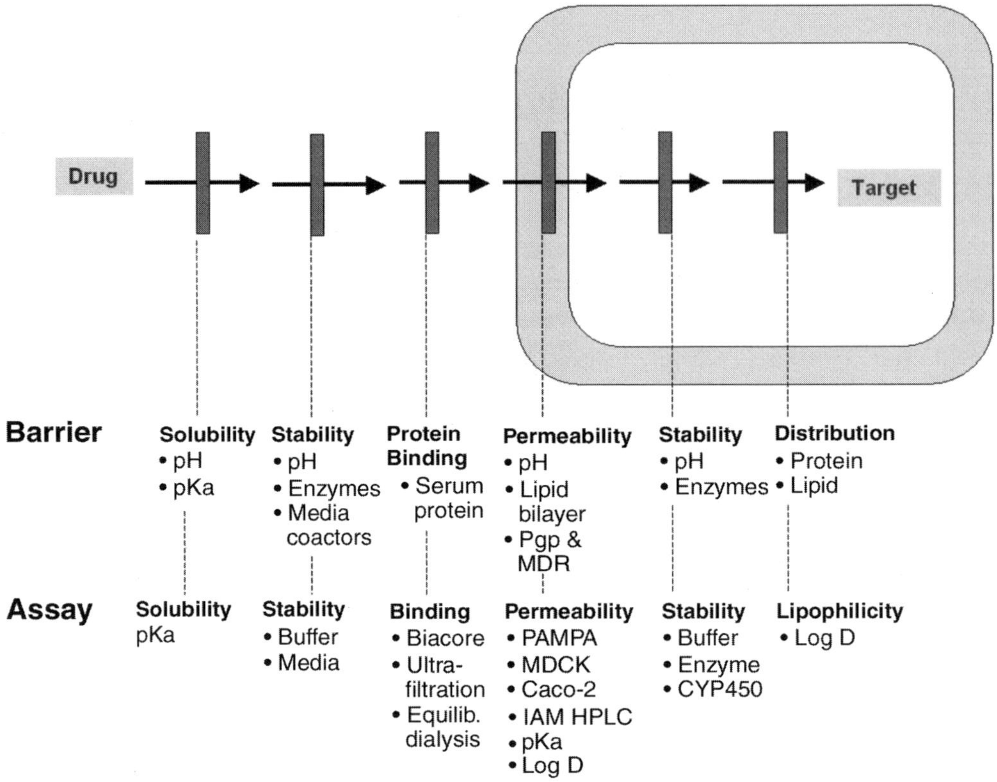

Fig. 1. Barriers to cellular drug delivery. Barriers are shown in bold. Detailed aspects of a barrier are listed as bullets. Pharmaceutical profiling assays that provide information on performance of compounds at the barriers are shown.

at these barriers and determine the ultimate concentration of compound reaching the target protein.

In this chapter, physicochemical and biochemical properties that affect cellular drug delivery will be discussed, along with automated assays that are used to measure them, and how this information may be used effectively.

2. Barriers to Cellular Drug Delivery

Several barriers exist between the drug that is in solution around the cell (in the interstitial fluid or cell-culture medium) and the target protein. These barriers are diagrammed in Fig. 1. The drug candidate first encounters the aqueous environment and its solubility at the pH of the solution depends on its intrinsic solubility and its pKa(s). For some compounds, poor solubility limits the concentration of compound surrounding the cells. Next the compound encounters the possibility of chemical conversion owing to chemical reactions. Its stability against hydrolysis by water, reactions with a substance in the solution, or catalysis by an enzyme (e.g., esterase) determines the dosed compound concentration available for cell penetration. Binding to proteins also limits the free compound in solution. Permeability of the cell membrane determines how much compound is accessible to intracellular targets. The total rate of permeability is a composite of several permeability mechanisms. The compound pKa and the pH of the solution determines the fraction of compound that is in the neutral form, that penetrates the lipid bilayer membrane. Passive diffusion through the lipid membrane is typically the predominant route of permeation. Active transport by membrane proteins can be a factor in the uptake of some compounds. Conversely, efflux by other membrane proteins (e.g., p-glycoprotein [P-gp]), removes drug candidate after it has gotten into the cell. Inside the cell, the compound may encounter other stability barriers, such as the intracellular pH and enzymes (e.g., esterase and cytochrome P450). Binding to other intracellular components owing to high lipophilicity distributes the compound and can limit the free movement of the compound to the target protein. Knowledge of these barriers and how the compound's properties perform at the barriers can help to interpret the net observed activity in cell culture experiments relative to noncellular in vitro enzyme and receptor-binding studies.

3. Pharmaceutical Profiling Assays for Cellular Delivery

As we have seen, compound properties can greatly affect cellular activity assays and should be taken into consideration in planning and interpreting experiments. In a later section, tactics for applying property information to cell-based research will be discussed. In this section, individual properties and how they can be screened for a large number of compounds is discussed *(1,4,9)*. Property screening can begin as soon as compounds are first studied, such as "hits" from high throughput screening, and carried through during the synthesis of analogs in a chemical series.

3.1. Integrity

For any compound that is new to a research project, it is first important to establish the integrity (identity and purity) of the material. A compound may have been synthesized several years before and could have been mis-identified or could have degraded during storage owing to oxidation or hydrolysis. Sometimes compounds are mislabeled. If an activity or property is attributed to the wrong structure, any subsequent SAR or structure-property relationships (SPR) will be incorrect. Liquid chromatography/ultraviolet/mass spectrometry (LC/UV/MS) methodology typically is used to check integrity. Recent methods *(16)* use a rapid (<5 min) high-performance liquid chromatography (HPLC) analysis with a wide solvent-polarity gradient to separate sample components, detection, and quantitation by UV, and alternating positive and negative ion mass spectrometry for the molecular weight of each component (Fig. 2). Typically, only compounds with identity verified by molecular weight and having high purity should be studied. Integrity profiling should be performed for all compounds tested in cell-based activity assays.

3.2. Solubility

Sufficient solubility is necessary for most experiments performed in drug discovery and development *(2,10,11)*. The maximum concentration at which a compound can be studied is determined by its solubility. Formulations can be developed to solubilize compounds at higher levels, if necessary, for in vivo dosing experiments. In drug discovery it is increasingly common for compounds to have low solubility. Solubility can affect activity as well as property assays *(10,11)*. Solubility experiments traditionally have been conducted by placing solid material into an aqueous buffer, followed by

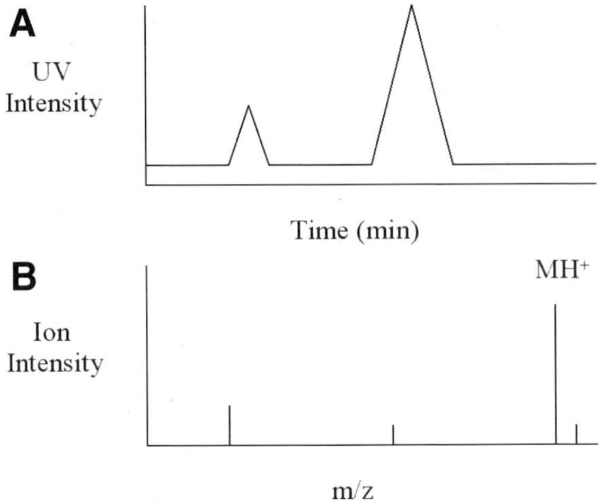

Fig. 2. Integrity profiling involves the estimation of sample purity from an HPLC chromatogram **(A)** and confirmation of identity from a mass spectrum **(B)**.

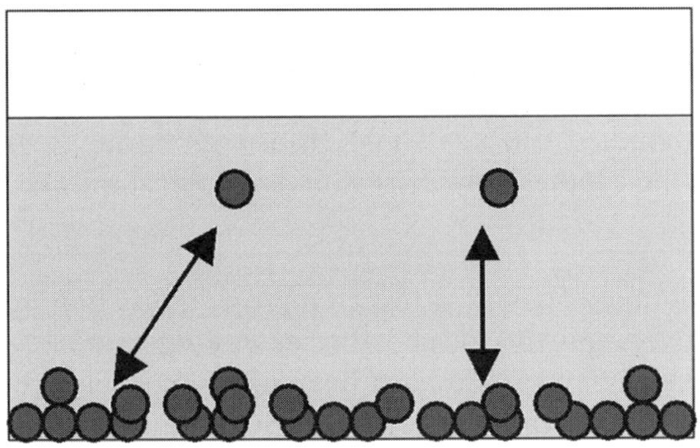

Fig. 3. Solubility profiling involves measurement of compound in aqueous buffer after a small volume of DMSO solution is added. Insoluble compound precipitates.

stirring and assaying using HPLC. This is typically called thermodynamic or equilibrium solubility measurement. However, for in vitro research it is more relevant to measure solubility when a dimethyl sulfoxide (DMSO) solution of the compound is added to an aqueous buffer (Fig. 3). This is because in vitro experiments are usually conducted by adding compound that has already been dissolved in organic solvent. This is a very different experiment and can produce very different solubility estimates. The experiment is typically called kinetic solubility measurement. Several automated assays for solubility have been described. These include turbidimetry *(2)*, nephalometry *(17)*, and direct UV *(18)* analysis. The turbidimetric and nephalometric methods involve the addition of DMSO compound solution to aqueous buffer. The solubility is determined from the appearance of precipitate, which is detected by light scattering. In the direct UV method, a 20 mg/mL solution of compound in DMSO is added to aqueous buffer in a 96-well plate. The final DMSO concentration is critical in any solubility assay because DMSO enhances aqueous solubility of organics. In this assay the final DMSO concen-

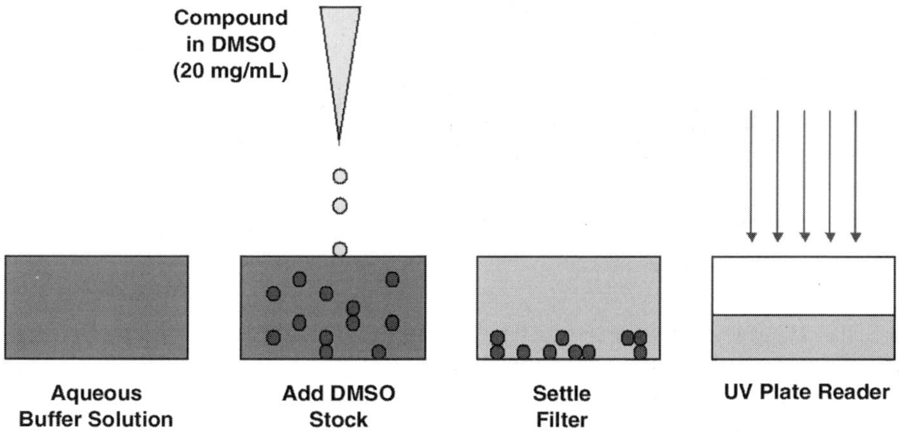

Fig. 4. Diagram of direct UV solubility assay. A DMSO solution of test compound is added to an aqueous buffer and allowed to equilibrate. The buffer is filtered to remove precipitate and the UV absorbance is obtained on a plate reader to measure concentration.

tration is less than 0.5% volume/volume. The solution equilibrates for a period of time (e.g., 18 h), the solution is filtered to remove precipitate, and the concentration of compound in solution is measured using a UV plate reader (*see* Fig. 4). Solubility measurement at the pH of the in vitro experiment is most valuable, because the compound's pKa and the solution pH can dramatically affect solubility. This is because compounds are most soluble in their charged form.

Lipinski *(10)* discussed how Caco-2 cellular experiments are affected by the compound solubility. For prediction of gastrointestinal (GI) drug absorption, Caco-2 should be run at a concentration of 100 μ*M*, because this is the approximate GI concentration for a 1 mg/kg orally administered drug. However, many labs perform the experiment at 10 μ*M*. A concentration of 100 μ*M* can saturate transporter proteins, which is not as likely at 10 μ*M*. Moreover, even at 10 μ*M*, up to one-third of compounds are not fully soluble, leading to inaccurate results.

Solubility should be screened as early as possible, owing to its great effect on activity assays. Currently, in silico methods for solubility are improving, but a measurement will assure greatest confidence for planning and interpreting experiments.

3.3. Permeability

In order for a compound to be active at intracellular therapeutic targets, it must permeate cellular lipid bilayer membranes. There are several major mechanisms of permeation into cells: passive diffusion, active uptake by membrane transporters, and active efflux by other membrane transporters. If the concentration of drug is high enough, protein transporters can become saturated, making passive diffusion the predominant permeation mechanism *(19)*. Currently, it is very difficult to predict by software whether or not a compound will be a substrate for an active transport protein.

Several assays have been described for predicting permeability by passive diffusion through lipid membranes (*see* Fig. 5). Early methods used liposomes *(20)*. Immobilized artificial membrane (IAM) chromatography has been used to simulate interactions of compounds with lipid membranes in a convenient, high-throughput format *(21–23)*. Cel-

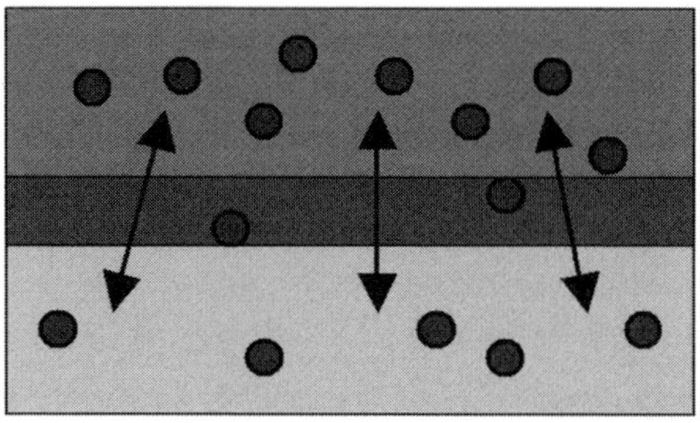

Fig. 5. Passive diffusion permeability profiling measures the rate of compound movement from the donor aqueous compound solution into the acceptor after passing through a membrane consisting of a cell monolayer or artificial lipid membrane.

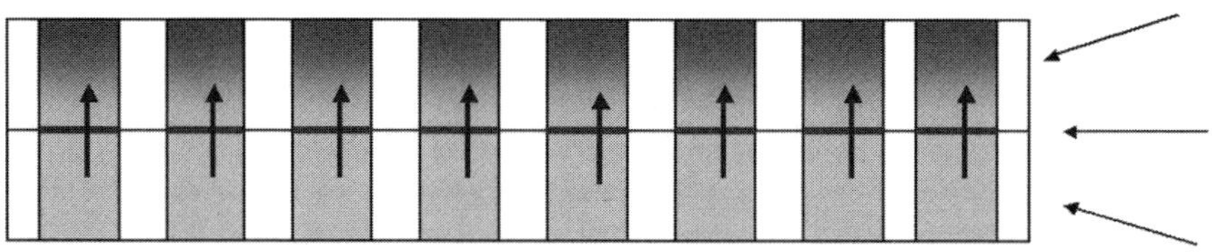

Fig. 6. Diagram of PAMPA assay. Test compound is placed in an aqueous buffer at 50 μ*M* concentration in the donor well of a 96-well plate. A porous filter plate is placed on top in contact with the water and the pours are filled with a solution of phoshpatidyl choline in dodecane. The test compound diffuses through the lipid and penetrates into the buffer in the acceptor wells.

lular-based models for permeability have been widely implemented using Caco-2 *(24,25)* and Madin-Darby canine kidney (MDCK) *(26)* cells. A monolayer of cells is grown to confluence on a porous filter surface and the test compound is placed in the buffer on one side of the monolayer (e.g., apical side). The rate of appearance of the compound on the other side of the monolayer (e.g., basolateral side) is measured using HPLC or LC/MS to calculate the permeability of the compound. Cellular models combine many mechanisms of permeability, so diagnosis of each mechanism for structural optimization requires additional studies. Expression of transporters varies with the lab or passage and should be monitored. Higher-throughput 96-well versions of Caco-2 have been reported *(27)*. Caco-2 cells are polarized, in that the transport proteins are typically found on the apical side. Thus, they may not accurately predict the intracellular concentration and may be most appropriate for studying mechanisms of GI absorption for a compound.

Parallel artificial membrane permeability assay (PAMPA) is receiving increasing attention in drug discovery *(28–31)* (*see* Fig. 6). In place of a cellular monolayer of

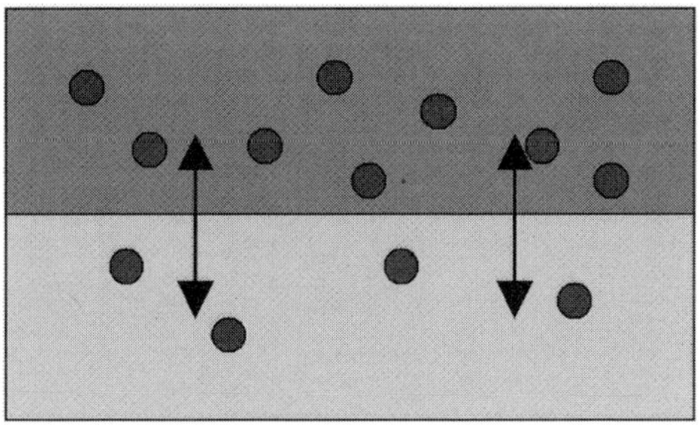

Fig. 7. Lipophilicity profiling measures the ratio of compound in the octanol and aqueous buffer layers after equilibration.

Caco-2, an artificial membrane is created using lipid in an organic diluent and supported within the pours of a 96-well filter plate. Because of the simpler aqueous matrix, a 96-well UV plate reader can be used for accelerated quantitation. The PAMPA assay is typically higher-throughput than Caco-2 and uses as little as 5% of the resources. A useful strategy for permeability is to use PAMPA for high-throughput assessment of passive diffusion for all compounds, then to use Caco-2 or another assay to probe other permeability mechanisms for a limited set of compounds. Permeability data has been used to correlate permeability to activity in cell-based activity assays and to diagnose mechanisms of permeability *(32)*.

Permeability through specialized in vivo membranes, such as the blood-brain barrier (BBB), may be predicted using specialized cell-culture models *(33)*, or with PAMPA using modified assay conditions *(34)*. Some in silico methods for permeability have been developed, including QMPRplus (Simulations Plus Inc.) for MDCK cell permeability.

3.4. Lipophilicity

Many properties are affected by the fundamental molecular property lipophilicity, the tendency of a compound to partition into nonpolar vs aqueous environments (*see* Fig. 7). Increasing lipophilicity generally has the following effects on barriers: increasing permeability, decreasing solubility, decreasing metabolic stability, increasing protein binding, and increasing distribution to lipophilic sites. The typical measure of lipophilicity is the log of the coefficient of partitioning between octanol and water, termed "Log P." Log P is used for the partitioning when all of the compound is in the neutral state. A related term, "Log D," is used to describe the distribution when the aqueous phase is at a pH where part or all of the solute is charged. Log P and Log D values have been compiled from the literature by Hansch *(35)*. For automation, the traditional "shake flask" method for Log P *(36–38)* was scaled down to 96-deep-well plate format *(39)*. In this assay, a sample in DMSO is added to a plate well containing octanol and aqueous buffer, equilibrium is established between the phases, and each phase is sampled and quantitated. Other methods for lipophilicity include the pH-metric method *(40,41)*, reversed-phase (RP)-HPLC *(16,42–48)* and microemulsion electrokinetic chromatography *(49)*. In the chromatographic methods, retention time is calibrated to Log P or Log D using standards and the lipophilicity of test compounds is

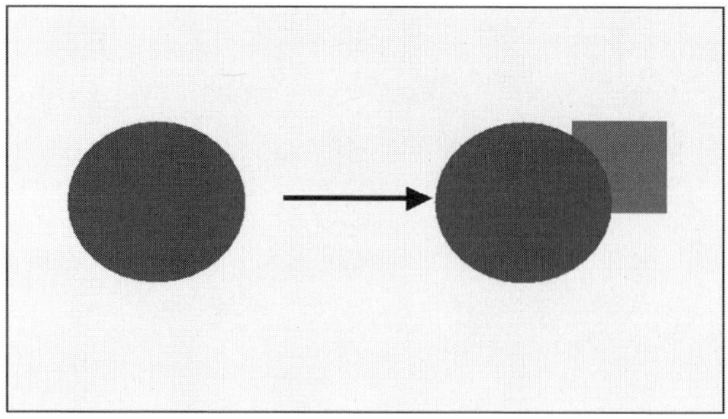

Fig. 8. Physicochemical stability measures the ratio of conversion product to unchanged compound after incubation in a physicochemical environment.

estimated from their retention time. Good calculations of Log D can be obtained from in silico programs such as Prolog D (CompuDrug Chemistry Ltd.).

3.5. Physicochemical Stability

Compounds can be chemically converted in aqueous solutions (*see* Fig. 8). These reactions can produce permanent products or reversible products in equilibrium with the starting material. Chemical conversion can cause variability in assay results. Conversion can be screened by incubating the compounds with buffer and testing for the percent of starting compound compared to an unincubated control. Components in some assay buffers, such as dithiothreitol (DTT), can react directly with compounds. Physical conditions in the laboratory can cause compound reactions also. These include basicity of some glassware, buffer pH, laboratory light exposure, solution temperature, and air oxidation or hydrolysis *(50–52)*. If compound degradation is a potential problem, compounds can be incubated under various pHs, high-intensity light, or elevated temperature to measure their rate of degradation. Quantitation typically uses LC/MS or HPLC.

3.6. pKa

Many compounds of interest in biomedical research have one or more pKas (the negative exponent of the equilibrium constant between the charged and uncharged forms of a particular functional group of the compound). The pKa and pH of the solution dictate the degree of compound ionization in solution (*see* Fig. 9). The ionized form is more soluble in water, because it is more polar. The neutral form is more permeable through lipid bilayer membranes. The pKa can also affect binding energies of a compound to the active site of the target protein. The pKa of a typical amine is in the range of 9–10; weaker bases have lower pkas. When the pH is below the pKa, the base will have more of its population in the protonated form. The pKa of a carboxylic acid is in the range of 3–4; weaker acids (e.g., phenols) have higher pKas. When the pH is below the pKa, the acid will have more of its molecules in the neutral form. A useful rule of thumb is that an equal concentration of the neutral and charged species is found in solution when the pH equals the pKa.

pKa has been measured in automated mode using techniques involving potentiometric titration *(40)*, capillary electrophoresis (CE) *(53)*, and spectral gradient analysis

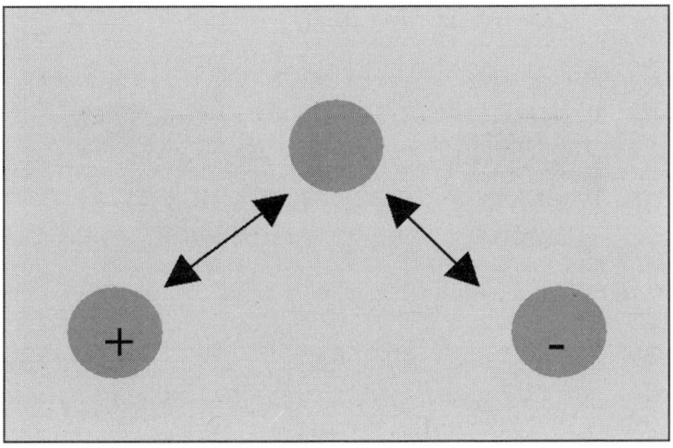

Fig. 9. pKa profiling measures the ratio of neutral compound to charged compound at a range of pHs.

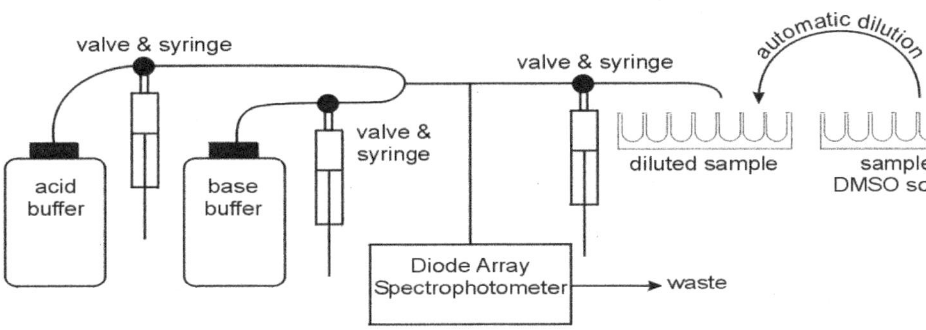

Fig. 10. Diagram of the SGA pKa analysis method. Sample in DMSO is diluted in aqueous buffer and injected into a flowing stream created by gradient mixing of acid and base buffers. pka is determined from the change in UV absorbance across the pH range.

(SGA) *(54)* (*see* Fig. 10). The SGA method creates a pH gradient in a flowing liquid stream over a 2 min period; the compound is injected continually into the stream and the UV absorbance of any chromophore in the near vicinity of the ionizable group changes with the ionization. This change is used to determine the pKa. pKa is calculated as the pH at which equal concentrations of charged and uncharged species are present. When the pKa is known, chemists can modify the ionizable functional groups on a molecule to change the solubility and permeability characteristics. Good calculations of pKa can be obtained from some in silico programs, such as Advanced Chemistry Development Inc.

3.7. Biochemical Stability

Many compounds in biomedical research are susceptible to biochemical conversion (i.e., metabolism) (*see* Fig. 11). Conversions can be caused by cytochrome P450 oxidations within the cell or other enzymatic reactions (e.g., phosphorylation, hydrolysis). Biochemical reactions reduce the concentration of compound. Significant instability

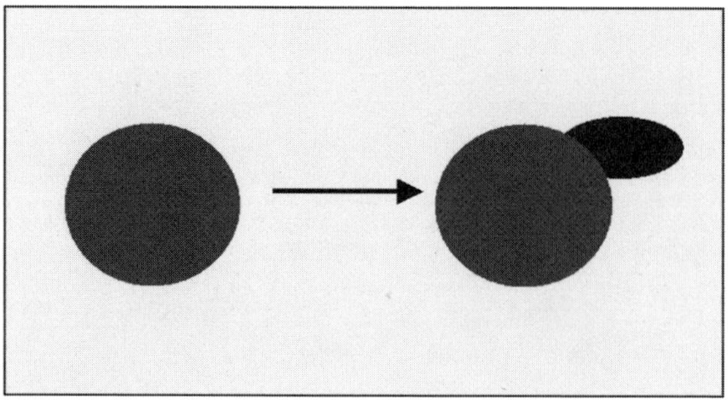

Fig. 11. Biochemical stability measures the ratio of metabolic product to unchanged compound after incubation in a biochemical environment.

compromises the proper interpretation of experiments and indicates a poor compound for in vitro studies. Metabolic stability assays are often conducted with liver microsomes containing the cytochrome P450 enzymes *(55–57)*, plasma *(58)*, or specific enzymes. Various animal species are often tested. It is useful to test stability in the same species as is being used in the cell culture or in vivo experiment. Stability assays are readily automated using standard laboratory robots (e.g., Packard, Tecan). Following incubation, the samples are typically quantitated using LC/MS.

Stability assays require high analytical sensitivity (owing to the low concentrations used to ensure that the enzymes are not saturated [e.g., 1 μM]) and selectivity to resolve the test compound signal from interfering signals of sample matrix components. Often high-throughput stability assays can only be achieved using mass spectrometry-based methods. Janizewski *(59)* described a customized LC/MS method having throughput on the order of 30 s/sample. Recently introduced instruments and software make implementation of high-throughput LC/MS/MS analysis easier for the average scientist *(60)*.

As with other profiling methods, the assay seems uncomplicated and easy to implement. However, experience has shown that results can vary widely, depending on the conditions and their day-to-day control. Factors such as DMSO concentration, sample concentration, species, and biochemical reagent (e.g., microsome) preparation can greatly affect the rate of biochemical reactions *(57)*. The data can be misleading if the assay is performed improperly.

3.8. Conversion Product Identification

When significant instability is indicated for a compound, useful information can be obtained by identifying the conversion products (*see* Fig. 12). This information allows redesign of structures in order to reduce reactions at the labile site of the molecule and reduce the conversion barrier. By identifying and isolating or synthesizing the conversion products, they can be tested for activity and/or toxicity. The use of LC/MS/MS for rapid structure elucidation has been described *(61–64)*. By using the MS/MS fragmentation of the parent compound as a template, the structures of major metabolites can be determined. MS techniques have the advantage of high throughput and low sample consumption (1–100 ng). Modern instrumentation and software allow the trained analytical chemist to rapidly identify the major conversion products of several compounds each day (*see* Fig. 13). When MS techniques do not indicate sufficient structural detail, nuclear magnetic resonance (NMR) or LC/NMR techniques *(65)* are

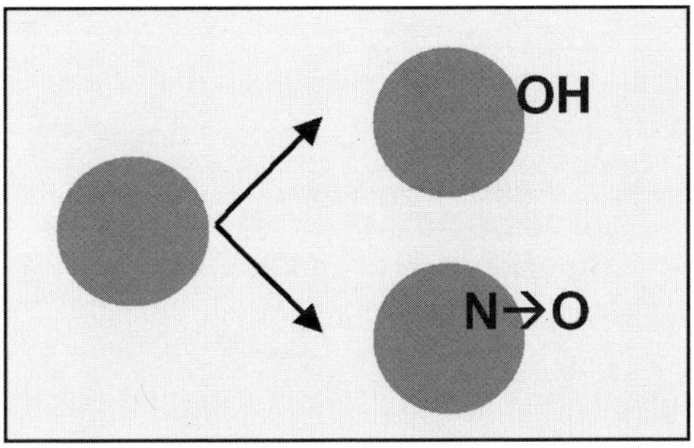

Fig. 12. Conversion product identification uses spectroscopic methods to rapidly identify the structure of the conversion product(s) from stability profiling.

used. This technique has made strides over the past 5 yr for trace compounds, but significantly higher amount of sample (0.1–10 μg) is needed and throughput is low compared to LC/MS/MS.

3.9. Active Transporters

The effect of active transport (*see* Fig. 14) on compound permeability is being studied more frequently in biomedical research. Cell-based assays can provide deeper understanding of transporter effects. Unknown affinity for transporters can cause errant prediction of membrane permeability if passive diffusion alone is considered. One of the most studied transporters, P-gp, is active in the intestine, BBB, and drug-resistant cancer cells. Polli *(66)* described and compared three methods for P-gp affinity: ATPase, Calcein AM, and MDR1-MDCK cell permeation. Many groups also use Caco-2 to study transport by use of apical to basolateral (A→B), basolateral to apical (B→A), and inhibitor studies (e.g., verapamil inhibits P-gp). Reduced P-gp affinity will likely improve performance in bioavailability, brain penetration, or activity in cancer cells.

Uptake transporters (e.g., OCT1, OATP1, OATP2, MOAT) and efflux transporters (MRPs) are under intensive study *(67)*. Not only will this research help to explain compound behavior in cell-culture experiments, but it will provide opportunities to enhance intracellular concentrations and permeation rates by reducing efflux or increasing uptake.

3.10. Cell Uptake

Because intracellular concentration often correlates with compound activity, assays for cell concentration are sometimes used (*see* Fig. 15). If the cellular activity is lower than expected from cell-free in vitro studies, the cellular concentration may be the cause, not the intrinsic activity. Low cellular concentration of drug may be owing to P-gp efflux, poor passive membrane permeation, or intracellular conversion/metabolism. In some cases, metabolic conversion may be necessary for activity and formation of the active species can be monitored. Cellular-exposure experiments have often utilized radiolabeled compounds for sensitive and rapid analysis. However, radiolabeling can be expensive and is not feasible in research involving the study of hundreds to thousands of compounds per year. Increasingly, LC/MS/MS techniques are being used to measure the

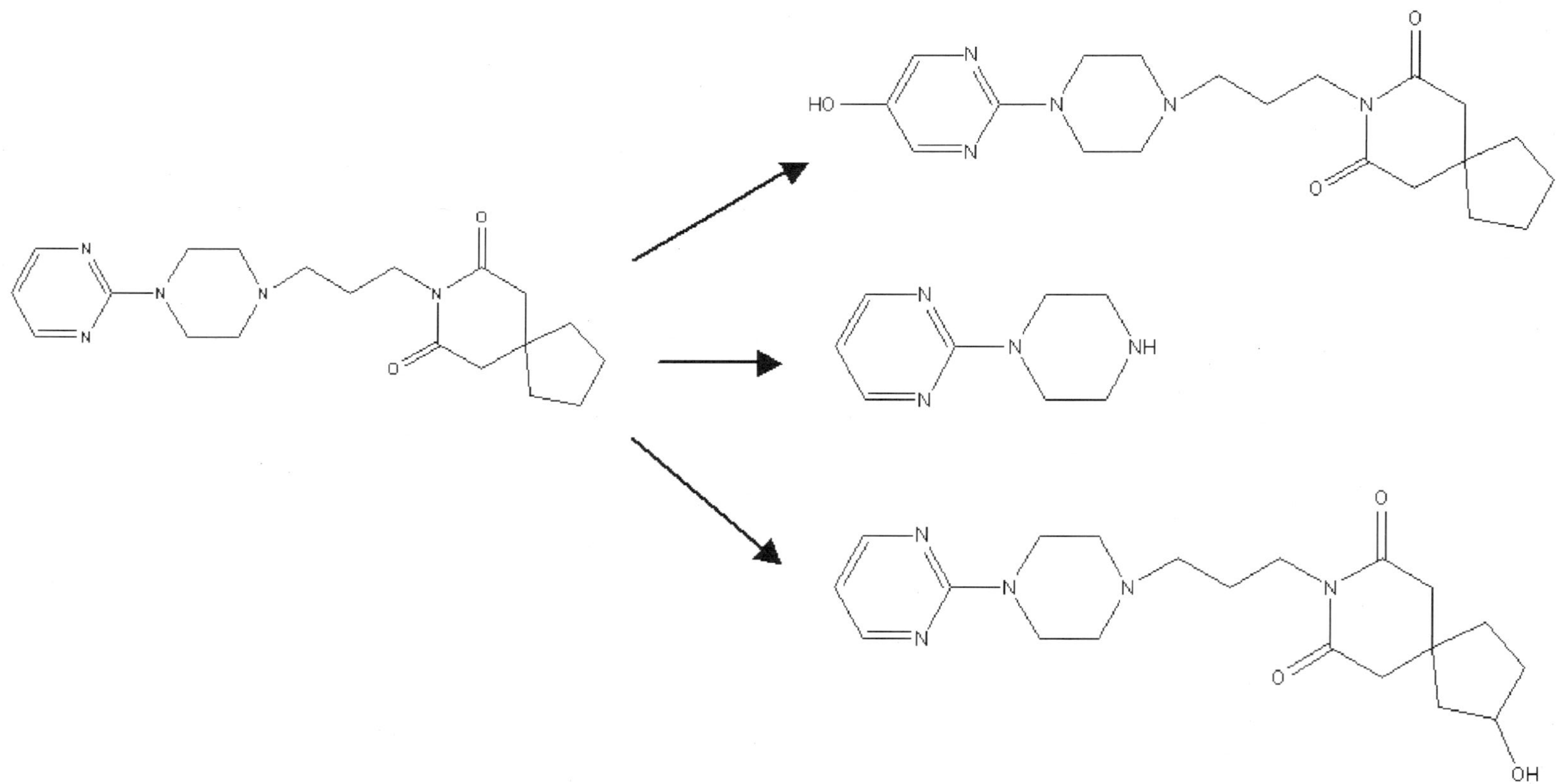

Fig. 13. Some common routes of metabolic degradation of busiprone catalyzed by cytochrome P450 enzymes in liver microsomes.

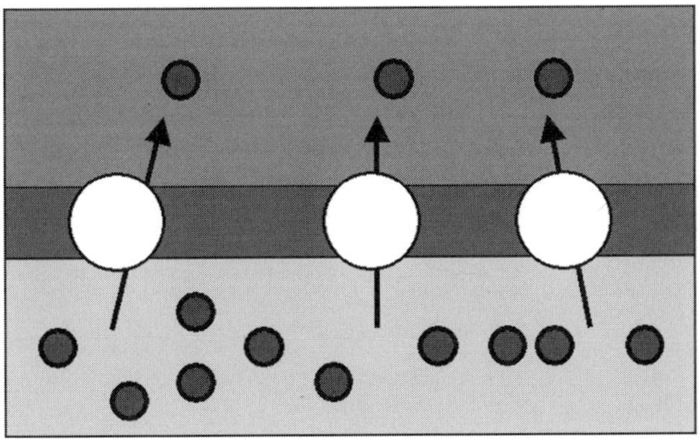

Fig. 14. Active transporter profiling indicates if cellular-membrane permeation is enhanced by active uptake or reduced by membrane efflux-transporter proteins.

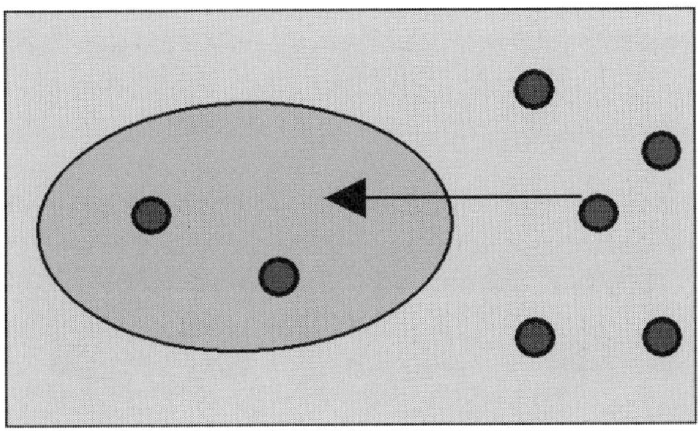

Fig. 15. Cell-uptake profiling measures the intracellular concentration of compound after it is added to the culture medium.

intracellular concentration of compounds, following incubation, washing, and harvesting of cells in culture plates *(68,69)* (*see* Fig. 16). Sensitive and selective LC/MS/MS methods can be rapidly and inexpensively developed, compared to radiolabeled detection schemes. The measured cellular concentration is the sum of compound bound to the cell membranes and proteins, as well as that dissolved in the cytoplasm.

3.11. Plasma Protein Binding

Binding of drug molecules to proteins (PPB) in plasma, cell-culture medium (e.g., serum albumin, α-acid glycoprotein), on the cell surface, or inside cells limits their free motion to the target protein (*see* Fig. 17). Several of the effects of binding offset each other, with regard to compound concentration at the therapeutic target, so application of PPB information can be confusing for chemists and pharmacologists and can be misinterpreted as a negative characteristic. PPB can be estimated using higher-throughput techniques, such as HPLC with HSA columns *(70)*, 96-well equilibrium dialysis *(71)*, Biacore *(72)*, or ultrafiltration (Millipore Microcon®).

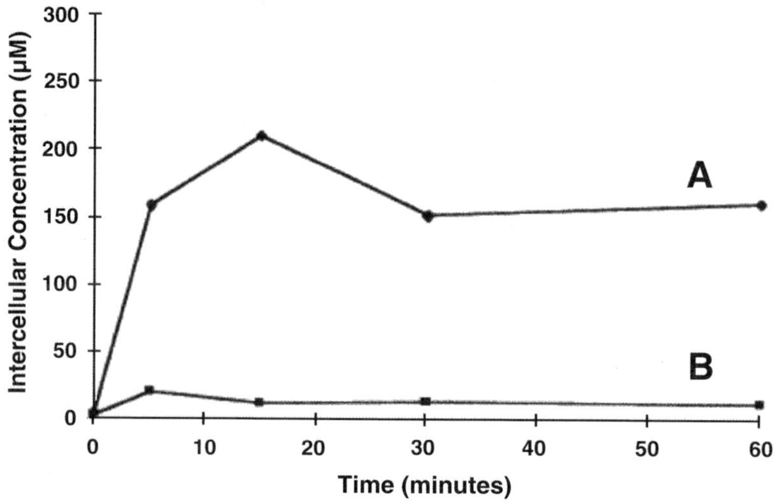

Fig. 16. Intracellular concentration of paclitaxel in normal cells **(A)** and multidrug resistant cells **(B)** in culture after dosing paclitaxel in cell-culture medium.

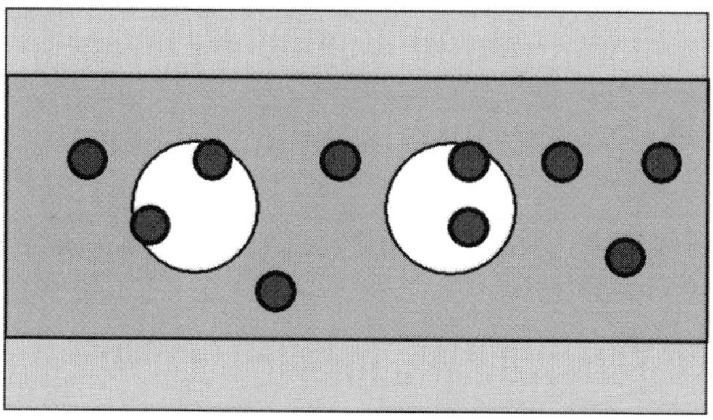

Fig. 17. Plasma protein-binding measures the ratio of compound that is bound to plasma proteins vs unbound in aqueous buffer solution.

4. Applying Pharmaceutical Profiling Data

In drug discovery, favorable properties are often termed "drug-like" and are predictive of potential drug-candidate success. This is because these properties are necessary for the delivery of compound to the target protein, which is often within the cell. Early knowledge of compound properties will aid any biomedical research because property information can be applied in several important ways.

There is a tradeoff between properties and potency. Lipinski *(2)* discussed the close relationship between properties and potency; poorer properties (solubility and permeability) can be tolerated for drugs with high potency (low mg/kg dose), but more favorable properties are necessary for drugs with lower potency. Van de Waterbeemd *(1)* provided a very informative discussion on "property-based design." This looks at optimizing the properties of compounds by structural modification, in order to improve the delivery of compound to the target protein. This can be done in parallel with structure-based design, in which SARs are a primary focus.

A key opportunity for the application of property information is better-informed research, by having more information to use in experimental design and interpretation. For example, solubility has a major effect on cellular experiments, so experiments can be planned at concentrations that are consistent with compound solubility. Unfavorable properties limit delivery of compound to the target protein, so properties greatly affect the perceived activity of compounds. Observed activity is a function of intrinsic target-protein binding, as well as delivery of compound to the intracellular target. Without property information, the cellular activity experiment may be incorrectly interpreted and limit research progress.

Complex processes, such as activity in cell-based assays, are functions of several properties. Thus, measurement of the individual properties that are components of a complex process allows the researcher to diagnose the limiting properties. Gan and Thakker *(12)* discussed how poor compound performance in complex systems can be diagnosed by examining individual properties. When this screening reveals the property responsible for the poor performance, then this property can be targeted for synthetic improvement. As new compounds are synthesized, the automated property profiling assays can be used to monitor improvement.

The dependence of a given biological activity assay on properties can be correlated using mathematical analysis techniques such as multivariate analysis (MVA). In one example *(9)*, MVA revealed that a cell-based potency assay correlated well with solubility and permeability, but not with stability. Once such correlations are recognized, then the properties most favorable to enhanced activity can be improved.

Early warning of major faults with a compound can save significant resources, time, and emotional investment. Research scientists can redirect their energies into more promising directions. If research focuses solely on SAR, time and resources can be wasted on compounds that have unfavorable properties. SAR is often favored by higher lipophilicity to promote better binding to the target protein, but such compounds can have low solubility and instability. An early alert to poor properties can help to avoid costly mistakes.

Often cellular experiments are performed in advance of in vivo experiments with sophisticated models such as transgenics or models that require weeks to complete. These models provide critical information, but are expensive and/or time-consuming. Property information can be included in the multifaceted information used to select compounds for advancement from cellular to sophisticated in vivo models.

If properties are well-characterized, they can also be used to guide optimization of compounds to improve cellular delivery. Sometimes the most promising leads have poor properties and this information helps to highlight the resources that will be needed to improve the properties. Drug-likeness has received increasing attention as an important predictor of success. The lack of drug-likeness reaches the attention of decision-makers who insure that a particular deficiency is addressed as the project proceeds. The SAR approach modifies the compound structure to make it bind better to the therapeutic enzyme or receptor. The SPR approach modifies the structure to improve properties that increase the delivery of the compound to the therapeutic target and sustain its concentration there. Thus, properties can help to guide synthetic chemistry efforts aimed at optimizing compounds during drug discovery. Examples of successful synthetic modifications include molecular size, hydrogen bonding, pKa, adding an ionizable group, and blocking substructures that are unstable. The optimum candidates are now optimized for both activity and properties.

At the same time, one must be aware that structural modifications that improve one property may diminish another property *(13)*. For example, increasing metabolic stability at one site may increase metabolism at another site, reduce permeability, or increase renal or biliary extraction. Also, in some cases, a property cannot be improved through structural modification without significantly reducing the activity. A rule of thumb is that if the poor properties result from a substructure that is critical for the activity, it will be difficult to both improve properties and maintain activity. Medicinal chemists can initiate property improvements by focusing on the modification of substructures that are shown, through SAR studies, to have little effect on activity.

Improvement of properties such as permeability through prodrugs is an established strategy. Profiling assays can be used to compare prodrug conversion rates. Specialized assays for the targeted conversion mechanism can be developed.

Activity response at the cellular level follows a dose-response relationship. Thus, a compound that is very active in an enzyme/receptor assay may not be active in cell culture if sufficient concentration is not established in the cells. A less potent compound may provide a better activity if its properties permit enhanced intracellular concentration over a longer time.

Lipinski discussed how solubility greatly affects activity assays *(10,11)*. In addition, if a compound has poor permeability it will not perform well in cell-based assays. Thus, the perceived activity of the compound in cells would be low, and the actual mechanism of the low activity, whether biochemical or physicochemical, would not be understood. In the past, property information was not available, but now researchers can take advantage of the improved methods for obtaining this information to enhance their research.

5. Method Selection Considerations

Several methods are available to researchers for assessing properties. The researcher should carefully examine how a method was developed and validated, such as the quality of the data set used and the quality of the correlation. Many methods are developed using a diverse set of compounds, but the compounds used may not be relevant to the researchers compound class and thus produce poor results. It is generally agreed that data and predictions are most accurate within a chemical series.

Rules and filters, such as the "rule of 5" *(2)*, rotatable bonds *(73)*, and polar surface area *(74)* provide a quick and easy guide for whether compounds are likely to have favorable properties for penetration into cellular targets *(10)*.

In silico models *(75)* are fast to use. They are inexpensive if the software cost can be spread over many compounds or research projects. They can be used in planning prior to chemical synthesis and to estimate the properties of compounds for which there is insufficient material for analysis, or when a particular property assay is not available.

Automated assays provide actual measurements of properties. The assays are rapidly performed and little material (mg level) is required. These assays are most efficient when a large number of compounds or several projects share the expense. Measurements provide an increased level of data assurance over in silico methods, but the assays need to be carefully developed, validated, and controlled for quality.

In-depth analyses provide the most definitive and reliable information. These assays require large quantities of material (10–100 mg), often require weeks to complete, and can only be performed on a limited number of compounds. Careful quality control (QC) must be performed.

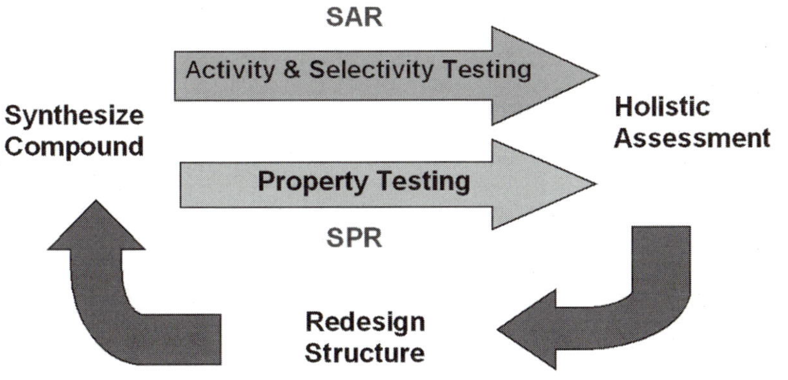

Fig. 18. Strategy for optimization involving parallel SAR and SPR studies, leading to holistic understanding of the compound and redesign of the structure for the next cycle of synthesis and testing.

The predictivity of in silico methods is rapidly improving as academic and commercial groups focus on their improvement. They rely on high-quality data from diverse compound structures. As in silico methods improve, they will provide sufficient information for early studies and measurement resources can be repositioned to assess other critical properties. Combination of in silico and measurement methods appears to provide the solution for obtaining the optimum property information with the available resources at this time.

Automated assays are often applied in a stepwise approach. Clearly, certain assays should be performed as early in research as possible: integrity to insure that the SAR is correct, and solubility and permeability to insure that biological activity measurements are not compromised by inappropriate assay conditions. Without this knowledge, active pharmacophores that were studied at levels above their actual solubility, may be eliminated prematurely from consideration. As research projects progress, additional property information can be obtained, or "full" profiles can be done on all new series compounds.

The effective application of pharmaceutical properties in drug discovery requires acceptance by discovery chemists and biologists. Several elements are necessary to gain buy-in. Pharmaceutical profiling must be as rapid as biological assays. Measurement of properties can occur in parallel with activity (Fig. 18) so that all the information is considered in redesigning structures for series optimization. Rapid and effective data communication to the researcher is critical. Tables of data that allow compound comparison are useful. Color coding of results, according to "bins" of favorable, moderate, and unfavorable property values allow researchers to rapidly review the data to prioritize compounds *(9)*. Visualization tools such as Spotfire *(76)* and multivariate analysis tools such as Simca *(77)* can provide additional insight that is not discerned by inspection of tables or single dimensional data. Lipinski *(11)* has discussed how simple graphical presentations are most effective for data communication. The research leader and scientists must be trained to understand the properties and their effects on the barriers encountered by the compound. Property data must be relevant to chemical structure, so that medicinal chemists can design new structures that improve lipophilicity, pKa, molecular weight (MW), hydrogen bonding, and stability. Effective automation of pharmaceutical profiling assays is necessary for efficiency, accuracy, and precision.

6. Conclusions

Pharmaceutical profiling is an emerging field in drug discovery. It complements the current SAR paradigm by providing SPR information. This is useful because properties have a major effect on the delivery of compounds to cellular therapeutic targets. Automated property assays have been developed for the critical properties so that data can be provided rapidly for researchers. The information allows researchers to plan and interpret their experiments in greater depth. It also allows the diagnosis of compound performance at barriers that limit cellular delivery. Compound properties can be improved via structural modifications guided by the assay data.

References

1. van de Waterbeemd H, et al. Property-based design: optimization of drug absorption and pharmacokinetics. J Med Chem 2001;44:1313–1333.
2. Lipinski CA, et al. Experimental and computational approaches to estimate solubility and permeability in drug discovery and development setting. Adv Drug Deliv Rev 1997;23:3–25.
3. Smith DA, van de Waterbeemd H. Pharmacokinetics and metabolism in early drug discovery. Curr Opin Chem Biol 1999;3:373–378.
4. Krämer S. Absorption prediction from physicochemical parameters. Pharm Sci Tech Today 1999;2:373–380.
5. Sinko JS. Drug selection in early drug development: screening for acceptable pharmacokinetic properties using combined *in vitro* and computational approaches. Curr Opin Drug Discov Dev 1999;2:42–48.
6. Venkatesh S, Lipper RA. Role of the development scientist in compound lead selection and optimization. J Pharm Sci 2000;89:145–154.
7. Kerns EH. High throughput physicochemical profiling for drug discovery. J Pharm Sci 2001;90:1838–1858.
8. Caldwell GW, et al. The new pre-preclinical paradigm: compound optimization in early and late phase drug discovery. Curr Topics Med Chem 2001;1:353–366.
9. Kerns EH, Di L. Multivariate pharmaceutical profiling for drug discovery. Curr Topics Med Chem 2002;2:87–98.
10. Lipinski CA. Avoiding investment in doomed drugs. Curr Drug Disc 2001;April:17–19.
11. Lipinski CA. Drug-like properties and the causes of poor solubility and poor permeability. J Pharmacol Toxicol Methods 2000;44:235–249.
12. Gan LSL, Thakker DR. Application of the Caco-2 model in the design and development of orally active drugs: elucidation of biochemical and physical barriers posed by the intestinal epithelium. Adv Drug Deliv Rev 1997;23:77–98.
13. Mandagere AK. Graphical model for estimating oral bioavailability of drugs in humans and other species from their Caco-2 permeability and *in vitro* liver enzyme metabolic stability rates. J Med Chem 2002;45:304–311.
14. Caldwell GW. Compound optimization in early- and late-phase drug development: acceptable pharmacokinetic properties utilizing combined physicochemical, *in vitro* and *in vivo* screens. Curr Opin Drug Disc Dev 2000;3:30–41.
15. Kerns EH, Di L. Pharmaceutical profiling in drug discovery. Drug Disc Today 2003;8:316–323.
16. Kerns EH, Di L, Petusky S, Kleintop T, McConnell OJ, Carter G. (2003) Pharmaceutical profiling method for lipophilicity and integrity using LC/MS. J Chrom B 2003;791:381–388.
17. Bevan C, Lloyd RS. A high-throughput screening method for the determination of aqueous drug solubility using laser nephelometry in microtiter plates. Anal Chem 2002;72:1781–1787.
18. Avdeef A. High throughput measurements of solubility profiles. In: Testa B, van de Waterbeemd H, Folkers G, Guy R, eds. Pharmacokinetic Optimization in Drug Research; Biological, Physicochemical, and Computational Strategies. Zurich, Switzerland, Verlag Helvitica Chimica Acta, 2001, pp. 305–326.

19. Artursson, P. Prediction of Drug Absorption: Caco-2 and Beyond. Book of Abstracts, PAMPA 2002 Conference, San Francisco, CA, July 14–17, 2002.
20. Balon K, et al. Drug liposome partitioning as a tool for the prediction of human passive intestinal absorption. Pharm Res 1999;16:882–888.
21. Ong S, et al. Immobilized-artificial membrane chromatography: measurements of membrane partition coefficient and predicting drug membrane permeability. J Chromatogr A 1996;728:113–128.
22. Caldwell GW, et al. Evaluation of the immobilized artificial membrane phosphatidylcholine drug discovery column for high-performance liquid chromatographic screening for drug-membrane interactions. J Chromatogr A 1998;800:161–169.
23. Masucci JA. Comparison of the retention behavior of β-blockers using immobilized artificial membrane chromatography and lysophospholipid micellar electrokinetic chromatography. J Chromatogr A 1998;810:95–103.
24. Artursson P, et al. Caco-2 monolayers in experimental and theoretical predictions of drug transport. Adv Drug Deliv Rev 2001;46:27–43.
25. Hidalgo IJ. Assessing the absorption of new pharmaceuticals. Curr Topics Med Chem 2001;1:385–401.
26. Irvine JD, et al. MDCK (Madin-Darby canine kidney) cells: a tool for membrane permeability screening. J Pharm Sci 88:28–33.
27. Menard CA, et al. Comparison of Caco-2 permeability data in 24- and 96-well formats, Book of Abstracts, LabAutomation 2001, January 2001, Palm Springs, CA.
28. Kansy M, et al. Physicochemical high throughput screening: parallel artificial membrane permeation assay in the description of passive absorption processes. J Med Chem 1998;41:1007–1010.
29. Avdeef A, et al. Drug absorption *in vitro* model: filter-immobilized artificial membranes. 2. studies of the permeability properties of lactones in *Piper methysticum* Forst. Eur J Pharm Sci 2001;14:271–280.
30. Wohnsland F, Faller B. High-throughput permeability pH profile and high-throughput alkane/water log P with artificial membranes. J Med Chem 2001;44:923–930.
31. Zhu C, et al. A comparative study of artificial membrane permeability assay for high throughput profiling of drug absorption potential. Eur J Med Chem 2002;37:399–407.
32. Kerns, EH et al. Combined application of PAMPA and Caco-2 in drug discovery. J Pharm Sci, in press.
33. Gumbleton M, Audus KL. Progress and limitations in the use of *in vitro* cell cultures to serve as a permeability screen for the blood-brain barrier. J Pharm Sci 2001;90:1681–1698.
34. Di L, et al. (2003) High throughput artificial membrane permeability assay for blood-brain barrier. Eur J Med Chem 2003;38:223–232.
35. Hansch C, et al. Exploring QSAR. Washington, DC, ACS Pub., 1995.
36. Danielsson L, Zhang Y. Methods for determining n-octanol-water partitioning constants. Trends Anal Chem 1996;15:188–196.
37. Buchwald P, Bodor N. Octanol-water partition: searching for predictive models. Curr Med Chem 1998;5:353–380.
38. Andersson JT, Schräder W. A method for measuring 1-octanol-water partition coefficients. Anal. Chem. 1999;71:3610–3614.
39. Gulyaeva N, et al. Relative hydrophobicity and lipophilicity of beta-blockers and related compounds as measured by aqueous two-phase partitioning, octanol-buffer partitioning, and HPLC. Eur J Pharm Sci 2002;17:81–93.
40. Avdeef A. pH-metric log P: II: refinement of partition coefficients and ionization constants of multiprotic substances. J Pharm Sci 1993;82:183–190.
41. Avdeef A, et al. pH-Metric logP 10. determination of liposomal membrane-water partition coeffients of ionizable drugs. Pharm Res 1998;15:209–215.
42. Minick DJ, et al. A comprehensive method for determining hydrophobicity constants by reversed-phase high-performance liquid chromatography. J Med Chem 1988;31:1923–1933.
43. Abraham MH. Connection between chromatographic data and biological data. J Chromatogr B 2002;745:103–115.
44. Valkó K. et al. Chromatographic hydrophobicity index by fast-gradient RP-HPLC: a high-throughput alternative to log P/log D. Anal Chem 1997;69:2022–2029.

45. Du CM; et al. Rapid gradient RP-HPLC method for lipophilicity determination: a solvation equation based comparison with isocratic methods. Anal Chem 1998;70:4228–4234.
46. Valkó K, et al. Rapid method for the estimation of octanol/water partition coefficient (log Poct) from gradient RP-HPLC retention and hydrogen bond acidity term ($\Sigma\alpha_2^H$). Curr Med Chem 2001;8:1137–1146.
47. Lombardo F, et al. ElogP_{oct}: a tool for lipophilicity determination in drug discovery. J Med Chem 2000;43:2922–2928.
48. Pagliara A, et al. Structural properties governing retention mechanisms on RP-HPLC stationary phases used for lipophilicity measurements. J Liquid Chromatogr 1995;18:1721–1745.
49. Poole SK, et al. Rapid method for estimating the octanol-water partition coefficient (log P_{ow}) by microemulsion electrokinetic chromatography. J Chromatogr B 2000;745:117–126.
50. Rourick RA, et al. Predictive strategy for the rapid structure elucidation of drug degradants. J Pharm Biomed Anal 1996;14:1743–1752.
51. Volk KJ, et al. Profiling degradants of paclitaxel using liquid chromatography-mass spectrometry and liquid chromatography-tandem mass spectrometry substructural techniques. J Chromatogr B 1997;696:99–115.
52. Kerns EH, et al. Stability profiling in drug discovery. Book of Abstracts, Symposium on Chemical and Pharmaceutical Structure Analysis. September, 2000.
53. Barták P, et al. Determination of dissociation constants of cytokinins by capillary zone electrophoresis. J Chromatogr A 2002;878:249–255.
54. Box KJ, et al. Rapid physicochemical profiling as an aid to drug candidate selection. In: Dixon GK, Major JS, Rice MJ, eds. High Throughput Screening: The Next Generation. Oxford, UK, BIOS Scientific Publishers Ltd., 2000; pp. 67–74.
55. Korfmacher WA, et. al. Development of an automated mass spectrometry system for the quantitative analysis of liver microsomal incubation samples: a tool for rapid screening of new compounds for metabolic stability. Rapid Comm Mass Spectrom 1999;13:901–907.
56. Xu R, et al. Application of parallel liquid chromatography/mass spectrometry for high throughput microsomal stability screening of compound libraries. J Soc Mass Spectrom 2002;13:155–165.
57. Di L, et al. (2003) High throughput microsomal stability screening assay for profiling drug discovery candidates. J Biomolec Screen 2003;8:453–462.
58. Wang G, et al. Semi-automated determination of plasma stability of drug discovery compounds using liquid chromatography-tandem mass spectrometry. J Chromatogr B 2002;780:451–457.
59. Janiszewski JS, et al. A high-capacity LC/MS system for the bioanalysis of samples generated from plate-based metabolic screening. Anal Chem 2001;73:1495–1501.
60. Kerns EH, et al. (2003) Integrated high capacity LC-MS-MS system for quantitation in drug discovery pharmaceutical profiling. Accepted for J Pharm Biomed Anal, in press.
61. Kerns EH, et al. Buspirone metabolite structure profile using a standard liquid chromatographic-mass spectrometric protocol. J Chromatogr B 1997;698:133–145.
62. Kim H, et al. Isolation of circulating metabolites in drug discovery using high-performance liquid chromatography, and their identification of liquid chromatography coupled with tandem mass spectrometry and nuclear magnetic resonance spectroscopy. J Sep Sci 2002;25:877–882.
63. Lopez LL, et al. Identification of drug metabolites in biological matrices by intelligent automated liquid chromatography/tandem mass spectrometry. Rapid Commun Mass Spectrom 1998;12:1756–1760.
64. Clarke NJ, et al. Systematic LC/MS metabolite identification in drug discovery. Anal Chem 2001;73:430A–439A.
65. Shockor JP. Application of directly coupled LC-NMR-MS to the structural elucidation of metabolites of the HIV-1 reverse transcriptase inhibitor BW935U83. J Chromatogr B 2002;748:269–279.
66. Polli JW, et al. Rational use of *in vitro* P-glycoprotein assays in drug discovery. J Pharmacol Exp Therapeut 2001;299:620–628.
67. Kim RB. Transporters and drug disposition. Curr Opin Drug Disc Devel 2000;3:94–101.

68. Kerns EH, et al. Cellular uptake profile of paclitaxel using liquid chromatography tandem mass spectrometry. Rapid Commun Mass Spectrom 1998;12:620–624.
69. Zhang Q, et al. Application of LC/MS/MS in the quantitation of SU101 and SU0020 uptake by 3T3/PDGFr cells. J Pharm Biomed Anal 2002,28:701–709.
70. Tiller PR, et al. Immobilized human serum albumin: liquid chromatography/mass spectrometry as a method of determining drug-protein binding. Rapid Commun Mass Spectrom 1995;9:261–263.
71. Kariv I, et al. Development of a high throughput equilibrium dialysis method. J Pharm Sci 2001;90:580–587.
72. Rich RL, et al. High-resolution and high-throughput protocols for measuring drug/human serum albumin interactions using biacore. Anal Biochem 2001;296:197–207
73. Veber DF, et al. Molecular properties that influence the oral bioavailability of drug candidates. J Med Chem 2002;45:2615–2623.
74. Palm K, et al. Evaluation of dynamic polar molecular surface area as predictor of drug absorption: comparison with other computational and experimental predictors. J Med Chem 1998;41:5382–5392.
75. Clark DE. Computational methods for the prediction of ADME and toxicity. Adv Drug Deliv Rev 2002;54:253–254.
76. Stoner CL, et al. (2002) The implementation of an ADME enabling selection tool for drug discovery. AAPS PharmSci 4, Abstract T3255.
77. Eriksson L, et al. Introduction to Multi- and Megavariate Data Analysis Using Projection Methods (PCA & PLS). Umea, Sweden, Umetrics AB Pub., 1999.

Index

About the Editors

D. Robert Lu received his PhD degree in Pharmaceutics from Purdue University in 1990. After working for 2 years (1990–1992) as an assistant professor at Idaho State University, he transferred to the College of Pharmacy, University of Georgia. In 2003, he moved to the School of Pharmacy at Temple University, where he holds a full professorship in pharmaceutics.

Dr. Lu's research focuses on drug targeting to cancers and cellular delivery of genetic materials using various delivery systems. He has published more than 60 peer-reviewed papers and an additional 70 meeting abstracts. In addition, he has served as the major professor for 16 graduate students, 14 of whom have accepted various academia and industry positions nationwide.

Svein Øie is professor and dean at the College of Pharmacy, University of Georgia. He received his pharmacy degree from the University of Oslo and served in the Norwegian Army as a lieutenant. He received his PhD in Pharmaceutics from the State University of New York at Buffalo in 1975 and spent the following 24 years at the University of California, San Francisco. In 2000, he joined the College of Pharmacy, University of Georgia as professor and dean.

Dr. Øie's research focuses on drug development in the area of oral absorption of peptides and cellular delivery, and on the effect of disease on drug distribution and elimination. His work includes more than 100 publications, and he has presented more than 80 papers and invited lectures at national and international meetings.